ASTROPHYSICS PROCESSES

Bridging the gap between physics and astronomy textbooks, this book provides step-by-step physical and mathematical development of fundamental astrophysical processes underlying a wide range of phenomena in stellar, galactic, and extragalactic astronomy. The book has been written for upper-level undergraduates and beginning graduate students, and its strong pedagogy ensures solid mastery of each process and application. It contains over 150 tutorial figures, numerous examples of astronomical measurements, and 202 exercises. Topics covered include the Kepler–Newton problem, stellar structure, binary evolution, radiation processes, special relativity in astronomy, radio propagation in the interstellar medium, and gravitational lensing. Applications presented include Jeans length, Eddington luminosity, the cooling of the cosmic microwave background (CMB), the Sunyaev–Zeldovich effect, Doppler boosting in jets, and determinations of the Hubble constant. This text is a stepping stone to more specialized books and primary literature. Password-protected solutions to the exercises are available to instructors at www.cambridge.org/9780521846561.

HALE BRADT is Professor Emeritus of Physics at the Massachusetts Institute of Technology. During his 40 years on the faculty, he carried out research in cosmic-ray physics and x-ray astronomy and taught courses in physics and astrophysics. Bradt founded the MIT sounding rocket program in x-ray astronomy and was a senior or principal investigator on three missions for x-ray astronomy. He was awarded the NASA Exceptional Science Medal for his contributions to HEAO-1 (High-Energy Astronomical Observatory) as well as the 1990 Buechner Teaching Prize of the MIT Physics Department and shared the 1999 Bruno Rossi prize of the American Astronomical Society for his contributions to the RXTE (Rossi X-ray Timing Explorer) program. His previous book, *Astronomy Methods – A Physical Approach to Astronomical Observations*, was published by Cambridge University Press in 2004.

Cover information

Views of the entire sky at six wavelengths in galactic coordinates; the equator of the Milky Way system is the central horizontal axis and the galactic center direction is at the center. Five of the views may be seen in greater detail on NASA's Web site "Astronomy Picture of the Day" (APOD). Except for the far-infrared and x-ray views, the colors represent intensity levels, with the greatest intensities lying along the equator. In all cases, the radiation shows an association with the galactic equator, the general direction of the galactic center, or both. The views are in frequency sequence as listed below and are located, respectively, as follows: background and upper and lower insets on the front cover; top, middle, and bottom insets on the back cover.

Radio sky at 408 MHz exhibiting a diffuse glow of synchrotron radiation from the entire sky. High-energy electrons spiraling in the magnetic fields of the Galaxy emit this radiation. Note the *North Polar Spur* projecting above the equator to the left of center. From three observatories: Jodrell Bank, MPIfR, and Parkes. [Glyn Haslam *et al.*; MPIfR, SkyView; APOD, 14 December 1997]

Radio emission at 1420 MHz, the spin-flip (hyperfine) transition in the ground state of hydrogen, which shows the location of clouds of neutral hydrogen gas. The gas is heavily concentrated in the galactic plane and manifests pronounced filamentary structure off the plane. [Courtesy of National Radio Astronomy Observatory/AUI/NSF; APOD, 13 January 2001]

Far-infrared (60–240 μm) sky from the COBE satellite showing primarily emission from small grains of graphite and silicates ("dust") in the interstellar medium of the Galaxy. The faint, whitish, large S-shaped curve (on its side) is emission from dust and rocks in the solar system; reflection of solar light from this material causes the zodiacal light at optical wavelengths. Color coding: 60 μm (blue), 100 μm (green), 240 μm (red). [E. L. Wright (UCLA); COBE; DIRBE; NASA; APOD, 17 May 2000]

Optical sky from a mosaic of 51 wide-angle photographs showing mostly stars in our (Milky Way) Galaxy, with significant extinction by dust along the galactic plane. Galaxies are visible at higher galactic latitudes, the most prominent being the two nearby Magellanic Clouds (lower right). [©Axel Mellinger; APOD, 2 February 2001]

X-ray sky at 1–20 keV from the A1 experiment on the HEAO–1 satellite showing 842 discrete sources. The circle size represents intensity of the source and the color denotes the type of object. The most intense sources shown (green, larger circles) signify compact binary systems containing white dwarfs, neutron stars, and black holes. Other objects are supernova remnants (blue), clusters of galaxies (pink), active galactic nuclei (orange), and stellar coronae (white). [Kent Wood (NRL); see *ApJ Suppl.* **56**, 507 (1984)]

Gamma-ray sky above 100 MeV from the EGRET experiment on the Compton Gamma-Ray Observatory. The diffuse glow from the galactic equator is due to the collisions of cosmic-ray protons with the atoms of gas clouds; the nuclear interactions produce neutral pions that decay to the detected gamma rays. Discrete sources include pulsars and jets associated with active galaxies ("blazars"). [The EGRET team; NASA; CGRO; APOD, 21 March 1998]

ASTROPHYSICS PROCESSES
The Physics of Astronomical Phenomena

HALE BRADT
Massachusetts Institute of Technology

CAMBRIDGE
UNIVERSITY PRESS

University Printing House, Cambridge CB2 8BS, United Kingdom

Published in the United States of America by Cambridge University Press, New York

Cambridge University Press is part of the University of Cambridge.

It furthers the University's mission by disseminating knowledge in the pursuit of education, learning and research at the highest international levels of excellence.

www.cambridge.org
Information on this title: www.cambridge.org/9781107677241

© H. Bradt 2008

First published 2008
Reprinted (with corrections) 2010
First paperback edition 2014

A catalogue record for this publication is available from the British Library

Library of Congress Cataloguing in Publication data

Bradt, Hale, 1930–
Astrophysics processes / Hale Bradt.
    p.   cm.
Includes bibliographical references and index.
ISBN 978-0-521-84656-1 (hardback)
1. Astrophysics.   I. Title.
QB461. B67   2008
523.01 – dc22   2007031649

ISBN 978-0-521-84656-1 Hardback
ISBN 978-1-107-67724-1 Paperback

To my wife Dottie, my daughters Elizabeth and Dorothy, and my sisters Val, Abby, and Dale Anne

They are my fans and I, theirs.

Contents

Figures

Tables

Preface

This volume is based on notes that evolved during my teaching of astrophysics classes for junior and senior physics students at MIT beginning in 1973 and thereafter on and off until 1997. The course focused on a physical, analytical approach to underlying processes in astronomy and astrophysics. In each class, I would escort the students through a mathematical and physical derivation of some process relevant to astrophysics in the hope of giving them a firm comprehension of the underlying principles.

The approach in the text is meant to be accessible to undergraduates who have completed the fundamental calculus-based physics courses in mechanics and electromagnetic theory. Additional physics courses such as quantum mechanics, thermodynamics, and statistics would be helpful but are not necessary for large parts of this text. Derivations are developed step by step – frequently with brief reviews or reminders of the basic physics being used because students often feel they do not remember the material from an earlier course. The derivations are sufficiently complete to demonstrate the key features but do not attempt to include all the special cases and finer details that might be needed for professional research.

This text presents twelve "processes" with derivations and focused, limited examples. It does not try to acquaint the student with all the associated astronomical lore. It is quite impossible in a reasonable-sized text to give both the physical derivations of fundamental processes and to include all the known applications and lore relating to them across the field of astronomy. The assumption here is that many students will have had an elementary astronomy course emphasizing the lore. Nevertheless, selected germane examples of the processes are presented together with background information about them. These examples cover a wide and rich range of astrophysical phenomena.

The twelve processes, with the principal applications presented, are the Kepler–Newton problem (*mass functions, exoplanets, galactic center orbits*); stellar equilibrium (*nuclear burning, Eddington luminosity*); stellar equations of state (*ideal and degenerate gases, the distribution function*); stellar structure and evolution (*normal and compact stars, binary evolution*); thermal bremsstrahlung (*clusters of galaxies*); blackbody radiation (*cosmological cooling*); synchrotron (*Crab nebula*) and curvature radiation (*pulsars*); 21-cm radiation (*galaxy rotation, dark matter, Zeeman absorption*); Compton scattering (*Sunyaev–Zeldovich effect*); relativity in astronomy (*jets, photon absorption in the cosmic microwave background*); dispersion (*interstellar medium*) and Faraday rotation (*galactic magnetic field*); and gravitational lensing (*quasars, Fermat approach, Hubble constant, weak lensing*). Cosmology as such is not systematically covered to limit the size of the text. Several related topics, however, are addressed: (*i*) the dark matter in galaxies and in clusters of galaxies, (*ii*) the cooling of the background blackbody radiation of the CMB, and (*iii*) determinations of the Hubble constant through both the S-Z effect and gravitational lensing.

Knowledge of the material in my previous textbook, *Astronomy Methods – A Physical Approach to Astronomical Observations* (AM), is not required for this text. The topics are largely complementary to those herein. I do, though, occasionally refer to it as an optional background reference. (The chapter numbers are those of the original edition.) The AM text, in Section 11.5, develops the formation of spectral lines in stellar atmospheres, one of the most basic processes in astronomy; hence, regrettably, this topic is not included in this book. Students and instructors are advised to download the PDF of this section from the resources link at www.cambridge.org/9780521846561. Other supplementary material and errata will also be posted at this site.

Again, SI units are used throughout to be consistent with most standard undergraduate science texts. Professional astronomers use cgs units. Unfortunately, this precludes progress in bringing the various science communities together to one system of units. It is also a significant hindrance to the student exploring astronomy or astrophysics. In this work I vote for ease of student access. Rather than use the customary and highly specialized astronomical unit of distance, the "parsec," I instead employ the better understood, but non-SI, unit, the "light year" (LY), which is the distance light travels in one year. This is a well-defined quantity if one specifies the Julian year of exactly 365.25 days each of exactly 86 400 SI seconds for a total for 31 557 600 s.

Other features of the book to note are as follows:

(*i*) Problems are provided for each chapter, and approximate answers indicated by the $\sim$ symbol are given when appropriate. Password protected solutions for instructors are available at the above mentioned resources link.

(*ii*) The problems are generally constructed to help carry the student through them and hence are mostly multipart.

(*iii*) Units are often given gratuitously (in parentheses) for algebraic variables to remind the reader of the meaning of the symbol.

(*iv*) For improved readability, equation, table, figure, and section numbers in the text do not carry the chapter prefix if they refer to the current chapter.

(*v*) For ease of referencing during class discussions, all equations are numbered, labels are provided for many of them, and important equations are each marked with a boldface arrow in the left margin.

(*vi*) Tables of useful units, symbols, and constants are given in the appendix.

(*vii*) A glossary of acronyms and abbreviations is provided just before the appendix.

(*viii*) Logarithms are base 10 if "log" and base e if "ln".

(*ix*) Quantitative information is meant to be up to date and correct but should not be relied upon for professional research. The goal here is to teach underlying principles.

In teaching this course from my notes, I adopted a seminar, or Socratic, style of teaching that turned out to be extremely successful and personally rewarding. I recommend this approach to teachers using this text. I sat with the students (up to about 20) around a table, or we would rearrange classroom desks and chairs in a circular or rectangular pattern so that we were all more or less facing each other. I would then have the students explain the material to their fellow students ("Don't look at me," I often said). One student would do a bit, and I would move on to another. I made my prompts easy and straightforward, avoided disparaging incorrect or confusing answers, and encouraged discussion among the students. I would synthesize arguments and describe the broader implications of the material interspersed by stories of real-life astronomy, personalities, discoveries, and so on.

The class would often become quite animated. During the discussions, the text was available to all and freely referenced. The students had to work hard to prepare for class, and thus they gained much from the class discussion. The course was also great fun for the teacher. In good weather, we would move outdoors and have our class on the lawn of MIT's Killian Court.

The author asks his readers' forbearance with the inevitable errors in the current text and requests to be notified of them. He also welcomes other comments and suggestions.

Hale Bradt
Salem, MA
USA
bradt@mit.edu

Note for 2010 printing: The only substantial corrections in this printing pertain to the Sunyaev-Zeldovich effect. These include (i) a revised Figure 9.7c with thanks to E. L. Wright of UCLA, (ii) a revised Eq. 9.26 and discussion pertaining thereto including clarification of the distinction between the "scattered spectrum" and the "modified" (observed) spectrum, with thanks to Alan Levine of MIT, and (iii) an addition to Problem 9.52. John Fulton of Austin Community College helpfully pointed out several other errors in both the text and the problem solutions, which are not cited here because they were not fundamentally misleading. I look forward to further comments from interested readers.

Hale Bradt
February 2010

Acknowledgments

I am indebted to many colleagues at MIT and elsewhere and to many students for their encouragement and assistance in hallway discussions, in class, and as readers of draft chapters over the course of the several decades that this work has been evolving. It is impossible to list fairly all those who helped in these ways, but I will mention those who come to mind. I apologize for omissions. It goes without saying that those mentioned are not responsible for errors; I assume that role.

Colleagues: Frederick Baganoff, John Bahcall (deceased), Marshall Bautz, Edmund Bertschinger, Kenneth Brecher, Robert Buonanno, Bernard Burke, Claude Canizares, Deepto Chakrabarty, George Clark, Sergio Colafrancesco, Angelica Costa-Tegmark, Alessandra Dicredico, Emilio Falco, Marco Feroci, Kathy Flanaghan, Peter Ford, Paolo Giommi, Mark Gorenstein, Marc Grisaru, Sebastian Heinze, Jacqueline Hewitt, Scott Hughes, Gianluca Israel, Paul Joss, Kenneth Kellermann, Alan Krieger, Pawan Kumar, Alan Levine, Walter Lewin, Alan Lightman, Herman Marshall, Jaffrey McClintock, Christopher Moore, James Moran, Edward Morgan, Philip Morrison, Stuart Mufson, Stanislaw Olbert, Dimitrios Psaltis, Saul Rappaport, Ronald Remillard, Harvey Richer, Peter Saulson, Paul Schechter, Rudolph Schild, Irwin Shapiro, David Shoemaker, David Staelin, Luigi Stella, Victor Teplitz, David Thompson, John Tonry, Wallace Tucker, Jan van Paradijs (deceased), Joel Weisberg.

Graduate and undergraduate students (at the time): Stefan Ballmer, David Baran, James "Gerbs" Bauer, Jeffrey Blackburne, Adam Bolton, Nathaniel Butler, Eugene Chiang, Asantha Cooray, Yildiz Dalkir, Antonios Eleftheriou, James Gelb, Karen Ho, Juliana Hsu, Tanim Islam, Rick Jenet, Jeffrey Jewell, Jasmine Jijina, Justin Kasper, Vishnja Katalinic, Edward Keyes, Janna Levin, Glen Monnelly, Matthew Muterspaugh, Tito Pena, Jeremy Pitcock, Philipp Podsiadlowski, Dave Pooley, Robert Shirey, Alexander Shirokov, Donald A. Smith, Mark Snyder, Seth Trotz. Student readers of recent proof chapters are Enrico Bozzo, Nicholas Dibella, Tamer Elkholy, David Ely, Elizabeth George, Dacheng Lin, Michael Matejek, Thanasin Nampaisarn, Stephen O'Sullivan, Alessandro Papitto, Rurik Primiani, Leslie Rogers, Madeline Sheldon-Dante, Sarah Vigeland, and Phillip Zukin.

I am especially grateful to colleagues Saul Rappaport and Stu Teplitz for their reading of the entire set of notes some years ago, to Alan Krieger for his more recent reading of the manuscript, to Saul for assistance on binary evolution (Section 4.5), to Stan Olbert for his suggested approach for the review of special relativity in Chapter 7 and for sensitizing me to the underlying role of the distribution function in astronomy (Section 3.3), and to Alan Levine for the derivation in Section 7.8 and many conversations and much encouragement.

Saul, Alan, and Stan have been especially generous with their time and counsel. In the days before personal word processors, secretaries Trish Dobson, Ann Scales, Patricia Shultz, and Diana Valderrama did yeoman's duty in typing revisions of the notes for my classes.

Much appreciated allowances have been made for my writing efforts by the Department of Physics at MIT, by my colleagues at the MIT Center for Space Research (now the MIT Kavli Institute for Astrophysics and Space Research), and by my associates in the Rossi X-ray Timing Explorer (RXTE) satellite program at MIT, the University of California at San Diego, and NASA's Goddard Space Flight Center. The hospitality of the Institute of Space and Astronautical Science (ISAS) in Japan and the Observatory of Rome (OAR) in Italy provided extended periods of quiet writing for which I am grateful. My hosts in Japan (Minoru Oda [deceased] and Yasuo Tanaka) and Italy (Roberto Buonanno, Emanuele Giallongo, and Luigi Stella) have my special thanks for these opportunities.

It has been a pleasure to work with the current and former staff of Cambridge University Press – in particular, Jacqueline Garget, Vincent Higgs, Jeanette Alfoldi, Lindsay Barnes, Anna-Marie Lovett, and especially Simon Mitton who guided and encouraged me from the beginning. Eleanor Umali, Donna Weiss, and their associates at Aptara Corp. – especially the skillful and ever so tolerant production/typesetting team in New Delhi – and copyeditor John Joswick have produced this book with great professionalism. All have been encouraging, creative, patient, and ever helpful.

Finally, I thank my wife Dottie and our daughters Elizabeth and Dorothy for their support over the years.

Also by the author

Astronomy Methods – A Physical Approach to Astronomical Observations
(Cambridge University Press, 2004)

Contents:

This text is an introduction to the basic practical tools, methods, and phenomena that underlie quantitative astronomy. The presentation covers a diversity of topics from a physicist's point of view and is addressed to the upper-level undergraduate or beginning graduate student. The topics include

- the electromagnetic spectrum;
- atmospheric absorption;
- celestial coordinate systems;
- motions of celestial objects;
- eclipses;
- calendar and time systems;
- telescopes in all wavebands;
- speckle interferometry and adaptive optics;
- astronomical detectors, including charge-coupled devices (CCDs);
- two space gamma-ray experiments;
- basic statistics;

- interferometry to improve angular resolution;
- radiation from point and extended sources;
- the determination of masses, temperatures, and distances of celestial objects;
- the processes that absorb and scatter photons in the interstellar medium together with the concept of cross section;
- broadband and line spectra;
- the transport of radiation through matter to form spectral lines; and, finally,
- techniques used to carry out neutrino, cosmic-ray, and gravity-wave astronomy.

1

Kepler, Newton, and the mass function

What we learn in this chapter

Binary star systems serve as laboratories for the measurement of star masses through the gravitational effects of the two stars on each other. Three observational types of binaries – namely, **visual**, **eclipsing**, and **spectroscopic** – yield different combinations of parameters describing the **binary orbit** and the **masses** of the two stars. We consider an example of each type – respectively, **α Centauri, β Persei (Algol)**, and **φ Cygni**.

Kepler described the orbits of solar planets with his **three laws**. They are grounded in **Newton's laws**. The **equation of motion** from Newton's second and gravitational force laws may be solved to obtain the **elliptical motions** described by Kepler for the case of a very large central mass, $M \gg m$. The results can then be extended to the case of **two arbitrary masses** orbiting their common barycenter (center of mass). The result is a **generalized Kepler's third law**, a relation between the masses, period, and relative semimajor axis. We also obtain expressions for the **system angular momentum** and **energy**. Kepler's laws are useful in determining the **orbital elements** of a binary system.

The generalized third law can be restated so that the measurable quantities for a star in a **spectroscopic binary** yield the **mass function**, a combination of the two masses and inclination. This provides a **lower limit to the partner mass**. Independent measures of the partner star's mass function and also of **orbital tidal light variations** or an **eclipse duration**, if available, can provide the information needed to obtain the masses of both stars and the inclination of their orbits. The track of one of the two stars in a **visual binary** relative to the other yields the **sum of the two masses** if the distance to the system is known. The tracks of both stars in inertial space (relative to the background galaxies) together with the distance yield the **two individual masses**.

Measurement of the optical mass function of the partner of the x-ray source Cyg X-1 revealed the first credible evidence for the **existence of a black hole**. Timing the arrival of pulses from a **radio or x-ray pulsar** provides information equivalent to that from a spectroscopic binary. Such studies have made possible the determination of the **masses of several dozens of neutron stars**. They have also provided the first evidence of **exoplanets**, which are planets outside the solar system. Almost **300 exoplanets** have now been discovered – most of them through optical radial velocity measures that detect the **minuscule wobble of the parent star**.

At the **center of the Galaxy**,[*] orbits of stars near the central dark mass have yielded the **mass of the central object**, $\sim 3 \times 10^6 \, M_\odot$, and give strong evidence that it is indeed a **massive black hole**. Spectral and imaging data from orbiting bodies, used together, have yielded the **distance to the center of the Galaxy** and to the nearby star cluster, **the Pleiades**.

---

[*] In this text, we use "Galaxy" for our own galaxy (the Milky Way) and "galaxy" for other galaxies.

1.1   Introduction

Between one-third and two-thirds of all stars are in *binary stellar systems*. In such a system, two stars are gravitationally bound to one another; they each orbit the common center of mass (*barycenter*) with periods ranging from days to years for normal stars and down to hours or less for systems containing a compact star. In this chapter, we examine the motions of the individual stars and describe how these movements can be deduced from observations. We then learn how to deduce the masses of the component stars. Finally, we examine some contemporary applications of Kepler's and Newton's laws.

The motions of stars in a binary system can be understood in terms of the second law ($F = ma$) of Isaac Newton (1643–1727). This is worked out initially for a massive star M orbited by a much smaller mass m (i.e., $M \gg m$). The motion of a body in a gravitational r^{-2} force field is found to follow an elliptical path. The derived motions satisfy *Kepler's laws*, which were empirically discovered by Johannes Kepler (1571–1630). Thereafter, the "two-body" problem is worked out for two bodies of arbitrary masses. The results for the $M \gg m$ case provide a useful shortcut to the solution of the more general case.

In many binary systems, the stars are so far apart that they evolve quite independently of each other. In this case, their binary membership is only of incidental interest. In many systems, however, the two stars are so close to each other that their mutual interactions greatly affect their structure and evolution through tidal distortion and interchange of matter. The creation of white dwarfs, neutron stars, and black holes can follow directly from the modified evolutionary paths; see Section 4.5. In this chapter our analyses pertain only to the effects of gravitational interaction of point masses.

1.2   Binary star systems

The binary systems observable in optical (visible) light are of three general types: *visual*, *eclipsing*, and *spectroscopic*. The classes are not mutually exclusive; for example, a system can be eclipsing *and* spectroscopic. The distinctions between the classes arise from the sizes of the two stars, their closeness to each other, and their distance from the observer.

An additional observational class is that of (*visual*) *astrometric binaries,* wherein only one star is detectable but is observed to wobble on the sky owing to its orbital motion about a stellar companion. An example is AB Doradus, which has been tracked to milliarcsecond precision with very long baseline (radio) interferometry (VLBI). It is now known to be a quadratic system of late-type stars.

Celestial laboratories

Binary stellar systems may be considered laboratories in space. One star interacts with the other, and its response to the environment of the other can be measured by the signals (photons) reaching us. One measurable quantity is the orbital period, and another is the line-of-sight velocity. If the orbit is viewed edge-on, the mass of a massive star can be determined by measurements of a much lighter companion. Similar information can be obtained from systems in which the masses of the stars have arbitrary values. Such studies have long been important in optical astronomy; since 1971, they have been important in x-ray astronomy.

There can, of course, be forces on the entire two-star binary system due to other (external) gravitational systems. These external forces will accelerate the system's center of mass. In this chapter, we assume that there are no significant external forces on the two-star system. In this case, the center of mass will be stationary or will drift through space with a constant velocity. Our attention will be focused on the motion of the two stars relative to each other and to their center of mass (barycenter).

The large proportion of stars in binary systems (about one-half) is one indication that stars are formed from the interstellar medium in groups. Triple systems are also common. Another indication is the existence of groups of stars in the Galaxy (open clusters) such as the Pleiades. The stars in such clusters were all formed at about the same time and probably condensed out of a single interstellar cloud. If two or more stars are formed sufficiently close to each other to be gravitationally bound, they will orbit each other and will thus be a binary or triple system.

The nature of the component stars in binary systems is as varied as the types of stars known to us. Almost any type of star can be in a binary system. Two main-sequence stars (Section 4.3) are common, but there are also highly evolved systems such as (*i*) *cataclysmic variables*, in which one component is a white dwarf; (*ii*) *neutron-star binaries*, in which one component is a neutron star; and (*iii*) *RS CVn binaries*, in which one component is a flaring K giant. (The latter class is named after the star RS in the constellation Canes Venatici = Hunting Dogs.) A binary system that includes a main-sequence star may contain a giant star because one of the stars will have moved off the main sequence to become a red giant; see Section 4.3.

In many of these cases, the close proximity of the two stars to each other leads to direct and complex interactions between them. For instance, as a star expands to become a giant, its gas envelope can overflow its potential well and flow onto a close partner. This process is called *accretion*. If the partner is a neutron star, the accretion leads to the emission of x rays. If it is a white dwarf, highly variable optical emission is seen as well as some x rays. The energy from the emission comes from the release of gravitational potential energy by the infalling material.

The close proximity of two stars can also disturb the atmosphere of a star, giving rise to turbulence and flaring (RS CVn binaries) caused by tidal effects. Their proximity can also distort the shape of a star in the same way that the moon changes the levels of the earth's oceans.

Accretion of gas from one star to another in a binary system can dramatically modify the evolution of the two stars. For example, accreted gas will increase the mass of a normal,

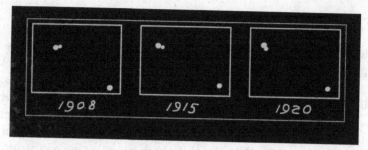

Figure 1.1. Kruger 60, a visual binary. The relative positions of the two stars (upper left) are seen to vary as they orbit each other with a period of 44.6 yr. The two stars are M stars with visual magnitudes 9.8 and 11.4. They are distant from the earth 12.9 LY with relative angular semimajor axis 2.4″ ($a_s = 9.5$ AU). [Yerkes Observatory]

gaseous star, leading to faster nuclear burning and a shorter life. The interactive evolution can lead to millisecond pulsars, which are neutron stars spinning with periods of a few milliseconds. The study of these objects can therefore provide great insight into the underlying physics of stars. Binary stellar systems are a diverse and fascinating breed of objects worthy of study in their own right; see Section 4.5.

Visual binaries

Visual binaries are systems that can be seen as two adjacent stars on an image of the sky, such as a photographic plate (e.g., Kruger 60 shown in Fig. 1 and Sirius). Over a period of some years, the two stars can be seen to orbit about each other. In such systems, the motion of one or both stars on the sky can be mapped to yield important parameters of the system.

An example of such mapping is α Centauri (Fig. 2). In this case, the asymmetry of the path is a consequence of an elliptical (eccentric) orbit. The orientation of the orbit to the line of sight gives it a strange appearance. The degree of eccentricity, the 80-yr period, and the angular size of the orbit (projected angular semimajor axis) provide information about the masses of the stars.

The inclination of the orbit relative to the line of sight is conventionally defined with the inclination angle i (Fig. 3a). If the observer is viewing the orbit face-on (i.e., normal to the orbital plane), the inclination is zero, $i = 0°$. If the observer is in the orbital plane, the inclination is $i = 90°$.

Eclipsing binaries

Eclipsing binaries are systems in which one star goes behind the other. This will happen if (*i*) the stars are very close to each other, (*ii*) one of the stars is sufficiently large, and (*iii*) the orbital plane is viewed more or less edge-on (inclination $\sim 90°$). Conditions (*i*) and (*ii*) can be summarized by requiring that the ratio of star size R to the distance s between the stars be of order unity:

$$R/s \approx 1.$$ (Approximate condition for eclipsing binary)   (1.1)

If this condition is met, the inclination need not be particularly close to $90°$.

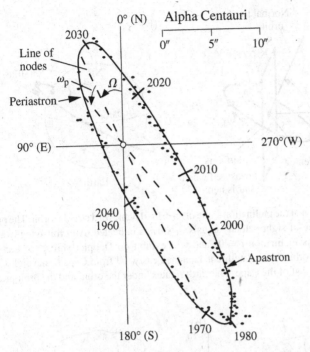

Figure 1.2. The α Centauri visual binary system. This is a plot of the positions projected onto the plane of the sky of one of the two stars relative to the other as a function of time (years). The origin is one of the components and the radial distances are in arcseconds (see scale). The track of the star in the plot is the projection of an ellipse, which is also an ellipse, but with shifted focus. The stars are at their smallest physical separation at the position marked *periastron* and at their largest at *apastron*. The *line of nodes* is the intersection at the focus (origin) of the plane of the orbit and the plane of the sky (see Fig. 11). The position angle Ω of the line of nodes and the (projected) longitude of periastron ω_p are indicated; they are defined in Fig. 11. The stellar components are bright main-sequence stars (G2 V and K IV) of visual magnitudes $m_V = 0.0$ and 1.36, respectively, with a period of 79.9 yr. This system is very close to the sun, 4.4 LY. (A faint, outlying additional companion, Proxima Cen, is the closest star to the sun.) [After D. Menzel, F. Whipple, and G. deVaucouleurs, *Survey of the Universe*, Prentice Hall, 1970, p. 467]

The stars in eclipsing systems are sufficiently close to each other that they can not be resolved on a traditional optical photograph; they appear to be a single star. The binary character is detected by the reduction of light emanating from the system during eclipse. Modern high-resolution imaging such as interferometry or adaptive optics, however, can sometimes resolve the two stellar components in these systems.

When the smaller of the two stars of a hypothetical binary (Fig. 4a) moves behind its companion, it is *occulted*, and only the light from the larger star reaches the observer. The *light curve* (flux density versus time) thus shows a reduction of light. The light also dims when the small star covers part of the big star. The edges of the dips in the light curve are not vertical; they show a gradual diminution of the light. This is due to the finite size of the small star. The two eclipses per orbit are of different depths; it is instructive to understand this (see

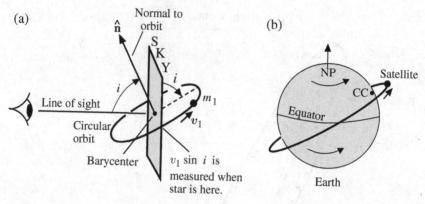

Figure 1.3. (a) Definition of the inclination angle of an orbital plane for a circular orbit. The position of the star when the line-of-sight velocity equals $v_1 \sin i$ is indicated. An orbit lying exactly in the plane of the sky has inclination $i = 0°$ and will exhibit no Doppler shifts. (b) Inclination of low-altitude satellite orbit. If the satellite is launched eastward from Cape Canaveral (CC), its greatest latitude will be that of the Cape. The earth rotates under the orbit, and the orbit precesses with a period of ~ 50 d.

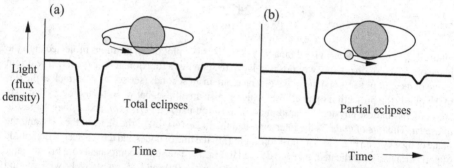

Figure 1.4. Schematic and hypothetical light curves of (a) a totally eclipsing binary and (b) a partially eclipsing binary. A light curve is a plot of flux density versus time. One eclipse is deeper than the other because the stars are assumed to have different surface brightnesses. Larger main-sequence (hydrogen-core-burning) stars are brighter per fixed solid angle than smaller main-sequence stars. The deeper eclipse occurs when the larger star is partially covered.

discussion immediately following). Sometimes the star merely grazes its companion, giving rise to *partial eclipses*. This case is shown schematically in Fig. 4b.

An actual (partial) eclipsing system, Algol, is shown schematically in Fig. 5. It contains one main-sequence star (B8 V) and one subgiant (K2 IV). They orbit each other with a period of 2.9 d. There is a third companion (not shown) at a greater distance that orbits the close pair in 1.9 yr. The system is ~ 100 LY distant, and the close pair are separated by 14 $R_\odot$. Their separation is thus only ~ 2 milliarcsec, and so they appear as a single star through most telescopes. Those now equipped with optical interferometry do resolve the components of this triple system.

The spectral type (Table 4.2) is a measure of the stellar color or temperature T, and this in turn determines the energy outflow per unit area from the stellar surface, which is approximately $\mathscr{F} = \sigma T^4$ (W/m^2), the flux density from a blackbody (6.18). Thus, in the

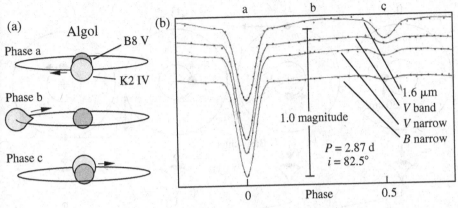

Figure 1.5. (a) Schematic drawing of the partially eclipsing system Algol (β Persei) approximately to scale. Darker shadings represent brighter surfaces. (b) Its light curve in four frequency bands. The designations a, b, c are used to associate portions of the light curve with particular phases of the orbit. The B8 V and K2 IV stars orbit one another with a period of 2.87 d and inclination 82.5° and are separated by 14 $R_\odot$. The K star fills and overflows its pseudopotential well (Roche lobe) and hence accretes matter onto the B star. The changing flux between the two eclipses is due to the changing aspect of the distorted K star and to backscattering of B-star light from the surface of the cooler K star. [(b) R. Wilson *et al.*, *ApJ* **177**, 191 (1972)]

visual band, the effect of partially covering the hotter B8 star is much more pronounced than is the effect of covering the same area of the cooler K2 star, as seen in Fig. 5.

The details of such light curves can tell astronomers a great deal about the stars in the binary system. The existence of the eclipse constrains the orbital plane to lie roughly in the line of sight; the duration and shape of the eclipse are related to the inclination, the separation of the stars, and their physical sizes. The changes of intensity are related to the surface brightnesses and hence the classes of the stars. Can you speculate about the cause of the gradual changes of light during the phases between the two eclipses of Fig. 5 (see caption)?

Spectroscopic binaries

Some close binaries do not eclipse each other because the orbit has low inclination, the stars are sufficiently separated, or both. In these cases, the binary nature of the stars can be identified only through the detection of periodic Doppler shifts in the spectral lines of one or both stars. These binaries are called *spectroscopic binaries*. The Doppler shifts are due to the motions about the system barycenter.

The radial velocity of the star must be great enough to be detected as a spectral Doppler shift, and it must be bright enough to yield sufficient photons for high-resolution spectroscopy. The motions are greatest for binaries of close separation (see (3) below); most known spectroscopic binaries have separations less than 1 AU. Not surprisingly, therefore, systems showing orbital spectroscopic variations often also exhibit eclipses; these are called *eclipsing spectroscopic binaries*.

The orbits and Doppler velocities of a hypothetical binary system are shown in Fig. 6. For simplicity, the orbits are circular and oriented such that the observer (astronomer) is in the plane of the orbit, $i = 90°$. Thus, once each orbit, each star approaches directly toward

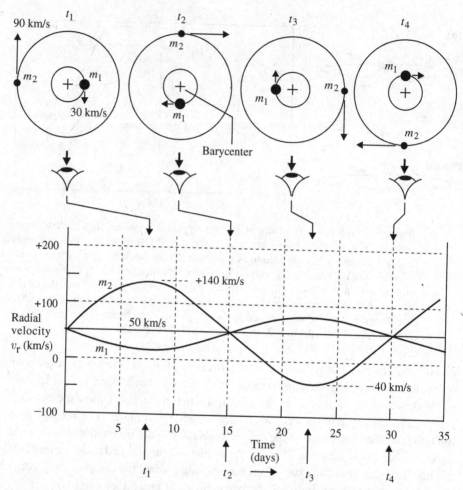

Figure 1.6. Hypothetical spectroscopic binary with circular orbits and a 30-d period shown at four phases of the orbit. The stationary observer is in the plane of the orbit. Star 1 is three times more massive than star 2. At any given time the two stars are on opposite sides of the barycenter, which is moving steadily away from the observer with a radial velocity component of +50 km/s. The star speeds relative to the barycenter are shown in the upper left. Star 2, with its smaller mass, is three times farther out from the barycenter than star 1; hence, it must travel three times faster to get around the orbit in the same time as star 1. The direction to the astronomer and the observed Doppler velocities are shown. [Adapted from G. Abell, *Exploration of the Universe*, 3rd Ed., Holt Rinehart Winston, 1975, p. 439, with permission of Brooks/Cole]

the observer, and a half-period later it recedes directly away. If the orbit were oriented such that it would lie in the plane of the sky, $i = 0°$, the stars would have no component of velocity along the line of sight. In this case, there would be no detectable Doppler shift.

The line-of-sight (radial) velocities are shown in Fig. 6 as a function of time for each star. They are not centered about zero radial velocity because the barycenter of the system is (in our example) receding from the observer with a radial velocity of $v_r = +50$ km/s. Star 1 is more massive than star 2 ($m_1 = 3m_2$). It is therefore closer to the barycenter (upper sketches) and is moving at a lesser velocity relative to the barycenter; see plots.

The radial-velocity curves are obtained spectroscopically from observations of the Doppler shifts of the frequency of stellar absorption (or emission) lines. The shift is toward higher frequency (blue shift) if the object approaches the observer and to lower frequency (red shift) if the object recedes. The usual sign convention for radial velocity in astronomy is "+" for a receding object and "−" for an approaching object.

The Doppler relation between the radial velocity v_r and the frequency shift $\Delta \nu$ for nonrelativistic speeds ($v \ll c$) is

$$\frac{\nu - \nu_0}{\nu_0} = -\frac{v_r}{c}, \qquad\qquad (v \ll c) \qquad (1.2)$$

where ν_0 is the *rest frequency* of the absorption line (as would be seen by an observer moving with the star), ν is the observed frequency at the earth, and v_r is the radial component of the star's velocity relative to that of the earth.

Additional features of Fig. 6 are as follows:

(*i*) zero Doppler velocity (relative to the barycenter) at t_2 and t_4 when the stars are moving at right angles to the line of sight;

(*ii*) sinusoidal radial-velocity curves as expected for projected circular motion at any inclination;

(*iii*) amplitudes that reflect the 3-to-1 mass ratio; and

(*iv*) phases that differ by exactly 180° owing to momentum conservation.

Data from an actual spectroscopic binary, ϕ Cygni, are shown in Fig. 7. Spectral lines from each of the two stars yield, from (2), the plotted radial velocity points. They show asymmetries introduced by the orientation of the elliptical orbits relative to the observer's line of sight and by the varying speeds of the stars as they move in their elliptical orbits. Note that the curves cross at a nonzero velocity. Again, this is due to the motion of the barycenter relative to the observer.

In many actual spectroscopic binary systems, astronomers obtain only one curve because one star is too faint – either in an absolute sense or because its light is swamped by its much brighter companion. These are called *single-line spectroscopic binaries*. If the brighter star is much more massive than its companion, as is likely (see Section 4.3), its motion may be too small to be measured. In this case only an upper limit to v_r is obtained. On the other hand, if the Doppler shifts of both stars are measurable, and if eclipses occur, a wealth of information is obtained. Such a system would be a *double-line eclipsing binary*.

1.3   Kepler and Newton

The laws of Kepler are described here together with an analysis of the ellipse and a presentation of Newton's equations of motion in polar coordinates. The latter lead to Kepler's laws, as we demonstrate in Sections 4 and 5. Here we also discuss how Kepler's laws govern the motions of earth-orbiting satellites.

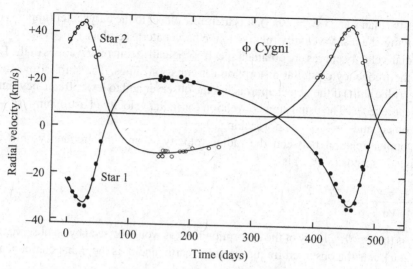

Figure 1.7. The radial velocities as a function of time for φ Cygni derived from the Doppler frequency shifts of the spectral lines. This is a double-line spectroscopic binary system consisting of two giants of about equal masses. The smooth curves are theoretical fits to the data points. The orientation of the elliptical orbit with respect to the line of sight and the nonconstant speeds in the orbit yield the strange shape. The barycenter recedes from the observer. [Adapted from R. Rach and G. Herbig, *ApJ* **133**, 143 (1961)]

Kepler's laws ($M \gg m$)

Kepler carried out a detailed analysis of the celestial tracks of the sun's planets as recorded with good precision by Tycho Brahe (1546–1601). The sun is so much more massive than the planets that it can be considered to be stationary (i.e., the condition $M \gg m$ holds). He discovered three simple laws that well describe the tracks of the planets, the speed variations of a planet in its orbit, and the relative periods of the orbits of the several planets. They are known as Kepler's laws and are as follows:

Kepler I. The orbital track of a given planet is an ellipse with the sun at one of the foci. (A circular orbit is a special case of an ellipse.)

Kepler II. The radius vector (sun to planet) sweeps out equal areas in equal time.

Kepler III. The square of the orbital period P^2 is proportional to the cube of the semimajor axis a^3 of the orbit. That is, $P^2 = c_1 a^3$, where c_1 is a constant *independent of the mass of the planet*. The physical constants that make up c_1 are now known (see (45) below), and so the law becomes

$$GMP^2 = 4\pi^2 a^3,$$
(Kepler III; $M \gg m$)   (1.3)

where M is the mass of the central object, if $M \gg m$.

The first law tells us that the orbits are elliptical (Fig. 8a) and that the sun is at one of the foci. It is remarkable, as we later demonstrate, that, according to a Newtonian analysis, an ellipse is precisely the expected track for an inverse-squared gravitational force law. Kepler was not just close; he was exactly right.

The second law (Fig. 8b) tells us that, as a planet traverses its orbit, it speeds up as it approaches the sun. It is fastest at the closest approach (*perihelion*) and slowest at its

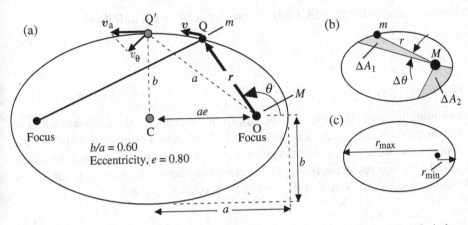

Figure 1.8. (a) Geometry of an elliptical orbit of a mass m showing the two foci, the semimajor axis a, the semiminor axis b, the radius vector r at the azimuthal angle θ, and the two radii whose summed lengths equal a constant, $2a$. The dashed radial line shows the special geometry when the orbiting body is on the semiminor axis at Q' and $r = a$. (b) Kepler's second law. The areas swept out in equal times are equal, $\Delta A_1 = \Delta A_2$. (c) Periastron (r_{min}) and apastron (r_{max}) for an elliptical orbit.

farthest distance (*aphelion*). This law is equivalent to the conservation of angular momentum, as we will find from Newton's second and gravitational laws. Again, Kepler was exactly right.

The third law moves on to compare the orbits of the several different planets in the solar system. The larger orbits have longer periods; the outer planets take longer to orbit the sun than do the inner planets. Again a Newtonian analysis yields the exact relation postulated by Kepler. The more remote the planet, the smaller are the forces and accelerations, which results in smaller angular velocities and hence longer periods.

Ellipse

An ellipse is a conic section that looks like a flattened circle. It can be constructed (Fig. 8a) by requiring that the sum of the two radii from the two foci to point Q on the ellipse be independent of the position of Q. An ellipse can be constructed with a pencil and a piece of string whose ends are anchored at the two foci; the fixed length of the string constitutes the constant sum of the two radii.

An ellipse is equivalently described mathematically with a function $r(\theta)$ that is the length of the vector r as a function of the angle θ defined in Fig. 8a,

$$r(\theta) = a\frac{1 - e^2}{1 + e\cos\theta},$$
(Equation of ellipse)   (1.4)

where a and e are constants called the *semimajor axis* and *eccentricity*, respectively. The eccentricity is restricted to values $0 \le e \le 1$; a larger value ($e > 1$) results in a hyperbola. A plot of (4) about the right-hand focus, with $e = 0.8$, yields the ellipse of Fig. 8a. One finds, in general, that the semimajor axis (1/2 the long dimension) is equal to the parameter a.

The minimum and maximum radii, from (4), are, for $\cos\theta = \pm 1$, as follows:

$$r_{\min} = a(1 - e), \qquad\qquad\qquad\qquad (\theta = 0^\circ) \quad (1.5)$$

and

$$r_{\max} = a(1 + e). \qquad\qquad\qquad\qquad (\theta = 180^\circ) \quad (1.6)$$

An object orbiting the earth is at the distance $r_{\min}$ from the center of the earth when it is at *perigee*, and the object is at *apogee* when it is at $r_{\max}$. If the object is orbiting a star, such as in a binary stellar system, these two parameters give the closest and farthest points and are called *periastron*, and *apastron*, respectively. For an object orbiting the sun, the corresponding terms are *perihelion and aphelion*. Note that the sum of (5) and (6) is (Fig. 8c)

$$r_{\min} + r_{\max} = 2a, \qquad\qquad\qquad\qquad\qquad (1.7)$$

which justifies our statement immediately above that a is the semimajor axis. The distance $2a$ is the length of the hypothetical string used to construct the ellipse graphically; to see this, visualize point Q to be at $\theta = 0^\circ$.

The string length may also be obtained from (4) at point Q′ midway along the ellipse (dashed line). First, note that the distance from origin O (at the focus) to C is, from (5),

$$a - r_{\min} = ae, \qquad\qquad\qquad\qquad\qquad (1.8)$$

and that, for this case,

$$\cos\theta = -ae/r, \qquad\qquad\qquad\qquad\qquad (1.9)$$

where $r = r(Q')$ is the unknown distance OQ′. Substituting (9) into (4) yields

$$r = a\frac{1 - e^2}{1 - (ae^2/r)}. \qquad\qquad\qquad\qquad\qquad (1.10)$$

Solving for r results in

$$r(Q') = a. \qquad\qquad\qquad\qquad\qquad (1.11)$$

The "string" has twice this length, or $2a$, which is the same as at $\theta = 0^\circ$ (7). This verifies that (4) is consistent with a constant-length string for a second point on the ellipse. The general proof of the constancy follows from (4), the ellipse equation (Prob. 31).

The half-width of the ellipse is specified as b, the semiminor axis. The triangle COQ′ and the Pythagorean theorem yield

$$b/a = (1 - e^2)^{1/2}. \qquad\qquad\qquad\qquad\qquad (1.12)$$

From this, we find that $(e = 0)$ corresponds to circular motion $(a = b)$ and that $e \to 1$ approaches straight-line motion $(b \to 0)$.

The Newtonian connection

Newton was born 12 years after Kepler died. He was able to show that the elliptical orbits of the planets deduced by Kepler could be understood in terms of a radial r^{-2} force proportional to the product of both masses (sun and planet). Such a simple force giving rise to elliptical orbits was an impressive theoretical result but did not, in itself, indicate the correctness of the theory. It was the fact that the planets do indeed follow elliptical orbits that validated Newton's theory. It is also a credit to Kepler, and to Tycho Brahe, whose data Kepler used, that Kepler's empirical laws proved to be precisely in accord with Newton's more fundamental theory.

The expression known as *Newton's law of gravitation* is

$$F_G = -\frac{GMm}{r^3} r, \qquad \text{(Newton's law of gravitation)} \qquad (1.13)$$

where F_G is the force on the planet, M and m are the masses of the sun and planet, respectively, and r is the radius vector directed from mass M (at focus O) toward mass m (at Q; Fig. 8a). The magnitude of the ratio r/r^3 is $1/r^2$; thus, (13) describes the familiar r^{-2} gravitational force. The negative sign indicates that the force on the planet is toward the sun. The vector force (13) is purely radial; it has no azimuthal component and is called a *central force*.

There is no torque N about the origin at mass M because F_G is radial and hence $N \equiv r \times F_G = 0$. The absence of torque implies that angular momentum J of the mass m (again taken about mass M) is conserved during the motion, $N = dJ/dt = 0$. A constant J vector during the orbit means that the motion is confined to a plane.

The expression known as *Newton's second law* $F = ma$ is, in differential form,

$$F = m\frac{d^2r}{dt^2}. \qquad \text{(Newton's second law)} \qquad (1.14)$$

For the gravitational force F_G (13), this yields the equation of motion

$$-\frac{GMm}{r^3} r = m\frac{d^2r}{dt^2}. \qquad \text{(Vector equation of motion)} \qquad (1.15)$$

The solution of (15), $r(t)$, should give the planetary motions described by Kepler.

The time-dependent radius vector $r(t)$ describes the motion of the mass m in the presence of the r^{-2} gravitational field. The vector can be described at any instant in terms of its polar components in the orbital plane – namely, radius r and azimuth θ. In our solution, the scalar function $r(\theta)$ will map out an ellipse (4), where θ, in turn, is a function of time, $\theta = \theta(t)$. The latter function describes the nonuniform speeds at which the planet moves around the ellipse.

The general method for finding a solution to an arbitrary differential equation is to guess possible solutions and to substitute them into the equation until one is found that satisfies it. (Certain classes of differential equations have well-known solutions, and others may be solved analytically by integration.) In our case, the elliptical motion suggested by Kepler in his laws will be used in Section 4 as the trial solution, and, of course, we will find that it satisfies the equation of motion (15). Several useful relations and insight into Kepler's laws will be obtained from this analysis.

Earth-orbiting satellites

We pause here to point out that Kepler's laws are a great help in the qualitative understanding of earth-orbiting satellite motion. For example, the second law, equal areas swept out in equal times, describes the rapid transit through perigee and the slower transit through apogee. X-ray astronomy satellites such as Chandra take advantage of this; it carries out observations far from the earth's radiation belts near apogee and spends relatively little time in the background-producing radiation belts near perigee.

The third law, $P^2 \propto a^3$, explains why earth-synchronous satellites that relay television signals with an orbital period of 24 h must be inserted into much higher orbits than low–earth-orbiting satellites that have periods of about 90 min.

Kepler's first law tells us that all such orbits should follow perfectly elliptical orbits in the presence of a perfect r^{-2} force (i.e., in the absence of any perturbing forces). In practice this condition is quite well satisfied for most satellites. In this approximation, the satellite follows the elliptical orbit perfectly and will always return to the point in inertial space where it had been inserted into the orbit. In extremely strong gravitational fields, however, orbits deviate from perfect ellipses according to Einstein's general theory of relativity.

Orbit change

Consider, for example, that mission controllers wish to change the orbit of an earth-orbiting satellite. The rocket on the satellite is commanded to give the satellite a momentum impulse with a brief "burn." The resultant new direction and velocity define the new orbit. If the impulse is directed in (or opposed to) the direction of the initial momentum, it will speed up (or slow down) the satellite, but the plane of the orbit will remain the same. If this impulse is in some other direction, not in the orbital plane, it will change the plane of the orbit.

The satellite will proceed along its new elliptical orbit, eventually returning through the same point where the impulse was applied. No matter how much energy you use (less than the escape energy), the satellite will always return to the firing point like a boomerang. You can not get rid of it! The new eccentricity and semimajor axis depend on the particular direction and energy the satellite has after the burn.

For example, one can raise a satellite from a low circular orbit into a high circular orbit with two rocket firings. The first burn imparts a momentum in the direction of the initial momentum, thus increasing the energy and leading to a higher apogee at the desired new altitude (say, synchronous altitude). The perigee, however, remains at the firing point according to the discussion just above; the intermediate orbit is therefore highly elliptical. A second firing when the satellite is at apogee, again in the direction of motion, will raise the perigee up to the desired final altitude, thus yielding the desired circular orbit.

Another example is the burn that gives the impulse leading to reentry of a manned spacecraft into the atmosphere. Assume an initial circular orbit. A burn is applied to slow down the spacecraft. It thus enters a new elliptical orbit with a lowered perigee that is within the upper atmosphere. Atmospheric friction then removes the additional energy required for a safe landing.

Launch inclination

This visualization is also useful in understanding the inclination of a satellite orbit relative to the earth's equatorial plane after the satellite is first launched into orbit. Consider the launch to be a single impulse given to the satellite directly above its launch site and high enough to be free of atmospheric drag. From Cape Canaveral, Florida, the impulse is generally eastward because the eastward rotation of the earth provides an additional thrust and larger weights can be launched. This impulse will generally be just sufficient to yield a circular orbit – that is, to lift the perigee (on the other side of the earth) to the appropriate altitude.

The satellite will then follow an orbit that is in the plane containing the earth's center and passing east–west through the impulse point (Fig. 3b). Thus, it moves first eastward, then southward, and then northward again, passing again through the impulse point going eastward. Because this is the most northern point in the orbit, the orbit inclination is the same as the latitude of Cape Canaveral, namely 28°.

This orbital track is almost stationary in inertial space; it is fixed relative to the distant galaxies, and the earth rotates under it. The northernmost point is thus fixed in inertial space, and Cape Canaveral passes under it once a day, during which time the satellite makes ~15 orbital passes. After a half-day, or 7–8 orbits, the satellite crosses the meridian of Cape Canaveral far to the south at latitude −28°.

In fact, the orbit is not exactly stationary. It actually precesses slowly in inertial space owing to gravitational torques applied by the bulge of the earth's equator and the sun. The time for the precession to complete a cycle is ~50 d for a low–earth-orbiting satellite.

One could launch into a more highly inclined orbit than 28° from Cape Canaveral by giving a northern component to the initial impulse with enough energy to yield a near circular orbit, but this would require a big energy expenditure. A southern component to the initial impulse would also increase the inclination; think about it! Furthermore, it is impossible to obtain an inclination less than 28° with a single impulse from Cape Canaveral; think about this too. Launch into an equatorial orbit (inclination 0°) from Cape Canaveral would require a second burn (impulse) and much energy. Where must that burn take place?

Finally, consider launch into a polar orbit, one that passes repeatedly over both the North Pole and South Pole. Can it be inserted with one burn from Cape Canaveral? Why are such launches always made from the west coast and not from Canaveral? (See Prob. 34.)

1.4   Newtonian solutions $M \gg m$

We now proceed to find the solution to the equation of motion (15). As noted after (15), we do this with a trial solution that describes the postulated elliptical motion.

Components of the equation of motion

The first step is to rewrite the vector equation (15) as two scalar equations, one for each component (radial and azimuthal), in polar (r, θ) coordinates as follows:

$$-\frac{GMm}{r^2} = m\frac{d^2r}{dt^2} - m\omega^2 r \qquad \text{(Radial equation of motion)} \qquad (1.16)$$

and

$$0 = mr\frac{d^2\theta}{dt^2} + 2m\omega\frac{dr}{dt}. \qquad \text{(Azimuthal equation of motion)} \qquad (1.17)$$

These equations follow from a description of acceleration for arbitrary differential motion of the vector r (Prob. 41). The angular velocity ω is shorthand for $d\theta/dt$; $\omega \equiv d\theta/dt$.

The left sides of (16) and (17) are the radial and azimuthal components of the force, respectively; because the vector force is totally radial, there is no azimuthal component. The right sides include the radial and azimuthal components of the vector acceleration. The radial acceleration consists of two terms: one incorporating acceleration in the radial direction (d^2r/dt^2) and the other radial acceleration arising from azimuthal motion $(\omega^2 r = v_\theta^2/r)$.

Similarly, the azimuthal expression (17) has a term that depends on the instantaneous radial velocity dr/dt. For circular motion, the radius r is fixed, and so this term is zero. In this case, the equation tells us that $d^2\theta/dt^2 = 0$ or that the angular velocity $\omega = d\theta/dt$ is constant as expected in the absence of any azimuthal force component.

Angular momentum (Kepler II)

Let us now show that the azimuthal equation of motion (17) directly yields angular-momentum conservation and Kepler's second law (equal areas swept out in equal times).

Rewrite (17) as follows:

$$\frac{1}{r}\frac{d}{dt}(mr^2\omega) = 0, \tag{1.18}$$

which you can verify by taking the derivative indicated and recalling that $\omega = d\theta/dt$. This tells us that the time derivative of $mr^2\omega$ is zero. Hence, this product must not depend on t; it is a constant of the motion.

Recall that the magnitude of the angular momentum vector $\boldsymbol{J} = \boldsymbol{r} \times \boldsymbol{p}$ may be written in scalar form several ways: $J = rmv_\theta$, where v_θ is the azimuthal component of the velocity, or as $J = mr^2\omega$ because $v_\theta = \omega r$ (from $\omega \equiv d\theta/dt$ and the definition of the radian). Thus, from (18), we find that angular momentum magnitude J is a constant of the motion:

$$J \equiv mr^2\omega = \text{constant.} \tag{1.19}$$

The differential area dA swept out as the mass m moves an angle $d\theta$ is indicated macroscopically by the triangular shaded area in Fig. 8b. The magnitude of the differential area is simply 1/2 the base r times the height $r\,d\theta$:

$$dA = (r\,d\theta\,r)/2 = (1/2)r^2\,d\theta. \tag{1.20}$$

The rate of area swept out is

$$\frac{dA}{dt} = \frac{1}{2}r^2\frac{d\theta}{dt} = \frac{1}{2}r^2\omega. \tag{1.21}$$

Comparison with (19) gives

$$➡\qquad \frac{dA}{dt} = \frac{J}{2m} = \text{constant.} \qquad \text{(Kepler's second law)} \tag{1.22}$$

This expression, $dA/dt = \text{constant}$, is a mathematical statement of Kepler II, which, as promised, we have now verified with Newton's laws.

Recall that angular momentum of a body or a system can not be determined unless the origin (or axis) about which the angular momentum is to be calculated is first specified. The convenient choice for the $M \gg m$ situation is to choose the (assumed stationary) mass M as the origin, as is done here. The force about a different origin would not be central, and angular momentum about that origin would not be conserved.

Elliptical motion (Kepler I)

We demonstrate here that the solution of the equation of motion (15) is an elliptical track. To do this, the elliptical trial solution $r(\theta)$ (4) and the radial equation of motion (16) are transformed to more convenient forms. The azimuthal equation of motion (17) is invoked through the conservation of angular momentum.

Trial solution transformed

Rewrite the ellipse expression $r(\theta)$ (4) making use of the ratio b/a (12) as follows:

$$r(\theta) = \frac{b^2}{a} \frac{1}{1 + e \cos \theta}. \qquad \text{(Trial solution; equation of ellipse)} \qquad (1.23)$$

Now define the variable $u \equiv 1/r$ because it will simplify the evaluation of the differential equation:

$$\Rightarrow \qquad u(\theta) \equiv \frac{1}{r} = \frac{a}{b^2}(1 + e \cos \theta). \qquad \text{(Trial solution; an ellipse; } u \equiv 1/r) \qquad (1.24)$$

This will be our trial solution. In using it, we will invoke the constancy of angular momentum described by

$$J = mr^2 \omega = m\omega/u^2 = \text{constant}. \qquad (1.25)$$

Radial equation transformed

Now rewrite the equation of motion (16) in terms of u and with the independent variable as the angle $\theta = \theta(t)$ rather than time t. The left side requires only the substitution $r = 1/u$:

$$GMm/r^2 \rightarrow GMmu^2. \qquad \text{(First term of radial equation of motion)} \qquad (1.26)$$

The rightmost term becomes, from (25),

$$m\omega^2 r \rightarrow m\left(\frac{Ju^2}{m}\right)^2 \frac{1}{u} = \frac{J^2}{m}u^3. \qquad \text{(Third term)} \qquad (1.27)$$

The $\mathrm{d}^2 r/\mathrm{d}t^2$ term on the right side of (16) is modified by expanding the derivatives:

$$\frac{\mathrm{d}r}{\mathrm{d}t} = \frac{\mathrm{d}r}{\mathrm{d}\theta}\frac{\mathrm{d}\theta}{\mathrm{d}t} = \frac{\mathrm{d}(1/u)}{\mathrm{d}\theta}\omega = -\frac{1}{u^2}\frac{\mathrm{d}u}{\mathrm{d}\theta}\frac{Ju^2}{m} = -\frac{J}{m}\frac{\mathrm{d}u}{\mathrm{d}\theta}, \qquad (1.28)$$

where we again use $\omega = Ju^2/m$ from (25). The second time derivative then similarly becomes

$$\frac{\mathrm{d}^2 r}{\mathrm{d}t^2} = \frac{\mathrm{d}}{\mathrm{d}\theta}\left(\frac{\mathrm{d}r}{\mathrm{d}t}\right)\frac{\mathrm{d}\theta}{\mathrm{d}t} = -\frac{J}{m}\frac{\mathrm{d}^2 u}{\mathrm{d}\theta^2}\omega = -\left(\frac{J}{m}\right)^2 \frac{\mathrm{d}^2 u}{\mathrm{d}\theta^2}u^2. \qquad (1.29)$$

The first term on the right side of (16) thus becomes

$$m\frac{d^2r}{dt^2} \rightarrow -m\left(\frac{J}{m}\right)^2\frac{d^2u}{d\theta^2}u^2 = -\frac{J^2}{m}u^2\frac{d^2u}{d\theta^2}.$$ (Second term)   (1.30)

Substitute (26), (27), and (30) into (16) to obtain the new version of the radial component of the equation of motion for $u(\theta)$:

$$\Rightarrow \qquad -GMmu^2 = -\frac{J^2}{m}u^2\left(u + \frac{d^2u}{d\theta^2}\right).$$ (Radial equation of motion)   (1.31)

Solution

Finally, the promised test of the trial solution (an ellipse) is at hand. Substitute the trial solution (24) into the radial equation of motion (31). (We reiterate that the azimuthal equation is taken into account through our use of $J =$ constant.) As the first step of the substitution, evaluate the parenthetical term in (31); take the second derivative of (24) to obtain $d^2u/d\theta^2 = -ae(\cos\theta)/b^2$. The parenthetical term in (31) becomes

$$\left(u + \frac{d^2u}{d\theta^2}\right) = \frac{a}{b^2}(1 + e\cos\theta) - \frac{a}{b^2}e\cos\theta = \frac{a}{b^2}.$$ (1.32)

The result of the substitution into (31), so far, is thus

$$-GMmu^2 \underset{?}{=} -\frac{J^2}{m}\frac{a}{b^2}u^2.$$ (1.33)

Because the variable terms u^2 cancel, there is no further need to invoke the trial solution (24). Our substitution is complete.

Return to the variable $r(\theta)$,

$$\Rightarrow \qquad -\frac{GMm}{r^2} \underset{?}{=} -\frac{J^2}{m}\frac{a}{b^2}\frac{1}{r^2},$$ (Test of trial solution)   (1.34)

where the "?" denotes that this is the test of a trial solution. The equality can indeed be satisfied because both sides of the equation vary as r^{-2}. The right side of (34) is proportional to the radial acceleration of elliptical motion defined in (16) with constant angular momentum. It varies as $1/r^2$, and this matches the $1/r^2$ variation of the Newtonian gravitational force (left side).

The trial solution satisfies the equation of motion (31) only if the equality (34) is satisfied. This provides a useful connection between the dimensions of the ellipse and the several constants of the system:

$$\Rightarrow \qquad \frac{b^2}{a} = \frac{J^2}{GMm^2}.$$ (1.35)

Substitute this into our initial trial solution (23) to obtain the elliptical motion in terms of the constants J and e:

$$\Rightarrow \qquad r(\theta) = \frac{J^2}{GMm^2}\frac{1}{1 + e\cos\theta}.$$ (Solution of equation of motion; $M \gg m$)   (1.36)

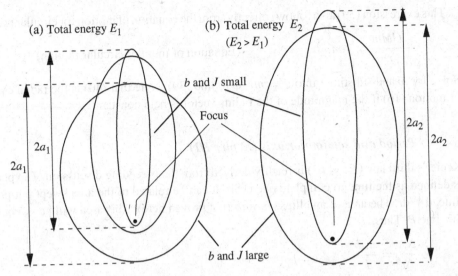

Figure 1.9. Total energy of elliptical orbits. (a) Three orbits of identical total energies E_1 (i.e., with identical semimajor axes a_1) but with differing semiminor axes or angular momenta J. A larger semiminor axis indicates a greater angular momentum. The focus is common to all three orbits. (b) Two orbits with the same total energy E_2 for $E_2 > E_1$ and a common focus. Again, they have different angular momenta.

This expression (36) is the desired solution of the radial and azimuthal equations of motion in terms of the given physical constants. Because it describes elliptical motion (by design of the trial solution), we have demonstrated that Kepler's first law (elliptical orbits) indeed follows from Newton's second and gravitational laws.

Angular momentum restated

The equality (35) may be solved for the angular momentum,

$$J = \left(\frac{GM}{a}\right)^{1/2} m\,b. \qquad \text{(Angular momentum of } m;\ M \gg m)\qquad (1.37)$$

This is the angular momentum of the mass m in an elliptical orbit of dimensions a and b about the mass M for $M \gg m$.

Consider the two sets of orbits in Fig. 9. The orbits of each set have the same semimajor axis a and hence the same total energy; see (52) below. Within each set, the orbits have differing semiminor axes b and hence differing angular momenta J. The limiting values of J, from (37), are

$$J \to 0 \qquad \text{(For } b \to 0;\ e = 1;\ \text{straight-line motion)}\quad (1.38)$$

and

$$J = m(GMr)^{1/2}. \qquad \text{(For } b = a = r;\ e = 0;\ \text{circular motion)}\quad (1.39)$$

The latter expression is the angular momentum of a mass m in a *circular* orbit of radius r about a mass M. It is also the maximum value of J for a given semimajor axis a because b can not exceed a.

This expression (39) also follows directly from the equation of motion for circular motion:

$$-\frac{GMm}{r^2} = -m\omega^2 r.$$ (Equation of motion; circular motion)   (1.40)

Solve for ω and substitute into $J = mr^2\omega$ to find (39). Note that (40) is the radial equation of motion (16) if the magnitude of the radius vector is held constant.

Period and semimajor axis (Kepler III)

Kepler's third law ($P^2 \propto a^3$) is readily derived from the preceding discussion. The period P is defined as the time to complete one orbit. It can be related to the area swept out per unit time, dA/dt. The area of an ellipse is πab (not proven here). This area will be swept out in the time P. Thus,

$$\frac{\Delta A}{\Delta t} = \frac{\pi ab}{P}.$$ (1.41)

Because the rate at which area is swept out is constant (on the basis of angular momentum conservation or Kepler II), the differential rate equals the average rate:

$$\frac{dA}{dt} = \frac{\pi ab}{P}.$$ (1.42)

The left term may be expressed in terms of angular momentum J (22) to yield

$$\frac{J}{2m} = \frac{\pi ab}{P}.$$ (1.43)

Solve this for P and eliminate J with (37),

$$P = 2\pi a^{3/2}(GM)^{-1/2},$$ (1.44)

and rewrite as

➡ $$GMP^2 = 4\pi^2 a^3.$$ (Kepler III; $M \gg m$)   (1.45)

This is Kepler III with the coefficients included, for the $M \gg m$ case, as stated in (3). It is a natural outcome of the application of Newton's second law to the gravitational problem. This law compares the orbit sizes of *different* planets. In contrast, Kepler I and Kepler II describe the orbit of *one* planet.

Note that the orbit sizes do not depend on the planetary mass. Acceleration by gravity is independent of mass; see (15) or remember the (probably false) legend of Galileo and the leaning tower of Pisa.

For a circular orbit of radius r, the semimajor axis is $a = r$, and (45) reduces to

$$GMP^2 = 4\pi^2 r^3.$$ (Kepler III for circular orbit)   (1.46)

This version may be derived in one step from the equation of motion (40) for a circular orbit. This is a quick way to obtain Kepler III if you remember to substitute a for r when the orbit is elliptical.

Total energy

The total energy (kinetic plus potential) of a mass m in an elliptical orbit about a massive object M can be expressed in terms of the parameters of the ellipse and the masses. For the $1/r^2$ gravitational force law, the total energy E_t of a mass m at position r and speed v is

$$E_t = \frac{1}{2}mv^2 - \frac{GMm}{r}, \tag{1.47}$$

where, as usual, the zero point of potential energy is set at $r \to \infty$. Because the total energy is a constant of the motion, it may be evaluated at any convenient place in the orbit such as the point Q' on the semiminor axis (Fig. 8a). At that position, the radius vector r has magnitude exactly equal to the semimajor axis a; see (11). This immediately gives us the potential term $(-GMm/a)$. Thus, the expression to evaluate is

$$E_t = \frac{1}{2}mv_a^2 - \frac{GMm}{a}, \tag{1.48}$$

where the speed at Q' (at distance $r = a$) is designated v_a.

At the position Q', the geometry of similar triangles (Fig. 8a) provides a relation between the quantity v_a and its azimuthal component v_θ:

$$\frac{v_\theta}{v_a} = \frac{b}{a}. \tag{1.49}$$

Write v_θ in terms of the angular momentum $J = mr^2\omega$,

$$v_\theta = \omega r = \frac{J}{mr^2}r \to \frac{J}{ma}, \tag{1.50}$$

and solve (49) to obtain v_a in terms of the orbital constants,

$$v_a = v_\theta \frac{a}{b} = \frac{J}{ma}\frac{a}{b} = \frac{J}{mb} = \left(\frac{GM}{a}\right)^{1/2}, \tag{1.51}$$

where J was eliminated with (37). Finally, substitute v_a into the expression (48) for E_t:

$$\rightarrow \quad E_t = -\frac{GMm}{2a}. \qquad \text{(Total energy; } M \gg m) \tag{1.52}$$

The total energy (52) turns out to be amazingly simple. Given the mass of the sun and a planet (M and m, respectively), the total energy (kinetic + potential) of the orbiting planet is completely determined by the size of the semimajor axis a. A larger a makes E_t less negative and hence greater in value. Compare the orbits of differing energies in Fig. 9a,b.

The size and shape of the orbit depend only on the values of E_t and J for given masses M and m. It is therefore possible to rewrite $r(\theta)$ (4) in terms of these constants rather than in terms of a, b, and e (Prob. 47). This turns out to be the expression (36) with eccentricity

$$e = \left[1 + \frac{2E_t J^2}{(GM)^2 m^3} \right]^{1/2}.$$

(Eccentricity)   (1.53)

The expressions developed here make it possible to gain an immediate qualitative feeling for planetary orbits. For example, an orbit with a long period will have a large semimajor axis (Kepler III (45)) and a large (meaning less negative) total energy (52). Also, the system has the maximum angular momentum (37) for a given semimajor axis a (or energy E_t) if the orbit is circular ($b = a$) as given in (39). Our expressions, so far, are valid only for the case of $M \gg m$. Finally, if the orbit is circular with radius r, our general expressions for J (37), Kepler III (45), and E_t (52) may be simplified with

$$a \to r; b \to r.$$

(Circular orbits)   (1.54)

We have now developed the connection between the Newtonian gravitational force and the elliptical orbits of Kepler. This has been done for the case $M \gg m$. The next level of generalization is to relax the latter restriction.

1.5   Arbitrary masses

For many stellar binary systems, the masses of the two stars are of comparable magnitude. Thus, a general solution must be sought for the motions of two gravitationally interacting pointlike objects of arbitrary masses. This is the *two-body problem*. It turns out that the motion of each star about the barycenter (center of mass) will again be elliptical. The definition of the barycenter then tells us that the motion of one star relative to the other is also elliptical.

The motion of two gravitationally bound bodies about their common barycenter can be determined from a joint solution of the (gravitational) equations of motion for the two bodies. Our task is simplified because the two equations of motion can be reduced to a single equation of form identical to the vector equation of motion (15) for the $M \gg m$ case. Hence, the solutions already obtained are applicable after some straightforward substitutions.

Relative motions

Here we cast the two-body problem in terms of relative coordinates and the reduced mass.

Relative coordinates: reduced mass

Consider two masses m_1 and m_2 with vector displacements r_1 and r_2 from their barycenter, as shown in Fig. 10a,b. Their separation is s, and the relative vector s is defined as the position of m_2 relative to m_1 by

$$s \equiv r_2 - r_1.$$

(1.55)

The same vector s is obtained for any choice of the origin from which r_1 and r_2 are measured or for any (nonrelativistic) choice of the observer's inertial frame of reference. It is convenient to choose the observer's frame of reference to be that in which the barycenter is at rest

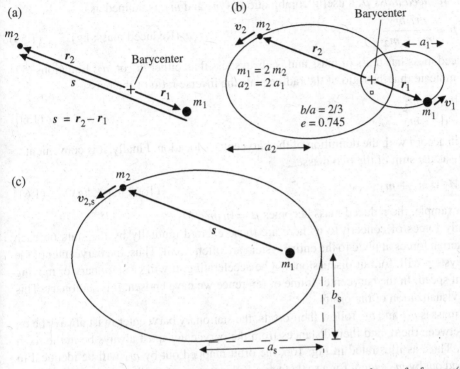

Figure 1.10. (a) Definitions of the radius vectors in the barycenter frame of reference (r_1, r_2) and in the frame of reference of $m_1(s)$. (b) Two masses with $m_1 = 2m_2$ orbiting their barycenter. Both ellipses have the same shape, and the ratio of their sizes is the inverse of the ratio of the masses. (c) Orbit of m_2 in the frame of reference of m_1. This *relative* orbit has the semimajor axis $a_s = 3a_1$; the star m_2 has relative velocity $v_{2,s}$.

(*barycenter system*) and further to choose the origin of the coordinate system to be at the barycenter, which lies between the masses (Fig. 10a). In this case, the magnitude of s is simply the sum of the magnitudes of the two vectors; that is,

$$s \equiv |s| \equiv |r_1| + |r_2|. \tag{1.56}$$

For our special case (origin at the barycenter), the definition of the position of the barycenter of two masses reduces to $m_1 r_1 = -m_2 r_2$. This and the expression for s (55) lead to the vector relations

$$r_1 = -\frac{m_2}{m_1 + m_2} s = -\frac{\mu}{m_1} s \tag{1.57}$$

and

$$r_2 = \frac{m_1}{m_1 + m_2} s = \frac{\mu}{m_2} s, \tag{1.58}$$

where the *reduced mass* μ, a useful combination of m_1 and m_2, is defined as

$$\mu \equiv \frac{m_1\, m_2}{m_1 + m_2}. \qquad \text{(Reduced mass; kg).} \qquad (1.59)$$

The reduced mass has units of mass and is always less than either m_1 or m_2. Equations (57) and (58) indicate that the ratio of the radii equals the inverse ratio of the masses,

$$\frac{|r_2|}{|r_1|} = \frac{m_1}{m_2}, \qquad (1.60)$$

which is in accord with the definition of the barycenter's location. Finally, it is convenient to define M_T as the sum of the two masses:

$$M_T \equiv m_1 + m_2; \qquad \text{(Total mass; kg)} \qquad (1.61)$$

thus, for example, the reduced mass becomes $\mu = m_1 m_2 / M_T$.

The only forces of concern to us here are those exerted mutually by the stars on each other. External forces applied to the entire system are often small. Thus, the barycenter of the two-star system will, for our discussion, not be accelerating; it will be stationary or moving at constant speed. In the barycenter frame of reference we have chosen, it is stationary. This eases the visualization of the orbits.

As the masses m_1 and m_2 follow their orbits, the stationary barycenter must always be on the line between them, and their distances from the barycenter must always be in the fixed ratio (60). Thus, as illustrated in Fig. 10b, the orbit mapped out by m_1 will be identical to that mapped out by m_2 except for a scale factor. If the motion of one is elliptical, the motion of the other is also elliptical. The shapes of the two ellipses (i.e., their eccentricities e) would be identical, but their sizes would differ by the ratio m_1/m_2.

In addition, the fixed ratio of the distances r_2 and s (57) indicates that the relative motion vector s would also map out an ellipse having a shape identical to that of r_2 but larger in scale. An observer on m_1 (Fig. 10c) would thus find that m_2 moves along an elliptical orbit. Note that this observer would be in a noninertial (i.e., accelerating) system and therefore that Newton's laws would not directly yield the motions.

Equation of motion

The equations of motion for each of the two masses are statements of Newton's second law with the appropriate gravitational forces applied in the inertial system at rest with respect to the barycenter. The force is Newton's gravitational force, which is proportional to the product $m_1 m_2$ and inversely proportional to the square of the separation of the stars s^2,

$$|F_G| = \frac{G m_1 m_2}{s^2}. \qquad \text{(Gravitational force)} \qquad (1.62)$$

Newton's second law for m_2 is the vector equation

$$F_{2,1} = m_2 \frac{d^2 r_2}{dt^2}, \qquad \text{(Newton's second law for } m_2) \qquad (1.63)$$

where $F_{2,1}$ is the force exerted *on* star 2 *by* star 1. The comparable equation for star 1 is

$$F_{1,2} = m_1 \frac{d^2 r_1}{dt^2}. \qquad \text{(Newton's second law for } m_1) \qquad (1.64)$$

Newton's third law states that, for two interacting bodies, the forces $F_{2,1}$ and $F_{1,2}$ are equal in magnitude but opposite in direction:

$$F_{1,2} = -F_{2,1}. \qquad \text{(Newton's third law)} \qquad (1.65)$$

Subtract (64) from (63) to obtain an equation in relative coordinates:

$$\frac{F_{2,1}}{m_2} - \frac{F_{1,2}}{m_1} = \frac{d^2(r_2 - r_1)}{dt^2}. \qquad (1.66)$$

Apply the definition of the relative vector s (55) and the relation between the two forces (65),

$$F_{2,1} \frac{m_1 + m_2}{m_1 m_2} = \frac{d^2 s}{dt^2}, \qquad (1.67)$$

and then invoke the reduced mass μ (59):

$$F_{2,1} = \mu \frac{d^2 s}{dt^2}. \qquad (1.68)$$

The gravitational force on m_2 can now be substituted into (68). It has the magnitude (62) and direction opposite to the vector s (Fig. 10a):

$$-\frac{G m_1 m_2}{s^3} s = \mu \frac{d^2 s}{dt^2}. \qquad (1.69)$$

Use the definitions of μ (59) and M_T (61) to rewrite the product $m_1 m_2$:

$$\rightarrow \qquad -\frac{G M_T \mu}{s^3} s = \mu \frac{d^2 s}{dt^2}. \qquad \text{(One-body equivalent equation of motion)} \qquad (1.70)$$

The expression (70) is very general. It does not depend on the speed of the (inertial) frame of reference (i.e., the speed of the observer relative to the barycenter) or on the location of the coordinate system's origin. The position vectors need not have originated at the barycenter. Nevertheless we will continue to use the barycenter frame with origin at the barycenter.

Equivalence to the $M \gg m$ problem

The two equations of motion for the two particles (63) and (64) thus reduce to one-body motion in (70). Furthermore, the relative coordinate s is governed by a differential equation identical in form to the vector equation of motion (15) used for the $M \gg m$ analysis. In fact, (70) has been arranged so that the masses M_T and μ play the same conceptual roles as did M and m in the former analysis.

Because $s = r_2 - r_1$, the solution $s(t)$ will map out the orbit of m_2 measured by an observer riding on m_1 (Fig. 10c). Comparison of (70) and (15) shows that the solution $s(t)$ must be exactly the one given previously for the $M \gg m$ case but with the substitutions specified in Table 1.

In effect, this amounts to solving the different problem of two hypothetical masses, M_T and μ, with separation $s(t)$ between them and with $M_T \gg \mu$. However, keep in mind that this is an artificial view of our problem. In actuality, the spacing s is between m_1 and m_2, not between "M_T" and "μ," and (70) was constructed with no restrictions on m_1 and m_2. The solution of (70) as specified here is exact and quite general for two arbitrary masses.

Table 1.1. *Conversion for two-body solutions*

M	$\rightarrow$	M_T	Total mass
m	$\rightarrow$	μ	Reduced mass
r	$\rightarrow$	s	Relative coordinate
r	$\rightarrow$	s	Relative coordinate; magnitude

Solutions

We now obtain the solutions of (70) from the expressions developed in Section 4 for the $M \gg m$ case. We know immediately that the result is that $s(\theta)$ sweeps out an ellipse. We found, in the $M \gg m$ case, that the size and shape of the ellipse depend on angular momentum and energy. We thus also examine the roles of total angular momentum and total energy in the two-body problem.

Angular momentum (Kepler II)

From our development in the earlier case ($M \gg m$), we recall that the azimuthal component of the equation of motion (17) indicated that the quantity $J = mr^2\omega$ is a constant of the motion; see (19). Similarly, our substitutions (Table 1) tell us that the quantity $\mu s^2 \omega$ is also a constant of the motion. It turns out that this is the total angular momentum J of the two-body system in the barycenter system:

➡ $$J = \mu s^2 \omega = \text{constant.} \quad \text{(Total angular momentum in barycenter system)} \quad (1.71)$$

We now demonstrate this to be the case. In the barycenter system, the angular momentum is measured relative to the barycenter to ensure zero torques and hence angular momentum conservation. Thus, we propose that

$$J = m_1 r_1^2 \omega + m_2 r_2^2 \omega \rightarrow \mu s^2 \omega. \quad (1.72)$$

Indeed, the rightmost term follows directly from the central terms with the aid of the transformations from r_1 and r_2 to s, (57) and (58), and the definition of μ (59) (Prob. 52a). Thus (71) does properly describe the (constant) angular momentum of the system in the barycenter system. Note that s and ω vary with time as the particles proceed around their orbits, whereas J remains fixed. The area swept out per unit time by the relative orbit is also constant; see (71) and (22). This is Kepler II for arbitrary masses.

The total angular momentum can also be written in terms of the parameters of the ellipse. In the process of satisfying the differential equation for $r(\theta)$, we found $J = (GM/a)^{1/2} mb$ (37), where a and b are, respectively, the semimajor and semiminor axes of the track defined by $r(\theta)$. Invoking our substitutions (Table 1), we find that the analog is

➡ $$J = \left(\frac{GM_T}{a_s}\right)^{1/2} \mu b_s, \quad \text{(Angular momentum in the barycenter system)} \quad (1.73)$$

where a_s and b_s are the semimajor and semiminor axes, respectively, of the ellipse that is swept out, in this case by the relative vector $s(\theta)$. Invoke again the fact, from (72), that J for the M_T, μ problem is equal to the system angular momentum in the barycenter system. Thus, (73) is another valid expression for the total angular momentum in the barycenter system.

Elliptical motion

The radial equation of motion (16) was used to solve for $r(\theta)$ under the assumption of constant J. The result is given in (36), and its analog, from invoking the substitutions, is

$$\Rightarrow \quad s(\theta) = \frac{J^2}{GM_T\mu^2} \frac{1}{1 + e\cos\theta}. \qquad \text{(Solution for relative motion)} \qquad (1.74)$$

This is the desired solution of the equation of motion (70). It gives the relative motion of m_2 with respect to the position of m_1, where $J = \mu s^2\omega$. The eccentricity e, as above (53), depends on the values of J and E_t. The motion $s(\theta)$ is demonstrably on ellipse; see (4).

Period and semimajor axis (Kepler III)

Kepler's third law was obtained by equating two versions of dA/dt – as functions of angular momentum (22) and of the period (42) – to obtain a relation (43) between the period P and angular momentum J. We then eliminated J with the relation $J(M, m, a, b)$ (37) and found $GMP^2 = 4\pi^2 a^3$. The more general analog from Table 1 is

$$GM_T P^2 = 4\pi^2 a_s^3 \qquad (1.75)$$

or

$$\Rightarrow \quad G(m_1 + m_2)P^2 = 4\pi^2 a_s^3, \qquad \text{(Kepler III)} \quad (1.76)$$

where again a_s indicates the semimajor axis of the orbit swept out by the relative vector s. The period P is the same for r_1, r_2, and s. This demonstrates that the sum of the masses and the period uniquely determines the semimajor axis of the relative orbit. For a circular orbit, one can make the substitution $a_s = s$ in (76).

Total energy

The final step in our $M \gg m$ analysis was the determination of the total energy. This was obtained by first writing down the total energy (47) at the arbitrary point Q in Fig. 8a, which becomes, for the M_T, μ problem,

$$E_t = \frac{1}{2}\mu v_s^2 - \frac{GM_T\mu}{s}, \qquad (1.77)$$

where v_s is the speed of the tip of the s vector at the time its magnitude is s. This energy is the same at any point Q because total energy is conserved in the hypothetical M_T, μ problem. The expression (47) was rewritten (48) for point Q′, where the separation $r \to a$, to obtain $E_t = -GMm/(2a)$ (52), which, for the M_T, μ problem, becomes

$$\Rightarrow \quad E_t = -\frac{GM_T\mu}{2a_s}. \qquad \text{(Total energy in the barycenter system)} \qquad (1.78)$$

The statement that the right-hand side of (77) or (78) is the total energy of our binary in the barycenter system must still be demonstrated.

In the barycenter system, the total energy is the sum of the kinetic energies of m_1 and m_2 and the potential energy of the two-mass system,

$$E_t = \frac{1}{2}m_1 v_1^2 + \frac{1}{2}m_2 v_2^2 - \frac{Gm_1 m_2}{s}, \qquad \text{(Barycenter system)} \qquad (1.79)$$

where v_1 and v_2 are the speeds of m_1 and m_2 in the barycenter system at some time and s is the separation of the masses at that time.

The question we ask is whether (79) is the same quantity as (77). Transformations between (v_1, v_2) and v_s may be obtained from an alternative definition of the barycenter – namely, that the total momentum measured in that system must equal zero,

$$m_1 v_1 + m_2 v_2 = 0. \qquad \text{(Momentum in barycenter system)} \qquad (1.80)$$

This states that the two velocity vectors are always in exactly opposing directions but with different magnitudes:

$$v_1 = -\frac{m_2}{m_1} v_2. \qquad (1.81)$$

The connection to the relative velocity v_s is obtained by taking the derivative of $s = r_2 - r_1$ (55):

$$v_s = v_2 - v_1. \qquad (1.82)$$

Together, (81) and (82) provide the conversion formulas:

$$v_1 = -\frac{m_2}{m_1 + m_2} v_s = -\frac{\mu}{m_1} v_s \qquad (1.83)$$

and

$$v_2 = \frac{m_1}{m_1 + m_2} v_s = \frac{\mu}{m_2} v_s, \qquad (1.84)$$

which are reminiscent of the expressions for r_1 and r_2 (57) and (58). Substitute these latter expressions into the expression for E_t (79) and invoke the equality $m_1 m_2 = M_T \mu$ from the definitions of μ (59) and M_T (61). The result is (77) (Prob. 52b), which is equivalent to (78). This demonstrates that (78) is indeed the total energy in the barycenter frame.

1.6   Mass determinations

The discussion in the previous section has established the elliptical motions of the two stars about their barycenter. Here these motions are related to the parameters that can be measured by astronomers. We then explore how the measurements can lead to physical properties such as the masses of the two stars.

Mass function

A very useful tool for the determination of masses is the *mass function f*. It follows directly from Kepler III (76), which we now restate,

$$G(m_1 + m_2)P^2 = 4\pi^2 a_s^3, \qquad \text{(Kepler III)} \qquad (1.85)$$

where a_s is the semimajor axis of the ellipse swept out by the relative vector s. Relative displacements (one star relative to the other) are appropriate for visual binaries in which one might track an orbit on a series of photographic plates, always measuring the distance of one star from its partner, as in Fig. 2.

In contrast, if one studies a spectroscopic binary, the frequency shifts represent line-of-sight velocities relative to the barycenter of the system. (There may be a constant offset due to barycenter motion; Fig. 6). In this case it is preferable to convert a_s to a function of the barycenter coordinate a_1 (or a_2), where a_1 is the semimajor axis of the ellipse swept out by m_1 with the barycenter as the origin (Fig. 10a,b). The ratio $|s/r_1|$ is a fixed value throughout the orbit (57) and hence must equal a_s/a_1. Thus, from (57),

$$a_s = \frac{m_1 + m_2}{m_2} a_1. \tag{1.86}$$

Substitute into (85) and rearrange terms:

$$GP^2 \frac{m_2^3}{(m_1 + m_2)^2} = 4\pi^2 a_1^3. \tag{1.87}$$

Spectroscopic data (e.g., the data of Fig. 7) yield neither a_1 nor the inclination i but only the product $a_1 \sin i$; see discussion under "Spectroscopic binary" following (102). To give $a_1 \sin i$ explicit visibility in the equation, both sides are multiplied by $\sin^3 i$. The two quantities that can be measured directly, P and $a_1 \sin i$, are then collected on the right side, leaving the unknown quantities m_1, m_2, and angle i on the left:

$$\frac{m_2^3 \sin^3 i}{(m_1 + m_2)^2} = \frac{4\pi^2}{GP^2} (a_1 \sin i)^3. \qquad \text{(Mass function equation for star 1)} \tag{1.88}$$

The left side of this expression is known as the *mass function*,

$$f_1 \equiv \frac{m_2^3 \sin^3 i}{(m_1 + m_2)^2}. \qquad \text{(Mass function for star 1)} \tag{1.89}$$

The subscript "1" in f_1 indicates that the Doppler velocity measurements yielding $a_1 \sin i$ were obtained from star 1 of mass m_1.

The successful measurement of both P and $a_1 \sin i$ from Doppler-shift studies of star 1 provides a numerical value for the right side of (88) and hence also of f_1. The equation (88) then contains the three unknowns m_1, m_2, and i. Two additional equations are needed if all three unknowns are to be determined.

A second and independent mass function equation will result if one is successful in measuring the Doppler shifts for star 2 to obtain $a_2 \sin i$. This can be used in the mass function equation for m_2 obtained by analogy to (88):

$$\frac{m_1^3 \sin^3 i}{(m_1 + m_2)^2} = \frac{4\pi^2}{GP^2} (a_2 \sin i)^3. \qquad \text{(Mass function equation for star 2)} \tag{1.90}$$

A third equation could describe the duration of an eclipse or the orbital brightness variations due to tidal distortion of the observed star by its binary companion. The fractional duration (relative to the orbital period) of an eclipse is directly affected by the inclination, eccentricity, and the sizes of the stars as well as their separation. The sizes and separation are related to the stellar masses for main-sequence stars (Table 4.2). Tidal light variations are a function of star spacing, masses, sizes, and the orbit inclination.

If these additional equations are not available, examination of (89) tells us directly that, for all possible values of i and m_1, the measured value of f_1 (kg) is the smallest mass m_2 that star 2 can have (Prob. 61).

The mass function equation, (88) or (90), is simply a rewrite of Kepler III in terms of barycentric parameters to make it useful for determining masses from *spectroscopic* binaries. Visual binary data are best addressed with the usual relative-parameter form of Kepler III (85).

Stellar masses from circular orbits

Here we explore the practicalities of obtaining masses for circular orbits with arbitrary masses m_1, m_2, but first we recall the $M \gg m$ case (AM, Chapter 9).

Massive central object

The mass determination is particularly straightforward in a system in which one star is known to be much more massive than the other, $M \gg m$. Consider a circular orbit of radius r. If the gravitational force, $F = -GMm/r^2$, and the centripetal acceleration, $a = -\omega^2 r$, are applied to Newton's second law, $F = ma$, one will find directly that

$$M = \frac{4\pi^2 r^3}{GP^2},$$

(1.91)

where $P = 2\pi/\omega$ is the orbital period and ω the angular velocity.

In this case, measurements of the radius and period of the low-mass object directly yield the mass of the central massive object. A prominent example is the sun-earth system. One can enter the radius of the earth's orbit (1.0 AU) and its orbital period (1 yr) into (91) to obtain the mass of the sun. Another example is the earth-moon system as well as most planet-satellite systems.

For the general case of two arbitrary masses, measurement of the parameters of one of them does not, in itself, give the mass of the other. However, it does place a lower limit on the mass of the other as noted above after (90).

Circular orbits

Circular orbits are important limiting examples of binary orbits. They are very common in systems that have small separations called *close binaries*. An example is an accreting *neutron-star binary* system in which gaseous matter from a normal star accretes onto the surface of its nearby partner, a neutron star (Section 4.4). Interaction between the objects dissipates energy, which tends to circularize an elliptical orbit.

Another example is the earth-moon system. Tidal interactions between the earth and the moon cause the moon's orbit to be quite circular with eccentricity only $e = 0.055$. This corresponds, from (12), to a ratio of semiminor to semimajor axes of nearly unity, $b/a = 0.998$.

One would not ordinarily assume a priori that the orbit is circular but would solve for all the orbital parameters. We do so here, however, to illustrate more simply the use of the mass function for the case of arbitrary mass values. In the circular case, the semimajor axis a_1 equals the radius r_1 of the orbit of $m_1(r_1 = a_1)$, and likewise for m_2.

Spectroscopic binary

Consider a hypothetical spectroscopic binary with circular orbits and arbitrary masses and focus on one of the two masses. If such an orbit has an inclination i, and if the orbiting mass m_1 has speed v_1 (Fig. 3a), the line-of-sight velocity v_r will vary sinusoidally with maximum value $v_1 \sin i$:

$$v_r(t) = v_1 \sin i \sin \omega t. \qquad \text{(Circular orbit)} \qquad (1.92)$$

The quantity $\sin i$ is a fixed value for a given system, whereas the $\sin \omega t$ term varies with the motion of the star around the circular orbit, as discussed previously regarding Fig. 6. The maximum value of $v_r(t)$ occurs when $\sin \omega t = 1$:

$$v_{r,max} = v_1 \sin i. \qquad (1.93)$$

This quantity is directly measurable from the Doppler data.

Rewrite the mass function equation (88) to incorporate the measured quantity $v_1 \sin i$. First apply $a_1 = r_1$ for a circular orbit and then eliminate r_1 with the relation

$$2\pi r_1 = v_1 P \qquad (1.94)$$

to obtain

$$\frac{m_2^3 \sin^3 i}{(m_1 + m_2)^2} = \frac{P}{2\pi G}(v_1 \sin i)^3. \qquad \begin{array}{l}\text{(f_1 equation in terms of $v_1 \sin i$;}\\\text{circular orbit)}\end{array} \qquad (1.95)$$

The right-hand side of this equation can be evaluated directly from Doppler-shift spectroscopy data such as those of Fig. 6. If the inclination of the binary orbit is not known, the maximum amplitude (relative to the barycenter velocity) of the curve in Fig. 6 would be noted (i.e., 30 km/s or 3×10^4 m/s for m_1). This is equal to $v_1 \sin i$ (93) and would be substituted into the right side of (95). The same plot would yield the orbital period P (30 d $\times$ 86 400 s/d).

This example (from Fig. 6) would thus yield the mass function

$$f_1 = \frac{m_2^3 \sin^3 i}{(m_1 + m_2)^2} = 1.7 \times 10^{29} \text{ kg} \rightarrow 0.084 \, M_\odot. \qquad (1.96)$$

This result has units of mass, but usually it is *not* the mass of either component. An exception occurs when $m_1 \ll m_2$ and $i = 90°$; then $m_2 = 0.084 \, M_\odot$. This is a lower limit for the mass m_2.

In Fig. 6, we place the observer such that $i = 90°$. In this case, the individual masses may be obtained if we make use of the Doppler curve for the other star m_2. From the ratio of amplitudes and the definition of the barycenter, the relative values of the masses are

$$m_2 = m_1/3. \qquad (1.97)$$

In this case, we have $i = 90°$ and $f_1 = 0.084 \, M_\odot$, and so the mass function equation (96) may be solved to yield

$$m_1 = 4.0 \, M_\odot; \quad m_2 = 1.3 \, M_\odot. \qquad (1.98)$$

If the inclination is known to be less than 90°, the masses will be substantially larger (Prob. 62).

Table 1.2. *Orbital elements*

Symbol	Unit	Name
a	m	Semimajor axis
e	—	Eccentricity
P	s	Period of revolution
T	date-time	Time of periastron passage
i	rad	Inclination of orbital plane
ω_p	rad	Longitude of periastron
Ω	rad	Position angle of node

A more general way to view such determinations is to evaluate the mass functions of both stars independently from the data of Fig. 6. This would give two equations. If the inclination is unknown, a third equation would be required to determine both masses and the inclination.

Stellar masses from elliptical orbits

The projection effects for elliptical orbits lead to rather strange stellar tracks and Doppler curves, as exhibited in Figs. 2 and 7. Visual and spectroscopic binaries provide quite different information. The former yield a geometric track for each star, whereas the latter yield line-of-sight velocities. The information one can extract from the data also differs. Here we qualitatively present some of the issues in retrieving the physical parameters from elliptical orbits.

Orbital elements

The seven quantities that describe the elliptical orbit of a star, including its orientation relative to the observer and the location of the star in the orbit, are given in Table 2. They are called the *orbital elements*. The orientation angles are further defined in Fig. 11 and its caption.

The origin of the ellipse is properly the barycenter of the two-star system, and the ellipse is the track of one of the two stars. Another set of elements would describe the orbit of the other star, but most of the parameters would be the same (Fig. 10b). The longitude of periastron would differ by π radians, and the semimajor axis would be different unless the two masses were identical. Often, data exist for only one of the stars. For visual binaries, the orbit is frequently plotted relative to the brighter star, as in Fig. 2. The parameters given thus refer to the relative orbit, but again, most of the elements are the same as those of one of the stars.

Visual binary: relative orbit

The relative orbit of a visual binary is often easier to obtain than independent absolute measures of the two stellar orbits against the background stars. In the latter case, the more massive star may move very little, and the reference stars may be relatively distant (in angle) from the binary and subject to their own and differing proper motions. Here we outline the analysis of a measured relative orbit.

The motion plotted by an observer is the projection of the orbit onto the x–y plane, the plane of the sky; see the lower ellipse in Fig. 11. It turns out that, in general, an ellipse projected onto a plane is also an ellipse but with a displaced focus. Think of a circular orbit with nonzero

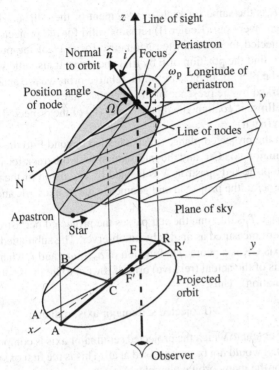

Figure 1.11. Elliptical orbit (shaded) and parameters that describe it viewed from above the sky plane or from outside the observer's celestial sphere. The origin is at the upper focus of the shaded ellipse, which would be at the barycenter of a binary system or at the location of one stellar component for a relative orbit (e.g., Fig. 2). The z-axis is the line of sight, and the x–y plane at $z = 0$ is the plane of the sky. The shaded portion of the ellipse and the observer are below the sky plane. The line of nodes is the intersection (at the focus) of the orbital plane and the sky plane; it is the axis of inclination rotation. The longitude of periastron ω_p is measured in the plane of the orbit from the line of nodes to the periastron in the direction of motion of the orbiting star as it recedes from the observer. The position angle of the line of nodes Ω is measured in the x–y plane eastward from north (N). The projection of this physical ellipse into the x–y plane is another ellipse (lower, unshaded ellipse) with (geometric) major axis A′R′ (light line) and focus F′. The major axis of the actual orbit projects to the (heavy) line ACFR. Because $\omega_p \approx 90°$, the projected major axis is approximately coincident with the geometric axis, but compare with Fig. 2. [In part from T. Swihart, *Astrophysics and Stellar Astronomy*, Wiley, 1968, pp. 65–66]

inclination and a massive star at its focus, the center of the circle. When projected, it becomes an ellipse with the central star at its center, but the mathematical "geometric" focus of the ellipse is not at the center. The projected (apparent) path of α Cen (Fig. 2) is an example of a projected elliptical orbit; it too is an ellipse.

The orbital elements (Table 2) refer to the physical orbit, and our data refer to the projected orbit. Thus we need the relationships between them. The projected ellipse for a visual binary has the relationships to the physical elliptical orbit (Fig. 11) listed below. They provide the expressions that permit the orbital elements to be deduced.

(*i*)   The *orbital period P* is the same for both physical and projected orbits. Thus the time for the star to complete a circuit of the projected orbit provides the actual period.

(ii)   The projection factor, $\cos i$, is the same for each area segment of the ellipse. Thus the constancy of the rate of area swept out (Kepler II) remains valid for the projected orbit when the origin is the projected focus F, not the geometric focus F′ of the projected ellipse. This may be used to find the position of F by trial and error. It also allows more accurate apparent orbits to be plotted. In the case of the relative orbit we are analyzing, the projected focus F is marked by the presence of the reference star.

(iii)   The center of the physical ellipse is coaligned with the center (C) of the projected ellipse for any possible orientation (try it).

(iv)   The line joining C and F in the projected plane, when extended beyond F to the orbital track, is the projected semimajor axis CR, and the intersection with the track locates the projected periastron R. The projected apastron A is located by extension of the line in the opposite direction. In general, the projected and geometric major axes AR and A′R′ are not coincident.

(v)   The time of periastron passage T is the time the star passes the projected periastron R.

(vi)   The projected distances are all measured as angles by an observer. The subtended angle $[CR]_{ang}$ of the semimajor axis CR is a geometric function of a_{ang}, ω_p, and i, where a_{ang} is the angular semimajor axis of the actual (relative) orbit if it were to lie in the plane of the sky (i.e., with zero inclination). Thus,

$$[CR]_{ang} = f(a_{ang}, \omega_p, i). \qquad \text{(Projected semimajor axis; radians)} \qquad (1.99)$$

The quantity ω_p controls the extent to which the projected semimajor axis is compressed by the inclination; at $\omega_p = 0$, it would not be compressed at all. This is the first of several equations needed to solve for the many orbital elements.

(vii)   The projected semiminor axis (CB) need not be perpendicular to the projected semimajor axis, but it must bisect line segments interior to the ellipse and parallel to the projected semimajor axis. It is therefore easily constructed; see segment CB in Fig. 11. The measured angle $[CB]_{ang}$ is thus, similarly, a function of orbital elements. Because the physical semiminor axis is a function of the semimajor axis a and the eccentricity e (12), the angular projected semiminor axis becomes

$$[CB]_{ang} = g(a_{ang}, e, \omega_p, i), \qquad (1.100)$$

where g is another geometrical function and again $[CB]_{ang}$ is an angular distance on the sky.

(viii)   The solid angle Ω_p (sr) of the projected ellipse can easily be measured. It is related to the solid angle of the actual ellipse (if it were at $i = 0°$) by the factor $\cos i$. The area of an ellipse of eccentricity e and semimajor axis a is $A = \pi ab = \pi a^2(1 - e^2)^{1/2}$. Apply the $\cos i$ factor and divide by the square of the distance to the binary to obtain

$$\Omega_p = \pi a_{ang}^2(1 - e^2)^{1/2} \cos i. \qquad (1.101)$$

(ix)   The projection factors for the angular distances $[CF]_{ang}$ and $[CR]_{ang}$ are the same because they lie along the same line. Thus their projected ratio is equal to that of the unprojected ratio. The center-to-focus distance for the physical ellipse is ae (Fig. 8a) giving a ratio of $ae/a = e$. Hence, for the projected ellipse,

$$\frac{[CF]_{ang}}{[CR]_{ang}} = \frac{a_{ang}\, e}{a_{ang}} = e. \qquad (1.102)$$

Measurement of the projected ratio therefore directly yields the eccentricity.

The four equations, (99) through (102), in principle allow one to solve for the unknown parameters e, a_{ang}, ω_{p}, and i because the quantities on the left side are all measurable. Of course, it would be necessary to have the actual forms of the functions f and g, which are not given here.

Another geometrical expression makes use of the quantities ω_{p} and i and the position angle of the projected major axis to produce the position angle of the node Ω. The final two orbital elements, period P and the time of periastron passage T, were determined as described in (*i*) and (*v*).

This might seem to complete our determination of the seven orbital elements. There is a missing element, however. Finding the star masses from Kepler III requires a physical length a_s for the relative semimajor axis, not the equivalent angular distance a_{ang} just determined. The quantity a_s follows directly from the angular value if the distance D from the observer to the binary is known, say, from parallax measurements. That is, $a_s = D a_{\text{ang}}$. Kepler's third law for the relative orbit (85) then yields the total mass $m_1 + m_2$.

Visual binary: two orbits

The two individual masses of a visual binary can be determined only by measuring the projected orbits of *both* stars in inertial space relative to the distant background stars and galaxies. The origins of the physical ellipses are at the barycenter of the binary. The barycenter must be located to proceed with the logic above, which leads to four of the orbital elements. As in item (*ii*) above, the barycenter can be located by trial-and-error applications of Kepler's second law (equal areas in equal times). Alternatively, one can deproject both orbits with trial values of i, ω_{p}, and Ω until the deprojected orbits have a common focus. This would be the barycenter.

One can then proceed as above with each orbit to obtain the two semimajor axes. Their sum is the relative semimajor axis a_s. (To see this, consider Fig. 10b when the radius vectors have lengths a_1 and a_2, as at Q′ in Fig. 8a.) This can be entered into Kepler III (85) to yield the sum of the masses. The ratio of the semimajor axes is the inverse ratio of the masses (60). These two expressions for the sum and ratio of the masses yield the individual masses.

Spectroscopic binary

In the case of spectroscopic binaries, one measures, in effect, the dimensions of the orbit along the line of sight. The Doppler shifts provide, at each instant, the line-of-sight component of the emitting star's speed. Integration yields the line-of-sight extent of the orbit. For an elliptical orbit, this is not necessarily $a \sin i$ because of the many possible orientations of the orbit relative to the line of sight.

The position angle of the node Ω is simply a rotation about the line of sight. Because a Doppler shift is a measure only of the line-of-sight component of the velocity, the shape and amplitude of the Doppler curve are not affected by a change in Ω. If the longitude of periastron ω_{p} is 90°, approximately as shown in Fig. 11, the line-of-sight extent of the orbit would be $2a \sin i$, but if $\omega_{\text{p}} = 0°$, it would be $2b \sin i$.

The Doppler curve for an elliptical binary in an arbitrary orientation (e.g., Fig. 7) may be solved for $a_1 \sin i$ by means of a nonlinear least-squares fit. The particular shape of the curve is also affected by the eccentricity and the (angular) location of periastron relative to the line of nodes (Fig. 11). The fit thus also yields e and ω_p. It can not, however, extract the inclination. The curve would look exactly the same for a large value of a_1 and a small inclination as it would for a smaller a_1 and a larger inclination, as long as the product $a_1 \sin i$, the eccentricity e, and the longitude of periastron ω_p were the same.

The value of $a_1 \sin i$ thus measured could be substituted into the mass function equation for star 1 (88). This expression is one of three needed to disentangle m_1, m_2, and i, as in the circular-orbit case. A fit to the data of star 2, if available, would lead to a second equation. Eclipse data, fits to the brightness changes owing to tidal forces, or both could provide the third.

Examples

Mass of a black hole in Cygnus X-1

The first strong case for the existence of a stellar black hole arose from spectroscopic observations of the optical binary companion of the bright celestial x-ray source Cygnus X-1. The optical partner is a quite massive supergiant star of spectral type O9.7 Iab (Table 4.2), which has evolved off the main sequence. The copious emission of x rays arises from gas that originates in the supergiant atmosphere and accretes onto, or into, a compact partner star – either a neutron star or a black hole; see Section 4.4.

The x rays arise from the release of gravitational energy as the gas descends into the deep potential well of the compact star. When the gas encounters a stellar surface, a shock, or an accretion disk, it becomes thermalized at x-ray temperatures $\sim 10^7$ K. The emission of copious hard (> 2 keV) x rays from a binary system is thus a sure indicator that the accretor is either a neutron star or a black hole.

Optical astronomers were able to obtain the mass function of the supergiant through Doppler shifts of its spectral lines. The orbital velocity variation is shown in Fig. 12. The variation fits a sine curve very well; hence, the orbit is quite circular. The period and amplitude of the velocity curve yield, from (95), a mass function of $f_{opt} \approx 0.23\ M_\odot$ (Prob. 64).

Unfortunately, the x rays do not provide a second mass function; these sources do not have sufficiently strong x-ray spectral lines. Some x-ray sources are pulsars that provide Doppler shifts (see below), but Cygnus X-1 is not one of them. Nevertheless, one can infer the mass of the optical star from its spectral type and luminosity – namely, that it is about $30\ M_\odot$. Consider the optical star to be star 1. Substitute $m_1 \approx 30\ M_\odot$ and $f_1 = f_{opt} = 0.23$ into the mass function (89) for arbitrary inclination to yield a mass limit for the compact object $m_2 > 6.8\ M_\odot$; uncertainties reduce this limit to $\gtrsim 6\ M_\odot$. This is substantially greater than the maximum theoretically expected mass for a neutron star, $\sim 3 M_\odot$.

One could well conclude therefore that the compact object in the Cygnus X-1 binary system is a black hole. The argument is not ironclad because the 3-$M_\odot$ limit is a theoretical limit dependent in part on untested physics.

Over the subsequent 30 years, Cygnus X-1 has failed to yield any hint that, instead, it is a neutron star. It is thus now widely accepted that Cygnus X-1 provided the first plausible evidence that a black hole could be an end point of stellar evolution. Now we have equivalent results for a few dozen sources in the Galaxy. In addition, the nuclei of active galaxies

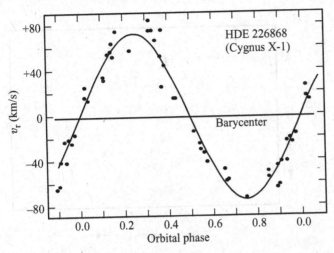

Figure 1.12. Doppler curve for spectral lines from the optical partner of the x-ray source Cygnus X-1. It is an O9.7 Iab supergiant of mass about 30 $M_\odot$ known as HDE 226868. The data, obtained from several observers over some months, are superimposed modulo the orbital period of 5.60 d. The best-fit curve shown is centered about -1.8 km/s and has amplitude 73.8 km/s. These data yield a mass function $f_{opt} = 0.23 \, M_\odot$, which, when combined with the supergiant mass, gives a mass $\gtrsim 6 \, M_\odot$ for the x-ray–emitting partner. This is a crucial element in the argument that the partner is a black hole. [R. Brucato and J. Kristian, *ApJL* **179**, L129 (1973)]

independently give strong evidence that they are *massive black holes* with masses of 10^6 to $10^8 M_\odot$. See more on black holes in Section 4.4.

Masses of neutron-star pulsars

A spinning neutron star can emit a pulse of radiation toward an observer one or more times each rotation. There are two general classes of such pulsars. Radio pulsars emit a beam of radiation that sweeps around the sky like a lighthouse beam. The large majority are isolated stars, but some are in binary systems with a compact companion.

X-ray pulsars are the other class. They are accreting binary sources as described just above for Cygnus X-1. In this case, though, the compact object is a neutron star; black holes are not expected to pulse. Most of the x-ray pulsing systems consist of a massive normal star and a neutron star of high magnetic field ($\sim 10^8$ T). In these cases, the accretion stream is likely to follow the magnetic field and impinge on the magnetic pole, creating an x-ray–emitting hot spot at the pole (Fig. 4.13). A distant observer might perceive the hot spot coming into and out of view as the neutron star spins; a pulse of x rays would thus be detected once for each rotation of the neutron star. See more on this in Section 4.4.

Spectroscopy in x rays is quite difficult to carry out and, as stated, the accreting sources are not strong x-ray line emitters. (The latter is also true of the radio pulsars.) Thus, spectroscopic Doppler shifts are not attainable for such stars. Instead, one can use the frequency of the pulsing itself. Pulse periods of x-ray pulsars range from milliseconds to about 1000 s. As the neutron star orbits the barycenter of the system, its line-of-sight velocity changes, giving rise to Doppler shifts of the pulsing frequency. These data can be treated exactly as Doppler variations of spectral lines obtained by optical astronomers.

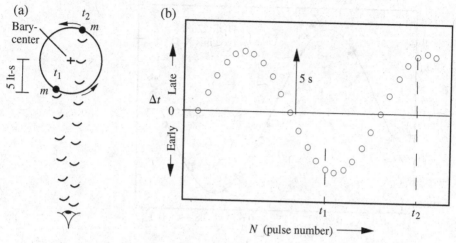

Figure 1.13. (a) Orbit of a pulsing neutron star of radius 5 light seconds showing the pulsar at two positions together with traveling pulses and observer. (b) Arrival time of pulses relative to the expected arrival time for a constant intrinsic period as a function of pulse number. Pulses are delayed when the pulsar is more distant than the barycenter and arrive early if closer than the barycenter. If the orbit is circular with $\sin i = 1$ as shown in (a), the plotted data will be sinusoidal with a half-amplitude that directly indicates the radius (semimajor axis) of the orbit.

In pulsar studies, observers actually measure the arrival time of each pulse and compare the times of arrival with the times expected if the pulse period were constant. As shown in Fig. 13, one expects the pulses to show varying delays that depend on the varying distance from the neutron star to the observer.

The pulse delays directly map the line-of-sight dimensions of the orbit. Fits yield the eccentricity and the product $a \sin i$, which, with the orbital period, provide a value for the mass function of the compact object (88). The optical Doppler data of the visible companion star can provide the second mass function. An eclipse duration can provide the final equation needed to solve for the two masses and inclination.

Binary radio pulsars consisting of two neutron stars in close orbits can exhibit further pulse-arrival-time changes caused by the emission of gravitational radiation. These can be fitted to the predictions of the general theory of relativity. The data from a single pulsing neutron star in such a system can yield all the orbital parameters, including both masses of the binary. This has been the case for the binary (Hulse–Taylor) pulsar, a system of two neutron stars (see AM, Chapter 12). A dozen or more such pulsar systems are now known.

Figure 14 shows a summary of mass measurements of neutron stars that are radio pulsars. They are all consistent with the relatively narrow range of masses, $1.35 \pm 0.04\ M_\odot$. The actual masses probably extend beyond this range because of differing accretion histories. Millisecond pulsars obtain their rapid spins from the torque applied by accreting matter from a gaseous binary partner during a previous stage of their evolution. The accreted matter is believed to be at least $0.1\ M_\odot$, and this could well differ from star to star.

The pulsars shown in the figure have spin periods ranging from 4 to 1000 ms. In general, the more rapidly spinning objects yield more precise mass estimates. Additional neutron-star mass measurements are obtained from *x-ray* binaries.

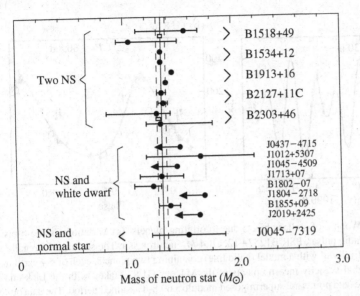

Figure 1.14. Neutron-star masses measured by pulse timing for 19 radio pulsars (neutron stars, NS) with 1σ error bars and one-sided 95% confidence limits. Those showing orbit decay due to gravitational radiation yield extremely precise masses. The top 10 NS are in 5 double NS systems, each of which yields masses of both NS. In two systems, the average of the two masses is obtained more precisely than that of either NS (open squares – size indicates error). The next eight are NS–white-dwarf systems, and the lowest is an NS–normal star system. The results match a relatively narrow range of masses, namely, $1.35 \pm 0.04\ M_\odot$ (vertical solid and dashed lines). X-ray pulsars also yield neutron-star masses. [Adapted from S. Thorsett and D. Chakrabarty, *ApJ* **512**, 288 (1999)]

1.7 Exoplanets and the galactic center

New lines of research are based directly on Newtonian binary orbits. Here we briefly describe two of them – namely the searches for exoplanets and the study of the galactic center region of the Galaxy.

Exoplanets

The search for planets outside the solar system, called *exoplanets*, is currently an active field of research. The number of such detected planets is now approaching 300. They are detected through several very different techniques and are found in diverse environments with differing masses and at various distances from the host star. For the most part they have been detected through the wobble of the parent star's position on the sky.

Exoplanets were first found in 1992 associated with a neutron star, the radio pulsar PSR B1257+12. This object is typical of most radio pulsars in that it is an isolated neutron star with no known binary companion – at least not one of stellar mass. One thus expects its pulsing to be quite uniform because it should be moving through space with a uniform velocity. This is not a normal gaseous star like our sun where one might expect to find a planet. It is an extremely compact neutron star with a large magnetic field that is spinning with a 6-ms period.

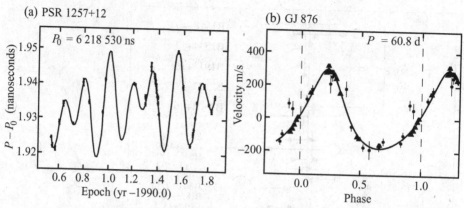

Figure 1.15. Wobble of stellar objects due to orbiting planets. (a) Variations in the 6-ms pulse period of the radio pulsar, PSR B1257+12, a 1.4-$M_\odot$ neutron star. The seemingly irregular variations are beautifully fit with a model (solid line) containing two planets, each of ~4 earth masses. (b) Optical radial velocity measurements of the M4 star GJ 876 taken with the Lick and Keck telescopes. The data points are superimposed modulo a 60.8-d assumed period. The data quite well match the variation expected for a ~2.1 Jupiter mass companion orbiting the star at this period with an eccentricity $e = 0.26$ (smooth curve). Better data and more refined analyses now indicate that both systems have additional, less massive planets. [(a) A. Wolszczan and D. Frail, *Nature* **355**, 145 (1992); (b) G. Marcy *et al.*, *ApJ* **505**, L147 (1998)]

Radio astronomers were able to measure the arrival times of the pulses from this pulsar with precisions of ~15 μs and so were able to look for minute discrepancies in the arrival times. An orbiting planet will cause the parent neutron star to wobble somewhat, and this will delay or advance pulses by small but detectable amounts. This motion of the parent star can reveal the presence of a planet that would not otherwise be detectable.

The pulse arrival times for PSR B1257+12 were found to have pronounced deviations up to ~2 ms from the expected arrival times for a steady period. These were found to fit quite well with a model in which two planets of 4.3 and 3.9 earth masses orbit a 1.4-$M_\odot$ neutron star in nearly circular orbits at 0.36 and 0.47 AU with periods of 67 and 98 d, respectively.

The arrival-time data for short intervals, $\gtrsim 2$ d, were used to calculate successive values for the 6-ms pulsar spin period. The variations in this period as first published are plotted in Fig. 15a. The variations in period are quite minute – only 0.03 ns compared with the 6-ms period – but are highly significant compared with the error bars of the data points.

Planets around normal stars (like the sun) are now being detected in abundance at optical wavelengths through the changing Doppler radial-velocity variations of optical absorption lines. High-dispersion spectrographs detect wavelength variations due to the wobble of the parent star. Precisions of ~3 m/s in the Doppler velocity are obtained; for comparison, the wobble velocity of the sun due to Jupiter is 12.5 m/s. Hence, Jupiter-mass planets orbiting solar-mass stars yield detectable Doppler shifts.

An example is the M4 star GJ 876 that has two planets of masses ~0.79 $M_{\rm Jup}$ and ~2.5 $M_{\rm Jup}$ and orbital periods of 30 and 61 d, respectively (1.0 $M_{\rm Jup} = 0.955 \times 10^{-3}\ M_\odot$). Figure 15b shows the first results for GJ 876. The radial velocities well match the variations expected for the 61-d orbit of a single planet of 2.1-$M_{\rm Jup}$. The second planet and a third of only 7.5 earth masses were discovered in subsequent data sets and analyses.

Radial velocity measurements of optical absorption lines have led to most of the $\sim$300 exoplanets now known. In a few cases, the planet has been observed transiting across the face of the star. This is observed as a decrease in the intensity of the star of $\sim$1% for a few hours each orbit. Several systematic searches for *planetary transits* are now being carried out. The detection or nondetection of a transit places limits on the eccentricity of the planetary orbit; detection yields the inclination and also the size of the planet.

Studies of a multiplicity of planetary systems of different characteristics (orbital radius, mass, inclination) are rich in information about the creation and evolution of planetary systems.

Galactic center

The center of the Galaxy is quite benign compared with activity seen in active galactic nuclei. Nevertheless, it has been a focus of scientific attention in recent years. New developments have intensified this interest.

Stellar orbits

High-resolution infrared detectors have made it possible to track the orbits of some dozen stars that lie very close to the radio source Sgr A*, the presumed center of the Galaxy. Over the course of $\sim$10 years, stars close to the center were found to follow projected elliptical orbits consistent with a single massive central object. The tracks of seven of them are shown in Fig. 16.

The innermost object in Fig. 16, S0-2, has the shortest period (15 yr) and thus has been tracked for most of its orbit at this writing. It passes only 120 AU from Sgr A* at its closest approach. The highly elliptical S0-16 with period 36 ± 17 yr passes only 45 AU from the central object.

Analysis of the projected orbits yields the mass of the putative black hole at the center, $\sim$3.7 $\times$ 10^6 $M_\odot$, for an assumed distance of 26 000 LY (8.0 kpc). The determination of the semimajor axis of the deprojected orbit and Kepler's third law (45) give the central mass directly.

Distance to the galactic center

The introduction of spectral data into the analysis can further provide a direct measure of the distance to the galactic center. Consider a simple constant-speed circular orbit with inclination 90° (observer in the plane of the orbit) at distance D from the earth. The linear velocity v_r may be measured via the Doppler shift of its absorption lines when the star directly approaches the observer. When the star is moving normal to the line of sight, multiple images will give its angular velocity $\omega = v_\theta/D$. Because $v_r = v_\theta$ for uniform circular motion, we have $v_r = \omega D$. If v_r and ω are measured from spectroscopy and imaging, respectively, the distance D follows.

In the present case, the motions of the stars are projected ellipses. Joint fits of both spectral and imaging data to an elliptical orbit necessarily involve the distance D, which is similarly forthcoming from the analysis. A recent result yields $D = 24.8 \pm 1.0$ kLY, which is consistent with the 25-kLY value adopted in this text or the commonly used value of 8.0 kpc (26 kLY). This same procedure has also been used to obtain a purely geometric distance to a binary system in the Pleiades, 430 ± 13 LY.

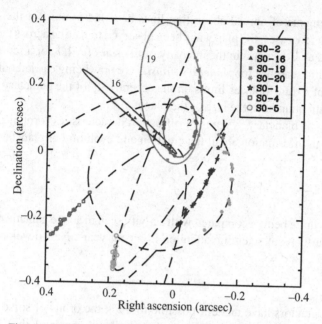

Figure 1.16. The tracks on the sky of seven "S stars" within 0.4″ of the Galaxy's central dark mass obtained with the Keck I 10-m telescope from 1995 to 2003. The origin of the coordinates is the dynamical center, which is coincident with the radio/x-ray source Sgr A*. The elliptical orbital fits to the data points are shown. At a distance of ~25 000 LY, the angular distance 0.1″ corresponds to 4.6 light days. The pericenter distance of the S0-16 orbit (labeled) is only 45 AU. The event horizon of the $3.7 \times 10^6 \, M_\odot$ central black hole is 0.074 AU [A. Ghez *et al.*, *ApJ* **620**, 744 (2005); see also F. Eisenhauer *et al.*, *ApJ* **628**, 246 (2005)]

Massive black hole

The elliptical orbit of S0-16 indicates that the radial extent of the central mass ($3.7 \times 10^6 \, M_\odot$) is no more than 45 AU. Although this is still 600 times the Schwarzschild radius (event horizon) of a nonrotating black hole of this mass (4.36), it still implies a huge mass density of at least $2 \times 10^{15} \, M_\odot / LY^3$. A cluster of dark objects (not black holes) of such average mass density would survive only ~10^5 yr owing to gravitational interactions that would eject its members, which is inconsistent with the ~10 GLY age of the Galaxy. Another possible model, the *Fermion ball*, wherein degeneracy pressure supports a massive object against gravitational collapse, also becomes much less tenable.

These data thus give one high confidence that the object at the center of the Galaxy is indeed a massive black hole exceeding $10^6 \, M_\odot$. Pronounced x-ray and infrared activity further signifies its unusual nature. This is a most persuasive case for the existence of such objects.

Problems

1.2 Binary star systems

Problem 1.21. Evaluate the destructive effect on a binary system by an external (point) gravitational body. Let both stars in the binary have one solar mass and be widely separated at 10 AU. (a) Consider two scenarios and for each find the ratio of the maximum possible external *force difference* on the two stars of the binary to the force exerted by the binary stars on each other. (*i*) The external force is due to a point source at the galactic center distant 25 000 LY and of mass $10^6 M_\odot$. Hint: consider the force gradient. (*ii*) The binary is in a globular cluster consisting of 10^6 stars in a sphere of radius 15 LY. The external force is due to a single nearby star at a typical or average separation distance. (b) In such a globular cluster, about how long would it take a given single star to experience an encounter with another single star within 10 AU? Hint: what does the virial theorem (2.14) tell you about the average stellar speed, and what is the cross section for collision? On average, what would be the interval between such collisions in the cluster as a whole? Compare both times with the $\sim 10^{10}$ yr age of globular clusters. [Ans. $\sim 10^{-18}$, $\sim 10^{-9}$; $\sim 10^9$ yr, $\sim 10^3$ yr]

Problem 1.22. (a) Look up the coordinates of the visual binary α Cen in a star catalog. When and from where on the earth could you see it? If it is available to you, go outside at an appropriate time and identify it to a friend. Use a small telescope to distinguish the two stars, which were separated by about $8''$ in 2005. (They are probably too close to resolve with binoculars, but the image might appear elongated.) Their types are G2 V (yellow/white) and K IV (redder) with V magnitudes $V = 0.0$ and 1.36, respectively. (b) If you can not get to where α Cen is visible (which is likely), look up and try to observe the northern-hemisphere visual binary, η Cas, with $V = 3.4$ and 7.2 and separation $\sim 10''$. (c) Also, try the binary 61 Cyg with $V = 5.2$ and 6.0 and separation $\sim 25''$ [Ans. Coords. J2000: $\alpha = 14$ h 39 m 36.5 s, $\delta = -60° 50' 02''$; $\alpha = 00$ h 49 m 06.3 s, $\delta = 57° 48' 55''$; $\alpha = 21$ h 06.9 m, $\delta = 38° 45'$]

Problem 1.23. Find the distances from which a spectroscopic binary can be detected also as a visual binary. Consider two 1-$M_\odot$ stars in a circular binary orbit viewed edge-on ($i = 90°$) from distance D. Assume that the two stars are bright enough to be detectable in both imaging and spectroscopy and that reliable binary detection requires image separation three times greater than the $1''$ resolution of a ground-based telescope. For spectroscopy, centroids of spectral lines can be determined to about a precision of $\Delta\lambda/\lambda = -\Delta\nu/\nu \approx 10^{-5}$. Here, a factor-of-three greater wavelength shift is required for confidence in the detection of a binary. Hint: use the Doppler relation (2) and Kepler's third law (76). Comment on the effect of imaging with optical interferometry with 1- milliarcsec resolution. [Ans. $\lesssim 20$ LY]

Problem 1.24. Consider the binary shown in Fig. 6. Note that the observer is in the plane of the orbit ($i = 90°$). (a) What is the total fractional frequency excursion, $\Delta\nu/\nu_0 = (\nu_{max} - \nu_{min})/\nu_0$, of a spectral line from star 1 as it proceeds around its orbit where ν_0 is the rest frequency? Use the classical Doppler shift. Repeat for star 2. (b) In the special theory of relativity (Section 7.4), there is a "second-order" Doppler shift (7.40),

$$\nu = \left(1 - \frac{v^2}{c^2}\right)^{1/2} \nu_0,$$

where v is the speed of the emitting object, ν_0 is the rest frequency, and ν is the detected frequency. This shift is most apparent when the object is moving normal to the line of sight when the "first-order" Doppler shift is zero. At time t_2, in Fig. 6, what is the fractional relativistic frequency shift $[(\nu - \nu_0)/\nu_0]_{rel}$ for star 2? Compare with your answers to (a). (c) At what inclination angle i of the orbit would $1/2$ the total fractional first-order frequency excursion for star 2 equal the magnitude of the second-order shift from part (b)? Comment on the likelihood of finding a system in which the second-order effect dominates throughout the orbit [Ans. $\sim 10^{-4}$, $\sim 10^{-3}$; $\sim -10^{-7}$; $\sim 30''$]

1.3 Kepler and Newton

Problem 1.31. Prove for any arbitrary point Q on an ellipse, defined by (4), that the sum of the two radii from the two foci is equal to $2a$, twice the semimajor axis. Refer to Fig. 8a.

Problem 1.32. (a) Calculate the ratio b/a and roughly sketch the ellipses that correspond to the following values of eccentricity e: 0, 0.2, 0.4, 0.6, 0.8, 0.9, 0.95, 0.99, 0.999, and 1.000. Make your sketches on a single figure where the long axis, $2a$, is common to all the ellipses. Tabulate your results and leave places for an additional number for each value of e. (b) Calculate and tabulate the periastron distance r_{min} (5) for each value of e; set $a = 1$. Plot on your sketch the location of a focus for each of the eccentricity values. Comment on how the eccentricity affects the shape and focus location of the ellipse for both high and low values of e. Does this change your thinking about the meaning of e? [Ans. for $e = 0.6$: 0.8 and 0.4]

Problem 1.33. Use the data for φ Cygni in Fig. 7 to answer the following. (a) What is the ratio of masses of the two stars in the φ Cygni system? (b) What is the period of the orbit? (c) What is the radial velocity of the barycenter? Is it approaching or receding from the observer? (d) What is the fractional frequency Doppler shift $(\Delta \nu / \nu)$ due to the barycenter motion? What is the minimum spectroscopic "resolution" $|\lambda/\Delta\lambda|$ required to detect this motion? (e) Can you extract the actual speed v (not the radial component only) of the barycenter from these data? (f) What do these data of Fig. 7 tell you about whether or not this object exhibits eclipses? (g) Why do the two curves look so much alike? Argue quantitatively from momentum conservation. (h) About when are the stars closest together? (*i*) From roughly what direction is the orbit of star 2 being viewed by the spectroscopist? Assume the observer–spectroscopist is in the plane of the orbit (i.e., $i = 90°$) rather than the actual $i = 78°$. Remember that receding velocities are positive. Illustrate your arguments with a sketch. Hint: consider the "zero" crossings, the extrema, and the asymmetries. (j) What is the approximate longitude of periastron ω_p (as defined in Fig. 11)? [Ans. ~1; ~400 d; ~5 km/s; $\lambda/\Delta\lambda \approx 10^5$; −; −; momentum; 20–40 d; −; ~30°] (Optional project: calculate and plot the Doppler curves for a star in an elliptical orbit for various view angles and eccentricities; see geometry of Fig. 11.)

Problem 1.34. NASA wishes to place a satellite into a circular orbit for the purpose of photographing the earth's *entire* surface over a period of ~12 h. (a) What inclination orbit would potentially accomplish this? In which direction, approximately, and from roughly where in the lower 48 states of the United States could NASA launch the satellite into this orbit? Assume the orbit is circular. Illustrate with a sketch. (b) If the field of view of the downward-looking camera is circular with angular radius $\theta_c = 30°$ and if the orbit is fixed in inertial space (i.e., it does not precess), at what altitude must the satellite be to just photograph completely the regions between the successive tracks of the satellite on the earth's surface? Take into account the spherical shape of the earth, recall that the period of the satellite orbit increases with altitude, and do not confuse altitude with distance to the earth's center. The geometry and mathematics are a bit involved. [Ans. 90° (polar); ~3400 km]

1.4 Newtonian solutions $M \gg m$

Problem 1.41 Show that the radial and azimuthal components of the acceleration vector in polar coordinates are as given in (16) and (17).

Problem 1.42. (a) Derive Kepler's third law for a circular orbit by rearranging the radial equation of motion for circular motion (40) for $M \gg m$. (b) What is the total energy of the orbiting body in terms of M, m, r (i.e., eliminating velocity v)? Demonstrate that your answers are consistent with those for an elliptical orbit, (45) and (52).

Problem 1.43. (a) Two different planets, m_1 and m_2, orbit a massive central object of mass M. The shapes of the two elliptical orbits are identical, but the orbit of m_2 is a factor of 9 greater in size than that of m_1. Can the angular momenta of the two objects be different even though the orbit shapes are identical? If so, what is the condition on the masses under which the angular momenta are equal? (b) Two objects of different masses are in the same orbit (same size and shape). How do their speeds at the same point in the orbit differ? Justify your answer with the Newtonian expression (37). What does Kepler have to say about this? (c) By what factor must the speed of the orbiting object change if the mass of the central object is doubled and, at the same time, the object is given the necessary additional velocity to maintain the same orbital track? [Ans. yes, $-$; $-$; $\sim$1.5]

Problem 1.44. The dwarf planet Pluto has a notably large eccentricity, $e = 0.250$. The semimajor axis of its orbit is 39.44 AU, and its mass is 0.17 m_{earth} ($m_{earth} = 6 \times 10^{24}$ kg). (a) What is the total (kinetic + potential) energy of Pluto; exclude internal and spin energies? (b) What is its angular momentum with respect to the sun? (c) How long does it take to orbit the sun? [Ans. $\sim 10^{31}$ J; $\sim 10^{40}$ kg m^2 s^{-1}; $\sim$250 yr]

Problem 1.45. A 200-kg satellite used for UV astronomy is in a highly elliptical orbit in the plane of the earth's equator with perigee (closest point to the earth) 400 km above the earth's surface and with apogee (farthest point) at the geosynchronous altitude. This altitude is defined as the one at which a satellite in a circular equatorial orbit would have a period equal to that of the earth's rotation (sidereal) period so that it remains over a fixed point on the equator. (a) What is the geosynchronous altitude r_s in earth radii measured from the *center* of the earth? (b) What is the eccentricity e of the orbit? (c) If a *circular* geosynchronous orbit were desired, when and in what direction would one give a rocket impulse to the satellite? How much energy is required? (Neglect the weight of the attached rocket and fuel, etc.) (d) If the rocket fails to ignite, frictional forces due to the tenuous atmosphere at perigee would gradually change the orbit. Consider that the friction simply imparts a momentary, small impulse to the satellite at each perigee passage. Describe how the orbit would change. To what semimajor axis and to what shape might it evolve before arriving at its eventual fate? What is that fate? [Ans. $\sim 10^8$ m; $\sim$0.7; $\sim 10^9$ J; $-$]

Problem 1.46. A mass m is in an elliptical orbit (semimajor axis a and eccentricity e) about a mass M, where $M \gg m$. Just as m reaches periastron, the central mass M is suddenly changed to mass fM, where f is the factor by which the mass increases or decreases. The orbiting mass m maintains its speed at this instant and hence also its kinetic energy; however, the change of mass causes its total energy to alter abruptly as do the orbit semimajor axis and eccentricity of its subsequent track. (a) Find an expression for the new semimajor axis a' in terms of a, f, and e. Hint: consider the total energy before and after the mass change. (b) Find expressions for the final eccentricity e' in terms of e and f; there are two cases to consider: (*i*) the periastron remains the periastron and (*ii*) it becomes the apastron. (c) Evaluate your expressions for a'/a and e' (Case (*i*)) for the conditions $f = 0.5, 0.9, 1.0, 1.1, 2.0$, and 100, each for $e = 0$, 0.5, 0.7, and 0.9. Tabulate the results and comment on the trends. What do negative values of a'/a and e' signify? Make a drawing approximately to scale showing the original orbit for $e = 0.7$ as a solid line and the new orbits for $f = 0.9$ and 1.1 as dashed or shaded lines. (d) Find the condition on f for circularization of the orbit, (i.e., find $f_c(e)$). Evaluate f_c for $e = 0, 0.5$, and 0.9 and find a'/a for each case. Tabulate your results and comment. (e) Find the expression $f_{u,per}(e)$ that is the value of f required for the mass m to just become unbound (for our periastron location case). Evaluate and tabulate $f_{u,per}$ for $e = 0, 0.5, 0.9$. By inspection of your derivation of a'/a, obtain the expression for $f_{u,ap}(e)$, the unbinding condition if the mass loss (by M) takes place when m is at apogee. Evaluate for $e = 0, 0.5, 0.9$ and add the results to your table. (f) Reconsider the unbinding condition for a system of two arbitrary masses (m_1, m_2) – that is, $M \gg m$ is not necessarily valid. With minimal or no further calculations, what can you say about this situation? How might this be relevant to a supernova explosion undergone by a star in a binary stellar system?

$$\left[\text{Ans. } a' = a\frac{f(1-e)}{2f-1-e}; \; e' = \frac{|f_c - f|}{f}; -; \; f_c = 1 + e; \; f_{u,per} = \frac{1+e}{2}, \; f_{u,ap} = \frac{1-e}{2}; - \right]$$

Problem 1.47. Find the expression (53) for the orbit eccentricity in terms of G, m, E_t, J, and numerical constants. Hint: begin with the definition of eccentricity (12) and use the expressions (35) and (52).

1.5 Arbitrary masses

Problem 1.51. (a) Demonstrate that (57) and (58) follow from the general definition of the position r_b of the barycenter of two point masses relative to an arbitrary origin: $r_b \equiv (m_1 r_1 + m_2 r_2)/(m_1 + m_2)$ if the origin is at the barycenter. (b) For a two-body gravitationally bound system, where $m_1 = 3m_2$, what are the relative sizes (e.g., the semimajor axes) of (*i*) the orbit of m_2 measured in the frame of reference of m_1, (*ii*) the orbit of m_1 in the frame of reference of m_2, (*iii*) the orbit of m_1 measured in the barycenter frame, and (*iv*) the orbit of m_2 in the barycenter frame? Make a simple sketch. [Ans. $-$; 4:4:1:3]

Problem 1.52. (a) Show that the total angular momentum magnitude J in the barycenter frame of reference is indeed the same as that inferred by analogy to the $M \gg m$ case. Follow the substitutions suggested in the text and fill in the missing steps to verify (72). (b) Repeat for the total energy E_t; that is, verify that (79) is equivalent to (77).

Problem 1.53. (a) Find the sum of the masses (in units of solar masses) in the binary system Kruger 60. Use the information in the caption to Fig. 1 and apply Kepler III. Is your result consistent with both stars' being spectral type M (see Table 4.2)? (b) Repeat this exercise for the hypothetical system shown in Fig. 6. The information in the figure is sufficient. What are the individual masses m_1 and m_2? (c) Repeat again for the system α Cen in Fig. 2. Use the angular scale in the figure and the distance to the system given in the caption. The line between periastron and apastron is foreshortened by the factor $\sim 2/3$ owing to the inclination of the orbit. Is your answer roughly consistent with the stellar types quoted in the caption? Refer to Table 4.2, but, if possible, use a reference that gives masses for additional intermediate classes. [Ans. total masses: $\sim 0.4\, M_\odot$; $\sim 5\, M_\odot$; $\sim 2\, M_\odot$]

Problem 1.54. The eccentricity of the moon's orbit is $e = 0.0549$ and its sidereal period is $P = 27.32$ d. Its mass is $m_M = 1/(81.301)$ of the earth mass, and its mean physical radius is $R_M = 1738$ km. The earth mass is $M_E = 5.974 \times 10^{24}$ kg, and its mean physical radius is $R_E = 6371$ km. In this problem, maintain fairly high numerical precision, to three or four places. (a) By what percentage does the ratio of the major to minor axes of the moon's orbit differ from unity? By what percentage does the ratio of the distances at apogee and perigee differ from unity? Comment. (b) Do these ratios refer to the orbit relative to the earth's center or to the orbit about the earth-moon barycenter? (c) What are the absolute values of the apogee and perigee distances in units of earth radii? Specify whether your answers are relative to the earth's center, moon's center, or to the barycenter. (d) What is the distance between the barycenter and earth's center in earth radii, at apogee $r_{E,a}$, and at perigee $r_{E,p}$? (e) What is the angle of the moon subtended by an observer when the moon is at apogee and directly overhead? Repeat when it is at perigee. Compare these with the sun's angular mean radius of 960″ at its mean distance. Comment on how this is pertinent to solar eclipses. [Ans. ~0.1%, ~10%; −; ~60 R_E; ~0.7 R_E; ~30′]

1.6 Mass determinations

Problem 1.61. (a) Write the equation for star 2 that is comparable to that for star 1 (88). (b) Explain why it is appropriate to call your equation the mass function equation for "star 2" rather than for "1" or "1 *and* 2." After all, it does contain *both* masses in it. (c) Show that the measured value of f_2 represents the lowest possible value for m_1.

Problem 1.62. An observer has only the data in the lower part of Fig. 6 and has no prior knowledge of the nature of the orbits (i.e., ignore the upper sketches). (a) By inspection only, what can you infer about the eccentricity of the orbit? Explain your reasoning. (b) Evaluate the two mass functions and determine if the limits to the masses they imply are consistent with the values given in the text in (98). (c) The absence of eclipses permits the observer to conclude, hypothetically, that the inclination is less than 30°. Does this change the constraints on the individual masses? If so, what are the new limits? (d) Assume the inclination is known to be exactly 30°. Find the two masses by solving the mass functions. [Ans. −; ≳2$M_\odot$, ≳0.1$M_\odot$; ≳20$M_\odot$, ≳1$M_\odot$; ~30$M_\odot$, ~10$M_\odot$]

Problem 1.63. The x-ray source A0620−00 is a compact star with an optical counterpart (V616 Mon) in a binary system. The x-ray source flared up for several months in 1975 and faded away to a very faint level whereupon the optical counterpart could be studied spectroscopically without contamination by florescence owing to x rays' impinging on the stellar atmosphere. The Doppler curve has been found to be sinusoidal, $\Delta\lambda/\lambda = A\sin(2\pi/P)t$ relative to the systemic barycenter Doppler shift with $P = 0.323\,014 \pm 0.000\,004$ d and $A = 1.523$ $(\pm0.027)\times10^{-3}$ where we quote the one-standard-deviation uncertainties. The mass m_{opt} of the optical star was determined from its spectral type to be no less than 0.7 $M_\odot$, and modeling of the changes in brightness due to tidal distortions indicated an inclination no more than 50°. Find the highest lower limit on the mass m_x of the x-ray star that you can claim *with high confidence*, taking into account the quoted errors at the two-standard-deviation level. Only a neutron star or a black hole could emit such copious x rays. A neutron star can not be more massive than ~3 $M_\odot$ according to theorists. Does your result allow you to exclude a neutron star and thereby claim it is a black hole? [Ans. ~7 $M_\odot$]

Problem 1.64. Find the mass function for the optical counterpart of Cygnus X-1 from the data of Fig. 12 together with the orbital period and mass of the optical star given in the caption. Use this to confirm the statements in the text and caption regarding the mass limit for the compact counterpart.

1.7 Exoplanets and the galactic center

Problem 1.71. (a) Consider a hypothetical star of $1\,M_\odot$ with a single earthlike planet in a circular orbit at 1 AU. Assume inclination 90° (observer in plane of orbit). If the star (not the planet) were emitting radio pulses, what would be the range of delays in the detected pulses as it orbits the barycenter of the two-body system? (b) What is the maximum radial (line-of-sight) velocity of the star? (c) Repeat (a) and (b) for a sunlike star with a single Jupiter-like planet at the Jupiter distance. Compare your answers with detectable limits for pulsing and radial velocity (i.e., spectroscopic) detections stated in the text. Useful values are $m_J = 318\,m_E$; $r_J = 5.2$ AU; $P_J = 11.86$ yr. [Ans. ~ 2 ms; ~ 0.1 m/s; ~ 3 s, ~ 10 m/s]

2

Equilibrium in stars

What we learn in this chapter

A normal star is basically a ball of hot gas held together by gravity. Processes that underlie the stability of a star begin when the stellar matter is still part of the diffuse interstellar medium (ISM). A portion of the ISM can not begin **condensation to higher densities** unless its size exceeds the **Jeans length**. Its gravitational potential must be sufficient to prevent the escape of individual atoms with thermal kinetic energies.

A star is in **hydrostatic equilibrium** when the **inward pull of gravity** on each mass element of the star is balanced by the upward force due to the **pressure gradient** at the location of the element. The **potential and kinetic energies** of the mass elements summed over an entire star in hydrostatic equilibrium yield the **virial theorem**. The theorem states that the sum of twice the kinetic energy and the (negative) potential energy equals zero. Its application to **clusters of galaxies** indicates they are bound by a preponderance of **dark matter**.

Several **time constants** characterize a star. A star would radiate away its current thermal content at its current luminosity in the **Kelvin–Helmholtz** or **thermal** time. In the **dynamical** time, a mass element at radius r without pressure support would fall inward a distance r under the influence of the (fixed) gravitational force at r. A photon will travel from the center of the star to its surface through many random scatters in the **diffusion** time.

Under stable conditions, the energy radiated from the stellar surface of normal (main-sequence) stars is replaced by **exothermic nuclear reactions** that convert **hydrogen to helium**. This occurs through the **proton–proton (pp)** chain of reactions dominant at the temperature of the sun's center. The **carbon–nitrogen–oxygen (CNO)** chain is important at somewhat higher temperatures. In later stages of stellar evolution, at even higher temperatures, **elements up to iron** can be created.

Stars of masses beyond $\sim 130\, M_\odot$ have such high luminosities that **radiation pressure** would **expel the outer layers of stellar material**. Such stars are not expected to exist. This upper limit of luminosity is called the **Eddington luminosity**. It is proportional to stellar mass and equal to $33\,000\, L_\odot$ at $1.0\, M_\odot$. A **neutron star** in a close binary system can accrete gaseous matter from its companion, and the infall energy gives rise to intense **x-ray radiation** that tends to inhibit the accretion. The **maximum rate of matter accretion** is that associated with the Eddington luminosity. The luminosities of **active galactic nuclei** indicate they have masses reaching to $10^8 \, M_\odot$.

At certain points in their evolution, stars can have atmospheres that are **unstable to pulsations**. They oscillate in radius, temperature, and luminosity. The oscillations are powered by the **conversion of heat to work** in gas elements of the star (as in the Carnot cycle) brought about by changing **levels of ionization** and hence **opacity**. Examples are the **cepheid variables** and the less luminous **RR Lyrae** stars.

2.1   Introduction

Stars such as our sun seem quite stable in their overall characteristics, and indeed they are. Here we examine the following seven issues of stability:

(*i*)   the size requirement for a portion of the interstellar medium (ISM) to begin condensation to higher densities that would lead eventually to star formation,

(*ii*)   the condition for a mass element of a spherical star to be in gravitational or hydrostatic equilibrium,

(*iii*)   the balance of kinetic and potential energies in such a system,

(*iv*)   the time constants that must govern changes of structure or energy content of the star,

(*v*)   the nuclear reactions that replace the radiant power emitted from the stellar surface,

(*vi*)   the limiting luminosity that places an upper limit to stellar masses, and

(*vii*)   an instability that yields stellar pulsations.

These processes provide insight into other phenomena in astronomy – for example, the dark matter in clusters of galaxies and the mass accretion rates in neutron-star binary star systems and in active galactic nuclei.

2.2   Jeans length

Stars are formed from condensations of the diffuse interstellar gas. The detailed physics of this process is a difficult theoretical problem. As the gas cloud contracts under the influence of its own gravity toward a density at which nuclear burning can begin, it must shed angular momentum and also overcome the pressure of the magnetic fields intrinsic to the ionized gases of the ISM. Just how all this takes place is still not well understood. Nevertheless, it does happen because stars do exist. The physics that must apply is interesting and well known; its application, however, is quite complex.

Portions of the ISM must fragment into individual clouds that will eventually become galaxies and stars. To collapse, the cloud must be of such a size that the magnitude of the gravitational potential energy of an atom in the cloud exceeds its kinetic energy. We calculate this size, which is known as the *Jeans length*. In clouds of smaller size, the atoms would escape the incipient cloud, and it would simply dissipate (Fig. 1).

Collapse criterion

The condition for cloud collapse in terms of the energies of an individual atom is

$$E_k \lesssim |E_p|$$

(2.1)

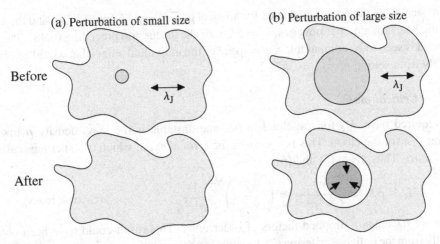

Figure 2.1. Density fluctuations in an interstellar cloud. (a) Perturbation to higher density of a cloud segment of size smaller than the Jeans length λ_J. The perturbation is not sustained. (b) Perturbation of size larger than λ_J. It will continue to contract.

or

$$\left(\frac{1}{2}m_H v^2\right)_{av} \lesssim \frac{GM m_H}{R}, \qquad \text{(Condition for fragmentation into cloud)} \qquad (2.2)$$

where M is the mass of the cloud and m_H is the mass of an individual hydrogen atom. The left side is the average kinetic energy of the atom, which is equal to $3kT/2$ for monatomic particles with a Maxwell–Boltzmann (thermal) distribution.

Let the mass of the cloud be approximately $M \approx \rho R^3$, where ρ is the mass density (kg/m³). Substitute into (2) and neglect factors of order unity to find

$$v^2 \lesssim GR^2 \rho. \qquad (2.3)$$

Solve for the radius of the cloud:

$$R \gtrsim \frac{v}{(G\rho)^{1/2}}. \qquad \text{(Critical size for collapse)} \qquad (2.4)$$

This is the critical size at which instabilities can develop so that the collapse can start. This size scale is known as the Jeans length after Sir James Jeans (1877–1946) and is expressed as

$$\lambda_J \approx \frac{v_s}{(G\rho)^{1/2}}, \qquad \text{(Jeans length; m)} \qquad (2.5)$$

where we write it in the usual form with the *speed of sound* v_s, which is, not surprisingly, close to the kinetic speed v, as we now demonstrate.

The speed of sound may be expressed in terms of the pressure P, the density ρ, and the ratio of specific heats γ as (not derived in this text)

$$v_s^2 = \frac{\gamma P}{\rho} = \gamma \frac{kT}{m_{av}} \rightarrow \frac{5}{3}\frac{kT}{m_{av}}, \qquad (2.6)$$

where we invoke the equation of state for an ideal gas, $P = \rho kT/m_{av}$ (3.39), and the ratio of specific heats for a monatomic gas, $\gamma = 5/3$; see the discussion preceding (6.48). The speed of sound is seen to be comparable to the speed of the individual atoms that would be obtained directly from setting $mv^2/2 = 3kT/2$.

Critical mass

The required mass M_c for the cloud to become unstable is the mass density ρ times the volume $\sim\lambda_J^3$ of the cloud. This is known as the *critical mass*, which is sometimes called the *Jeans mass*. Thus, from (5) and (6),

$$\blacktriangleright \qquad M_c \approx \rho\lambda_J^3 = \frac{v_s^3}{G^{3/2}\rho^{1/2}} \approx \left(\frac{kT}{Gm_{av}}\right)^{3/2}\frac{1}{\rho^{1/2}}, \qquad \text{(Critical mass)} \qquad (2.7)$$

where we have again dropped factors of order unity. This result could have been obtained directly from the collapse criterion (2), together with the relations $mv^2 \approx kT$ and $\rho \approx M/R^3$, without defining the Jeans length and speed of sound (Prob. 21).

As an example, we calculate the critical mass for the cold neutral component (hydrogen clouds) of the ISM described in Table 10.2. The hydrogen number density is $4 \times 10^7 \mathrm{m}^{-3}$ at a temperature of $T \approx 100$ K. The expression (7) yields a mass of $M_c \approx 3000\,M_\odot$ and a Jeans length of $\sim$60 LY. This is somewhat larger than the cloud sizes listed in the table. We would thus conclude that the clouds of Table 10.2 are unlikely to collapse further. Local regions of substantially higher density, however, surely could.

If our hypothetical cloud further collapses, say by a factor of 100 in size, and hypothetically manages to cool itself by radiation so as to remain at the 100-K temperature, the critical mass would be 1000 times less, or $M_c \approx 3\,M_\odot$. This suggests fragmentation to smaller clouds; the initially large low-density clouds might be expected to fragment one or more times as they contract. The group of cloudlets from a single large cloud thus could become a cluster of hundreds to thousands of newly formed individual stars.

On another scale, the rarified low densities and moderately high temperatures of some regions of intergalactic space could well lead to critical masses comparable to the masses of galaxies. From absorption lines in quasar spectra, one can infer the existence of clouds in intergalactic space with particle number densities of $\sim 10^2 \mathrm{m}^{-3}$ and temperatures of $\sim 3 \times 10^4$ K. This leads to a critical mass of $M_c \approx 10^9\,M_\odot$, the mass of a small galaxy. Even more rarified regions could result in even larger critical masses that could lead to larger galaxies and clusters of galaxies.

Our calculations here assume a smooth, homogenous medium when in fact the interstellar medium is highly irregular with density fluctuations and also magnetic fields and angular momenta. We will find in Section 5 that the dynamical infall time is $(G\rho)^{-1/2}$ (37). Hence, in our simplistic scenario, the regions of high density will collapse rapidly and, in turn, become the centers of further fragmentation and increasingly rapid collapse.

2.3   Hydrostatic equilibrium

An element of the gas in a star is attracted by means of gravity to all other elements of the star. For a spherically symmetric mass distribution, the total gravitational force on the element is directed toward the center of the star. If each mass element of the star has no net force on it, it is said to be in *hydrostatic equilibrium*. In this case, the inward pull of gravity is exactly

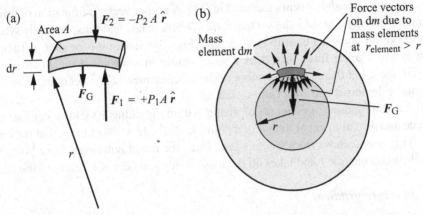

Figure 2.2. Hydrostatic equilibrium for a mass element of area A and thickness dr at radius r in a star. (a) Force balance. The upward force due to the differential pressure, $(P_1 - P_2)A$, is balanced by the downward gravitational force F_G on the element. (b) Gravitational forces on a mass element in a spherical star for mass elements at greater radii than dm (thin arrows) and for the sum of all mass elements interior to dm (thick arrow). The former sum to zero.

balanced by the upward force due to the gradient of the gas pressure. (Radiation pressure can also play a role in the most massive stars.) Here we find the differential equation that represents this balance of forces.

Balanced forces

Consider an element of gas of thickness dr and area A at a distance r from the center of the star, where the mass density is $\rho(r)$ (Fig. 2a). The element is a segment of a spherical shell of radius r with its center at the center of the star. The pressure P_1 at the bottom of the element will exert an upward force on the element, whereas the lesser pressure P_2 at the top will exert a downward force that is smaller than the upward force.

The net upward force due to the pressure differential, $dP = P_2 - P_1$, is

$$F_P = -A\,dP. \tag{2.8}$$

We choose the "+" direction to be upward (increasing radius r), and so in this case the pressure differential is negative, $dP < 0$, and the associated force is in the positive (outward) direction. The condition for hydrostatic equilibrium is

$$F_G + F_P = 0, \tag{2.9}$$

where F_G is the gravitational force. The condition thus becomes

$$\Rightarrow \qquad F_G = A\,dP. \qquad \text{(Hydrostatic equilibrium)} \tag{2.10}$$

The force due to gravity on our element of gas, $dm = \rho A\,dr$, is described by Newton's gravitational law (1.13)

$$F_G = \frac{-G\mathfrak{M}(r)\,\rho(r)A\,dr}{r^2}, \qquad \text{(Newton's gravitational law)} \tag{2.11}$$

where $\mathfrak{M}(r)$ is *included mass*, the portion of the stellar mass interior to radius r.

The mass outside radius r is not included in (11). A proper vector summation of forces due to the mass elements beyond r shows that, for a $1/r^2$ force law and for spherically symmetric mass distribution, their net contribution is zero (Fig. 2b). Each bit of matter just above our element exerts a larger force than does a bit of matter in a distant (lower) part of the star (the $1/r^2$ effect), but there are many more of the latter elements. The net effect is zero force, which can be demonstrated with Gauss's law.

A similar summation for elements of matter within the radius r yields a net force with a magnitude and direction equal to that of a pointlike mass $M = \mathfrak{M}(r)$ located at the center of the star. This, too, follows from Gauss's law. Thus, the net gravitational force given in (11) ignores the mass outside r and takes all the mass inside r to be at the center of the star.

Pressure gradient

Substitute the gravitational force (11) into (10) to obtain the equation of hydrostatic equilibrium in its usual form as

➡
$$\frac{dP}{dr} = \frac{-G\mathfrak{M}(r)\rho(r)}{r^2}$$
$$= -\rho(r)\,g(r). \qquad \text{(Hydrostatic equilibrium)} \qquad (2.12)$$

This differential equation indicates how, at radius r, the gradient of pressure (dP/dr) depends on the total stellar mass at lesser radii, $\mathfrak{M}(r)$, and on the density ρ at the radius r if the inward and outward forces on the elements of the star are in balance.

The acceleration due to gravity g is introduced in (12), where

$$g \equiv \frac{G\mathfrak{M}(r)}{r^2} \xrightarrow{r=R} \frac{GM}{R^2}. \qquad \text{(Gravitational acceleration; N/kg or m/s}^2\text{)} \qquad (2.13)$$

At the surface of a star of radius R and mass M, the acceleration becomes equal to GM/R^2, and thus the force on a test mass m becomes the familiar $F = mg$.

The reader is cautioned that, in our derivation, the area A is tacitly taken to have the same value at the top and bottom of the mass element. This is not precisely correct for spherical geometry. Our assumption of equal areas does, in fact, give the correct final result, and the essential physics is illustrated. A proper derivation makes use of the gradient of pressure ∇P in spherical coordinates (Prob. 31).

The equation of hydrostatic equilibrium is one of the essential elements required in the modeling of a star's structure (Section 4.3).

2.4   Virial theorem

A global way to understand the equilibrium state of a star is provided by the *virial theorem*. It is an energy argument that makes no reference to the detailed internal structure of the star. The theorem is important in other contexts also – notably in revealing dark matter in clusters of galaxies.

Potential and kinetic energies

The virial theorem is a useful statement about the relative magnitudes of the total kinetic and potential energies of a system of particles in *stable equilibrium bound by gravity*.

The theorem states that twice the *total* kinetic energy of the particles, $2\Sigma E_k$, added to the *total* potential energy ΣE_p must equal zero:

$$2\Sigma E_k + \Sigma E_p = 0. \qquad \text{(Virial theorem)} \qquad (2.14)$$

The summation is over all particles in the system. Keep in mind that, in a bound system, the potential energy will always be negative for the usual convention of $E_p = 0$ at infinite separation.

The virial theorem tells us that the total (kinetic plus potential) energy of the system is equal to half the total potential energy or to the negative of the kinetic energy; that is,

$$E_{tot} = \Sigma E_k + \Sigma E_p = (1/2)\Sigma E_p = -\Sigma E_k. \qquad (2.15)$$

The virial theorem (14) is valid for potentials that vary inversely with distance as does the gravitational potential.

The virial theorem is obeyed by a satellite of mass m in a circular earth orbit at radius r about the much more massive earth of mass M. Substitute into $F_r = ma_r$ the gravitational force GMm/r^2 and also the radial acceleration $-v^2/r$. Solve for $E_k = mv^2/2$ to find $E_k = GMm/(2r)$, which is just one-half the magnitude of the potential energy of the satellite, $-GMm/r$. This is in agreement with (14).

If the satellite loses total energy because of gradual atmospheric drag, it must move to smaller radii (lower altitudes) to lose potential energy and hence total energy. Because the potential becomes more negative, increasing in magnitude, the kinetic energy must *increase* according to the virial theorem. This increase in E_k is only one-half the loss of potential energy, and so total energy indeed decreases. The atmospheric drag tries to slow the satellite, but instead it falls to a lower orbit and speeds up!

The virial theorem is not used explicitly in calculations of stellar structure. Nevertheless, any solution of the basic equations of stellar structure must be checked for stable equilibrium with it. This is acknowledgment of the central role of the virial theorem.

Derivation

The virial theorem (14) may be derived from first principles with the aid of the hydrostatic equilibrium equation (12). The latter is appropriate because a collection of particles in equilibrium will have zero net force on each and every gas element of its interior.

Define a volume function $\mathcal{V}(r)$, which is the spherical volume of a star enclosed within the radius r, and multiply it by dr:

$$\mathcal{V}(r)\,dr = \frac{4}{3}\pi r^3 dr. \qquad (2.16)$$

Multiply the left and right terms by, respectively, the left and right terms of the equation of hydrostatic equilibrium (12):

$$\mathcal{V}(r)dP = -\frac{1}{3}\frac{G\mathfrak{M}(r)}{r}[4\pi r^2\rho dr]. \qquad (2.17)$$

The quantity in square brackets is the mass of a shell of matter $d\mathfrak{M}$, and so

$$\mathcal{V}(r)dP = -\frac{1}{3}\frac{G\mathfrak{M}(r)}{r}d\mathfrak{M}. \qquad (2.18)$$

The right side of (18) is the potential energy of the shell $d\mathfrak{M}$ at radius r divided by 3.

Integrate this equation over the entire volume of the star:

$$\int_{\text{star}} \mathcal{V}(r)dP = -\frac{1}{3}\int_{\text{star}} \frac{G\mathfrak{M}(r)}{r}d\mathfrak{M}. \tag{2.19}$$

The integral on the right side is the sum of the potential-energy magnitudes of all the shells. With the minus sign, the right side of the equation is thus the potential energy of the entire star divided by 3; that is, $\Sigma E_{\text{P}}/3$.

For the integral of the left side, consider the differential of the product $P\mathcal{V}$:

$$d(P\mathcal{V}) = Pd\mathcal{V} + \mathcal{V}dP. \tag{2.20}$$

Integrate over the volume of the entire star,

$$(P\mathcal{V})|_0^R = \int_{\text{star}} Pd\mathcal{V} + \int_{\text{star}} \mathcal{V}dP, \tag{2.21}$$

where R is the radius of the star. The left side equals zero because the volume function $\mathcal{V}(r)$ vanishes at $r = 0$ and the pressure P vanishes at the surface of our idealized star. Substitute this result into the left side of (19) to obtain

$$\blacktriangleright \qquad -3\int_{\text{star}} Pd\mathcal{V} = \Sigma E_{\text{p}}. \qquad \text{(Virial theorem; general form)} \tag{2.22}$$

This is the *general form* of the virial theorem.

Evaluate the integral on the left side of (22) for a perfect nonrelativistic gas. From kinetic theory (3.34), we have

$$P = \frac{2}{3}n\left(\frac{1}{2}mv^2\right)_{\text{av}} = \frac{2}{3}u_{\text{k}}, \tag{2.23}$$

where u_{k} is the kinetic energy density (J/m^3). Integrate this expression over the star to obtain

$$\int_{\text{star}} Pd\mathcal{V} = \int_{\text{star}} \frac{2}{3}u_{\text{k}}d\mathcal{V} = \frac{2}{3}\Sigma E_{\text{k}}. \tag{2.24}$$

Substitute this result into the left side of (22) to obtain

$$-3\frac{2}{3}\Sigma E_{\text{k}} = \Sigma E_{\text{P}}, \tag{2.25}$$

which is the virial theorem (14).

In this derivation, we required stability in (17) through the condition of hydrostatic equilibrium. In turn, this invoked Newton's r^{-2} gravitational law against which a support pressure must exist. We also assumed nonrelativistic particles.

For a relativistic gas with a blackbody distribution (e.g., photons), the pressure from (6.25) and (6.43) is $P = u_{\text{k}}/3$. Substitution into (22) yields

$$\Sigma E_{\text{k}} + \Sigma E_{\text{p}} = 0. \qquad \text{("Virial theorem"; relativistic gas particles)} \tag{2.26}$$

Note that in this case, $E_{\text{tot}} \equiv \Sigma E_{\text{k}} + \Sigma E_{\text{p}} = 0$, rather than $E_{\text{tot}} = -\Sigma E_{\text{k}}$ for the nonrelativistic case (15). This indicates that a gas of highly relativistic particles (which behave as photons) would be just critically (gravitationally) bound if the gas is in hydrostatic equilibrium. Hydrostatic equilibrium is the basis of the virial theorem (17). Such a gas would be in unstable equilibrium.

Stars

If a star had no interior nuclear energy source, it would behave like the earth-satellite system discussed above after (15). With the loss of energy through radiation from the surface, the star would gradually shrink. At any given stage, it would be in (quasi) stable equilibrium, and the virial theorem would be valid. As the star shrinks, the potential energy becomes more negative and the total kinetic energy increases.

Increased kinetic energies mean higher temperatures. As the star loses total energy and shrinks, it becomes hotter. Viewed another way, the star gives up more potential energy than necessary to compensate for the lost radiation, and the extra potential-energy decrease goes into heating the gas. The star exhibits a *negative specific heat*; the removal of heat results in a higher temperature T!

The virial theorem is relevant to gas clouds that are in quasi-stable equilibrium while collapsing to form stars before internal nuclear burning commences. The cloud heats up as it shrinks; the decreasing potential energy is converted to heat, in part, and the resultant pressure inhibits further shrinkage. Further collapse is possible only with the continuing loss of energy (e.g., by radiation at infrared wavelengths). Centrifugal forces due to bulk motions and magnetic pressures also inhibit shrinkage.

Toward the end of a star's normal life, when the hydrogen fuel in the core is completely expended, the core shrinks and heats up according to the virial theorem. Eventually, the kinetic energies of helium nuclei (the ashes of the hydrogen burning) become sufficiently energetic to overcome their mutual Coulomb repulsion and helium burning commences. When the helium is expended, shrinkage and heating again take place until, for a sufficiently high stellar mass, even higher-mass elements begin to burn.

Nuclear burning at the center of a star in equilibrium produces just enough energy in a given year to replace that lost from the surface during that year. This amounts to only about one part in 10^7 for the sun; see the discussion of the thermal time scale in Section 5. A star is thus, to good approximation, in hydrostatic equilibrium, and the virial theorem properly describes it.

The virial theorem for relativistic particles (26) underlies the collapse of a white dwarf star when its mass exceeds the famous Chandrasekhar limit of $\sim 1.4 \ M_\odot$ (4.26). At the point of collapse the electrons in the core of the star have become relativistic (Section 4.4 under "Chandrasekhar mass limit"). The electrons are degenerate (Section 3.6) and hence provide effectively all of the pressure; the pressure from the nondegenerate nucleons is negligible (3.57). Hence, the relation (26) as derived from (22) applies, and unstable equilibrium exists.

A slight bit of additional mass accreted onto a white dwarf in this condition from a companion star tips the white dwarf toward collapse. In the case of a white dwarf, the kinetic energy term in (26), being derived from the pressure, arises largely from the electrons while the potential energy term arises largely from the much more massive nuclei. All this is discussed in more depth in the above referenced sections.

Clusters of galaxies

Galaxies are distributed in space in a highly nonuniform manner, exhibiting "walls" of galaxies, voids with few galaxies, and distinct clusters consisting of tens to hundreds, or even a few thousand, galaxies. An important application of the virial theorem pertains to clusters of galaxies.

Spatial distribution

The galaxies in a cluster of galaxies, in many instances, are distributed in space in such a way that they appear to be in stable equilibrium. The masses of the constituent galaxies may be estimated from their luminosities. The measured velocity dispersions together with the masses provide an estimate of the total kinetic energy ΣE_k of all the galaxies in a cluster. The masses and the positions of the galaxies in the cluster provide the total potential energy ΣE_p.

One finds generally that these values do not satisfy the virial theorem; there is an excess of kinetic energy. Given these values, the galaxies would not be bound together; they would be in the process of flying apart.

Nevertheless, the appearance of many clusters of galaxies argues strongly that they are not dispersing but rather are in a stable configuration. It is thus now widely believed that clusters contain an unknown type of matter that is invisible (*dark matter*) in addition to the visible galaxies. The virial theorem can be satisfied if the dark matter has 10 to 50 times the visible mass in clusters. In other words, the gravitational matter holding the cluster together is 90–98% dark.

Virial mass

The virial theorem may be used to find the total mass of a stable collection of particles if one has measures of particle masses and speeds. We derive this *virial mass* here and also explore the observational approach to the determination of dark matter in clusters of galaxies.

For a collection of particles interacting solely with r^{-2} gravitational forces, the virial theorem (14) may be expressed as a summation over all the individual particles and pairs of particles,

$$2 \sum_i \frac{1}{2} m_i v_i^2 - \sum_{\text{pairs}} \frac{G m_i m_j}{r_{ij}} = 0, \qquad \text{(Virial theorem)} \qquad (2.27)$$

where v_i is the speed of particle i and r_{ij} is the separation distance between the ith and jth particles.

Consider the elementary case of N identical galaxies in a cluster of galaxies, where each galaxy has mass m and N is a large number. Multiply the first term of (27) by N/N and the second by N^2/N^2,

$$Nm \frac{1}{N} \sum_i v_i^2 - G \frac{(Nm)^2}{2} \frac{1}{N^2} \sum_i \sum_{j \neq i} \frac{1}{r_{ij}} = 0, \qquad (2.28)$$

where the factor $1/2$ avoids double counting of the pairs. The single summation (first term) has N terms, whereas the double summation (second term) has $N(N-1)$ terms because those with $i = j$ are excluded. For large N, one can make the approximation $N(N-1) \approx N^2$. The total mass may be written as $M = Nm$, and the summations may be expressed in terms of average values to give

$$M \langle v_i^2 \rangle_{\text{av}} - G \frac{M^2}{2} \langle r_{ij}^{-1} \rangle_{\text{av}} = 0, \qquad (2.29)$$

where the averaged quantities are the speed squared and the inverse separation. The total mass required to hold the cluster in stable equilibrium, the *virial mass*, is therefore

$$M = \frac{2\langle v_i^2 \rangle_{av}}{G\langle r_{ij}^{-1} \rangle_{av}}. \tag{2.30}$$

The velocities v_i are obtained from the Doppler shifts of the spectral lines of the individual galaxies in the cluster. Such measurements yield, after correction for the overall recession of the cluster, the individual line-of-sight components of the velocity, $v_{i,los}$. If all directions of motion are equally probable, the other two components will, on the average, have the same magnitude; thus,

$$\langle v_i^2 \rangle_{av} = 3\langle v_{i,los}^2 \rangle_{av}. \tag{2.31}$$

Similarly, galaxy separations obtained from telescopic measures must also be corrected for projection effects.

The mass obtained from (30) is that required to provide stability of the cluster. In typical clusters, it is much greater than the luminous mass, which is the mass inferred from the visible galaxies in the cluster. The extra *dark matter* implied by this need not be, and probably is not, contained solely in the individual galaxies, as might have been implied here. Dynamical studies of galaxy rotation (Section 10.4) reveal dark matter associated with individual galaxies that exceeds luminous matter by a factor of a few or at most ~10. The larger factors found in clusters of galaxies (up to ~50) indicate that large amounts of dark matter must be distributed throughout the intergalactic medium within the cluster.

2.5 Time scales

Three time scales are pertinent to a star: the thermal or Kelvin, the dynamical, and the diffusion time scales. These provide understanding, respectively, of how rapidly a given star (*i*) might evolve without a nuclear energy source, (*ii*) might collapse inward given a sudden lack of support as in a supernova collapse, or (*iii*) would transfer radiant energy from its center to its surface.

Thermal time scale

The *Kelvin–Helmholtz time scale* τ_K, or simply the Kelvin or thermal time scale, is the time it takes for a star to radiate away an energy that approximates its total current kinetic energy content ΣE_k (J) at its current luminosity L (W):

$$\tau_K \approx \frac{\Sigma E_k}{L}. \tag{2.32}$$

The summation is over all particles in a star.

According to the virial theorem, the kinetic energy is half the magnitude of the total potential energy. For a spherical mass distribution of radius R and total mass M, the total potential energy ΣE_p can be found by summing the potential energies between all pairs as in (27). The result is $\Sigma E_p \approx -GM^2/R$, where the missing coefficient, which depends on the radial density distribution, is of order unity. For our purposes, set $\Sigma E_k \approx |\Sigma E_p|$ to obtain

$$\tau_K \equiv \frac{GM^2}{RL}, \qquad \text{(Kelvin–Helmholtz time scale)} \qquad (2.33)$$

which is taken to be the definition of the Kelvin contraction time scale.

Substitute the solar values into (33) to find

$$\tau_K \approx 3.0 \times 10^7 \text{yr}. \qquad \text{(Solar Kelvin time)} \qquad (2.34)$$

If the nuclear energy source in the sun were to turn off today, it would take about 10^7 years for the sun to lose a substantial fraction of its current energy content. This does not mean that all its energy would be expended. Shrinkage of the sun to smaller radii converts gravitational energy into additional kinetic energy. As we have seen, this raises the temperature. In turn, this raises the luminosity and thus shortens the Kelvin time scale. Without replenishment of energy from hydrogen burning, the sun's lifetime at (roughly) its present luminosity would be 10^7 to 10^8 yr.

This age was one of the biggest puzzles of astrophysics for many years. It is much less than the known ages of the earth and solar system. Studies of radioactivity in rocks on the earth indicate that its age and hence that of the solar system is $\geq 3.8 \times 10^9$ yr. Studies of radioactive elements in meteorites indicate that the material in the solar system condensed into solid bodies about 4.5×10^9 years ago; this latter age is taken as the age of the earth and roughly the age of the sun in its current state (luminosity and temperature).

The solar Kelvin time of $\sim 3 \times 10^7$ yr (34) is the future lifetime of the sun in its current state in the absence of any nuclear energy source. This is also the maximum time that it could have been in its current state; the thermal energy would not support a longer life. The thermal energy content of the sun is thus insufficient by a factor of ~ 100 to have provided the observed luminosity for the apparent age of the solar system (4.5×10^9 yr). Planetary evidence precludes substantial changes in the sun's energy output during this period.

One might have argued that the sun was larger (and cooler) in earlier times in accord with the virial theorem and is just now passing through its current state. The times at which it would have been within acceptable sizes and temperatures would still be far short of the required 4.5×10^9 yr. One therefore concludes that another source of energy must be present. We now know that to be nuclear fusion.

Dynamical time scale

The *dynamical time scale* τ_{dyn} is the time for a star to collapse inward under the influence of gravity with no opposing forces such as pressure. A brief dimensional argument provides an approximate magnitude for this quantity.

Consider a star of mass M and radius R and find the (approximate) time for a test mass m at or near the surface to fall a distance R if the surface were not there and if the gravitational force on it were to remain constant at the surface value (which of course is not the case). The acceleration of the test mass m is

$$a = \frac{F}{m} = \frac{GM}{R^2}. \qquad (2.35)$$

The time to fall a distance R follows from the familiar constant-force expression $s = at^2/2$, where $s = R$. We drop the factor of 2, solve for $t \equiv \tau_{\text{dyn}}$, and invoke (35) to obtain

$$\tau_{\text{dyn}} \equiv \left(\frac{R}{a}\right)^{1/2} = \left(\frac{R^3}{GM}\right)^{1/2}. \qquad \text{(s)} \qquad (2.36)$$

The factor M/R^3 is the approximate density ρ of the matter giving rise to the gravitational force. The dynamical time constant thus becomes

➡ $$\tau_{\text{dyn}} = (G\rho)^{-1/2}. \qquad \text{(Dynamical time scale)} \qquad (2.37)$$

The mean mass density of the sun is 1400 kg/m^3, and so (37) yields

$$\tau_{\text{dyn},\odot} = (G\rho)^{-1/2} = 3300\,\text{s} = 55\,\text{min}. \qquad \text{(Sun dynamical time)} \qquad (2.38)$$

The higher densities in the interior of the sun yield a somewhat smaller value of $\sim$20 min, which is sometimes quoted.

The present sun, under free-fall conditions, would take less than an hour to collapse to a small fraction (say $\sim$1/3) of its current radius. At this more dense condition, it would have a new shorter collapse time according to (37). The characteristic time would become shorter and shorter as the matter collapsed in on itself.

A white dwarf of 1 $M_\odot$ and 0.01 $R_\odot$ would have an average density $\sim$10^6 times greater than that of the sun and hence a dynamical time constant 10^{-3} that of the sun, or

$$\tau_{\text{dyn,wd}} = (G\rho)^{-1/2} = 3.3\,\text{s}. \qquad \text{(White dwarf)} \qquad (2.39)$$

The density for this case is 1.4×10^9 kg/m^3. As the collapsing matter approaches nuclear densities of 10^{17} kg/m^3, the dynamical time constant is on the order of a millisecond:

$$\tau_{\text{dyn, nuclear matter}} = (G\rho)^{-1/2} = 0.4\,\text{ms}. \qquad \text{(Nuclear matter)} \qquad (2.40)$$

These latter two times play roles in the inward collapse of the degenerate core of a star toward neutron-star densities or further into a black hole.

Diffusion time scale

The third time scale of interest is the time it takes for photons to work their way out of the sun via many, many scatters or absorption–reemission processes in a "random walk." This is the *diffusion time*.

One-dimensional random walk

Consider a one-dimensional photon random walk that proceeds as follows. Start the photon at $x = 0$; flip a coin to decide whether to step it a length ℓ ahead (heads for $\Delta x = +\ell$) or a step backward (tails for $\Delta x = -\ell$); step the photon as indicated; flip the coin again to obtain the direction of the second step; take the second step; flip the coin again, and so forth, until the photon has made $N = 100$ steps.

When finished with the 100 steps, one might naively expect the photon to be at the origin because the *expected* numbers of heads and tails would be equal. However, for a *single* 100-step trial, they are not equal because of statistical fluctuations in the numbers of heads N_{H}

and tails N_T. After the 100 steps, the net number of steps is $N_{net} = N_H - N_T$ and the distance from zero is

$$x = N_{net}\ell = (N_H - N_T)\ell. \tag{2.41}$$

This will generally have a positive or negative value and is rarely exactly zero.

If one averages over *many* 100-step trials, an *average* final displacement $\langle x \rangle$ that approaches zero will indeed be found because, in the limit of an infinite number of trials, half of them will have final displacements in the positive direction and half in the negative direction:

$$\langle x \rangle = \langle N_H - N_T \rangle \ell \approx 0. \qquad \text{(Expected value; many trials)} \tag{2.42}$$

The brackets indicate the average value of many N-step trials.

Now, consider fluctuations. Let the total number of steps in a single trial be fixed at $N = N_H + N_T = 100$. Write the net number of steps as $N_{net} = N_H - N_T = 2N_H - N$. The binomial distribution describes random trials with fixed probabilities such as this. In our case with probability 1/2 of a head, the mean number of heads (in many trials) turns out to be $m_H = N/2 = 50$ (as expected) with standard deviation $\sigma_H = N^{1/2}/2 = 5$ (not intuitively apparent). This leads us to the standard deviation of N_{net},

$$\sigma_{net} \equiv \sigma(2N_H - N) = 2\sigma(N_H) \equiv 2\sigma_H = \sqrt{N}, \tag{2.43}$$

where we noted that the total number N is fixed and therefore does not contribute to the fluctuations. (Do not confuse the two uses of σ: standard deviation and cross section.) The uncertainty in the net number turns out to be the square root of the total number of steps. (See AM, Chapter 6 for a discussion of statistics.)

This uncertainty indicates that the actual values of N_{net} will be distributed about the "expected" value of $N_{net} = 0$ with standard deviation $N^{1/2}$. The end points of many 100-step trials will thus be distributed as a Gaussian function centered on zero (Fig. 3). (The Gaussian has the form $\exp(-ax^2)$.) Its standard (root-mean-square or rms) deviation from zero, after N steps each of length ℓ, is, from (43),

➡ $$x_{rms} = N^{1/2}\ell. \qquad \text{(Root-mean-square displacement)} \tag{2.44}$$

For many 100-step trials, one finds the end points at a typical (rms) distance 10ℓ from the origin, and for many 1000-step trials, a distance of 32ℓ. The final positions will lie in both the positive and negative directions, and thus the *average* position will be close to zero in accord with (42).

If many photons are started at the origin at the same time, their distribution in x at a later time (after N steps each) will be the aforementioned Gaussian. As time proceeds and N increases, the distribution becomes progressively spread out. The photons "diffuse" out to larger and larger distances according to (44). This gradual spreading (diffusion) of photons along the x-axis is shown in Fig. 3 as a widening of the Gaussian function.

Let us now turn the question around. How many steps does it take per photon for a substantial number of photons in a sample to reach a distance X from their point of origin? It follows from (44) that the number of steps required, in our one-dimensional problem, is about

$$N \approx \left(\frac{X}{\ell}\right)^2. \qquad \text{(Average number of steps for photon to reach } X\text{)} \tag{2.45}$$

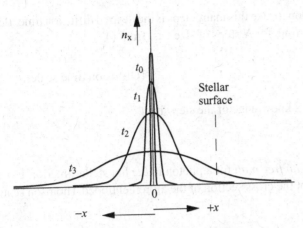

Figure 2.3. Diffusion (random walk) of photons along one axis. The Gaussian distribution is shown for four different times. The number of photons in distance interval dx is $n_x dx$. A large number of photons start from the origin (center of star). As time progresses, the distribution of photon positions gradually widens until substantial numbers reach and escape from the stellar surface.

Three-dimensional walk

The diffusion in a star is three-dimensional. The transport of photons in three dimensions within the star is well approximated with a random walk process. A photon is Thomson scattered (absorbed and reemitted) by electrons many times during its passage. Although Thomson scattering is not isotropic, it is appropriate to assume so when averaging over a nearly isotropic distribution of incident directions. One can therefore argue that, with each collision, photons will scatter along any one of three axes (x, y, z) with equal probability. (See AM, Chapter 10, regarding Thomson scattering.)

The progress of a given photon along any one arbitrarily chosen axis will therefore be slowed. After N steps, only $N/3$ steps will be along the x-axis. The rms distance along the x-axis (44) thus becomes $x_{\text{rms}} = (N/3)^{1/2}\ell$; distances along the other axes will be the same. We are interested in the number of steps it takes for the photon to reach any point on the surface of the star – a distance R from the center. Steps in any of the three directions can contribute motion toward the surface. The mean-square distance from the center after N steps is thus

$$R_{\text{rms}}^2 = x_{\text{rms}}^2 + y_{\text{rms}}^2 + z_{\text{rms}}^2 \qquad (2.46)$$

$$= \left(\frac{N}{3} + \frac{N}{3} + \frac{N}{3}\right)\ell^2 = N\ell^2.$$

The required number of steps to reach the surface of the star at radius R is then

$$N = \left(\frac{R}{\ell}\right)^2, \qquad \text{(Steps to reach star surface)} \qquad (2.47)$$

where ℓ is the step size.

The time required for a photon to take this many steps is the desired diffusion time. Because the speed of a photon is c, the time for N steps of size ℓ is, from (47),

$$\tau_{\text{dif}} = \frac{N\ell}{c} = \frac{R^2}{c\ell}. \qquad \text{(Diffusion time scale)} \qquad (2.48)$$

Evaluation of this time requires knowledge of the step size ℓ.

Mean free path

The average step size is the *mean free path* for a photon in the hot stellar interior. It is a reasonable approximation to adopt the cross section of the aforementioned Thomson scattering of photons by free electrons,

$$\sigma_{\text{T}} = \frac{8\pi}{3} r_{\text{e}}^2 = 6.6525 \times 10^{-29}\,\text{m}^2; \qquad \text{(Thomson cross section)} \qquad (2.49)$$

$$r_{\text{e}} = \frac{1}{4\pi\varepsilon_0} \frac{e^2}{m_{\text{e}}c^2} = 2.8179 \times 10^{-15}\,\text{m},$$

where r_{e} is the classical radius of the electron. This cross section applies to interactions with photon energies substantially less than the rest energy of the electron, $h\nu \ll m_{\text{e}}c^2$.

The relation between cross section and mean free path ℓ is (AM, Chapter 10)

$$\ell = (\sigma_{\text{T}} n_{\text{e}})^{-1}, \qquad (2.50)$$

where n_{e} is the number density of scatterers (usually electrons). For a completely ionized hydrogen gas, n_{e} is equal to the number density of protons that carry most of the mass. Hence, $n_{\text{e}} = (M/m_{\text{p}})/V \approx M/(m_{\text{p}}R^3)$, and

$$\ell \approx \frac{m_{\text{p}}R^3}{\sigma_{\text{T}}M} = 4 \times 10^{-3}\,\text{m} = 4\,\text{mm}, \qquad \text{(Photon mean free path in sun)} \qquad (2.51)$$

where we substituted solar values for M and R. This is the value for the average solar mass density. At the sun's center, the density is ~ 100 times greater and the mean free path correspondingly less.

Substitute (51) into (48) to obtain the approximate diffusion time,

$$\tau_{\text{dif}} \approx \frac{\sigma_{\text{T}}}{cm_{\text{p}}} \frac{M}{R}, \qquad \text{(Diffusion time)} \qquad (2.52)$$

which for solar values becomes

$$\tau_{\text{dif}} \approx 3 \times 10^{11}\,\text{s} = 10\,000\,\text{yr}. \qquad (2.53)$$

The solar diffusion time is actually somewhat larger because absorption and emission processes further delay the photons on their way to the surface. The value usually quoted is

$$\tau_{\text{dif}} \approx 20\,000\,\text{yr}. \qquad \text{(Diffusion time; sun)} \qquad (2.54)$$

Solar luminosity

The solar luminosity may be estimated from the diffusion time if the sun is taken to be a ball of hot gas with uniform internal temperature $T_\odot \approx 5 \times 10^6$ K. The solar diffusion time just obtained tells us that the entire energy content of the *photons* in the sun, at some given time, will be carried to the surface in 20 000 yr. The energy density of photons in blackbody radiation of temperature T is aT^4 (J/m^3), where $a = 7.566 \times 10^{-16}$ J m^{-3} K^{-4} (6.25). This times the solar volume is the total photon energy content. Divide this by the diffusion time to obtain the luminosity estimate

$$L_\odot \approx \frac{aT^4 \frac{4}{3}\pi R_\odot^3}{\tau_{\text{dif}}} = 1 \times 10^{27} \text{ W}, \tag{2.55}$$

which is within a factor of a few of the actual value, 4×10^{26} W.

The diffusion time of 20 000 yr is much shorter than the thermal time scale of 2×10^7 yr. It turns out that the photons in the sun contain only $\sim 10^{-3}$ the thermal energy present in the particles. It would thus take ~ 1000 sun loads of photons to remove most of the energy from the sun, and this would take $\sim 1000 \times 20\,000$ yr $= 2 \times 10^7$ yr, which is the thermal time scale.

2.6 Nuclear burning

The power source for most stars is the burning of hydrogen in the core of the star. The pressures and temperatures there are sufficient to allow the hydrogen nuclei to undergo fusion reactions that lead to helium. Such reactions are exothermic; they give up mass and release energy in the form of kinetic energy of the reaction products. This provides the power to replace the energy being radiated from the surface of the star. The result is that the star remains in a fairly stable state for much of its active life – some 10^{10} yr in the case of the sun. The basics of the important nuclear reactions are described here.

Stable equilibrium

The stable equilibrium of nuclear-burning stars is maintained by a negative feedback mechanism. If the star is perturbed to smaller size, the densities and temperature at the core increase owing to the greater gravitational force. This leads to more nuclear reactions because the particle fluxes and velocities are greater. The increased energy released into the core causes the star to expand, thus returning it to its original state. Similarly, if the star is perturbed to a larger size, the reduced densities and temperatures at the core diminish the energy output, and the star will shrink back to its original stable state.

Coulomb barrier

One might be inclined to think that nuclear burning would not take place at all. The dominant element in the sun is hydrogen, and it is completely ionized throughout most of the solar volume. It is thus proton–proton interactions that yield the energy release. For this interaction to take place, the protons must come within the short range of the nuclear forces, and this requires that the kinetic energies be great enough to overcome the huge Coulomb repulsive

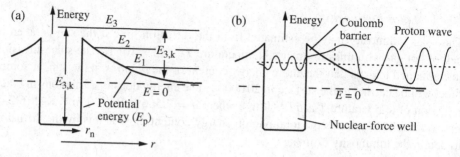

Figure 2.4. Potential (dark curve) of a proton showing the combined square-well nuclear and Coulomb r^{-1} potential. The negative gradient (slope) of the potential gives the direction and magnitude of the force. The steep sides of the nuclear well represent a strong attractive force, and the sloping sides further out represent the repulsive Coulomb force for an approaching proton. (a) Three (total) energy levels for approaching protons. Classically, protons with energies E_1 and E_2 are repulsed. The third, with E_3, has sufficient energy to override the potential barrier to come within range of the nuclear force. Kinetic energies, $E_{3,k} = E_3 - E_p$, are shown at two locations. (b) Incoming proton treated as a wave. It can, with low probability, tunnel through the Coulomb potential barrier. Nuclear reactions can thus occur at much lower particle energies (i.e., temperatures) than would otherwise be possible.

force at these short distances. In fact, the average kinetic energy of protons at the center of the sun is about a factor of 1000 less than required (Prob. 61). Stated otherwise, the average proton energy is insufficient to overcome the *Coulomb barrier*.

This problem is surmounted by the wave nature of particles that allows them to penetrate some distance into potential barriers (Fig. 4). If the barrier is sufficiently narrow, a particle can leak (tunnel) through it into the nuclear potential well even if, classically, it has insufficient energy to overcome the barrier. The leakage probability through a Coulomb barrier increases rapidly with particle energy because the barrier narrows with increasing energy. There are sufficient numbers of particles in the high-energy tail of the Maxwell–Boltzmann distribution at 10^7 K to provide the required leakage into the nuclear well and hence the required nuclear reactions.

A slight change in temperature will substantially increase proton numbers in the tail and will also raise the average proton energy. The reaction rates are thus highly temperature sensitive. A modest temperature rise will markedly increase the rate of nuclear interactions. This is a crucial aspect of the stability feedback just described.

Nuclear warmer

Only a tiny fraction of the stellar thermal energy content of a star is radiated away from the stellar surface each year – only about 1 part in 20 million for the sun (see discussion of the thermal time scale above). The nuclear energy that must be supplied each year to offset this loss is thus only a very small part of the total thermal energy content of the sun.

One should therefore not think of the sun as a raging nuclear furnace like a basement oil burner that is expected to bring a house up to temperature in an hour or two. Rather, think of it as a (huge) ball of hot gas with a low-powered nuclear "warmer." Nonetheless, in the case of the sun, the warmer puts out 4×10^{26} W; the sun is a very big house with large thermal content.

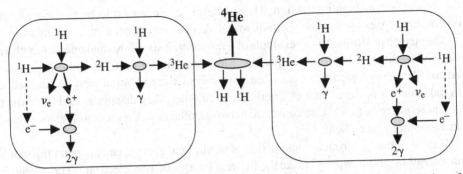

Figure 2.5. Schematics of the pp series of nuclear reactions that dominate the conversion of hydrogen into helium in the sun. The net effect is that a helium nucleus, ^{4}He, is created from four hydrogen nuclei, ^{1}H. The shaded ovals are intermediate states before the decay (or annihilation) to the final products. The dashed arrow indicates that the annihilated e$^-$ can be considered to have been associated with one of the input protons in its neutral atomic state.

An elementary model of a normal star can thus treat the star simply as a gravitationally bound, stable ball of hot gas. At the next level of sophistication, though, the model would include the effects of a distributed source of energy in the central regions and the propagation of this energy toward the surface, where it is radiated into space.

Proton–proton (pp) chain

The dominant chain of nuclear interactions in the sun is known as the *proton–proton chain* (*pp*). The reactions of the pp chain can take place at temperatures above about 5×10^6 K, which are found in the central regions of the sun. (The core temperature of the sun is 1.6×10^7 K; Table 4.1.) In these regions, the gases are totally ionized. At the beginning of hydrogen burning, the hydrogen content was 71% by mass, the helium content 27%, and heavier elements 2% (see AM, Chapter 10). These are the so-called *solar-system abundances*. At present the hydrogen content at the sun's center has been reduced to ~36% by the hydrogen burning described here.

The series of reactions in this chain converts four protons to a helium nucleus. The latter is known to particle physicists as an *alpha particle*. There are several alternate pathways in the pp chain; we first describe the most probable.

Nuclear interactions

The most probable sequence of nuclear reactions in the pp process is illustrated in Fig. 5. It begins when two hydrogen nuclei (protons), each of unit mass number (^{1}H $+^1$H), interact and momentarily form an intermediate state. This immediately decays to a hydrogen isotope of mass number 2 (^{2}H, a *deuteron*), an electron neutrino ν_e, and a *positron* (e$^+$, a positively charged electron). The first reaction is thus

$$^1\text{H} + {}^1\text{H} \rightarrow {}^2\text{H} + \nu_e + e^+. \tag{2.56}$$

A deuteron consists of a proton and a neutron; a proton is converted to a neutron in this interaction. The neutrino is a neutral particle that easily traverses matter; it will most likely escape the star without interaction. The subscript "e" denotes it to be an electron neutrino, one of three types of neutrino, each of which also has an antineutrino counterpart.

The ejected positron e^+ is an example of *antimatter*. It soon finds and interacts with a nearby electron (ordinary matter) in the plasma. This electron, for particle counting purposes, can be considered to have been associated with one of the two input protons (dashed arrow). In the interaction, the e^+ and e^- annihilate each other; they disappear, and their kinetic and rest-mass energies ($2mc^2$) are converted to two gamma rays. Caution: never shake hands with an antimatter space alien!

If the e^+ emerges from the interaction with minimal energy, each gamma ray would have an energy of about $m_e c^2$ (511 keV), the rest energy of one electron. (The e^- has kinetic energy of only ~ 1 keV at 10^7 K.) In this case the energy of the two gamma rays would be

$$E(2\gamma) \approx 2m_e c^2 = 1.022 \text{ MeV}. \tag{2.57}$$

Because the gamma rays will quickly interact and share their energy with electrons in the surrounding plasma, this 1.0 MeV – or more if the e^+ had significant kinetic energy – contributes to the star's internal thermal energy. Also contributing is the kinetic energy of the deuteron created in the reaction (56). The neutrino most probably escapes from the sun, and so its energy is lost.

Subsequently, another proton (^{1}H) collides with the deuteron (^{2}H) to give an isotope of helium (^{3}He) and a gamma ray γ:

$$^2\text{H} + {}^1\text{H} \rightarrow {}^3\text{He} + \gamma. \tag{2.58}$$

The ^{3}He nucleus consists of two protons and one neutron. Its kinetic energy and that of the γ ray contribute additional energy to the plasma. Keep in mind that, although we write our elements with atomic notation, the primary reactions are between nuclei, not atoms.

The next step requires that another set of reactions (56) and (58) take place in such a way that another three protons produce a second ^{3}He nucleus, as shown in the right-hand box of Fig. 5. The two ^{3}He nuclei then interact to give the stable isotope of helium (^{4}He) and two free protons as follows:

$$^3\text{He} + {}^3\text{He} \rightarrow {}^4\text{He} + 2\,{}^1\text{H}. \tag{2.59}$$

The ^{4}He nucleus (alpha particle) consists of two protons and two neutrons. All in all, six protons are consumed to create a helium nucleus and two free protons. The net effect is that one helium nucleus is created from four protons.

Baryon, lepton, and charge conservation

Several conservation laws must be obeyed in nuclear interactions. The number of baryons (protons and neutrons) must be preserved, as must the number of leptons (electrons and neutrinos). Electric charge and energy must also be conserved. Antiparticles such as e^+ count negatively for lepton number conservation. To keep track of these, one must keep in mind the electrons associated with the interacting nuclei of (56), (58), and (59).

Referring to Fig. 5, we therefore find that six hydrogen "atoms" are introduced into the interactions. These consist of six protons (^{1}H) and six electrons, all of which were initially free of one another in the ionized plasma. The input constituents thus have zero total charge.

After the sequence, we have one helium nucleus (^{4}He) consisting of two protons and two uncharged neutrons, two free electrons associated with the helium nucleus, and also the two free protons (^{1}H) and their two associated electrons. In addition, two ν_e are created in the interactions and also six gamma rays.

The six baryons are conserved in the interactions as already noted. The proton charge drops from $+6e$ to $+4e$ because two protons become neutrons in the formation of the deuterons, and the electron charge increases from $-6e$ to $-4e$ owing to the two e^+e^- annihilations. Thus, total electric charge is maintained at zero. The reduction in electron number is a reduction in lepton number from 6 to 4, but that is made up by the two ν_e created in the interactions.

The three conservation laws (baryon, lepton, and charge) are each thus satisfied in reactions (56), (58), and (59).

Energy conservation

Energy conservation underlies the exothermic nature of the reactions. In special relativity, a mass m has an energy equivalence of mc^2 known as the *rest energy* (Section 7.3), and this must be accounted for in the interactions. A decrease of the total mass of the constituent particles appears as increased kinetic energies of the interaction products relative to the input kinetic energies. One must take care to include the electrons in this accounting.

In our accounting of electrons and nucleons (protons and neutrons), the conversion is, in essence, from four neutral hydrogen *atoms* to one neutral helium *atom*:

$$4\,H_{atom} \rightarrow 1\,He_{atom}. \qquad \text{(Atoms)} \qquad (2.60)$$

The energy released is thus that associated with the difference in mass of four hydrogen atoms and one helium atom. We calculate this mass decrease and also the associated energy release below; see (62).

pep, hep, and Be reactions

There are several alternative paths in the pp chain that will take place with somewhat lower probabilities than the primary chain just described. Each produces a helium nucleus from four protons. Because the emitted neutrino energies vary with the path, more or less energy may be lost from the star owing to neutrino escape. The principal reactions are shown in Fig. 6 with their relative likelihood of contributing to a given ^{4}He termination. Other pathways are possible but highly improbable.

There are two branches that produce the deuteron, ^{2}H: the pp process and a less probable *pep process* that entails an input electron in the interaction. The pep process is involved in only 0.4% of the ^{4}He terminations. It does emit a relatively high-energy neutrino (1.44 MeV) that has been more easily detectable by neutrino astronomers than the ≤ 0.42 MeV neutrino of the pp reaction.

The nucleus ^{3}He created in the second step (58) can be transformed in three ways. It can interact with ^{3}He as given above (59) to yield ^{4}He directly, or it can interact with a ^{4}He or with a ^{1}H. The latter "hep" reaction is very rare; it produces ^{4}He directly together with a high-energy neutrino. The former reaction, ^{3}He $+$ ^{4}He, takes place 15% of the time and produces Be^7, which in turn has two branches. The less probable of these leads to ^{8}Be and a high-energy neutrino.

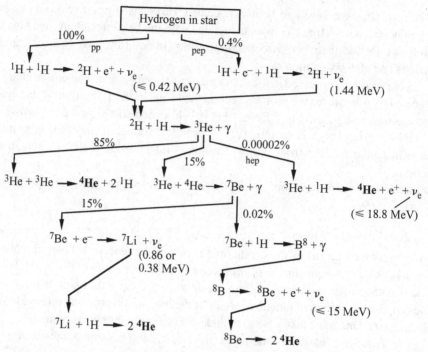

Figure 2.6. Hydrogen pp burning showing the nuclear reactions for all significant branches. The percentages of ^{4}He terminations that pass through the several paths are shown. The neutrino energies are also given: a fixed energy for the two-body final states and a distribution of energies with a maximum energy for the three-body final states. [J. N. Bahcall, *Neutrino Astrophysics*, Cambridge Univ. Press, 1989, Table 3.1]

The several neutrino-emitting reactions in the chain provide neutrino astronomers with a view of the interior of the sun. Neutrinos from the sun have been detected by large neutrino detectors on the earth. The fluxes and energies are generally in accord with our understanding of the sun. A long-standing ~50% deficit of detected electron neutrinos is now understood as being due to transformation of electron neutrinos to other types during their passage through the sun (AM, Chapter 12).

CNO cycle

Another set of nuclear interactions becomes important at somewhat higher temperatures. It is called the CNO cycle because carbon, nitrogen, and oxygen nuclei are involved in the reactions. Nevertheless, this cycle creates a helium nucleus from four protons just like the pp chain.

The CNO process makes use of the occasional carbon nucleus in the core of a star that was formed from the debris of previous generations of stars. It consists of six sequential interactions, as listed in Table 1.

First, a carbon nucleus in the plasma undergoes a fusion reaction with a hydrogen nucleus. The result is the heavier nucleus ^{13}N. This then spontaneously undergoes a radioactive beta decay reaction. The decay products are a positron e^+, an electron neutrino ν_e, and a heavy

Table 2.1. *CNO cycle*

$^{12}C + {}^1H$	$\rightarrow$	$^{13}N + \gamma$	
^{13}N	$\rightarrow$	$^{13}C + e^+ + \nu_e$	(β decay)
$^{13}C + {}^1H$	$\rightarrow$	$^{14}N + \gamma$	
$^{14}N + {}^1H$	$\rightarrow$	$^{15}O + \gamma$	
^{15}O	$\rightarrow$	$^{15}N + e^+ + \nu_e$	(β decay)
$^{15}N + {}^1H$	$\rightarrow$	$^{12}C + {}^4He$	

isotope of carbon ^{13}C. This reaction is called β *decay* because electrons were called *beta rays* in the early days of radioactivity studies. In effect this reaction converts a proton in the nucleus to a neutron. Another proton interaction then yields the stable nucleus ^{14}N.

Similar steps sequentially yield ^{15}O and ^{15}N, but instead of the expected ^{16}O the final fusion yields the energetically favored $^{12}C + {}^4He$. The net result is the formation of one ^4He from four ^1H. The initial ^{12}C has been replaced; it simply plays the role of a catalyst. The energy released is identical to that of the pp chain.

Energy production

Both the pp and CNO cycles yield one helium *nucleus* from four input protons. Counting electrons as well as nuclei, we found (60) that the net effect is the conversion of four hydrogen *atoms* to one helium *atom* with all their electrons.

Yield per cycle

The *atomic* masses of hydrogen and helium are, respectively, 1.007 83 and 4.002 60 atomic mass units (amu), where 1.0 amu $= 1.660\,53 \times 10^{-27}$ kg. An He atom has significantly less mass than four of the hydrogens (4.031 32 amu); hence, mass is lost. Because the energy equivalence of mass m is mc^2, the energy release E_r from a single set of interactions (60) is

$$E_r = -\Delta E_{\text{rest}} = (4M_H - M_{He})c^2 \qquad (2.61)$$
$$= 4.29 \times 10^{-12}\,\text{J}$$
$$= 26.75\,\text{MeV}$$
$$= 0.0071 \times 4M_H c^2.$$

The yield thus turns out to be 27 MeV per chain, which is a modest 0.71% of the rest-mass energy of the initial four hydrogen atoms.

In absolute terms, the yield of energy per kilogram is huge:

$$\frac{E_r}{4\,M_H} = 0.0071c^2 = 6.4 \times 10^{14}\,\text{J/kg}. \qquad (4{}^1H \rightarrow {}^4He) \qquad (2.62)$$

This would satisfy my personal overall energy needs for about 10 000 yr. Ten grams would take care of my lifetime. Of this energy, 2% is carried away by the neutrinos. Finally, we note that nuclear burning is a factor of $\sim 10^7$ more efficient than chemical burning in which only the electronic bonds of molecules of order eV are involved.

Sun lifetime

The energy yield from the pp chain per kilogram of material (62) allows us to calculate a lifetime or maximum age for the sun taking into account the amount of nuclear fuel. We estimate that only about 10% of the mass is in the hot central region of the sun where it can be burned in fusion reactions. Taking into account the 2% loss to neutrinos, we find that the total available energy from hydrogen burning is approximately, from (62),

$$E_{\text{total}} \approx 0.98 \times 6.4 \times 10^{14} \text{ J/kg} \times 2 \times 10^{30} \text{ kg} \times 0.1$$
$$\approx 1.3 \times 10^{44} \text{ J}. \tag{2.63}$$

At the present solar luminosity, the fuel supply would last for the time

$$\tau_{K,\odot} \approx \frac{E_{\text{total}}}{L_\odot} = \frac{1.3 \times 10^{44} \text{ (J)}}{4 \times 10^{26} \text{ (J/s)}} = 3 \times 10^{17} \text{s} \tag{2.64}$$

or

$$\tau_{K,\odot} \approx 1 \times 10^{10} \text{ yr}. \tag{2.65}$$

Thus, the nuclear energy source will sustain the sun for a period comparable to or longer than the age of the earth, $\sim 4.5 \times 10^9$yr. We saw (34) that the thermal energy content of the sun would be expended in about 10^7yr, which is much less than the demonstrated age of the solar system. This discrepancy was a major puzzle until the physics of nuclear reactions came to be understood.

Energy-generation function

The energy-generation rates, ϵ (W/kg), for these hydrogen-burning processes are proportional to the numbers of interactions that take place per second. The flux of "projectile" particles and the "target" particle density are both proportional to the mass density ρ (kg/m^3). Hence, the interaction rate per cubic meter is proportional to ρ^2. If only a fraction X of the mass is hydrogen, then the number of pp interactions is proportional to $X^2\rho^2$. Divide this by density ρ (kg/m^3) to get the interaction rate per kilogram. The rate is also a strong function of the temperature.

For the case of hydrogen burning, the energy generation rate $\epsilon_{pp}(\rho, T)$ can thus be parameterized approximately as

$$\epsilon_{pp} = \epsilon_0 X^2 \left(\frac{\rho}{10^5 \text{ kg/m}^3} \right) \left(\frac{T}{10^7 \text{ K}} \right)^\beta, \qquad \text{(Hydrogen burning; W/kg)} \tag{2.66}$$

where ϵ_0 and β may be taken to be constants but only within modest ranges of temperature. The numerical parameter β depends on the type of nuclear interactions involved. It is about 4 for the burning of protons to helium with the pp chain and about 15 with the CNO chain. For the pp chain and the 1.6×10^7 K temperature of the solar center, the leading coefficient is $\epsilon_0 \approx 2 \times 10^{-3}$ W/kg. The solar system (or galactic) fractional abundance of hydrogen is $X = 0.71$ by mass, and at the center of the sun it has been reduced to $X \approx 0.36$ by nuclear burning. At the solar center, with $\rho_c = 1.5 \times 10^5$ kg/m^3, $T_c = 1.6 \times 10^7$ K, and $\beta = 4$, we find from (66) that $\epsilon_{pp} = 2.5 \times 10^{-3}$ W/kg.

Table 2.2. *Hydrogen-burning reactions*

Chain/ cycle	Temperature where dominant (K)	Power per unit mass (W/kg)[a]	Stars where reaction dominates
pp	5 to ~15 × 10^6	$\epsilon \propto \rho T^4$	Sun and less massive
CNO	≳ 20 × 10^6	$\epsilon \propto \rho T^{15}$	Type A and more massive

[a] These expressions are approximations of the actual formulas that apply for temperature regions of interest. As a function of temperature, the CNO reaction turns on gradually in the region of its threshold. The net power output as a function of temperature is a combination of the pp and CNO processes and will be a continuous function of temperature. This could be characterized with a variable exponent β in (66).

The onset of the CNO cycle occurs at a somewhat higher temperature than the pp cycle. Massive stars have higher core temperatures than less massive ones. In sufficiently massive stars, the CNO cycle will dominate the pp process because of its strong temperature dependence of $\sim T^{15}$. In the sun, the pp cycle dominates, but the CNO cycle provides about 10% of its energy.

The two hydrogen-burning cycles are summarized in Table 2. In practice, one would use an *effective generation function* $\epsilon(\rho, T)$ that takes into account both processes, pp and CNO. This function is fundamental to the modeling of stars.

2.7 Eddington luminosity

There is a practical maximum to the luminosity of a normal star. At this luminosity, the radiation pressure will eject the outer layer (photosphere) of the star. Any attempt to increase hydrogen burning and hence luminosity by adding mass will be defeated by this radiation pressure.

We demonstrate here that the limiting luminosity, known as the *Eddington luminosity* after Sir Arthur Eddington (1882–1944), increases linearly with the mass of the star, $L_{Edd} \propto M$. The luminosity of normal (main-sequence) stars also increases with the mass of the star, but more rapidly – roughly as $L_{star} \propto M^3$ (Table 4.3). Thus, as mass increases, the stellar luminosity can reach the limiting luminosity. This yields an upper limit to the mass of a stable star.

The limit occurs when the momentum p of the outward-moving photons is transferred to a photospheric particle at a rate dp/dt that exactly cancels the inward force of gravity on the particle. The forces balance if the rate of collisions between the photons and the particle is sufficiently great. The Eddington "limit" refers to the limiting luminosity for spherical symmetry in which the radiation emerges in all directions from the star with equal intensity.

Forces on charged particles

Consider for simplicity that the photospheric gases are a plasma of ionized hydrogen consisting solely of electrons and protons. Gravity pulls inward on a proton with a force proportional to the mass of the particle or 1836 times greater than the force on an electron. In contrast,

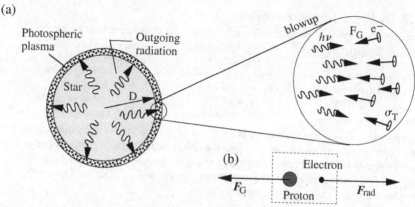

Figure 2.7. Principle of Eddington limit to the luminosity for isotropic conditions (an idealized case). (a) Photons streaming out from the star into the photospheric plasma of electrons and protons (hydrogen plasma). The photons transfer their momenta to the electrons via Thomson scattering. (b) Forces on a "system" of one proton and one electron. The inward gravitational force is primarily applied to the proton, whereas the radiation pressure is applied to the electron. The Eddington luminosity is that which balances the two forces.

the outward radiation force on the plasma is applied primarily to the electrons (Fig. 7a) because the cross section is much greater than it is for a proton. Thus the number of collisions with, and the force upon, the electrons are much higher.

These two effects would tend to separate the electrons and protons, but the strong electrostatic forces prevent substantial separation of the oppositely charged particles. We thus may consider each electron-proton system as a single unit (Fig. 7b) upon which both gravitational and radiation forces are exerted.

Radiative force

Consider an electron just above the surface of a star bathed in the flux of photons emerging from it. Find the average radiation force on an electron. The energy $E = h\nu$ carried by an individual photon is related to its momentum p according to $E = pc$ (7.21), or

$$p = E/c. \qquad \text{(kg m s}^{-1}) \quad (2.67)$$

Thus, an energy flux $\mathscr{F}$ (W/m^2 = J s^{-1}m^{-2}) carries momentum $\mathscr{F}/c$ per unit area and time. Because momentum per second is a force ($F \equiv dp/dt$) and force per unit area is a pressure, $\mathscr{F}/c$ is the *radiation pressure*,

$$P_{\text{rad}} = \frac{\mathscr{F}}{c}. \qquad \text{(Pressure; momentum m}^{-2}\text{ s}^{-1}) \quad (2.68)$$

The force on a single electron is simply the pressure times the cross section,

$$F_{\text{rad,e}} = P_{\text{rad}}\, \sigma_{\text{T}} = \frac{\mathscr{F}\sigma_{\text{T}}}{c}, \qquad (2.69)$$

where σ_{T} is the Thomson cross section (49), which, again, is appropriate for a photon of energy $h\nu \ll m_e c^2$ interacting with a free electron. Because the electron rest energy is 511 keV, the requirement on $h\nu$ is well satisfied for optical photons ($\sim$2 eV) and even for moderately energetic x rays of a few tens of keV.

The relation between the luminosity L (W) of an isotropically radiating, spherical star and the flux $\mathscr{F}$ (W/m^2) at radius D from its center is

$$\mathscr{F} = \frac{L}{4\pi D^2}. \tag{2.70}$$

Substitute into (69) to obtain

$$\Rightarrow \quad F_{\text{rad,e}} = +\frac{L\,\sigma_T}{4\pi D^2 c}. \qquad \text{(Force on election, directed outward)} \tag{2.71}$$

This is the outward radiative force experienced by a single electron at distance D from a star of luminosity L.

Balanced forces

The gravitational force on an electron depends on the mass associated with it, which in this case is primarily that of a proton. The dimensionless *electron molecular weight* μ_e is defined as the mass associated with each free electron of a medium in units of the proton mass m_p. For a totally ionized hydrogen, $\mu_e \approx 1$; that is, there is one nucleon per free electron. For a gas of completely ionized heavy elements, there are about two nucleons per free electron, or $\mu_e \approx 2$. The mass (kg) associated with each free electron is $\mu_e m_p$. The inward force of gravity on a single e$^-$p pair at the distance D from the center of a star of mass M is therefore

$$\Rightarrow \quad F_G = -\frac{GM\mu_e m_p}{D^2}. \qquad \text{(Gravitational force, directed inward)} \tag{2.72}$$

The condition for the Eddington limit is that the net force on the electron-proton system be zero:

$$F_G + F_{\text{rad,e}} = 0. \tag{2.73}$$

Substitute from (71) and (72) and solve for L to obtain the Eddington luminosity:

$$\Rightarrow \quad L_{\text{Edd}} = \frac{4\pi GM_\odot \mu_e m_p c}{\sigma_T}\frac{M}{M_\odot}. \qquad \text{(Eddington luminosity; W)} \tag{2.74}$$

Because both forces (71) and (72) vary as D^{-2}, the expression for L_{Edd} is independent of distance from the star. If the forces balance at one radius, they will do so at all other radii (for our spherically symmetric case). Also, the more massive the star, the greater this limiting luminosity will be because the relation (74) is linear with mass of the star.

Substitute numerical values into (74) to find

$$\Rightarrow \quad L_{\text{Edd}} = 1.26 \times 10^{31}\mu_e\left(\frac{M}{M_\odot}\right)\text{W}, \tag{2.75}$$

which, for $M = 1$ kg (as a reference value only), yields a modest 6.32 W for $\mu_e = 1$ (pure hydrogen). Keep in mind that our derivation was for the ideal, spherically symmetric situation.

For 1 $M_\odot$, divide both sides by the solar luminosity, $L_\odot = 3.85 \times 10^{26}$ W, to obtain

$$L_{\text{Edd}}/L_\odot = 3.27 \times 10^4\,\mu_e\,\frac{M}{M_\odot}. \tag{2.76}$$

The Eddington luminosity for a 1-$M_\odot$ star is 33 000 times greater than the actual solar luminosity. Gravity will therefore, for the most part, keep the solar plasma confined to the sun.

In fact, minute amounts of plasma flow outward from the sun to the earth and beyond. This *solar wind* is an extension of the extremely hot ($\sim 10^6$ K) *solar corona*, which is probably heated by dissipation of currents associated with magnetic fields that thread through it. At the high coronal temperatures, the particle velocities are so great that they are not contained by the sun's gravity. Except in regions where it is confined by the magnetic fields, the coronal gas expands into interplanetary space to become the solar wind observed near the earth.

Maximum star mass

The Eddington limit places an upper bound on the mass of a gravitationally bound star. For the pp process, the luminosity scales roughly as the mass to the power 16/5 (Table 4.3), or

$$\frac{L_{star}}{L_\odot} = (M/M_\odot)^{16/5}, \tag{2.77}$$

whereas the Eddington luminosity increases only as the mass to the power 1.0. If the stellar mass were sufficiently large, its luminosity would reach L_{Edd}; that is, $L_{star} = L_{Edd}$. Equate (76) and (77) and solve for the associated mass to obtain

$$\left(\frac{M}{M_\odot}\right)_{max} = 115. \qquad \text{(Eddington stellar mass limit for } \mu_e = 1\text{)} \tag{2.78}$$

This is the maximum mass a hydrogen-burning star can have without exceeding the Eddington luminosity and thereby ejecting its outer layers. This limit applies to a pure hydrogen gas for which $\mu_e = 1$. The associated luminosity, from (77), is

$$\left(\frac{L_{star}}{L_\odot}\right)_{max} = 3.9 \times 10^6. \qquad (\mu_e = 1) \tag{2.79}$$

If the surface gases were, for example, pure helium ($\mu_e = 2$), the limits would become 157 $M_\odot$ and 1.06×10^7 $L_\odot$. Observational studies are in general accord with these upper limits. The most massive stellar objects convincingly shown to be single (not multiple) stars have masses no greater than about 130 $M_\odot$.

Massive, luminous, hydrogen-burning stars can be variable owing to unstable atmospheres and mass ejections. Such stars are known as *luminous blue variables* (LBVs). They lie at the upper limit of luminosity of hydrogen-burning stars and are erratically variable on time scales of months to years. Some, such as η Car and P Cyg, have experienced huge outbursts in historic times. The LBVs typically exhibit nebulae of ejected material that can be attributed, at least in part, to ejections propelled by the stellar radiation.

Mass accretion rate

The Eddington limit plays another major role in stellar astrophysics. Radiation pressure limits the rate at which matter can be accreted onto a star. This is important, for example, in the formation of stars (protostars) and in the accretion of matter onto x-ray–emitting neutron stars. In the latter case, the x-ray luminosity arises directly from the gravitational infall energy of the plasma accreting from a companion star; see Fig. 4.13. The radiation pressure thus directly limits the rate of matter accretion and hence the observed luminosity.

Neutron-star accretion

Neutron stars are compact final states of relatively massive stars. They have densities comparable to the nucleus of an atom and are typically of mass $1.4 M_\odot$ and radius 10 km (Section 4.4). They thus have deep gravitational potential wells and high gravitational fields. Gaseous material accreting from a close companion star gains tremendous energy as it falls toward the neutron-star surface.

When the gas impinges on the surface, its kinetic energy will be sufficient to heat the surface to x-ray temperatures of $\sim 10^7$ K. Alternatively, such temperatures may occur at the inner edge of an accretion disk or in a shock lying just off the surface. Such sources are observed by x-ray astronomers with instruments in orbit above the earth's atmosphere.

For a 1.4-$M_\odot$ neutron star, the maximum possible luminosity from (75) is 1.8×10^{31} W or 46 000 $L_\odot$ for $\mu_e = 1$. This is comparable to the x-ray luminosity of the most luminous stellar-accreting x-ray sources. This is, in itself, strong evidence that the gravitational field of a neutron star is the energy source of the luminosity.

Accretion luminosity

The luminosity (in all wavelengths) from the release of gravitational energy by gas falling onto a star of mass M and radius R is related to the mass inflow rate dm/dt (usually called the *accretion rate*) through the expression

$$\Rightarrow \qquad L \approx \frac{GM(dm/dt)}{R}. \qquad \text{(Accretion luminosity; W)} \quad (2.80)$$

This expression follows directly from the potential energy lost by an element of mass dm as it infalls from "infinite" radius to the radius R – namely, $GMdm/R$. Division by the time interval dt yields the rate at which this potential energy is given up. This is the maximum luminosity one would expect under the assumption of spherically symmetric, steady-state radial infall with all energy being reradiated in the waveband observed.

The approximate rate of mass accretion, $\dot{m} \equiv dm/dt$, that corresponds to the Eddington luminosity is obtained by setting the luminosity (80) equal to the Eddington luminosity (75) for $\mu_e = 1$:

$$\frac{GM\dot{m}_{\mathrm{Edd}}}{R} \approx 1.26 \times 10^{31} \frac{M}{M_\odot}. \qquad (2.81)$$

Solve for $\dot{m}_{\mathrm{Edd}}$ to obtain

$$\Rightarrow \qquad \dot{m}_{\mathrm{Edd}} \approx 1.26 \times 10^{31} \frac{R}{GM_\odot}, \qquad \begin{array}{l}\text{(Accretion rate to yield the} \\ \text{Eddington luminosity; kg/s)}\end{array} \quad (2.82)$$

where R is the radius of the star.

Note that $\dot{m}_{\mathrm{Edd}}$ depends only on the radius of the recipient star, not on its mass! The Eddington luminosity increases linearly with mass, but so does the energy given each infalling proton. Thus, for a star of radius R, the rate of infalling protons required to yield the Eddington luminosity is independent of the star's mass. On the other hand, if the star is larger at the same mass, there is less potential energy loss per infalling proton, and so more matter must be accreted to yield the Eddington luminosity.

For a neutron star ($R = 10$ km),

$$\dot{m}_{Edd} \approx 9.5 \times 10^{14} \text{ kg/s} \qquad \text{(Maximum accretion rate; } R = 10 \text{ km)} \qquad (2.83)$$
$$\approx 1.5 \times 10^{-8} \, M_{\odot}/\text{yr}.$$

This is the maximum accretion rate that will be accepted by the neutron star. Any additional mass trying to accrete will be blown off by the outgoing radiation for our simplified, spherically symmetric situation. In practice, magnetic fields and accretion disks will significantly modify these values, but they should still be valid within a factor of order unity.

Massive black holes

Active galactic nuclei are compact masses in the centers of galaxies exhibiting intense non-thermal emission that is often variable, which indicates small sizes (light months to light years). The luminosities range up to 10^{39} W for a bright quasar. If the luminosity stems from gravitational potential loss by material accreting onto (or into) the central object, it should not exceed the Eddington luminosity. The latter must therefore be at least equal to 10^{39} W, which in turn requires, from (75), a central mass of $\geq 10^8 \, M_{\odot}$. These objects are most likely massive black holes. Independent upper limits to their sizes render other scenarios unlikely – for example, a cluster of millions of luminous stars (Section 1.7).

The accreting material probably consists of stars or stellar debris spiraling in toward the black hole. The Schwarzschild radius, or event horizon, of a (nonrotating) black hole is at radius $R_S = 2GM/c^2$ (4.36), which is 2 AU for $M = 10^8 M_{\odot}$. Substitute this into (82) to obtain the value of dm/dt required to reach the Eddington luminosity at this mass. One finds about half a solar mass per year. On a galactic scale, that is a rather modest appetite. See Section 4.4 for more on black holes.

2.8   Pulsations

Most normal stars emit a relatively steady flow of radiant energy. The visible sky is quite stable for the most part, although dramatic events such as dwarf novae, supernovae, binary eclipses, and so on do occur. It turns out that all stars are subject to variability at some level. Stars are prone to physical oscillations as is any physical object, and at a low level they all oscillate. Such oscillations can be detected through brightness and atmospheric velocity oscillations. This field is rightly called *asteroseismology* and, for solar studies, *helioseismology*. Such studies probe the interior structure of stars or the sun.

Under certain conditions, the oscillations can be driven to high amplitudes. Such stars are known, generally, as *pulsating variables*. They include the quasi-periodic *cepheid variables* and *RR Lyrae* stars that are used as distance indicators (AM, Chapter 9). Here, we present some of the basic physics that underlies the pulsations of these two types. The evolutionary states of these and other types of variable stars are discussed briefly in Section 4.3.

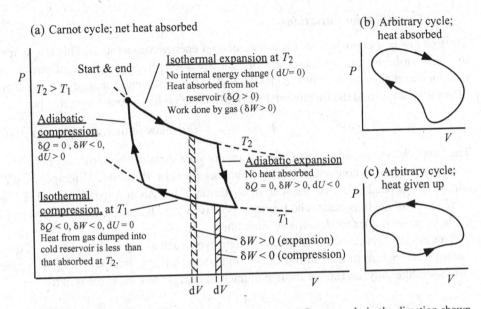

Figure 2.8. (a) The Carnot cycle. A gas taken through the Carnot cycle in the direction shown undergoes two isothermal state changes ($\Delta T = \Delta U = 0$) and two adiabatic state changes $\delta Q = 0$. The gas, in this example, does net work on its surroundings because it expands at a higher pressure than when it is compressed. The gas absorbs heat from the hot reservoir at temperature T_2 and dumps less heat into the cool reservoir at T_1. The net heat absorbed from the entire circuit is converted to the work done. (b, c) Arbitrary cycles. Clockwise motion yields net work done on the surroundings, as in (a), whereas counterclockwise motion results in negative work. Pulsations are possible in cases (a, b) but not (c).

Heat engine

The pulsating variable stars discussed here oscillate in radius, temperature, and luminosity. The oscillations require an energy source because otherwise they would quickly be quenched by dissipation in the gas. The energy source is the abundant heat (radiant energy) being transported through the star. In the study of thermodynamics, one encounters the *Carnot cycle* in which a volume of gas is carried ("reversibly") through a series of four changes of state and brought back to its initial state (Fig. 8a). The net effect of the four-step cycle is that heat is absorbed and the gas does net work on its surroundings.

This is evident if one recalls that $\int P \, dV$ is the work done by the gas and that this is the area under a curve on a P-V plot such as any one of the four tracks of the path in Fig. 8a. The work done is positive if the gas sample moves to the right ($dV > 0$) and negative if it moves to the left ($dV < 0$). The work done by the gas during its transit along the upper track in the figure is thus greater than the negative work done on the lower track. Any cycle that runs clockwise (Fig. 8b) does net work on its surroundings. For a counterclockwise cycle (Fig. 8c), the net work by the gas is negative; the surroundings do work on the gas.

In a pulsating star, elements of gas that constitute the star will cycle through a series of states during each oscillation. A necessary requirement for oscillations is that the elements of gas do net mechanical work on the star. The required energy must be provided by a net absorption of heat. We now formalize these statements.

Condition for pulsations

The *first law of thermodynamics* is a statement of energy conservation. This law states that, in a *reversible process*, an increment of heat δQ absorbed by a sample of gas during an incremental change of state must equal the sum of the incremental change of internal energy dU of the sample and the incremental work δW done on its surroundings; that is,

$$\delta Q = dU + \delta W. \qquad \text{(First law of thermodynamics)} \qquad (2.84)$$

The Greek deltas remind us that Q and W are not state variables because they depend on the history of the gas. In contrast, the state variables – pressure P, volume V, temperature T, and entropy S – depend solely on the state or condition of the gas at a specified time. The signs of δQ and δW are important. The former is positive when the gas element absorbs heat as is the latter when it does work on its surroundings.

If a gas sample is carried through a complete cycle such as any of those of Fig. 8, the state variables will return to their initial values. In an ideal gas, the internal energy is a function of temperature only, and so the change in internal energy over the cycle is zero:

$$\oint dU = 0. \qquad (2.85)$$

The net work done by the gas over the entire cycle equals the total heat absorbed and is thus, from (84),

$$W = +\oint \delta Q. \qquad (2.86)$$

The sign of W must be positive for pulsations to occur.

The entropy change is defined as $dS \equiv \delta Q / T$. Because entropy S is a state variable, the integral of dS over an entire cycle must be zero (Prob. 81):

$$\oint dS \equiv \oint \frac{\delta Q}{T} = 0. \qquad (2.87)$$

The temperature of a gas sample cycles as the star goes through its pulsing cycle. It is reasonable for our purpose to assume that the cyclic variation in temperature $\Delta T(t)$ is a small fraction of a mean temperature T_0. The condition we seek applies to small oscillations that might then grow to become large pulsations. Thus, we write

$$T(t) = T_0 + \Delta T(t) = T_0 \left(1 + \frac{\Delta T(t)}{T_0} \right). \qquad (2.88)$$

Substitute this into (87) and expand the denominator for the condition $\Delta T / T \ll 1$:

$$\oint \frac{\delta Q}{T_0} \left(1 - \frac{\Delta T(t)}{T_0} \right) \approx 0. \qquad (2.89)$$

Rearrange to isolate the integral of δQ,

$$\oint \delta Q \approx \oint \frac{\Delta T(t)}{T_0} \delta Q, \qquad (2.90)$$

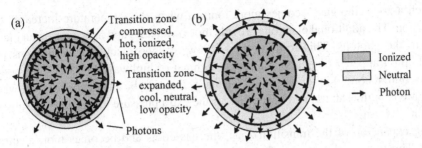

Figure 2.9. Stellar atmosphere undergoing pulsations. (a) Compression of the ionization transition zone. Heating and the resulting ionization increase the opacity, thus trapping photons (and hence heat) in the stellar gases. (b) Expansion of the ionization transition zone. During expansion, the temperature drops and the opacity decreases, thus releasing photons.

and substitute into (86) to obtain

$$W \approx \oint \frac{\Delta T(t)}{T_0} \delta Q. \qquad (2.91)$$

This shows that if work is to be done on the surroundings (W positive), then δQ should be positive when the temperature deviation ΔT is positive and negative when ΔT is negative. That is, heat should be absorbed when the temperature is high ($T > T_0$) and discharged when the temperature is low ($T < T_0$), as in the examples of Fig. 8a,b.

The standard internal combustion engine is an example of such a heat engine; heat is introduced (burning fuel) when the temperature is high from compression, and heat is ejected in part when the gas has cooled as the result of expansion. In the case of a star, the *net work* done by *all* of the gas elements through one cycle must be positive. (Some elements might contribute negative work.) Thus, one integrates (91) over the entire mass M of the star:

$$\Rightarrow \qquad W_{\text{star}} \approx \int_M \oint \frac{\Delta T(t,m)}{T_0(m)} \delta Q(m)\, dm > 0. \quad \text{(Condition for pulsations)} \qquad (2.92)$$

The loop integral is carried out for each mass element dm in the star, and the results are summed over the entire mass of the star. If the result is positive, pulsations can, in principle, develop.

Ionization valve

What is the mechanism that drives the heat engine in a pulsating variable? In the surface layers of cepheid and RR Lyrae variables, the operative mechanism is the "valving" of the heat by the changing opacity of the gas in the star's ionization transition zone. This is the level in the star at which the gas density and temperature are close to those required for ionization.

Transition zone

In the transition zone, a modest temperature or density change can markedly alter the ionization state of the atoms. For hydrogen and helium, the transition zone lies relatively near the surface of the star. Inward of the zone, the gas will be hot and ionized, and outward it will be cool and neutral (Fig. 9).

In such a zone, the gas becomes more ionized when the temperature increases during compression. The additional electrons provide more scatterers and hence create an increased opacity to the photons. This restricts the outflow of energy, and so the photons remain bunched up ("trapped") like cars on a rough segment of highway (Fig. 9a). This conforms with the required condition for pulsations – namely, that heat be absorbed, $\delta Q > 0$, when the temperature is high. The trapped heat provides the extra pressure that drives the subsequent expansion.

During expansion of the transition zone, the gas cools and becomes more neutral (on average). This reduces the opacity (fewer electrons) and allows photons to escape; the highway becomes smooth and the cars (photons) rush ahead. Heat is released, $\delta Q < 0$, while the temperature is low – again in accord with the condition for pulsations. At the lower temperature and expanded volume, the pressure is reduced according to the ideal gas law $PV = \mu RT$ for μ moles. The outer layers of the star then lack pressure support, and so they fall rapidly owing to the self-gravity of the star. The cycle then repeats again and again. The continuous valving allows the thermal energy to overcome the dissipation.

Application of the equations of stellar structure to narrow layers of the star allows one to determine the P-V trace of individual layers. Regions that do positive work will drive the pulsations. These turn out to be the outer layers where, not surprisingly, hydrogen and helium are partially ionized. The inner layers are mostly dissipative.

The oscillations can grow to large amplitudes if the net work is positive. The large temperature and density excursions are mostly in the outer regions where the valving takes place.

The results of such calculations indicate the existence of a region on the luminosity-temperature plot (the Hertzsprung–Russell diagram) in which a star is not stable; it will pulsate. This region is called the classical instability strip and is illustrated in Fig. 4.9. The figure also illustrates numerous other types of pulsational variables we have not discussed here.

Variable stars as distance indicators

Certain types of variable stars, known as cepheid variables and RR Lyrae stars, exhibit a *period-luminosity relation*. The measurement of the period of such a star yields its luminosity through this relation. The luminosity can thus be used as a standard candle in the determination of distances. Cepheid variables are quite luminous, 300–30 000 $L_\odot$, and thus can be used out to much larger distances than can most main-sequence stars. For more on this, see Section 4.3 and AM, Chapter 9.

Problems

2.2   Jeans length

Problem 2.21. (a) Find the expression (7) for the critical (Jeans) mass directly from (2) and from the relation between temperature and atomic speeds ($mv^2/2 = 3kT/2$) without bothering to define the Jeans length and the speed of sound. Neglect factors of order unity. (b) What is the critical mass in units of solar mass for a hypothetical region of the interstellar medium of our Galaxy with particle density of 0.5×10^6 H atoms/m^3 at $T = 100$ K? Compare with the masses of giant molecular clouds that range from $10^3 \, M_\odot$ to $10^5 \, M_\odot$.

(c) What is the size (radius) in light years of the region occupied by this matter? [Ans. −; ∼10^4 $M_\odot$; ∼500 LY]

Problem 2.22. A cloud of mass M_0 that is about equal to the critical mass has uniform mass density ρ and consists solely of identical particles of mass m. It separates itself from the general ISM and commences to shrink slowly and uniformly by radiating energy. Assume that it continues to remain at the Jeans critical condition (2) by changing its temperature as it shrinks. Ignore any other perturbing effects such as density fluctuations, angular momentum, and magnetic fields. (a) Find an expression for the gas temperature T of the cloud in terms of M_0, the mass density ρ, the particle mass m, and physical constants. Neglect factors of order unity. As the size of the cloud decreases by a factor of 10, by what factor and sign does the temperature change? (b) Find approximate expressions for, compare, and comment on the total kinetic and potential energies in the cloud before and after this decrease. [Ans. $T \approx (Gm/k)M_0^{2/3}$ $\rho^{1/3}$;−]

2.3 Hydrostatic equilibrium

Problem 2.31. (a) Explain the role of the minus sign in the equation of hydrostatic equilibrium (12); does it make physical sense? (b) Derive the equation of hydrostatic equilibrium in spherical coordinates. Set up the equilibrium condition in vector notation, making use of the gradient of pressure ∇P, which, in our spherically symmetric case in spherical coordinates is $\nabla P = (\partial P/\partial r)\hat{r}$, where $\hat{r}$ is a unit radial vector. [Ans. −; $\nabla P = -g(r)\rho(r)\,\hat{r}$]

2.4 Virial theorem

Problem 2.41. A hot ball of gas (similar to the sun) consisting totally of ionized hydrogen and having mass M and radius R and no nuclear energy source is in gravitational equilibrium; it thus obeys the virial theorem. It radiates energy from its surface with a blackbody spectrum at the rate $L = 4\pi R^2 \sigma T_{\text{eff}}^4$ (W) (6.20), where T_{eff} is the effective surface temperature, which is some constant fraction f of the virial temperature, $T_{\text{eff}} = fT_{\text{v}}$. (This is an ad hoc way to crudely simulate the effect of opacity.) As the gas loses energy, the ball gradually shrinks, giving up potential energy while continuing to satisfy the virial theorem. In applying the virial theorem, consider the gas ball at any time to be isothermal at "virial" temperature T_{v} and the potential term to be $E_{\text{p}} = -GM^2/R$. (a) Determine how each of the following depend on f, M, R, and other constants: (*i*) the thermal energy content E_{k}, (*ii*) the temperature T_{v} required by the virial theorem, and (*iii*) the luminosity L. (b) Find an expression for $R(t)$ in terms of physical constants, M, the initial radius R_0, and f. (It is convenient to express a combination of constants with a single parameter.) (c) According to your result, how long does it take for a ball of gas with solar values $M = M_\odot$ and $R_0 = R_\odot$ to decrease to zero radius if $f = 1$? Comment on your answer. Hint: what is T_{v} at radius $R_\odot$? (d) Find the value of f that gives the correct effective temperature (5800 K) for the sun. (e) With this value and your solution for $R(t)$, (*i*) how long would it take for the sun to shrink to zero diameter, and (*ii*) how long would it take for its angular *diameter* to shrink 0.01″ from its current value of 1920″? Comment on both answers. (f) Project this model backward in time and find when the sun would have been three times its current size. How would its effective temperature and luminosity compare with today's values in our model? Comment. [Ans. $\propto M^2/R$, $\propto M/R$, $\propto f^4 M^4/R^2$; $R = R_0 - \alpha t$, where $\alpha \approx 10^{-55} f M^2$; ~ 40 min.; $\sim 10^{-3}$; $\sim 10^7$ yr, ~ 80 yr; $\sim 10^7$ yr]

Problem 2.42. An astronomer notes a large number of galaxies in a tight cluster, suggesting strongly that they are in equilibrium in a gravitational potential. That is, they should obey the virial theorem. For simplicity consider the cluster to consist of only three galaxies, each of visible mass $m = 10^{11} M_\odot$. Fortuitously, at this time, their spatial (x, y, z) positions form an equilateral triangle on the plane of the sky with sides $d = 500\,000$ LY (a few times the diameter of our Galaxy). The measured radial (line-of-sight) components of their velocities are $+150$, -200, and $+240$ km/s, respectively, where "$+$" indicates recession; the other components are unknown. (a) Find values of $2\Sigma_k$ and $-\Sigma_p$ in joules and calculate their ratio. What might you conclude from this result? (b) Derive an expression from the virial theorem for the *virial mass* M_v required for equilibrium for our three-mass system in terms of the average line-of-sight velocity squared, the spacing d, and the gravitational constant G. Assume that all the virial mass is equally divided among the three galaxies and thus that your expressions from (a) apply. (c) Substitute into your expression the measured line-of-sight velocities. By what factor does the virial mass M_v exceed the visible mass $M = 3m$ of the three galaxies? [Ans. ~ 40; $M = 9d \langle v_{i,\mathrm{los}}^2 \rangle_{\mathrm{av}} / G$; ~ 40]

2.5 Time scales

Problem 2.51. Find the thermal (Kelvin–Helmholtz) time scale of a hypothetical white dwarf star of mass $1.0 M_\odot$ and radius $10^{-2} R_\odot$. It has a thin, high-opacity, nondegenerate shell at effective (surface) temperature $33\,000$ K as compared with the solar effective temperature of 5800 K. The luminosity is $L = 4\pi R^2 \sigma T_{\mathrm{eff}}^4$ (6.20). Its interior consists of high-momentum (degenerate) electrons and nondegenerate carbon nuclei at temperature 3×10^7 K. The carbon nuclei carry almost all of the available thermal energy. [Ans. $\sim 5 \times 10^7$ yr]

Problem 2.52. (a) Find an expression for the time it takes a satellite to orbit just above the surface of a spherical celestial body of mass M and radius R in terms of the average density ρ_{av} of the celestial body. Compare this with the dynamical time constant (37) for that body. (b) What is the dynamical time τ_{dyn} for (i) the central region of the sun within $\sim 0.1 R_\odot$, where the average density is about 100 times that of water; (ii) the earth ($M = 6 \times 10^{24}$ kg, $R = 6400$ km); (iii) a white dwarf star with the mass of the sun and the size of the earth; (iv) a neutron star of mass $1.4 M_\odot$ and radius 10 km; and finally (v) a proton ($m = 1.7 \times 10^{-27}$ kg, $R = 1.5 \times 10^{-15}$ m)? For parts (ii)–(v), assume constant density. What does the latter result (v) tell you about a neutron star? [Ans. comparable; ~ 5 min, ~ 30 min, ~ 5 s, ~ 0.2 ms, ~ 0.3 ms]

Problem 2.53. Derive from first principles, without reference to the text if possible, the diffusion time for particles traveling at speed v in a two-dimensional (x, y) space to reach a distance R from their release point if the scatterers have a density n (m^{-2}) and linear cross section for scatter σ (m). Let each particle scatter with equal probability into the $\pm x, \pm y$ directions. [Ans. $R^2 \sigma n / v$]

2.6 Nuclear burning

Problem 2.61. (a) What is the magnitude of the electrostatic force between two protons that approach to within 1.4×10^{-15} m, center to center, of each other, the approximate distance to which the nuclear force reaches? (b) At about what temperature must a Maxwell–Boltzmann gas be such that particles of the average kinetic energy, $\sim 3kT/2$, can approach to within this distance? How does this temperature compare with the temperature at the center of the sun? [Ans. ~ 100 N; $\sim 10^{10}$ K]

Problem 2.62. Answer the following regarding the pp chain: (a) How much mass of hydrogen undergoing the pp interaction would be required to power a 1-kW heater for 1 yr and (b) to power the sun for 1 μs? (c) Show that electric charge is conserved in each of the several nuclear interactions illustrated in Fig. 6 for the pp chain. (d) How much energy (eV units) is released per nucleon (proton or neutron) in the chemical burning of oil that yields 140 000 BTU/gallon? (1 BTU $\approx$ 1055 J and 1 gal of water weighs 3.78 kg. The oil density is about that of water.) Compare this energy release with that for the pp chain. Look up the wood-burning yield and repeat. [Ans. ~ 50 mg; ~ 600 metric tons; –; ~ 0.4 eV/nucleon, ~ 0.2 eV/nucleon]

Problem 2.63. Make a diagram similar to Fig. 5 to illustrate the CNO cycle.

Problem 2.64. (a) Estimate, from the solar luminosity, the value of the proportionality constant ε_0 in the expression (66) for nuclear power generation through the pp process, where $\beta = 4$. Assume all the luminosity originates within radius 0.1 $R_\odot$, where $T = 1.6 \times 10^7$ K, $\rho = 1.5 \times 10^5$ kg/m^3, and hydrogen has been depleted to a fraction $X = 0.36$ of the mass. The solar luminosity is 3.8×10^{26} W. Hint: what is the energy-generation rate ϵ_{pp} (W/kg) at the sun's center? (b) Find the hypothetical power output (W) from (*i*) 1 m^3 of water and (*ii*) a swimming pool of dimensions 25 m $\times$ 15 m $\times$ 2 m if the water would generate power at the solar rate ϵ_{pp} (W/kg). Comment on your answers. [Ans. $\sim 10^{-3}$ W/kg; ~ 2 W, ~ 1 kW]

2.7   Eddington luminosity

Problem 2.71 Consider the massive black hole of $\sim 3 \times 10^6$ $M_\odot$ at the center of the Galaxy at distance $\sim 25\,000$ LY from the earth. (a) What is its Eddington luminosity ($\mu_e = 1$)? Compare this with its actual x-ray luminosity during a flare that reached a luminosity of 4×10^{28} W. (b) What is the approximate accretion rate that would result in this flare and what would it be at L_{Edd}? Adopt Newtonian energies and assume that all the potential energy loss down to the Schwarzschild radius, $2GM/c^2$, appears as radiation. Give your results in kilograms per second and also in solar masses per year. (c) What is its flux (W/m^2) at the earth and what would it be at L_{Edd}? Compare the latter with the brightest persistent celestial x-ray source, Sco X-1, at 4×10^{-10} W/m^2, with the solar x-ray flux during bright flares at $\sim 10^{-6}$ W/m^2, and with the (mostly optical) total flux from the sun (i.e., the solar constant at 1365 W/m^2). [Ans. $\sim 10^{38}$ W; $\sim 10^{-11}$ $M_\odot$/yr, $\sim 10^{-2}$ $M_\odot$/yr; $\sim 10^{-13}$ W/m^2, $\sim 10^{-4}$ W/m^2]

Problem 2.72. (a) What is the Eddington luminosity for 1 kg of material for $\mu_e = 1$? (b) Does this suggest that a liter of water in an open pan would start ejecting material if it were radiating at this luminosity? Discuss why or why not. (c) Suppose you took the water into the vacuum of space and released it so it would be free of earth's gravity. How might it behave? What does the Jeans criterion tell you (Section 2) about the temperature that would be required for self-gravity to dominate the thermal motions? State your assumptions. [Ans. ~ 5 W; –;–]

2.8   Pulsations

Problem 2.81. (a) Demonstrate that the entropy change ΔS integrated around a complete Carnot cycle (Fig. 8) is zero as stated in (87). Note the quantities that are zero on each leg, the first law of thermodynamics (84), and the relation between internal energy and temperature, $dU = C_V\, dT$, where the constant C_V (J/K) is the specific heat at constant volume. (b) Use your calculations to demonstrate that $PV^\gamma = \text{constant}$ describes a gas element as it expands or contracts adiabatically. Here $\gamma \equiv C_P/C_V$ is the ratio of specific heats at constant pressure and volume, which are related as $C_P = C_V + R$. The gas constant R is that found in the ideal gas law (for one mole), $PV = RT$.

3

Equations of state

What we learn in this chapter

An **equation of state** (EOS) of a gas is the pressure as a function of density and temperature, $P(\rho, T)$. It is fundamental to the understanding of stellar interiors. For an ideal gas, the gas particles have a distribution of momenta described by the **Maxwell–Boltzmann** (M-B) **distribution**.

The distribution of gas particles is most generally described as a six-dimensional (6-D) **phase-space density**, $f(x, y, z, p_x, p_y, p_z, t)$, known as the **distribution function**. It is directly related to **specific intensity**. The phase-space density can vary with time, but, according to **Liouville's theorem**, it is conserved in a frame of reference that travels in phase space with the particles – under certain conditions. The **propagation of cosmic rays** in the Galaxy generally satisfies these conditions.

From the M-B distribution, one finds the pressure, and hence the EOS, $P = (\rho/m_{av})kT$ of an "ideal" particulate gas, which, rewritten, is the ideal gas law, $PV = \mu RT$. This EOS applies to most stellar interiors. A **photon gas** in equilibrium with its surroundings, **blackbody radiation**, has an EOS that depends solely on temperature, $P = aT^4/3$, as developed in Chapter 6. This pressure plays a dominant role in the centers of the most massive stars and did so in the early universe.

Highly dense stars such as **white dwarfs** and **neutron stars** have very different equations of state. The former, and in part the latter, are supported by **degeneracy pressure**, a quantum mechanical phenomenon. On a microscopic scale, 6-D **phase space** is **partitioned into states** each of volume h^3, where h is the **Planck constant**. Electrons and protons (both spin 1/2 **fermions**) have the restriction that no more than two of them can occupy a given phase-space state.

A gas of fermions becomes completely **degenerate** when it is so cold or compressed that the lowest-energy **phase-space states are completely filled**. This forces particles up to the **Fermi momentum or energy**, which is higher than that expected for the thermodynamic temperature of the gas. This results in an **electron degeneracy pressure** that can support a **white dwarf** star against gravitational collapse. The **Fermi distribution** of energy of the particles is used to calculate the **EOS of a degenerate gas** in the limit of complete degeneracy. To first order, the pressure depends only on matter density ρ; it is independent of temperature. For **nonrelativistic particles**, we find $P \propto \rho^{5/3}$ and, for **relativistic particles**, $P \propto \rho^{4/3}$. The latter "softer" EOS can lead to white dwarf collapse (Section 4.4).

3.1   Introduction

An equation of state (EOS) gives the pressure at a position in space as a function of mass density ρ (kg/m^3) and temperature T; that is, $P(\rho, T)$. This is a statement of how the pressure of matter responds to changes in ρ and T. The EOS is an essential element in the modeling of star structure.

A gas of temperature T contains particles traveling in many different (random) directions and with a wide range of speeds. The most basic distribution of velocities is that of a gas in thermal equilibrium wherein collisions of the particles have allowed the energy to be shared appropriately among all particles. In particular, the *Maxwell–Boltzmann* (M-B) distribution presented here applies to a gas of pointlike and nonrelativistic particles that have negligible interparticle forces. Its EOS, when rewritten, is known as the *ideal gas law*.

Many gases encountered in astrophysics obey the ideal gas law (or its equivalent EOS) quite well. In the interiors of normal (nondegenerate) stars, the gases are mostly ionized, and so interparticle (electrostatic) forces are present. Typically they yield potential energies of only ~10% of the kinetic energy densities. Thus, the ideal gas law is a satisfactory approximation for use in understanding the broad properties of stellar interiors, but detailed models must take into account the electrostatic potential energies.

Matter becomes *degenerate* at extremely high densities or low temperatures such as found in white-dwarf and neutron-star interiors. In this state, particles, for the most part, fill all the available quantum energy states up to well beyond thermal energies. Thus, when the temperature rises, particles do not acquire additional kinetic energy.

Such a gas does not obey the ideal gas law; instead it is described by a different relation, $P(\rho, T)$, which turns out to be independent of temperature in the limit of perfect degeneracy. Such a gas is not called "ideal." However, because it has no interparticle forces in the simple case, one may still call it a *perfect gas*.

The particle speeds in a degenerate gas are described by the *Fermi–Dirac* (F-D) distribution for a gas of half-integer spin particles and by the *Bose–Einstein* (B-E) distribution for a gas of integer-spin particles. For sufficiently high temperatures, low densities, or both, the F-D and B-E distributions reduce to the classical M-B distribution.

In this chapter, we first examine the velocity distribution in an ideal gas to arrive at the M-B distribution of momenta in three-dimensional (3-D) space. The six-dimensional (6-D) generalization of this distribution, the *distribution function*, is introduced. The EOS of an ideal gas is then derived from the M-B distribution. The EOS of a photon gas ($P = aT^4/3$) is also presented; it is derived in Section 6.2; see (6.43).

Finally, a degenerate gas of fermions is introduced through a graphic representation of phase space for a one-dimensional (1-D) gas. The Fermi momentum and energy are defined, and the Fermi–Dirac distribution function is presented. This leads to two temperature-independent equations of state, one for nonrelativistic particles ($P \propto \rho^{5/3}$) and the other for relativistic particles ($P \propto \rho^{4/3}$).

The fundamental and secondary equations of stellar structure presented in Section 4.2 will include the equations of state presented herein.

3.2 Maxwell–Boltzmann distribution

We present and discuss, but do not derive, the distribution of momenta (or equivalently velocities) of particles in an ideal gas known as the Maxwell-Boltzmann (M-B) distribution. The concept of *momentum space* is presented.

The particles of a gas have a range of momenta with an average value that depends on the temperature. The functional form of the distribution of speeds follows when three constraints on a gas of particles in thermal equilibrium are applied: (*i*) all magnitudes of velocity component v_x are, a priori, equally probable and the same is true of v_y and v_z, (*ii*) the total energy is constrained to a fixed value, and (*iii*) the total number of particles is constrained to a fixed value. One further assumes that the temperatures are not so high as to drive the particles to relativistic speeds.

The net effect of items (*ii*) and (*iii*) is that velocities are actually limited; they are not in fact equally probable. One could quickly use up the available energy with only a few excessively high-velocity particles or with too many low-velocity particles. Item (*i*) ensures that particles do have a wide distribution of speeds, whereas items (*ii*) and (*iii*) result in a variation of the probability of different speeds.

One-dimensional gas

The statistical-mechanics calculation based on the preceding principles is, in fact, a maximization of entropy, or "randomness," under the constraint of fixed particle number and fixed total energy. The resultant distribution for a 1-D gas with positions and motions limited to one dimension (e.g., the x direction) is the Gaussian

$$\mathcal{P}(v_x)dv_x = \left(\frac{m}{2\pi kT}\right)^{1/2} \exp\left(-\frac{mv_x^2}{2kT}\right) dv_x, \quad \text{(Probability of } v_x \text{ in } dv_x) \quad (3.1)$$

where $\mathcal{P}(v_x)dv_x$ is the one-dimensional probability of finding a given particle at speed v_x in the speed interval dv_x. The integral of (1) over all speeds equals unity in accord with the usual meaning of probability.

Visualize a one-dimensional gas for which (1) describes the distribution of velocity v_x, which can take on values $-\infty < v_x < +\infty$. The function is symmetric about $v_x = 0$ and falls off symmetrically on either side; the exponential becomes e^{-1} when the translational energy $mv_x^2/2 = kT$. The average value of v_x is obviously zero, but the average kinetic energy ($\propto v_x^2$) will have a finite value.

This result can also be expressed in terms of the particle momentum

$$p_x = mv_x \quad (3.2)$$

rather than the particle speeds. To change the variable in (1), impose the usual correspondence

$$\mathcal{P}(p_x)\,dp_x = \mathcal{P}(v_x)\,dv_x, \quad (3.3)$$

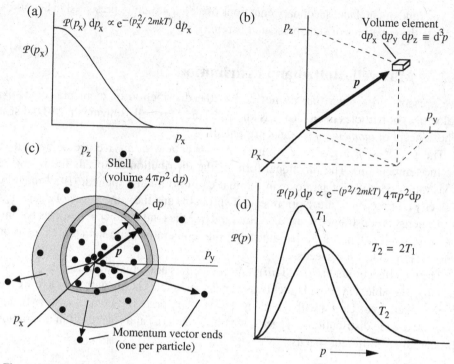

Figure 3.1. Nondegenerate gas obeying Maxwell–Boltzmann (M-B) distribution in momentum space. (a) Gaussian probability function $\mathcal{P}(p_x)$ of finding a particle with momentum p_x per unit interval p_x for $p_x \geq 0$. (b) Volume element in 3-D momentum space; the momentum vector p has components p_x, p_y, and p_z. (c) Momentum space with dots representing individual particles; the number of vector ends within a shell includes all particles with momentum of magnitude, p in dp. The density of particles in momentum space is greatest at the origin. (d) Probability $\mathcal{P}(p)$ (11) of finding momentum *magnitude* p in unit momentum interval versus p, in a 3-D gas, for two temperatures, where $T_2 = 2T_1$. The function goes to zero at the origin and also at high momenta.

where $\mathcal{P}(p_x)dp_x$ is the probability of finding the particle at momentum p_x in the momentum interval dp_x. Equations (2) and (3) yield $\mathcal{P}(p_x) = \mathcal{P}(v_x)/(dp_x/dv_x) = \mathcal{P}(v_x)/m$. Thus, the converted equation is again a Gaussian:

$$\mathcal{P}(p_x)\,dp_x = \left(\frac{1}{2\pi mkT}\right)^{1/2} \exp\left(-\frac{p_x^2}{2mkT}\right) dp_x. \qquad \text{(Probability of finding } p_x \text{ in } dp_x) \qquad (3.4)$$

This function is plotted in Fig. 1a. The probability of finding a given particle with momentum p_x is maximum at $p_x = 0$ and decreases as the magnitude of p_x increases. The probabilities for all $p_x(-\infty < p_x < +\infty)$ again sum to unity (Prob. 21a). The average translational kinetic energy of these particles, $kT/2$, can be obtained from this distribution (Prob. 42).

Three-dimensional gas

Consider now a 3-D gas. If collisions between particles randomly change their directions, the three motions (along the x-, y-, and z-axes) are independent of one another, and the same momentum distribution applies to each. The *isotropy of space* demands that the three axes be equivalent. We will now obtain the probability of finding a particle with a given (previously specified) *vector* momentum with components p_x, p_y, and p_z in the intervals dp_x, dp_y, and dp_z.

Maxwell–Boltzmann distribution

The combined probability for the occurrence of an event that depends on independent events is simply the product of the independent probabilities. For example, the probability of obtaining "heads" in five successive coin flips is the product of the five individual probabilities $(1/2)^5 = 1/32$.

Similarly, the probability of finding a particle with vector momentum p in some volume interval,

$$d^3p \equiv dp_x \, dp_y \, dp_z, \tag{3.5}$$

is the product of the independent probabilities for obtaining p_x, p_y, and p_z in those intervals:

$$\mathcal{P}(p) \, d^3p = \mathcal{P}(p_x) \, dp_x \, \mathcal{P}(p_y) \, dp_y \, \mathcal{P}(p_z) \, dp_z \tag{3.6}$$

and, from (4),

$$\mathcal{P}(p) \, d^3p = \left(\frac{1}{2\pi mkT} \right)^{3/2} \exp\left(-\frac{p_x^2 + p_y^2 + p_z^2}{2mkT} \right) dp_x \, dp_y \, dp_z. \tag{3.7}$$

Because $p_x^2 + p_y^2 + p_z^2 = p^2 \equiv p \cdot p$, we have

$$\Rightarrow \quad \mathcal{P}(p) d^3p = \left(\frac{1}{2\pi mkT} \right)^{3/2} \exp\left(-\frac{p \cdot p}{2mkT} \right) d^3p. \quad \text{(Maxwell–Boltzmann}$$
$$\text{distribution in momentum space)} \tag{3.8}$$

This is the *Maxwell–Boltzmann distribution* in 3-D momentum space. It is dimensionless, and thus $\mathcal{P}(p)$ has the units $(\text{momentum})^{-3} = (\text{N s})^{-3}$.

The function $\mathcal{P}(p)$ depends only on the magnitude of the momentum squared, p^2. It is thus independent of the particle motion direction, as it should be. As before, the distribution is a function of the kinetic energy of a gas particle E because $E = mv^2/2 = p^2/2m$. (In this chapter, we use E, not E_k, to represent the kinetic energy of a particle or photon.)

The distribution (8) then takes the form

$$\mathcal{P}(p) \propto \exp(-E/kT). \tag{3.9}$$

The kinetic energy at which the exponential has the value e^{-1} is thus $E = kT$. As in the 1-D case, the vector momentum with the highest probability occurs at $E = 0$ (i.e., at $p = 0$). The function $\mathcal{P}(p)d^3p$ is the probability of obtaining a particle with vector momentum p in d^3p – that is, a particle moving with specified momentum magnitude in a specified direction.

Momentum space

A 3-D *momentum space* (Fig. 1b) has axes that represent the components p_x, p_y, and p_z. In this space, a particle with a vector momentum p is represented as a point at the position corresponding to its components p_x, p_y, and p_z. If the vector p is drawn from the origin to this point, it will have the correct magnitude and components. This is the exact counterpart of a radius vector in x, y, z space. The infinitesimal volume element d^3p in this space is given in (5) and shown in Fig. 1b.

A collection of gas particles of all possible momenta can be represented by a collection of dots throughout the space (Fig. 1c). One then imagines the vectors, one for each dot. The probability $\mathcal{P}(\mathbf{p})$ is proportional to the density of dots at the position $\mathbf{p}$. In fact, our formalism (8) specifies $\mathcal{P}(\mathbf{p})$ to be the probability of finding a particle with vector momentum $\mathbf{p}$ in unit volume of momentum space.

Distribution of momentum magnitude

The probability $\mathcal{P}(p)$ of finding a particle with a given magnitude of the momentum p in dp, including all directions of travel, is a sum of the probabilities for all volume elements at magnitude p in momentum space. These elements occupy a spherical shell of radius p and radial thickness dp in 3-D momentum space. Their total volume is $4\pi p^2 dp$ (Fig. 1c):

$$d^3p = 4\pi p^2 dp. \qquad \text{(Spherical shell in isotropic momentum space)} \qquad (3.10)$$

The total probability $\mathcal{P}(p)dp$ of finding a particle with a given *magnitude* of momentum p in dp is thus the expression (8) with $4\pi p^2 dp$ substituted for d^3p:

$$\mathcal{P}(p)dp = \left(\frac{1}{2\pi mkT}\right)^{3/2} \exp\left(-\frac{p^2}{2mkT}\right) 4\pi p^2 dp. \qquad \begin{array}{l}\text{(Probability}\\ \text{of } p \text{ in } dp)\end{array} \qquad (3.11)$$

The exponent again carries the kinetic energy of the particle, $E = p^2/(2m)$. The function $\mathcal{P}(p)$ is one form of the M-B distribution for a 3-D gas. The function itself is 1-D in the variable p (Fig. 1d). The area under the curve is again unity (Prob. 21b) as expected for a probability function:

$$\int_0^\infty \mathcal{P}(p)\,dp = 1. \qquad (3.12)$$

The shape of the function (Fig. 1d) is determined by the product of the exponential and p^2 terms. Because the p^2 term goes to zero at $p = 0$, the function goes to zero there. At high p, when $E \gg kT$, the exponential overcomes the p^2 term and drives the function toward zero. In between, the function has a maximum that defines the most probable value of p. The most probable momentum magnitude is thus not zero. The distribution in speed v has the same functional shape as (11) because $p = mv$.

Compare the distributions for the same gas at two temperatures, $T_2 = 2T_1$, as shown in Fig. 1d. The $T^{-3/2}$ dependence in (11) causes the higher temperature curve to have less amplitude at low momentum, but the $\exp(-p^2/2mkT)$ term extends the curve to higher p before forcing it toward zero.

3.3   Phase-space distribution function

Particle positions and momenta are completely described by the 6-D *distribution function*

$$f(x, y, z, p_x, p_y, p_z, t) \equiv f(\mathbf{x}, \mathbf{p}, t). \qquad \begin{array}{l}\text{(Distribution function;}\\ \text{particles m}^{-3} \text{ (N s)}^{-3})\end{array} \qquad (3.13)$$

This is the particle density in a 6-D *phase space* as a function of the six position and momentum coordinates at time t. The distribution function is the underpinning of the EOS of any material.

We first write the distribution function for M-B statistics. Then, in a diversion from our main task of finding the EOS for the M-B distribution, we present other useful aspects of the distribution function. We show its direct relation to specific intensity, the fundamental measurement quantity of astronomy, and its relation to other measurable quantities. Interestingly, it underlies the conservation of specific intensity through Liouville's theorem. It is also a Lorentz invariant, as we demonstrate in Section 7.8.

Maxwell–Boltzmann in 6-D phase space

The quantity $\mathcal{P}(\mathbf{p})$ is the probability per unit momentum volume of finding a particle with momentum $\mathbf{p}$. If N particles are distributed uniformly in volume V of physical (x, y, z) space, the number of particles per unit momentum volume is $N\mathcal{P}(\mathbf{p})$ with units $(\text{mom.})^{-3} = (\text{N s})^{-3}$. Divide this by volume V to obtain the number of particles per unit six-dimensional volume (units $\text{m}^{-3}\,(\text{N s})^{-3} = (\text{J s})^{-3}$), namely, $f = N\mathcal{P}(\mathbf{p})/V$. Because the density of particles in x, y, z space is $n = N/V$, the distribution function becomes

➡ $$f = n\mathcal{P}(\mathbf{p}) \qquad\qquad \text{(Distribution function for}$$
$$= \frac{N}{V}\left(\frac{1}{2\pi mkT}\right)^{3/2}\exp\left(-\frac{p^2}{2mkT}\right), \qquad \begin{array}{l}\text{M-B statistics; J}^{-3}\,\text{s}^{-3})\end{array} \qquad (3.14)$$

where we have substituted in $\mathcal{P}(\mathbf{p})$ for M-B statistics (8).

This is the particle density in 6-D (x, y, z, p_x, p_y, p_z) *phase space*. The spatial and momentum components of each particle specify its coordinates in this space. Think of a Cartesian coordinate system with six orthogonal axes. In this space, every particle can be plotted as a point just as was done in 3-D momentum space (Fig. 1c).

For unit volume, $V = 1$, the distribution function becomes $N\mathcal{P}(\mathbf{p})$. Thus the 3-D momentum-space scatter plot, Fig. 1c, may be viewed as a representation of phase space if the physical particles are distributed uniformly in physical x, y, z space. Think of each element d^3p as carrying with it unit x, y, z volume.

The distribution function describes the complete dynamical state of a collection of particles from which all measurable quantities (e.g., number density, particle flux density, and bulk velocity) can be obtained by appropriate integration of f.

Measurable quantities

All possible measurements of numbers, energies, and directions of travel of an incoming flux of particles can be described as an integral over the distribution function. We cite two examples. Their justification is left to the reader.

The simplest is an integration over all momenta. This yields the *number density* n (particles/m^3 at position x, y, z and time t) in physical space:

$$n(x, y, z, t) = \iiint f\,\text{d}^3p. \qquad \text{(Number density in } x, y, z \text{ space; m}^{-3}) \qquad (3.15)$$

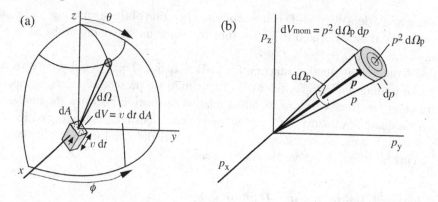

Figure 3.2. Volume element (shaded) in (a) physical (x, y, z) space and (b) momentum (p_x, p_y, p_z) space.

The net *particle flux density* (number $s^{-1} m^{-2}$) at position or time (x, y, z, t), a vector quantity, is

$$\mathscr{F}_p(x, y, z, t) = \iiint vf\,d^3p, \qquad \text{(Particle flux density; m}^{-2}\,s^{-1}) \qquad (3.16)$$

where v is the velocity vector associated with a particular cell. The quantity $\mathscr{F}_p$ is the net number of particles that would cross unit surface in unit time. The integral sums over all momentum elements at the position x, y, z and hence over all vector velocities v. If the particles at some position move chaotically and isotropically such that the velocity vector sum is zero, there will be no net flux density at that position.

Specific intensity

Knowledge of the distribution function is equivalent to knowledge of the specific intensity that is the basis of most measurements in astronomy. We first demonstrate the relation of f to *particle specific-intensity J* and then find its relation to *energy specific-intensity I*. Our expressions will be relativistically correct.

Particle number

Define the *number specific-intensity $J(U)$* to be the number of particles crossing unit surface per unit time into unit solid angle within unit interval of particle energy U:

$$J(U) : \left(\frac{\text{Particles}}{\text{s m}^2\,\text{J sr}} \right). \qquad \text{(Particle specific-intensity)} \qquad (3.17)$$

Define the energy U to be the total relativistic energy, the sum of the rest and kinetic energies of a given particle, $U = E + mc^2$. For high-energy cosmic ray sources, where $U \gg mc^2$, the distinction between total and kinetic energies is moot. For lower-energy particles, an experiment might measure the fluxes at different kinetic energies E, and the spectrum would most likely be plotted as a function of E.

Consider now the particles passing through surface dA (Fig. 2a) with energy between U and $U + dU$ and direction normal to the surface within the solid angle $d\Omega$ in the time interval dt. It follows from the definition of J (17) that the number of such particles is

$$dN = J\,dA\,d\Omega\,dU\,dt. \qquad \text{(Number of particles in specified differential intervals)} \qquad (3.18)$$

This sample of particles may also be described in terms of the phase-space density f (13). The number dN is the product of f and the relevant 6-D phase-space volume element. In physical space (Fig. 2a), the volume element that contains the particles is $v\,dt\,dA$, where v is the particle speed. In momentum space (Fig. 2b), the element is $p^2\,dp\,d\Omega_p$, where $d\Omega_p$ is the solid angle in momentum space. The 6-D volume element is the product of the momentum and physical elements. Thus,

$$dN = f\,p^2\,dp\,d\Omega_p\,v\,dt\,dA. \qquad (3.19)$$

Equate (18) and (19) because both describe the same sample of particles:

$$J\,dA\,d\Omega\,dU\,dt = f\,dA\,p^2\,dp\,d\Omega_p\,v\,dt. \qquad \text{(Conservation of particles)} \qquad (3.20)$$

The axes in the spatial and momentum spaces are coaligned (p_x-axis coaligned with x-axis, etc.). Thus, the solid angles are equivalent, $\Omega = \Omega_p$.

In special relativity, the total energy U of a particle is the sum of its kinetic and rest energies, $U = E + mc^2$ (7.16). The momentum p and total energy U are related by $U^2 - (pc)^2 = (mc^2)^2$ (7.20). Differentiation of the latter yields the relation between the intervals dU and dp:

$$U\,dU = c^2\,p\,dp. \qquad (3.21)$$

The velocity and momentum are related because $p = \gamma m v$, where $\gamma = U/mc^2$, from (7.14) and (7.17). Thus,

$$v = \frac{c^2}{U}p. \qquad (3.22)$$

Substitute (21) and (22) into the equality (20) to obtain

$$\blacktriangleright \qquad J = p^2 f, \qquad \text{(Relation between particle specific-intensity and phase-space density)} \qquad (3.23)$$

where the units of f and J are given in (13) and (17), respectively. This expression is relativistically correct. It is an important and very general relation that tells us the phase-space density and an intensity are essentially the same quantity.

Energy and photons

The energy specific-intensity $I(\nu)$ (W m^{-2} Hz^{-1} sr^{-1}) often used by astronomers for photon fluxes and the particle specific-intensity $J(U)$ (s^{-1} m^{-2} J^{-1} sr^{-1}) may both be applied to photons. The quantities differ in two ways. First, $I(\nu)$ describes the energy flux rather than the particle flux. Secondly, it gives the flux per unit frequency interval (Hz) rather than per unit energy interval (J). The one is *energy* flux density per unit *frequency* interval, whereas the other is *particle* flux density per unit *energy* interval.

The energy flow per (s m^2 sr) in a flux of photons may be expressed in terms of both I and J and the expressions equated to find a relation between them as follows:

$$I(\nu)\,d\nu = U\,J(U)\,dU. \qquad\qquad \text{(W m}^{-2}\text{ sr}^{-1}) \qquad (3.24)$$

For photons, the energy U and frequency ν are related because $U = h\nu$ and the differential is $dU = h\,d\nu$, giving

➡ $$I(\nu) = J(U)\,h^2\nu. \qquad\qquad \text{(Conversion, } J \text{ to } I) \qquad (3.25)$$

This is the conversion from J to I in terms of frequency.

Finally, eliminate $J(U)$ in (25) with the relation $J = p^2 f$ (23) and then eliminate p with $p = h\nu/c$ to obtain

➡ $$I(\nu) = \frac{h^4\nu^3}{c^2}f. \qquad\qquad \text{(Relation between } I \text{ and } f;\ \text{W m}^{-2}\text{ Hz}^{-1}\text{ sr}^{-1}) \qquad (3.26)$$

This expression is the relation between the distribution function and the energy specific-intensity and is the equivalent of (23).

The specific intensity at a given location, $I(\nu, \theta, \phi, t)$, is a function of frequency, time, and direction of propagation. As was the case for the distribution function, I can be integrated over any one (or more) of the several variables to obtain other measurable quantities; see appendix, Table A4. Examples are the spectral flux density $S(\nu)$ (W m^{-2} Hz^{-1}), the flux density $\mathscr{F}$ (W/m^2), and energy fluence $\mathscr{E}$ (J/m^2). With (26), these can also be written as integrals over the phase-space density f.

Liouville's theorem

Consider a selected group of particles with a modest spread of momenta and positions about central values of these quantities. The particles flow through phase space as time proceeds, subject only to "smooth" forces (no collisions). Let an observer travel alongside them as they propagate through space. *Liouville's theorem* (not derived here) tells us that the distribution function f in the observer's frame of reference is conserved. Furthermore, at each point along the track, a stationary (laboratory) observer would measure the same value of f for these (non-relativistic) particles because the spreads of positions and momenta are invariant for Galilean transformations.

This theorem remains valid for charged particles in the presence of magnetic fields, which change trajectories but not momentum magnitudes. It is not valid if the particles suffer collisions or encounter momentum-dependent forces such as dissipation and radiation.

Cosmic rays streaming through the Galaxy obey the theorem quite well. A measurement of the phase-space density f, or equivalently, the specific intensity J (23), at the earth thus gives the value of f or J of the particles when they were in the far reaches of the Galaxy.

Conservation of specific intensity

The invariance of specific intensity I in astronomical measurements is a direct consequence of Liouville's theorem. The invariance of f and the equivalence of f and I (26) tell us that I is also conserved if one follows a group of photons of some frequency ν through space. This is valid given negligible absorption, scattering, or redshift.

An important ramification of this is that the emitted surface brightness $B(\nu, \theta, \phi)$ of an astronomical body and the detected specific intensity $I(\nu, \theta, \phi)$ from that same body are, in effect, the same quantity (AM, Chapter 8); that is,

$$B(\nu, \theta, \phi) = I(\nu, \theta, \phi).$$ (Equality of surface brightness and specific intensity) (3.27)

The brightness B is the power emitted from unit projected area of the surface into the unit solid angle at θ, ϕ, and the specific intensity I is the power received at the detector per unit detector area and solid angle at θ, ϕ. Both quantities have the same units (W m^{-2} Hz^{-1} sr^{-1}).

The invariance of I with the distance to a diffuse celestial object is a related example. The measured specific intensity (power per steradian) of the sun's surface would not change if the sun were to be moved to twice its current distance. This is usually explained as the cancellation of the increased size (as distance squared) of the patch within the observer's beam and the decreased radiation (inverse square of distance) received from each element of the patch (AM, Chapter 8). Nevertheless, it is a consequence of Liouville's theorem.

Relativity connection

The observer following the particles does so even as they are guided along curved paths by magnetic fields. This observer would thus be in a noninertial (i.e., accelerating) frame of reference to which the Lorentz transformations of special relativity do not apply. However, one can adopt an inertial (constant velocity) frame of reference that, at some moment, is instantaneously at rest with respect to the particles. In this case, the Lorentz transformations would apply.

One may then ask how phase-space density f transforms from this moving inertial frame to the laboratory (stationary) frame if the particles are moving at a speed approaching the speed of light c. We demonstrate in Section 7.8 that f has the same value in two such frames; that is, it is a Lorentz invariant. It then follows from (26) that I/ν^3 is also a Lorentz invariant. This will prove useful in our discussion of the intensity of radiation from jets in Section 7.6.

Now, the frequency ν of photons is not a Lorentz invariant according to the relativistic Doppler shift (Section 7.4). Hence, because I/ν^3 is a Lorentz invariant, the specific intensity I is not. We will find in Section 7.6 that the specific intensity emanating from a celestial object will be greatly modified if the source (e.g., a knot of radiation in a jet) is moving rapidly toward or away from the observer.

3.4 Ideal gas

We now return to the determination of the EOS for a perfect gas that obeys M-B statistics.

Particle pressure

Pressure is the force per unit area that gas particles exert on a wall of the gas container. It can be expressed in terms of the mass m, speed v, and number density n of the particles of which it consists. For now, assume that the gas consists of identical particles, each of mass m, and that all of them move at the same speed v directed toward a wall. A cloud of such particles is shown in Fig. 3.

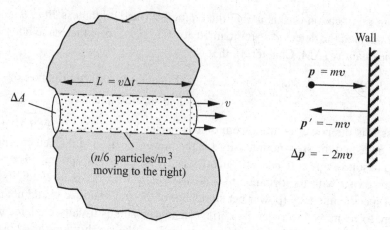

Figure 3.3. Pressure arising from atoms striking a surface. In this approximation, one-sixth of the atoms in the cylinder move to the right and strike area ΔA of the wall in time Δt. The pressure turns out to be 2/3 times the kinetic-energy density in the gas.

Momentum transfer

The momentum of one of the particles in Fig. 3 is

$$\boldsymbol{p} = m\boldsymbol{v},$$
(3.28)

where $\boldsymbol{p}$ and $\boldsymbol{v}$ are the vector momentum and velocity, respectively. When the particle hits the wall head-on and bounces off with an equal speed (assume an elastic collision), the new momentum is

$$\boldsymbol{p}' = -m\boldsymbol{v},$$
(3.29)

and the momentum change of the particle is

$$\Delta\boldsymbol{p} = \boldsymbol{p}' - \boldsymbol{p} = -2m\boldsymbol{v}. \qquad \text{(Momentum change)}$$
(3.30)

The negative sign indicates that the $\Delta\boldsymbol{p}$ vector is opposed to the direction of the $\boldsymbol{v}$ vector. From momentum conservation, a positive momentum of $+2m\boldsymbol{v}$ is transferred to the wall.

The momentum transfer per second to the wall yields the force on it by definition, $\boldsymbol{F} \equiv d\boldsymbol{p}/dt$. The force per unit area on the wall is the desired pressure. The pressure is thus the momentum transferred by one particle times the number of particles that strike 1 m^2 of the wall in 1 s.

The particles that strike area ΔA of the wall in time Δt are those that, at a given instant, are traveling toward it within the imaginary cylinder shown in Fig. 3. This cylinder has base area ΔA and length $v\Delta t$, the distance the particles travel in time Δt. The volume of the cylinder is

$$V = v\Delta t \, \Delta A,$$
(3.31)

where v is the magnitude of the velocity. A given particle can move in any given direction. Thus, one can simplistically assume that only one-sixth of them are moving toward any one of the six walls of the box. In this approximation, the number N of particles in the cylinder heading toward the right wall is

$$N = (n/6)V = (n/6)\,v\,\Delta t\,\Delta A, \tag{3.32}$$

where n is the particle number density.

The total momentum transfer per unit time and per unit area (the pressure P) requires that we multiply N (32) by the momentum transfer of a single particle, $+2mv$ (30), and divide by ΔA and Δt:

$$P = \frac{N\,\Delta p}{\Delta t\,\Delta A} = \frac{nv}{6}\,2mv = \frac{nmv^2}{3} = \frac{2}{3}\,n\,\frac{mv^2}{2}. \qquad (\text{N/m}^2) \tag{3.33}$$

The pressure is proportional to the product of the number density n and the kinetic energy $E = mv^2/2$ of a single particle. In other words, it is proportional to the *kinetic energy density*, nE.

This result is correct even if the particles travel in random directions. A particle approaching a wall at angle θ to the normal of the wall will impart momentum $2mv\cos\theta$ to it upon reflection, and the number of particles striking unit area of the wall per unit time is $nv\cos\theta$. The product of these two terms is the rate of momentum transfer. Integration over the hemisphere of angles yields the factor of $1/3$ in (33) (Prob. 41).

If the particles further have a distribution of speeds, the contribution to the pressure from n_i particles (per unit volume) in an energy interval ΔE_i at E_i is proportional to $n_i E_i$, where E_i is the kinetic energy of one such particle. The total pressure is simply $P = (2/3)\Sigma n_i E_i = (2/3)n(\Sigma n_i E_i/n) = (2/3)\,n E_{av}$, where the summation is over the several energy bins and $n = \Sigma n_i$ is the total number density of particles. In general,

$$P = \frac{2}{3}n\left(\frac{mv^2}{2}\right)_{av} = \frac{2}{3}\,n\,E_{av}. \qquad (\text{N/m}^2 \text{ or J/m}^3) \tag{3.34}$$

It thus turns out that pressure is a measure of the *kinetic energy density* of the gas. In fact, pressure has the same dimensions (units) as kinetic energy density. Except for the factor of $2/3$, the two quantities would be numerically equal for our ideal gas.

Finally, we caution the reader that, most generally, pressure is neither a vector nor a scalar; it is a *tensor*. The flow of material at a given point in space has both direction and magnitude; hence, it is a vector. When a flow of material strikes a surface obliquely, it produces a vector force normal to the surface. These two vector directions require one to define pressure as a tensor. Within an isotropic gas in equilibrium, however, the tensor pressure becomes a scalar (it has no direction). It is this scalar quantity that we derived.

Average kinetic energy

The average kinetic energy E_{av} in the pressure expression (34) may be expressed as a function of temperature if the gas particles obey the M-B distribution. Take the product of the kinetic energy at momentum p and the 3-D probability $\mathcal{P}(p)\,dp$ (11) integrated over all p to obtain (Prob. 42)

$$E_{av} = \left(\frac{p^2}{2m}\right)_{av} = \int_0^\infty \frac{p^2}{2m}\mathcal{P}(p)\,dp = \frac{3}{2}kT, \quad \begin{array}{l}\text{(Average particle energy for} \\ \text{ideal monatomic gas; J)}\end{array} \tag{3.35}$$

where k is the Boltzmann constant and T is the absolute temperature. This result, $E_{av} = 3kT/2$, tells us that the average translational kinetic energy of the particles is a direct measure of the temperature of the gas.

It is convenient to recall that $kT = 1$ eV at $T = 11\,600$ K (AM, Chapter 2). Scaling from this, we find that a gas with $kT = 1$ keV has a temperature $T = 12 \times 10^6$ K. A 12-million K monatomic gas of ionized hydrogen will thus have an average particle energy of $E_{av} = 3kT/2 = 1.5$ keV.

It is not unreasonable to expect that collisions of electrons and protons in a gas of this temperature would lead to the emission of ~ 1 keV photons (x rays) if the electrons were prone to emit a large fraction of their kinetic energy in the form of a single photon (i.e., $h\nu \approx kT$). This is indeed the case for gases in thermal equilibrium such as for optically thin thermal bremsstrahlung (Chapter 5) and for blackbody radiation (Chapter 6). In brief, the temperature dictates the nature of the radiation. Gases at 6000 K emit optical light, and those at $\sim 10^7$ K emit primarily at x-ray frequencies.

The result (35) is valid for a monatomic gas. A monatomic gas is said to have three *degrees of freedom*, one each in the x, y, and z directions. Each degree of freedom will have an average kinetic energy per particle of

$$\frac{E_{av}}{3} = \frac{1}{2}kT.$$

(Average energy per degree of freedom) (3.36)

Diatomic molecules (like dumbbells), have *rotational kinetic energy* about two axes. (Rotation about the third axis, the axis of symmetry, has negligible moment inertia and carries no energy.) Thus, with the five degrees of freedom (three translational and two rotational), the average energy of a diatomic molecule is

$$E_{av} = \frac{5}{2}kT.$$

(Diatomic gas) (3.37)

Molecules are easily disassociated owing to their weak bonds. Thus they do not enter into our discussion of EOS in the hot interiors of stars.

Equation of state

The EOS for an ideal gas, $P(\rho, T)$, follows directly from the preceding calculations of pressure and average kinetic energy.

Physical form

The pressure may be obtained as a function of temperature simply by substituting the relation $E_{av} = 3kT/2$ (35) into the expression for pressure (34):

➡ $$P = nkT$$

(EOS; ideal gas) (3.38)

or

➡ $$P = \frac{\rho}{m_{av}}kT.$$

(3.39)

This is the desired EOS. The *average* particle mass, m_{av}, is the mass per unit volume ρ divided by the total number n of particles per unit volume, $m_{av} = \rho/n$. For a hydrogen plasma of protons and electrons, $m_{av} \approx m_{proton}/2$. If heavier elements are present, the total number density n (or the average mass) of all ions, atoms, and electrons must be used. The pressure P in (39) is found to be proportional to the mass density and to the temperature.

Equation (39) is valid if the densities are sufficiently low that the gas effectively consists of "pointlike" particles; that is, their interaction energies are small compared with their kinetic energies. This is typically the case for the gas in normal stars – both in the upper layers with low density and in the deeper denser and hotter layers.

In the latter case, the atoms are nearly completely ionized. As noted in Section 1, the electrostatic potential energies are modest compared with kinetic energies. Also, a proton or nucleus is many times smaller than the neutral atom (a factor of $\sim 10^5$ for hydrogen), and so the particle sizes remain largely pointlike relative to their separations. For these reasons, the EOS (39) remains approximately valid for particulate pressure even at the centers of sunlike stars.

Macroscopic form (ideal gas law)

Equation (39) is, in fact, the *ideal gas law*. If the gas contains μ moles, or μN_0 atoms (where N_0 is Avogadro's number), in volume V, then the number density is

$$n = \mu N_0 / V, \tag{3.40}$$

and, from (38),

$$P = nkT = (\mu N_0 / V)kT \tag{3.41}$$

$$PV = \mu N_0 kT. \tag{3.42}$$

Invoke the definition $N_0 k \equiv R$, the *universal gas constant*, to obtain

➡ $$PV = \mu RT, \qquad \text{(Ideal gas law)} \tag{3.43}$$

which is the usual form of the ideal gas law. The ideal gas law is simply the EOS for a gas of noninteracting particles that obey M-B statistics. In astrophysics, the commonly used forms are (38) and (39).

3.5   Photon gas

Consider a gas of photons in thermal equilibrium with the walls of the container or with the gas particles in its midst. In this case, photon momenta obey a "blackbody" distribution (6.6) rather than the M-B distribution, and the EOS, $P(\rho, T)$, is (6.43)

➡ $$P = \frac{aT^4}{3}, \qquad \text{(Pressure of a photon gas; N/m}^2) \tag{3.44}$$

where a is a constant, $a = 7.566 \times 10^{-16}$ N m^{-2} K^{-4}. The pressure depends only on the temperature; there is no density dependence. Also, the energy density u_{rad} of the photon gas is three times greater than P (6.25):

$$u_{\text{rad}} = aT^4. \qquad \text{(Energy density of a photon gas; J/m}^3) \tag{3.45}$$

A photon gas is fundamentally different from a classical particulate gas in that the number of photons is not conserved. It can absorb a large amount of energy for a given temperature rise, as if it had many degrees of freedom, because an increase in temperature results in the creation of large numbers of additional photons.

3.6   Degenerate electron gas

The concept of degeneracy in a very dense gas of Fermi–Dirac particles (fermions) is presented here, beginning with a graphical representation of a one-dimensional gas. The Fermi–Dirac distribution function is introduced. The equations of state for both nonrelativistic and relativistic degenerate gases are derived.

Fermions and bosons

Fundamental (elementary) particles may be grouped into two categories, called fermions and bosons, based on their quantum mechanical spin. On the quantum level, they behave quite differently, as we describe here.

Spin

Fundamental particles always carry an intrinsic angular momentum that can be zero or of magnitude $\gtrsim \hbar$, where $\hbar \equiv h/(2\pi)$ and h is the Planck constant. The value is the same for all particles of a given type but can differ for different types of particles. The quantum number that specifies the magnitude of the angular momentum is called *spin*; it approximates the angular momentum in units of $\hbar$. If the spin quantum number is S, the square of the angular momentum magnitude is $S(S+1)\hbar^2$. The spin S of a particle can only have a value that is zero or a multiple of $1/2$.

In addition, the projection of an angular momentum onto some defined axis is also quantized. Spin $1/2$ particles have two possible projections, $\pm\hbar/2$, referred to as "spinup" and "spindown." We discuss particle spin further in Section 10.3.

Particles with half-integer spins, $1/2, 3/2, 5/2, \ldots$, are Fermi–Dirac (F-D) particles or *fermions*. Electrons, protons, and neutrinos are examples, each of which has spin $1/2$. Nuclei with an odd number of *nucleons* (protons and neutrons) such as H^1 (the proton), ^{3}He, ^{7}Li, and ^{35}Cl are also fermions.

Particles with integer spin (e.g., 0, 1, 2, ...) are called *bosons*. Examples are photons with spin 1 and the nuclei ^{4}He and ^{12}C, each with spin zero. Fermions and bosons obey very different statistical rules on how they occupy quantum mechanical states.

Pauli exclusion principle

The state of a free particle in a one-dimensional space may be described by its position x and momentum p_x. This is a position in a two-dimensional phase space with coordinates x, p_x (Fig. 4); recall our discussion of phase space in Section 3. An element of area in this two-dimensional phase space has the dimensions of momentum times position (N s m); this is equivalent to energy times time (J s).

Quantum mechanics tells us that, on tiny scales, this phase space is "quantized"; it consists of cells or states, each of area equal to the Planck constant (Fig. 4); that is,

$$\Delta x \Delta p_x = h,$$

$$(3.46)$$

where $h = 6.63 \times 10^{-34}$ J s. The *Pauli exclusion principle* dictates that no more than one fermion of a given spin state can occupy a given state in our x, p_x space. Fermions of spin $1/2$ have two spin states. Thus, no more than two particles can occupy a given state of area h, and the two must have opposite spins.

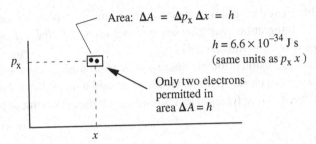

Figure 3.4. Two-dimensional phase space (x, p_x) for a one-dimensional gas showing an area of $h = 6.6 \times 10^{-34}$ J s (Planck constant). This area is a quantum state that can contain no more than two electrons.

In a 3-D space, phase space has six dimensions, x, y, z, p_x, p_y, p_z. A phase-space quantum state has a 6-D volume

$$\Delta x\, \Delta y\, \Delta z\, \Delta p_x\, \Delta p_y\, \Delta p_z = h^3 \qquad \text{(Volume of single quantum state in 6-D phase space; } \text{J}^3\, \text{s}^3) \qquad (3.47)$$

with units of $\text{J}^3\, \text{s}^3$. The Pauli exclusion principle allows only two half-spin particles in each such volume element – one with spinup and the other with spindown.

A familiar application of the exclusion principle is the population of the electron states of an atom. Only two electrons may occupy any state of an atom. Because the ground ($n = 1$) state of an atom has only one angular momentum state (s state), it contains only two electrons, one of each spin. The $n = 2$ level has one s and three p angular-momentum states, and so it can accommodate eight electrons.

Degeneracy

A gas of fermions with a sufficiently low temperature, or with a sufficiently high density, will fill all the lowest momentum states. Some of the fermions thus find themselves in higher momentum states than they would normally occupy. The momentum (or energy) distribution therefore differs markedly from the M-B distribution, for which the particles occupy only a very small portion of the lowest states. A gas filling all of the lowest energy states is called a degenerate gas.

The high momenta of particles in a degenerate gas result in an abnormally high pressure. It is this *degenerate pressure* of the constituent electrons that supports a white dwarf star from gravitational collapse. White dwarfs would not exist if this nonintuitive quantum mechanical pressure were unavailable.

Statistics and distribution functions

Particles that obey the Pauli exclusion principle are said to obey *Fermi–Dirac* (*F-D*) statistics. For an assembly of such particles, one can (with difficulty and not here) calculate the most probable arrangement of particles in the several states to obtain the expected distribution function $f(x, y, z, p_x, p_y, p_z)$ for an arbitrary degree of degeneracy. In contrast, the distribution for complete degeneracy is quite easily obtained. We will quote the general expression below, in (51), but will derive and use the simpler limiting expression to find the EOS.

We learned above (13) that the distribution function is the number of particles per unit phase-space volume, $(J\ s)^{-3}$. Integration over the space variables yields the momentum distribution needed for the pressure and thus the EOS.

Bosons do not obey the Pauli exclusion principle; there is no a priori limit on the number of bosons that may be placed in a given phase-space state. They are said to obey *Bose–Einstein* (B-E) statistics. The distribution function for massless bosons (photons) is presented, but not derived, in Section 6.1.

One-dimensional degeneracy

Here we present a graphical view of a 1-D Fermi–Dirac gas to illustrate the concept of degeneracy.

Plots of 2-D phase space

Consider a hypothetical 1-D gas of 37 electrons in which the particles move back and forth along the *x*-axis. At a given instant, the momenta and positions of the particles may be represented in phase space, as in Fig. 5a. Cells or states that particles may occupy are indicated with square regions, each of area 1.0 *h*. Each such cell may contain only 0 or 1 or 2 electrons. Each electron of the gas is indicated with a small circle.

The gas in Fig. 5a is in a nondegenerate condition as for an M-B gas. The greatest density of electrons is in the lowest-momentum state in accord with the one-dimensional Gaussian M-B distribution given in (4) and plotted in Fig. 1a. This distribution is also shown to the right of Fig. 5a.

Also indicated in Fig. 5a is the momentum p_x at which the kinetic energy is at its average value, $p_x^2/2m = kT/2$. The key to the nondegenerate character is that the lowest-momentum cells are not all fully occupied – some are empty and some contain only one electron. Ten electrons occupy the lowest layer of states in the figure, although these states have room for fourteen. In a truly nondegenerate situation, the electrons would occupy only a tiny fraction of these states – for example $<1\%$. The electrons would be at higher energies (momenta) owing to their thermal content.

Fermi momentum

Now let us move to degenerate conditions. If the gas of electrons in Fig. 5a loses or conducts heat away to other particles or to radiation and has no internal energy source, it will gradually cool down. This means that the electrons gradually decrease their kinetic energies and hence their momenta. They will settle down into the lower-momentum states and eventually, when cool enough, will pack themselves solidly into the lowest-momentum states (Fig. 5c).

The dashed line labeled $(E = kT/2)$ in Fig. 5c shows the momentum corresponding to the average kinetic energy that the electrons would ordinarily have at this low temperature. It is lower than before the cooling and is so low that many electrons can not find enough low-momentum states to represent the temperature properly. They thus find themselves stacked up into higher states with abnormally high momenta for the temperature. The momentum at the upper end of this distribution is called the *Fermi momentum*, p_F.

A gas in this state is highly *degenerate*. Note that the momentum distribution $\mathcal{P}(p_x)$ versus p_x is no longer a Gaussian function but is approximately rectangular, as shown to the right of

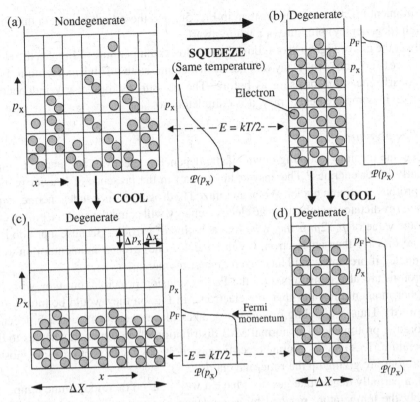

Figure 3.5. The two dimensions of phase space, p_x and x, for a one-dimensional gas of fermions containing 37 particles (shaded circles) for different temperatures and physical lengths ΔX. (a) Nondegenerate gas. (b, c, d) Completely degenerate gas. Shown are the distribution probabilities $\mathcal{P}(p_x)$ or, equivalently, the distribution function f from (4) and (52). The space is divided into individual quantum states $\Delta x \, \Delta p_x$, each of which may contain zero, one, or two particles of spin 1/2. The horizontal dashed lines indicate the momentum corresponding to particle energy $kT/2$, the average energy for a 1-D ideal (M-B) gas (i.e., for $p_x^2/2m = kT/2$). The electrons in a degenerate gas are forced to much higher average energies than $kT/2$ because of a shortage of available states at lower energies.

Fig. 5c. Only a few of the electrons, those at the upper (slightly rounded) end of the momentum distribution, are free to lose or gain energy as the temperature changes. At sufficiently low temperatures approaching absolute zero, the distribution would become quite rectangular with the upper momentum limit at p_F. The gas would be *completely degenerate*. Small temperature changes would have no effect on the particle distribution nor, consequently, on the pressure of the gas.

Compression and cooling

A compression of the gas, from Fig. 5($a \rightarrow b$), also yields degeneracy. The gas is squeezed by reducing the length of physical space Δx available to it. In this case the "temperature" is held constant. In the compressed state, very few particles can occupy the lowest-momentum states because there are not many of them. Particles are therefore forced to abnormally high energies for this (relatively high) temperature. Note again that they are packed solidly up to a Fermi momentum, which is higher than in Fig. 5c. It is these high momenta that give rise to the high (degeneracy) pressure of a white dwarf.

Finally, if the gas of Fig. 5b "cools down" to the lower temperature of Fig. 5c, the electrons have no place to go, and so they stay very nearly in the same states of high momentum (Fig. 5d), exerting nearly the same pressure as before. The pressure is thus nearly independent of temperature. In complete degeneracy, it is completely so.

Temperature

What does one mean by "cooling down" if the degenerate electrons are unable to move to significantly lower energies? The answer lies in part in the presence of heavier particles – namely, protons or atomic nuclei. When an ionized hydrogen gas is *non*degenerate, *equipartition of energy* dictates that the average electron energy will equal the average proton energy. The average value of $p_x^2/2m$ is the *same* for each kind of particle. Because a proton is 1836 times more massive than an electron, it will have 43 times greater momentum on average. Heavier nuclei, if present, would have even greater momenta.

The protons are also constrained by the Pauli exclusion principle, but they will occupy much higher momentum states than the electrons; in Fig. 5a they would be mostly off the page (upward). Thus, as cooling or squeezing takes place, the electrons reach degeneracy long before the protons do. The normal M-B distribution of proton energy serves to define the temperature. To say a degenerate electron gas cools down can mean, for the most part, that the protons are giving up their thermal energy.

In fact, a partially degenerate gas does have a well-defined thermodynamic temperature. It is simply the temperature reached by a nondegenerate gas (a "thermometer"), such as the aforementioned protons, that is in contact with it. The protons and electrons exchange energy through collisions. A collision can not occur if the momentum state of the electron after the postulated collision is already occupied. Because most such states are occupied in a gas approaching complete degeneracy, collisions between the electrons and protons are vanishingly rare.

The degenerate electrons thus can not easily share their energies with the proton thermometer. Hence, the thermometer will indicate a thermodynamic temperature much lower than the high electron momenta would otherwise imply.

Three-dimensional degeneracy

These ideas are extended to a 3-D degenerate gas, which is properly described in a 6-D phase space.

Fermi momentum

When a 3-D electron gas is completely degenerate, the particles are in the lowest possible momentum states, as in the 1-D case. The three momentum coordinates of 6-D phase space are shown in Fig. 6a. Recall that, in this representation, the momentum vector for any particle is a radial vector. Particles in a degenerate gas will seek out the unoccupied cells of lowest-momentum magnitude and hence will tend to arrange themselves in a sphere surrounding the origin, as shown.

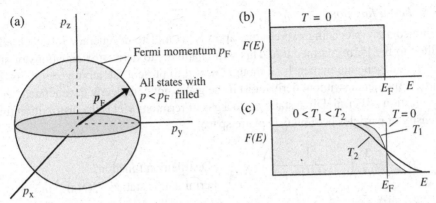

Figure 3.6. (a) Distribution of a completely degenerate three-dimensional gas in three-dimensional momentum space. The maximum (Fermi) momentum is p_F. The particles occupy the lowest possible momentum states (i.e., those in a sphere surrounding the origin). The physical (x, y, z) dimensions of phase space are not shown. (b) The Fermi function (52) for a completely degenerate gas $(T = 0)$ as a function of kinetic energy E. The Fermi energy is E_F. (c) The same function for two nonzero temperatures T_1 and T_2, where $0 < kT_1 < kT_2 \ll E_F$.

The radius of the sphere represents the momentum of the most energetic electrons in the degenerate gas; its magnitude is the Fermi momentum p_F. This radius is the 3-D analog to the heights of the occupied cells in the 1-D examples of Figs. 5b,c,d. The three spatial coordinates x, y, z of phase space are not shown in Fig. 6a because it is quite difficult to draw a 6-D figure in our 3-D world; it could take all morning.

The magnitude of the Fermi momentum of a completely degenerate gas of density n_e follows from writing the total number N_e of electrons of a gas in terms of p_F. The volume of the sphere in Fig. 6a is $(4/3)\pi p_F^3$. The 6-D phase-space volume is the product of this and the physical volume V. Thus, $V_{\text{phase}} = (4/3)\pi p_F^3 V$. (For example, if the physical volume is a sphere of radius R, then $V = 4\pi R^3/3$.) The number of allowed states is V_{phase} divided by the quantum volume element h^3 (47). Each such state can hold two electrons, and so

➡ $$N_e = \frac{2}{h^3} \frac{4}{3} \pi p_F^3 V, \qquad \text{(Number of electrons; complete degeneracy)} \quad (3.48)$$

where V is the physical volume. The electron number density n_e (in physical space) is

➡ $$n_e = \frac{N_e}{V} = \frac{2}{h^3} \frac{4}{3} \pi p_F^3. \qquad (3.49)$$

Solve (49) for p_F in terms of n_e:

➡ $$p_F = h \left(\frac{3}{8\pi} n_e \right)^{1/3}. \qquad \text{(Fermi momentum; N s)} \quad (3.50)$$

The Fermi momentum p_F is completely specified by the electron density n_e when temperatures are sufficiently low that degeneracy is nearly complete.

In defining n_e, we have assumed a uniform degree of degeneracy – namely, that p_F is the same at all positions of concern such as those in the central region of a star. A higher region in the star with lesser mass and electron densities would have a smaller Fermi momentum.

Fermi function

The immediately preceding expressions apply to complete degeneracy in which all states are filled to the Fermi momentum. This will enable us to derive the equations of state for completely degenerate matter. It is, though, useful to present the more general distribution of particle momenta (without derivation) if the gas is not completely degenerate.

Application of Fermi–Dirac statistics to a gas of fermions yields the distribution function, the density of particles per unit 6-D phase space,

$$f_{\text{F-D}} = \frac{2}{h^3} \left(\frac{1}{e^{\alpha + (E/kT)} + 1} \right),$$

(Distribution function;
Fermi–Dirac statistics; $(\text{N s})^{-3}\,\text{m}^{-3}$)   (3.51)

where E is the particle kinetic energy and $\alpha(n, T)$ is a complicated function of density n and temperature T. When considering the variation of $f_{\text{F-D}}$ with E for given n, T, the parameter α serves as a constant.

The distribution function $f_{\text{F-D}}$ must become the M-B distribution in the limit of nondegenerate conditions. Compare (51) with the distribution function for an M-B gas (8). The latter is proportional to $\exp(-E/kT)$ and is lacking the α parameter and also the "1" in the denominator. In the limit of nondegenerate conditions and given knowledge of $\alpha(n, T)$, the function does, in fact, become equal to the M-B distribution function.

The quotient $2/h^3$ in (51) is demonstrably the number of particles per unit phase-space volume expected for complete degeneracy. The parenthetical factor is thus an adjustment to this number. If the gas is almost completely degenerate, one can associate the parameter α in (51) with the negative of the *Fermi energy* E_F divided by kT. The factor in brackets is then called the *Fermi function* $F(E)$,

➡ $$F(E) \equiv \frac{1}{e^{(E - E_F)/kT} + 1},$$

(Fermi function; $kT \ll E_F$;
nearly complete degenerate)   (3.52)

which contains the energy E and temperature T dependence of the distribution function. For nearly complete degeneracy, the distribution function thus becomes

$$f_{\text{F-D}} = \frac{2}{h^3} F(E).$$ (Distribution function; nearly complete degeneracy)   (3.53)

Consider the Fermi function $F(E)$ (52) for $T = 0$. As E increases, the fraction $(E - E_F)/kT$ abruptly changes from $-\infty$ to $+\infty$ at $E = E_F$. The function $F(E)$ thus jumps from unity to zero at the Fermi energy; it is rectangular, as shown in Fig. 6b. The function exhibits the sharp cutoff expected for complete degeneracy.

The Fermi function is sketched for temperatures T_1 and T_2, both near zero, in Fig. 6c. The region occupied by particles is shaded for the $T = T_1$ case. The distribution (52) reaches 50% when the exponential equals unity at $E = E_F$. The lower states are fully occupied, and so only the very few electrons near E_F are free to change their energies. As temperature changes (in this regime of high degeneracy), both the distribution and average energy change only slowly.

Fermi energy

The Fermi energy corresponds to the Fermi momentum p_F. The nonrelativistic relation between the Fermi (kinetic) energy and momentum is $E_F = p_F^2/(2m)$. Relativistically, it is the total energy associated with the Fermi level U_F less the electron rest-mass energy mc^2. Use (7.20) to express U_F in terms of p_F and m:

$$E_F = U_F - mc^2 = \left(p_F^2 c^2 + m^2 c^4\right)^{1/2} - mc^2, \quad \text{(Fermi (kinetic) energy)} \qquad (3.54)$$

which reduces to the nonrelativistic value in the limit of low p_F. The Fermi energy is simply the highest particle energy encountered in a completely degenerate gas – namely, that of, a particle on the outer surface of the sphere of Fig. 6a. At extremely high energies, $E_F \approx p_F c$.

Pressures of electrons and protons

How do the pressures due to the electrons and protons in a hydrogen plasma compare? The pressures in the gases we have encountered are generally comparable to the kinetic energy densities. This is generally true regardless of whether the gas is degenerate or relativistic. Thus,

$$P \approx n E_{av}, \qquad \text{(N/m}^2) \qquad (3.55)$$

where n is the particle number density (m^{-3}) and E_{av} is the average of the translational kinetic energy E of a particle. The numerical coefficient in (55) would be 2/3 for a nonrelativistic gas of particles (34) and 1/3 for a relativistic gas of photons or particles; compare (44) and (45).

Consider a nondegenerate hydrogen plasma. Because the electrons and protons are in the same plasma, equipartition of energy dictates that the average kinetic energies of the electrons and the protons be equal. If the number densities n for each component are also equal, then, according to (55), the pressures due to the two components are equal. Thus, for the *non*degenerate hydrogen plasma,

$$P_{electrons} = P_{protons}. \qquad \text{(Nondegenerate hydrogen plasma)} \qquad (3.56)$$

As the electrons become degenerate, such as through squeezing (Fig. 5b), they are forced up to abnormally high momenta or energies. Because they started with the same energies as the protons, they now attain much higher energies than the protons on the average. Given that the pressure is generally about equal to kinetic energy density, the electrons in their degenerate state exert *greater* pressure than the (still nondegenerate) protons; that is,

$$P_{electrons} \gg P_{protons}. \qquad \text{(Hydrogen plasma; electrons degenerate;}$$
$$\text{protons nondegenerate)} \qquad (3.57)$$

This is generally true of electron-degenerate matter.

It is thus the pressure of the electrons that keeps a white dwarf from collapsing inwardly despite the tremendous inward pull of its own gravity. The pressure from protons and nuclei provides only insignificant support against collapse. Rather, it provides almost all of the inward gravitational pull.

Nonrelativistic EOS

The EOS, $P(\rho, T)$, for a degenerate electron gas of nonrelativistic particles is obtained here with a semiquantitative derivation.

Average kinetic energy

In the case of a nondegenerate, nonrelativistic gas, we found (34) that the pressure is 2/3 the kinetic energy density. We restate this here in terms of momentum,

$$P_e = \frac{2}{3} n_e E_{av} = \frac{2}{3} n_e \left(\frac{p^2}{2m_e} \right)_{av}, \qquad \text{(Nonrelativistic pressure)} \qquad (3.58)$$

where we have added the subscripts "e" to represent the electron, the primary source of degenerate pressure in white dwarfs.

Examination of the derivation of (58) in the discussion before (34) shows that this result is valid for any distribution of momenta of nonrelativistic particles. The coefficient 2/3 follows from geometric considerations that are independent of the momentum distribution. Thus, we obtain the EOS, $P(\rho, T)$, for a degenerate nonrelativistic gas simply by calculating the average of $(p^2/2m)$ for the appropriate momentum distribution – in this case the F-D distribution.

The average kinetic energy is obtained from an integration over all momenta. The volume of a momentum shell in phase space at momentum p in volume V of physical space is $4\pi p^2 dp V$. In the limit of complete degeneracy, the number of particles per unit volume of phase space is constant at $2/h^3$, which follows from our earlier arguments about quantum states (47) or from the distribution function (51).

The product of these two terms, $2/h^3$ and $4\pi p^2 dp V$, is the total number of electrons in a momentum-space shell. Divide by the total electron number N_e to obtain the fraction of electrons in the shell. Use this as the weighting factor in the calculation of the average, $(p^2/2m_e)_{av}$:

$$E_{av} = \frac{1}{N_e} \int_0^{p_F} \frac{p^2}{2m_e} \frac{2}{h^3} V 4\pi p^2 dp = \frac{1}{N_e} \frac{4\pi V}{m_e h^3} \frac{p_F^5}{5}. \qquad (3.59)$$

The integration is from zero momentum to the Fermi momentum p_F because, in complete degeneracy, there are no electrons at higher momentum than p_F. Eliminate $n_e = N_e/V$ with (49) to obtain

$$E_{av} = \frac{3}{5} \frac{p_F^2}{2m_e}. \qquad \begin{array}{l} \text{(Average kinetic energy; nonrelativistic,} \\ \text{complete degeneracy)} \end{array} \qquad (3.60)$$

The average kinetic energy is simply 3/5 of the kinetic energy associated with the maximum possible momentum, p_F.

Pressure

Substitution of (60) into (58) would yield the pressure P_e in terms of p_F and n_e, but these are related through (50). Eliminate p_F in (60) with (50) and then substitute the resultant E_{av} into the expression for P_e (58):

$$P_e = \frac{1}{20}\left(\frac{3}{\pi}\right)^{2/3}\frac{h^2}{m_e}n_e^{5/3}. \qquad \text{(Electron pressure)} \qquad (3.61)$$

The electron number density n_e can be expressed in terms of the total mass density ρ (kg/m^3) if we make use of the *electron molecular weight* discussed in Section 2.7, before (2.72), and formally defined here by

$$\mu_e \equiv \frac{1}{N_e}\sum_i \frac{m_i}{m_p}. \qquad \text{(Electron molecular weight; dimensionless)} \qquad (3.62)$$

This is the total mass per free electron in units of the proton mass, or, roughly, the number of nucleons per free electron. The summation is over all particles (atoms, electrons, protons, and heavier nuclei) in a given system, and N_e is the total number of free electrons in the system. For a completely ionized hydrogen plasma, the mass associated with each electron consists of one proton and one electron, and so $\mu_e \approx 1$.

The electron number density n_e may thus be written as the mass density ρ divided by $\mu_e m_p$, the mass per electron:

$$n_e = \frac{\rho}{\mu_e m_p}. \qquad \text{(m}^{-3}) \qquad (3.63)$$

Substitute (63) into (61) to obtain the (nonrelativistic) degenerate EOS:

$$\Rightarrow \quad P_e = \frac{1}{20}\left(\frac{3}{\pi}\right)^{2/3}\frac{h^2}{m_e}\left(\frac{\rho}{\mu_e m_p}\right)^{5/3}. \qquad \begin{array}{l}\text{(Nonrelativistic EOS;}\\ \text{complete degeneracy; N/m}^2)\end{array} \qquad (3.64)$$

Here, ρ is the total mass density (kg/m^3) of the matter, including nondegenerate protons and nuclei, and $\mu_e m_p$ is the mass per electron. This expression (64) is not valid if the electrons reach relativistic velocities (i.e., speeds approaching the speed of light $v \approx c$).

The result (64) may be summarized as

$$\Rightarrow \quad P_e \propto (n_e)^{5/3} \propto (\rho/\mu_e)^{5/3} \propto \rho^{5/3}. \qquad \text{(Nonrelativistic; } \mu_e \approx \text{constant)} \qquad (3.65)$$

In the last term just above, we assume that $\mu_e \approx$ constant for the conditions under consideration. The factor μ_e reflects the composition and ionization state of a gas and is thus relatively constant for a completely ionized gas of fixed composition. In this approximation, the pressure depends on the matter density but not on the temperature. This is in accord with our discussions above in which we noted that cooling a degenerate gas does not significantly reduce the average momentum of the particles; see Fig. 5b,d.

Relativistic EOS

If the electrons are forced up to such high momenta that they become relativistic, yet another EOS applies. This may be obtained in a manner similar to that for the nonrelativistic case. We express the pressure in terms of the average kinetic energy density and find the latter by integration over the momentum distribution.

For highly relativistic particles (i.e., for $v \approx c$), the kinetic energy and momentum are related as they are for a photon, namely $E \approx pc$. (In this case, the rest energies are negligible, and so the total and kinetic energies are comparable, $U \approx E$.) It is thus appropriate to use the relation between pressure P and energy density u for photons. From (44) and (45), we have

$$P = u/3. \qquad \text{(Relativistic particles)} \qquad (3.66)$$

As for an ideal gas, the factor $1/3$ follows from geometric considerations (Prob. 41); it is independent of the distribution of momenta. The electron pressure may thus be written

$$P_e = \frac{u}{3} = \frac{n_e E_{av}}{3} \approx \frac{n_e (pc)_{av}}{3}. \qquad (3.67)$$

The average electron energy $(pc)_{av}$ is obtained from an integral similar to that for the nonrelativistic case (59), and again, we eliminate $n_e = N_e/V$ with (49):

$$pc_{av} = \frac{1}{N_e} \int_0^{p_F} pc \frac{2}{h^3} V 4\pi p^2 dp = \frac{3}{4} cp_F. \qquad \text{(Average energy; J)} \qquad (3.68)$$

Again eliminate p_F in favor of n_e with (50) and substitute this result into the pressure equation (67). Finally, write n_e in terms of the mass density ρ and the electron molecular weight (63) to obtain the desired EOS:

➡ $$P_e = \frac{1}{8} \left(\frac{3}{\pi} \right)^{1/3} ch \left(\frac{\rho}{\mu_e m_p} \right)^{4/3} \qquad \begin{array}{l} \text{(Relativistic EOS;} \\ \text{complete degeneracy; N/m}^2) \end{array} \qquad (3.69)$$

➡ $$P_e \propto (\rho/\mu_e)^{4/3} \propto \rho^{4/3}. \qquad (\mu_e \approx \text{constant}) \qquad (3.70)$$

This expression is valid for highly relativistic and completely degenerate electrons. Again, if μ_e is independent of temperature, the EOS is independent of temperature. The pressure is maintained even as the gas cools.

The relativistic EOS carries a lower power of the mass density, $\rho^{4/3}$, than does the nonrelativistic case, $\rho^{5/3}$. It is said to be a "softer" EOS than the nonrelativistic expression. If the matter is squeezed to a higher density, the pressure rises less rapidly in the $\rho^{4/3}$ case than for the $\rho^{5/3}$ case. This can lead to the collapse of a white dwarf star, as described in Section 4.4.

Summary of EOS

The equations of state derived above are summarized here. In each case the pressure is given as a function of the local mass density and temperature. For complete degeneracy, the pressure turns out to be quite independent of temperature. The EOS for a *non*degenerate (ideal) gas of particles obeying Maxwell–Boltzmann statistics is, from (39),

$$P = \frac{\rho}{m_{av}} kT, \qquad \text{(Nondegenerate particle pressure; Pa)} \qquad (3.71)$$

where m_{av} is the average particle mass, ρ is the mass density (kg/m^3), and the pressure P may be expressed in pascals (Pa = N/m^2).

A gas of photons has EOS (44),

$$P = \frac{aT^4}{3}, \qquad \text{(Photon gas; Pa)} \qquad (3.72)$$

where a is a constant, $a = 4\sigma/c = 7.566 \times 10^{-16} \, \text{N m}^{-2} \, \text{K}^{-4}$. Note that in the latter case there is no dependence on mass density ρ; the pressure depends only on temperature. If both photons and particles are present (as they surely will be), and if both contribute significantly to the pressure, the proper expression for the nondegenerate case is the sum

$$P = \frac{\rho}{m_{av}} kT + \frac{aT^4}{3}. \qquad \text{(Photons and nondegenerate particles; Pa)} \qquad (3.73)$$

The equations of state for completely degenerate electrons in the nonrelativistic and relativistic cases, are, from (64) and (69),

$$P_e = \frac{1}{20} \left(\frac{3}{\pi}\right)^{2/3} \frac{h^2}{m_e} \left(\frac{\rho}{\mu_e m_p}\right)^{5/3} \qquad \text{(Nonrelativistic;}$$
$$\qquad\qquad\qquad\qquad\qquad\qquad\qquad\qquad \text{complete degeneracy; Pa)}$$
$$= 9.92 \times 10^6 \left(\frac{\rho}{\mu_e}\right)^{5/3} \qquad\qquad\qquad\qquad\qquad (3.74)$$

and

$$P_e = \frac{1}{8} \left(\frac{3}{\pi}\right)^{1/3} ch \left(\frac{\rho}{\mu_e m_p}\right)^{4/3} \qquad \text{(Totally relativistic;}$$
$$\qquad\qquad\qquad\qquad\qquad\qquad\qquad\qquad \text{complete degeneracy; Pa)}$$
$$= 1.232 \times 10^{10} \left(\frac{\rho}{\mu_e}\right)^{4/3}. \qquad\qquad\qquad\qquad\qquad (3.75)$$

The relativistic equation is softer than the nonrelativistic equation. If the electrons at the core of a white dwarf reach relativistic velocities, and hence obey (75), they become incapable of supporting the white dwarf.

The regions of temperature and mass density of stellar matter in which the several equations of state (71)–(75) are generally operative are illustrated in Fig. 7. At high temperatures and low densities, radiation dominates the pressure. At intermediate regions, particles provide most of the pressure. At high densities and low temperatures, the matter becomes degenerate and electron pressure becomes important. At extremely high densities, the electrons become relativistic, and the softer equation of state dominates.

Each point in the diagram represents a given density and temperature, and for each there is an appropriate EOS. The equations presented here and in the figure are for ideal limiting cases. In intermediate regions of partially relativistic particles or partial degeneracy, other intermediate forms are required. Keep in mind that the EOS within a star depends on the local density and temperature. Thus the EOS will vary from point to point within a star. For a spherically symmetric (nonrotating) star, the variation would be in the radial direction only.

The nondegenerate equation (71) applies in the inner regions of many normal stars, including the sun, which is represented by the fine dashed line in Fig. 7. Radiation pressure must be taken into account (73) in some of the hottest, most massive stars. Electron degeneracy

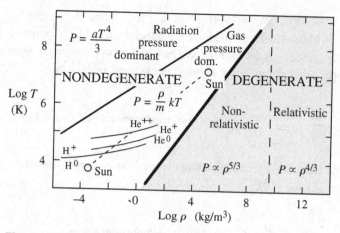

Figure 3.7. Temperature–density log-log plot (base 10) showing the regions where the various equations of state (71)–(75) in stellar material are operative. The boundaries between the several regions are lines of equal pressures. Their locations depend somewhat on the chemical composition through the electron molecular weight μ_e. The equations of state given are limiting cases; see text. The locus of points for the sun (dashed line) from its surface (lower left) to its center (upper right) are indicated together with the boundaries between regions of neutral and ionized hydrogen and between regions of neutral, singly ionized, and doubly ionized helium. [Adapted from M. Schwarzschild, *Structure and Evolution of the Stars*, Princeton University Press 1958]

pressure (74) supports white dwarf stars. The EOS at the center of a white dwarf star approaches the relativistic equation (75) as the star's mass approaches the Chandrasekhar limit of $\sim 1.4\,M_\odot$. Neutron stars are supported primarily by repulsive nuclear forces, but the degeneracy pressure of neutrons plays a role. The failure of all support mechanisms leads to the black hole phenomenon.

Problems

3.2   Maxwell–Boltzmann distribution

Problem 3.21. (a) Verify that the one-dimensional M-B probability $\mathcal{P}(p_x)\mathrm{d}p_x$ (4) integrated over all possible p_x is unity. Note that p_x can take on negative as well as positive values. (b) Similarly, demonstrate that the integrated probability of obtaining any given (3-D) momentum *magnitude* p is unity, as specified in (12). Why are the limits of integration different from those applied for part (a)?

Problem 3.22. Find the most probable momentum (magnitude) p of a gas that obeys the M-B distribution in terms of the temperature and other constants. Compare your answer with the momentum corresponding to the average kinetic energy of particles in a monatomic gas. [Ans. $\sim (2mkT)^{1/2}$]

3.3   Phase-space distribution function

Problem 3.31. (a) Confirm that the dimensions of the left and right sides of (23) agree, and do the same for (26). (b) The distribution function for a sample of photons is $f = Ap^{-4}$, where A is a constant and the photon momentum is $p = h\nu/c$. What is the directional energy flux $\mathcal{F}$(W m^{-2} sr^{-1}) over the frequency range ν_1 to $10\nu_1$? Hint: start with (26). [Ans. –; $Ac^2 \ln 10$]

3.4 Ideal gas

Problem 3.41. Calculate the pressure of a gas in terms of particle density n, particle mass m, and speed v to obtain the result (34). Assume the particles move in all directions with equal probability but that all have the same speed v. Set up and evaluate an integral that takes into account the contributions to the pressure by particles approaching the wall at all angles of incidence. Comment on how to take into account a distribution of speeds. Hint: what is the momentum transferred to the wall by a particle of momentum p impinging elastically at angle θ from the normal?

Problem 3.42. (a) Demonstrate that the average kinetic energy associated with one of the degrees of freedom of a monatomic gas such as motion in the x direction is $kT/2$. Use the one-dimensional probability function to calculate the average of $p_x^2/2m$. (b) Argue from this result that the average of the total energy $p^2/2m$ should be $3kT/2$ for a monatomic gas. (c) Demonstrate this directly by evaluating the integral (35).

Problem 3.43. (a) The mass density of the sun at its center is 1.5×10^5 kg/m^3, and the central temperature is 1.6×10^7 K. What is the pressure at the center? Assume a gas of pure ionized hydrogen. Express your answer in pascals (Pa or N/m^2) and in (earth) atmospheres. The atmospheric sea-level pressure is 1.01×10^5 Pa. (b) The earth's atmosphere consists mostly of N$_2$ and O$_2$, ∼80% and ∼20%, respectively, by number. Given the atmospheric temperature (∼300 K) and the pressure at sea level just given, what is the sea-level mass density (kg/m^3) of the atmosphere? What is the ratio of the mass density at the center of the sun to that at the bottom of the earth's atmosphere? [Ans. ∼10^{12} atm; ratio ∼10^5]

Problem 3.44. Demonstrate that the units of pressure (N/m^2) and kinetic energy density (J/m^3) are identical.

Problem 3.45. Derive the EOS for an ideal gas of diatomic particles – molecules with two identical atoms, like dumbbells – where each atom has, on average, an additional $1/2\,kT$ of rotational kinetic energy for each of the two ways it can rotate about its center of mass with significant kinetic energy. Be sure that you can justify any point at which your derivation differs from that for a monatomic gas as given in the text; think carefully about the meanings of temperature and pressure. [Ans. $P \propto nkT$]

3.5 Photon gas

Problem 3.51. (a) Show that the energy density $u(R)$ (J/m^3) of a photon gas expanding adiabatically (with no loss or gain of heat) in a spherical mirrored room of radius R decreases as R^{-4}. Consider a photon of initial frequency ν_0 that crosses back and forth across the full diameter $2R$. Let the mirrored walls of the container move outward at speed $v \ll c$ relative to the center. Start by finding the frequency change of the photon due to a single reflection; consider the Doppler effects for an observer at rest on the receding wall and another at rest with respect to the center of the sphere. From this, find how the frequency changes with R, and then find $u(R)$. (b) How does the temperature of the radiation depend on R? This is one way of understanding the cooling of the cosmic microwave background. [Ans. –; R^{-1}]

3.6 Degenerate electron gas

Problem 3.61. (a) Qualitatively discuss the ideas underlying the two relativistic expressions in (54) for kinetic energy E_F. If necessary, refer to Chapter 7. (b) Demonstrate that, at $p_F \ll mc$, the expression (54) yields the classical result $E_F = p_F^2/2m$.

Problem 3.62. What is the approximate Fermi energy E_F for the electrons in a white dwarf star of radius $R = 10^{-2} R_{\odot}$ and $M = 1 M_{\odot}$, in units of MeV? Assume the star consists of heavy elements so that there is about one electron for every two nucleons. Use a whole-star (average) value for the electron density. Are the particles relativistic (i.e., is $E_F \gtrsim mc^2$)? Hint: first, find p_F. [Ans. $\sim$0.2 MeV]

Problem 3.63. Derive, with closed book, the nonrelativistic EOS (64) for a degenerate electron gas starting with the pressure expression (58).

Problem 3.64. Derive the relativistic EOS (69) starting with pressure-to-energy-density relation (66); include the steps skipped in the text.

Problem 3.65. Confirm that (64) and (69) have units of pressure (N/m^2).

Problem 3.66. The vertical dashed line in Figure 7 at $\log \rho = 9.58$ (kg/m^3) represents the place where the nonrelativistic and relativistic electron pressures, P_N and P_R, are equal. Find the value of μ_e adopted for this figure. What can you say about the composition of the gas? [Ans. $\sim$2]

4

Stellar structure and evolution

What we learn in this chapter

The structure of a star may be modeled with the aid of the four fundamental equations of **stellar structure** – namely, **mass and luminosity distributions**, **hydrostatic equilibrium** and **radiative** (or **convective**) **transport** – together with the **equation of state** (EOS) and several secondary equations. Each of these describes some essential applicable **physics that must be obeyed at each point of the stellar interior**. Given a stellar mass and elemental composition, these equations can be used to create a **model of a star** that yields the radial distribution of mass density, pressure, temperature, and luminosity within the star. The **adiabatic constraint** leads to a condition for **convective energy transport** in stars.

Hydrogen-burning stars of different masses lie along the **main sequence** on a **Hertzsprung–Russell (H-R) diagram**, which is a plot of luminosity versus temperature. Stars of high mass burn their hydrogen rapidly, in millions of years, whereas stars of low mass can burn hydrogen stably for billions of years. After leaving the main sequence, a star becomes a **red giant** or a **supergiant** and eventually evolves to become a **white dwarf, neutron star**, or **black hole**. A group of stars created together from a single collapsing cloud at a given time is known as a **globular cluster** or an **open cluster**. H-R diagrams of cluster stars dramatically exhibit the **differing ages** of clusters and the results of **stellar evolution**. **Scaling laws** or **homology transformations**, derived from the fundamental equations, yield the variation of temperature, pressure, and luminosity as a function of stellar mass and radius.

Compact stars have very different characteristics. The approximate **mass-radius relation** of **white dwarf stars** follows directly from the nonrelativistic EOS and the equation of hydrostatic equilibrium. The relativistic degenerate EOS yields the maximum possible mass of such a star, the **Chandrasekhar limit**. **Neutron stars** are found as **radio pulsars** and in accreting, x-ray–emitting **neutron-star binaries**. **Black holes** are also believed to reside in some **x-ray binaries** and in the cores of **active galaxies**. The properties of black holes are few: mass, charge, and angular momentum. The **mass** of a compact object yields an **event horizon** (**Schwarzschild radius**), and the object's **angular momentum** determines the radius of the **innermost stable orbit** of matter orbiting it. As matter collapses to a singularity, the **Planck length** defines the size scale at which **gravity confronts quantum mechanics**.

The **evolution of a binary stellar system** is affected by the **evolution of the constituent stars, gravitational radiation, accretion** of matter from one star to the other, **stellar winds**, and the **supernova** of one or both of the partners. The **effective equipotentials** in the rotating frame of reference define the **Roche lobes** and the **Lagrangian points** that control the accretion process. Accretion with mass and angular momentum conservation leads to a **changing separation** of the stars. **Sudden mass loss** due to a supernova changes the semimajor axis and eccentricity of the orbit. A possible evolutionary path of two normal stars to a **high-mass x-ray binary** (HMXB) and thereafter to a **binary radio pulsar** is presented. Similarly, the path to a **low-mass x-ray binary** (LMXB) is given. The **spinup of a neutron star** to millisecond periods occurs in an LMXB, as evidenced by the **accreting millisecond x-ray source** SAX J1808-3658.

4.1 ' Introduction

The equations of state discussed in Chapter 3 describe the relations between pressure, temperature, and matter density at any given point in a star. This in itself does not indicate how a star is constructed. There are four fundamental equations and several auxiliary "secondary" equations based on physical laws that must also be satisfied throughout the star. One of these is the equation of hydrostatic equilibrium presented in Chapter 2. With these, a model star can be constructed given only a total mass and chemical composition.

We present the physical equations as a way of broadly illustrating the physics underlying the structure of a star. We indicate briefly how they lead to an actual model of a star such as the sun and how such stars evolve. The scaling of star parameters for stars of different masses is presented. Physical laws are applied to degenerate stars to give sizes and maximum masses, and characteristics of black holes are described. Finally, fundamentals of binary-system evolution are developed.

For most of this chapter, we consider spherically symmetric and nonrotating stars with uniform composition throughout. Nevertheless, it is important to remember that stellar structure and evolution can be affected significantly by star rotation, a binary companion, and the depletion of elements at the core caused by nuclear burning.

4.2 Equations of stellar structure

We present here the four fundamental equations of stellar structure and the secondary equations that must be satisfied in a stellar model. One of the fundamental equations has two forms determined by how energy is transported through the inner layers of the star. The secondary equations pertain to the state of the stellar gas.

Fundamental equations

Hydrostatic equilibrium

The first of the fundamental equations is the equation of *hydrostatic equilibrium* (2.12), which was developed in Section 2.3:

$$\frac{\mathrm{d}P(r)}{\mathrm{d}r} = -\frac{G\mathfrak{M}(r)\rho(r)}{r^2} \qquad \text{(Hydrostatic equilibrium)} \qquad (4.1)$$
$$= -g(r)\rho(r).$$

This is a mathematical statement that, at any point in the star, the net force on a volume element of gas must be zero. Here, $P(r)$ is the pressure (N/m^2) at radius r in the star, $\rho(r)$ the mass density (kg/m^3), and $\mathfrak{M}(r)$ the mass interior to the radius r. The negative sign tells us that, as expected, the pressure decreases with increasing radius within the star.

Mass distribution

The definition of $\mathfrak{M}(r)$ just stated gives the mass distribution equation. Consider a spherical shell of stellar material of radius r and thickness dr and write the associated element of included mass, $d\mathfrak{M} = 4\pi r^2 \rho \, dr$. Divide by dr to obtain

$$\frac{d\mathfrak{M}(r)}{dr} = 4\pi r^2 \rho(r). \qquad \text{(Mass distribution)} \qquad (4.2)$$

This is the second fundamental equation. It describes the distribution of mass within the star.

Luminosity distribution

Consider the energy flow from the center of the star to its surface. Let $\mathscr{L}(r)$ be the luminosity (W) at radius r – that is, the power (W) passing through a mathematical spherical surface of radius r. According to this definition, the stellar luminosity at the outer surface of a star of radius R is $L = \mathscr{L}(R)$.

At each position in the star, or in any radial shell, there is the possibility of power generation by nuclear reactions. This is described with the energy-generation function $\epsilon(\rho, T)$, which gives the power generated per kilogram (W/kg); see (2.66). Because ρ and T are both functions of radius r in the star, the parameter may be written as a function of r – namely, $\epsilon(r)$.

Write the change in luminosity $d\mathscr{L}(r)$ across a spherical shell of thickness dr due to nuclear burning therein and divide by dr to obtain

$$\frac{d\mathscr{L}(r)}{dr} = 4\pi r^2 \rho(r) \, \epsilon(r). \qquad \begin{array}{l}\text{(Luminosity distribution;}\\ \text{conservation of energy)}\end{array} \qquad (4.3)$$

This is a statement of energy conservation. It is the third of the four fundamental equations. As stated previously (Table 2.2), the energy source function can be parameterized as $\epsilon(\rho, T) \propto \rho T^\beta$, where $\beta \approx 4$ for the pp process and $\beta \approx 15$ for the CNO process.

Radiation transport

The fourth fundamental equation is an expression for the radial gradient of the temperature $T(r)$, namely, $dT(r)/dr$. The content of the equation depends on whether the transport of energy toward the surface is by the transport of radiation (photons) or by convection. The former takes place through many absorptions and reemissions as the photons work their way toward the stellar surface and the latter through bulk motions of gas. We first address the former.

Consider the fluxes from two spherical surfaces separated by one mean free path, $\Delta r = (\kappa \rho)^{-1}$ (Fig. 1), where κ is the opacity (m^2/kg). One surface is at position r and the other is at $r + \Delta r$. The two have slightly different temperatures, the outer being cooler, and they

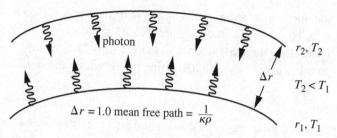

Figure 4.1. Radiation transport in the interior of a star. Two spherical segments of imaginary surfaces radiate toward each other as blackbody surfaces, $\mathscr{F} = \sigma T^4$. If the inner surface is hotter than the outer surface (as in a star), there is a net flow of radiation in the outward direction.

radiate toward one another as blackbodies with fluxes $\mathscr{F} = \sigma T^4$ (W/m^2) (6.18). The photons arriving at each layer are absorbed or scattered there (on average) because the two layers are separated by one mean free path. The lower, hotter layer radiates power upward and receives somewhat less power back from the upper, cooler layer.

There remains, therefore, a net outward flow of energy from the lower layer, $r = r_1$. The net outward flux from 1 m^2 at radius r is

$$\mathscr{F}_{\text{net}} = \sigma T_1^4 - \sigma T_2^4, \qquad\qquad \text{(W/m}^2\text{)} \qquad (4.4)$$

and the flux over the entire spherical surface is the luminosity at radius r,

$$\mathscr{L}_{\text{rad}}(r) = -4\pi r^2 \left(\sigma T_2^4 - \sigma T_1^4\right). \qquad\qquad (4.5)$$

Rewrite (5) for an arbitrary r in differential form as

$$\mathscr{L}_{\text{rad}}(r) \approx -4\pi r^2 \frac{\mathrm{d}(\sigma T^4)}{\mathrm{d}r} \Delta r, \qquad\qquad (4.6)$$

where Δr is the distance between the surfaces.

Finally, substitute the mean free path, $\Delta r = (\kappa\rho)^{-1}$, into (6), take the derivative, and solve for the temperature gradient. Our simplifications cause the result (Prob. 22) to be in error by $\sim$30%. The correct result is

➡ $$\left(\frac{\mathrm{d}T(r)}{\mathrm{d}r}\right)_{\text{rad}} = -\frac{3}{64\pi\sigma} \frac{\kappa(r)\rho(r)}{(T(r))^3 r^2} \mathscr{L}(r), \qquad \text{(Radiation transport; K/m)} \qquad (4.7)$$

where $\sigma = 5.67 \times 10^{-8}$ W m^{-2} K^{-4} is the Stefan–Boltzmann constant, and κ is the opacity, which is the cross section per kilogram (m^2/kg) for absorption, scattering, or both.

The radiation transport equation tells us that a high luminosity $\mathscr{L}(r)$ at some r requires a large temperature gradient to drive the energy across the layer Δr. A high opacity, κ, also leads to a high temperature gradient for a given luminosity. This is similar to an electrical circuit that, from Ohm's law, produces a high electric potential gradient if either the resistance ("κ") or the current ("$\mathscr{L}$") is high. (See AM, Chapter 10 for more on opacity.)

Convective transport

The alternative energy transport mechanism, convection, entails energy transfer by means of rising bubbles of hot gas. Upon reaching the surface, the bubbles lose energy via radiation,

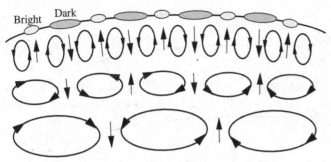

Figure 4.2. Sketch showing an idealized view of several layers of convective loops in the convective region of the sun. The gas radiates energy rapidly upon reaching the surface. The surface therefore exhibits hot and cooler regions at the positions of rising and descending gas, respectively. These regions appear as bright and dark granulations in photographs (Fig. 3). The structure is highly chaotic, like boiling cereal.

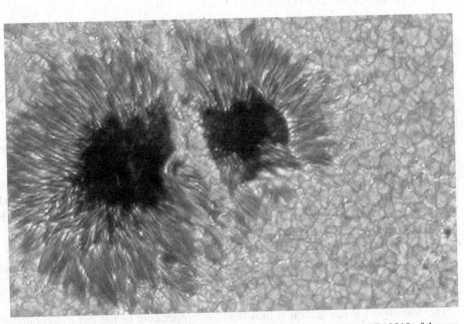

Figure 4.3. High-resolution photograph of 49 × 77″ portion of active region AR10019 of the sun taken in the G band (581 nm) on 3 July 2002. It shows the dark (cooler) sunspots, penumbral filaments surrounding them, and the churning, boiling granulations due to convective bubbles. The granules are about 1 to 2″ in size corresponding to ∼1000 km, which is comparable to the size of Texas. The patterns change on time scales of a few minutes. [Dutch Open Telescope, DOT]

cool rapidly, and then sink to lower levels, where they again absorb heat and rise again. This leads to circulations such as those illustrated in Fig. 2, which yield hot and cool regions on the surface. Convection cells are visible in high-resolution photographs of the solar surface (Fig. 3).

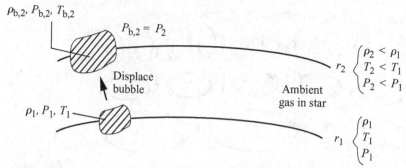

Figure 4.4. Displacement of gas bubble in a star. When an element of gas undergoes a fluctuation upward to a level of lower pressure and lesser temperature, it will quickly adjust to the lower pressure of its new position, $P_{b,2} = P_2$, but heat is not readily removed from it. The expansion is thus adiabatic. If the new bubble density $\rho_{b,2}$ is less than the density ρ_2 of the surrounding ambient gas, the bubble continues to float upward and convection takes place.

Condition for convection

Consider a star governed by radiation transport in which the temperature decreases with height (radius r); that is, $(dT/dr)_{\text{rad}} < 0$. Consider a small volume ("bubble") of gas at radius r_1 (Fig. 4). Let it be in equilibrium with its surroundings with the same mass density ρ_1 and temperature T_1.

Now let the bubble be perturbed upward (outward to r_2) to a slightly cooler level in the star while remaining in pressure equilibrium with its surroundings. As it rises to a level of lower pressure, it will expand and cool. If the rise is sufficiently rapid, negligible heat will leave or enter it, and the expansion, by definition, will be adiabatic, $\delta Q = 0$. We previously encountered adiabatic state changes in the Carnot cycle (Fig. 2.8).

The bubble and its surroundings are both cooler at the higher level than they were at the lower level. The fate of the bubble depends on which is cooler at the higher level – the bubble or the surroundings. If the bubble is cooler than its surroundings, it will be denser than the surroundings (given equal pressures) according to the nondegenerate EOS, $P \propto \rho T$. Being denser, it will sink back to its equilibrium position, and convection will not occur.

If, on the other hand, the temperature of the bubble at level r_2 is greater than that of its surroundings, it will be less dense and will continue to float upward. Convection does take place.

The condition for convection is thus that the temperature of the gas in the bubble under adiabatic expansion must decrease less with increasing stellar radius than the surroundings do. We take the surroundings to be governed by radiation transport (7), the competitive process. The condition for convection just stated reads, right to left, as

$$\left|\frac{dT}{dr}\right|_{\text{rad}} > \left|\frac{dT}{dr}\right|_{\text{adiab}}. \qquad \text{(Condition for convection)} \qquad (4.8)$$

The adiabatic temperature gradient is independent of luminosity at a given location r, as we demonstrate below; see (13). In contrast, the radiative gradient (7) increases with luminosity. Thus, according to (8), reading left to right, convection will occur if the luminosity $\mathcal{L}(r)$ being driven through the star at r is so large that it would require a radiative temperature gradient of magnitude greater than that of the adiabatic gradient.

Convection is a much more efficient process for transferring energy than radiation transport. Thus, if (8) is satisfied at some r, convection will become the dominant energy transfer mechanism in that region, and the temperature gradient will have the adiabatic value. In fact, it turns out that the gradient near the surfaces of stars like the sun, with convection in their outer parts, can actually exceed the adiabatic value (i.e., the gradient is *super-adiabatic*).

Adiabatic temperature gradient

Here we calculate the temperature gradient for a rising gas bubble undergoing adiabatic expansion while in pressure equilibrium with its surroundings; its pressure thus conforms to hydrostatic equilibrium (1). The temperature gradient dT/dr may be expanded as follows:

$$\left(\frac{dT}{dr}\right)_{adiab} = \left(\frac{\partial T}{\partial P}\right)_{adiab}\frac{dP}{dr} + \left(\frac{\partial T}{\partial S}\right)_{adiab}\frac{dS}{dr}$$

$$= \left(\frac{dT}{dP}\right)_{adiab}\frac{dP}{dr}, \tag{4.9}$$

where S is another state variable, the *entropy*. The definition of entropy change is $dS \equiv \delta Q/T$, and the adiabatic condition is $\delta Q = 0$; the rightmost fraction is thus zero, $(dS/dr)_{adiab} = 0$. This statement is implicit in the fact that T is a function of only one state variable, namely P, the other being eliminated by the adiabatic constraint, $PV^{\gamma} = $ constant. The partial derivative $\partial T/\partial P$ (9) is thus equivalent to the total derivative dT/dP, as given in the second line of (9).

We now evaluate the $(dT/dP)_{adiab}$ term in (9). It can be shown (Prob. 2.81) that a gas sample that expands or contracts with no heat input or removal ($\delta Q = 0$) obeys the constraint

$$PV^{\gamma} = c_1; \quad \gamma \equiv \frac{C_P}{C_V}, \qquad \text{(Adiabatic constraint)} \tag{4.10}$$

where c_1 is a constant and γ is the ratio of the specific heats for state changes with fixed pressure and with fixed volume:

$$C_P \equiv |\delta Q/dT|_P; \quad C_V \equiv |\delta Q/dT|_V. \qquad \begin{array}{l}\text{(Specific heat definitions;}\\ \text{J K}^{-1}\text{ mole}^{-1}\text{)}\end{array} \tag{4.11}$$

Eliminate V in (10) with the ideal gas law for one mole, $PV = RT$ (3.43), take the derivative dT/dP, and then eliminate c_1 with (10) to find (Prob. 23)

$$\left(\frac{dT}{dP}\right)_{adiab} = \left(1 - \frac{1}{\gamma}\right)\frac{T}{P}. \tag{4.12}$$

The dP/dr term in (9) has the hydrostatic equilibrium value (1). Substitute (1) and (12) into (9) to find the equation for convective transport,

$$\begin{aligned}\Rightarrow \quad \left(\frac{dT(r)}{dr}\right)_{adiab} &= -\left(1 - \frac{1}{\gamma}\right)\frac{T}{P}\frac{G\mathfrak{M}(r)\rho(r)}{r^2}\\ &= -\left(1 - \frac{1}{\gamma}\right)\frac{m_{av}}{k}g(r), \qquad \text{(Convective transport)}\end{aligned} \tag{4.13}$$

where γ is the ratio of specific heats (10), the product ρT was eliminated with the ideal gas law, $P = \rho kT/m_{av}$ (3.39), and the gravitational acceleration $g(r) \equiv G\mathfrak{M}/r^2$ (2.13) was invoked. This is the gradient that must be compared with that of radiation transport (7) in determining the nature of the energy transfer. As anticipated in the discussion after (8), it is independent of luminosity.

The condition for convection may be satisfied in some regions of a star and not in others. For example, energy transfer in the sun is radiative in the inner $0.7\,R_\odot$ and convective in the outer $0.3\,R_\odot$. In the outer parts of the sun, where there is no nuclear burning, $\mathcal{L}(r)$ does not vary with radius but temperature is decreasing. The strong temperature dependence, $\propto -T^{-3}$, of the radiative temperature gradient (7) becomes sufficient at some radius to drive the gradient magnitude above that of the adiabatic gradient magnitude (13). At this point, the convection condition (8) is satisfied, convection occurs, and the gradient is held at or near the adiabatic value.

Secondary equations

Secondary equations of stellar structure describe the states of a star's interior gases. They too must be satisfied at each point in the star. One is the EOS for sunlike stars when radiation pressure is negligible (3.39),

$$\Rightarrow \qquad P = \frac{\rho}{m_{\rm av}}kT. \qquad\qquad \text{(Equation of state)} \qquad (4.14)$$

In addition, the following quantities are also needed, each as a function of ρ, T, and the chemical composition of the gas: the average particle mass $m_{\rm av}(\rho, T, {\rm comp})$, the energy source function $\epsilon(\rho, T, {\rm comp})$ (W/kg), the opacity $\kappa(\rho, T, {\rm comp})$ (m^2/kg), and the ratio of specific heats, $\gamma(\rho, T, {\rm comp})$. Keep in mind that the EOS (14) is also a function of composition through $m_{\rm av}$.

The Saha equation is required for the calculation of opacities contributed by different elements in the star. The opacity to photons varies with ionization state. The Saha equation gives the degree of ionization of a given atomic element in thermal equilibrium as a function of temperature and electron density:

$$\Rightarrow \qquad \frac{n_{r+1}n_{\rm e}}{n_r} = \frac{G_{r+1}\,g_{\rm e}}{G_r}\frac{(2\pi\,m_{\rm e}kT)^{3/2}}{h^3}\exp\left(-\frac{\chi_r}{kT}\right). \qquad \text{(Saha equation)} \qquad (4.15)$$

Here n_r and n_{r+1} are the number densities of atoms in the ionization state r (e.g., with four electrons missing) and the ionization state $r+1$ (five electrons missing) of a given element. Also, $n_{\rm e}$ is the number density of free electrons, G_r and G_{r+1} are the partition functions of the two states, $g_{\rm e} = 2$ is the statistical weight of the electron, $m_{\rm e}$ is the electron mass, and χ_r is the ionization potential from state r to $r+1$. For hydrogen, under most astrophysical conditions, $G_r \approx 2$ and $G_{r+1} \approx 1$. For the neutral ground state of hydrogen, $\chi_r = 13.6\,{\rm eV}$. See AM, Chapter 9 for somewhat more on this.

4.3   Modeling and evolution

The fundamental equations, (1), (2), (3), and (7) or (13) together with the secondary equations and appropriate boundary conditions, allow one to construct theoretical models of a star. These can be compared with the observable parameters of real stars to gain understanding of their interior structure.

Approach to solutions

Such a model, in the simplest case, can be constructed if only two parameters are specified: the mass M and the chemical composition of the gaseous material of which the star is formed. For example, if one starts with the solar mass and solar chemical composition, a modeling program will yield a star with radius, luminosity, and surface temperature that approximates the actual sun when it first started burning hydrogen on the main sequence. The model would demonstrate that energy is propagated via radiation transfer in the central regions of the sun and via convection in the outer parts. In addition, it would yield the interior radial distributions of pressure, mass, mass density, luminosity, and temperature.

How is the model for a spherical (nonrotating) star obtained? The opacity function $\kappa(\rho, T, \text{comp.})$ and the energy-generation function $\epsilon(\rho, T, \text{comp.})$ are determined independently – either theoretically or experimentally – and substituted into (7) and (3), respectively. Consider the composition to be uniform throughout the star. The EOS, $P(\rho, T, \text{comp})$ is then used to eliminate the density ρ in the four fundamental equations.

The result is four differential equations in four unknown functions – namely, $P(r)$, $\mathfrak{M}(r)$, $\mathfrak{L}(r)$, and $T(r)$ – each of which is a function of r. The four boundary conditions are $\mathfrak{M}(0) = 0$, $\mathfrak{L}(0) = 0$, $T(R) = 0$, and $P(R) = 0$. The latter condition is appropriate for stars with radiative envelopes; another form is required for convective envelopes. The independent variable in the fundamental equations is the radius r. In principle, this boundary value problem will yield a unique solution for the four unknown functions if a stellar radius R is specified. The stellar mass, $M = \mathfrak{M}(R)$, follows from the solution. It is possible to reconfigure the problem so that, instead, one specifies M and the solution yields R.

The solution of the four equations must be carried out numerically, and this would be tedious work without the availability of electronic computers. Solutions for stars of different initial masses and composition enlarge the computational task. Each such star will exhibit structural changes as the nuclear burning gradually depletes and finally exhausts the hydrogen at its core. Models must also take into account the varying composition with radius. Modeling must continue as the star proceeds through its evolutionary stages.

Sun

The results of models can be compared with observations of stars. If, for example, a model successfully matches the radius and mass of a given star, the entire interior structure of the star is known in principle. The understanding could be in error if the processes are not sufficiently well understood.

The structure of the sun is quite well understood at present with modeling playing a central role. The model has been checked with *neutrino studies*, which probe the nuclear reactions at the core of the sun. *Helioseismology* probes the overall mass structure of the sun through measures of its mechanical (acoustic) oscillations.

Distributions of mass, energy generation, temperature, and density of the sun are shown in Fig. 5. Interior parameters are given in Table 1 as are other primary characteristics of the sun.

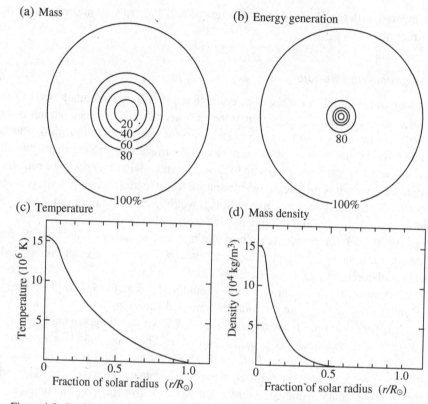

Figure 4.5. Profiles of (a) mass, (b) energy generation, (c) temperature, and (d) mass density for the sun. The labels on the circles in (a) and (b) indicate the percentages of mass and energy generation within the spherical region shown. The central density in (d) is 150 times that of water, but the average density is only a factor of 1.4 greater (Table 1). [Adapted from G. Abell, *Exploration of the Universe*, 3rd Ed., Holt, Rinehart Winston, 1975, Fig. 23.7, with permission of Brooks / Cole. Solar model by R. Ulrich]

The central temperature is 16 million K, and the mass fraction of hydrogen relative to the mass of all elements in the center has been depleted down to 36% by nuclear burning. This is to be compared with 69% for the photosphere and 71% for the solar system as a whole.

Estimates (Prob. 31) of the central pressure of a star in terms of its mass and radius can be obtained by integrating the equation of hydrostatic equilibrium under the simplifying assumption of a constant density equal to the average density, $\rho_0 = M/(4\pi R^3/3)$. The EOS at the center can then be used to estimate the central temperature.

For the sun, the nondegenerate EOS is used to find a central temperature that is a factor of a few lower than the actual value. This calculation underestimates the central pressure by about two orders of magnitude. The discrepancies are not surprising given the large underestimate of the central density in this approximation.

Note that the average mass density of the sun is just slightly greater than water and less than that of the earth. The sun's great mass relative to the earth is due mostly to its being larger by a factor of ~ 100 with volume $\sim 10^6$ times greater.

Table 4.1. *Solar quantities*[a]

Quantity	Value
Mass, M	1.989×10^{30} kg
Radius, R	6.955×10^8 m
Luminosity, L	3.845×10^{26} W
Spectral type and class	G2 V
Absolute bolometric magnitude, M_{bol}	4.74
Absolute visual magnitude, M_V	4.82
Photosphere temperature[b]	6520 K
Effective temperature	5777 K
Central temperature, T_c	1.57×10^7 K
Average density, ρ_{av}	1.41×10^3 kg/m³
Central density, ρ_c	1.51×10^5 kg/m³
Central pressure, P_c	2.33×10^{16} N/m²
Central hydrogen mass fraction[c]	0.355
Radius of convection zone base	$0.71\, R_\odot$

[a] Demarque and Guenther in *Allen's Astrophysics Quantities*, 4th Ed., ed. A. N. Cox. AIP Press 1999.
[b] At unit optical depth at 500 nm.
[c] The surface hydrogen mass fraction is 0.6937. Solar system abundances by mass are $X = 0.71 \pm 0.02$ (hydrogen), $Y = 0.27 \pm 0.02$ (helium); $Z = 0.019 \pm 0.002$ (heavier than helium); see AM, Chapter 10.

Main-sequence stars

Models of stars burning hydrogen at their cores with differing total masses yield differing radii and luminosities. Stars less than $\sim 0.06\, M_\odot$ will not burn hydrogen because the densities and temperatures at the core are too low. Stars above about $120\, M_\odot$ can not exist because they will blow away their outer layers (see Eddington limit; Section 2.7.) Over this range of masses, the stellar radii vary from 0.1 to 15 $R_\odot$, the latter being the most massive. The radii vary by a factor of only 150 for a mass range of ~ 2000.

The luminosity is much more sensitive to the mass. The preceding range of masses yields bolometric luminosities from 0.011 $L_\odot$ for the least massive stars (spectral type M8) to $8.6 \times 10^5\, L_\odot$ for the massive spectral type O5 stars – a range of almost a factor of 10^8. Hydrogen-core-burning stars are called *main-sequence stars*.

Spectral types

The characteristics of a range of stellar spectral types are summarized in Table 2. The spectral types O, B, A, F, G, K, M run from the hot and massive O stars with helium absorption lines to the cool M stars with metallic absorption bands. The sequence has been remembered by many generations of students with the phrase "O, Be A Fine Girl (Guy), Kiss Me." The types are based on spectral line features but generally can be associated with a color or surface temperature. The types are subdivided into 0–9 with B0 being hotter than B1.

Table 4.2. *Stellar spectral types and characteristics*[a]

Type	M_V[b]	BC[c]	T_{eff}^d	$M/M_\odot$[e]	$R/R_\odot$[f]
Main sequence (Class V)					
O3				120	15
O5	−5.7	−4.4	42 000	60	12
B0	−4.0	−3.16	30 000	17.5	7.4
A0	+0.65	−0.30	9 790	2.9	2.4
F0	+2.7	−0.09	7 300	1.6	1.5
G0	+4.4	−0.18	5 940	1.05	1.1
K0	+5.9	−0.31	5 150	0.79	0.85
M0	+8.8	−1.38	3 840	0.51	0.60
M5	+ 12.3	−2.73	2 880	0.21	0.27
Giants (Class III)					
B0				20	15
A0				4	5
G5	+0.9	−0.34	5 050	1.1	10
K5	−0.2	−1.02	4 050	1.2	25
M0	−0.4	−1.25	3 690	1.2	40
Supergiants (Class I)					
O5				70	30
B0	−6.5	∼ −2.6	∼28 000	25	30
A0	−6.3	−0.41	9 980	16	60
G0	−6.4	−0.15	5 370	10	120
M0	−5.6	−1.29	3 620	13	500

[a] Drilling and Landolt in *Allen's Astrophysical Quantities*, ed. A. N. Cox, AIP Press, 1999, p. 389.
[b] Absolute visual magnitude.
[c] Bolometric correction: $M_{bol} = M_V + BC$, where M_{bol} is the bolometric magnitude, a measure of total flux at IR–opt–UV wavelengths. Its relation to luminosity is $M_{bol} = -2.5 \log(L/L_\odot) + 4.74$; AM, Chapter 8.
[d] Effective temperature; see (16).
[e] Ratio of mass to solar mass.
[f] Ratio of radius to solar radius.

Stars are also categorized into *luminosity classes* I–V to differentiate giants from main-sequence stars (see "Giants and supergiants" below). Main-sequence stars are luminosity class V. The sun is thus classified as a G2 V star.

Convective regions

The models also reveal the regions of convective and radiative transport. As found previously (8), convection occurs if the magnitude of the radiative temperature gradient $|dT/dr|_{rad}$ would otherwise exceed the adiabatic gradient. Convection, as noted in the discussion after (8), maintains the adiabatic gradient and carries energy upward much more efficiently than does radiation.

As illustrated in Fig. 6, the lowest-mass stars are completely convective. Stars of mass ∼0.4 $M_\odot$ develop a radiative core that reaches to greater and greater radii for more massive stars until, at ∼1 $M_\odot$, only the outer envelope is convective (e.g., the sun). At ∼1.5 $M_\odot$,

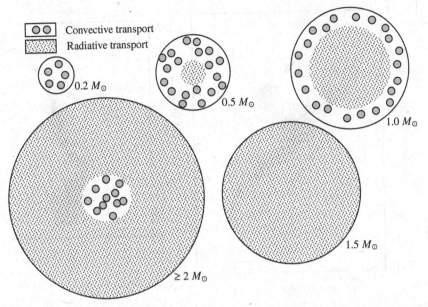

Figure 4.6. Zones of convection and radiation in stars of various masses. The lowest-mass stars are completely convective, a radiative core develops at ~0.4 $M_\odot$, the star is completely radiative at ~1.5 $M_\odot$, and the core region is again convective at ≳ 2 $M_\odot$. The relative sizes of the stars shown here are approximately correct.

the star is completely radiative, but a new convective core is about to develop. Stars of mass ≳ 2 $M_\odot$ have convective cores and radiative envelopes. In these stars, the convective core is in the region where energy is being produced and encompasses 10–20% of the stellar mass.

Hertzsprung–Russell diagram

Stars can be plotted on a *Hertzsprung–Russell (H-R) diagram*, which nominally is a plot of log luminosity L versus log temperature T with temperature increasing from right to left. In practice, though, the observational quantities that represent L and T are plotted instead. Two H-R diagrams are shown in Fig. 7a,b. The former is for a sample of stars in the solar neighborhood at a wide range of distances, whereas the latter is for stars in a single globular cluster; thus, they are all at about the same distance from the sun.

Color-magnitude diagram

The vertical axis in Fig. 7a gives the absolute magnitude in the visual band M_V instead of log luminosity. (The absolute magnitude is defined as the apparent magnitude, $m_V \equiv V$, corrected to distance 10 pc; AM, Chapter 8). Because the magnitude scale is logarithmic, M_V is a logarithmic measure of the luminosity in the visual band. Lower absolute magnitudes represent greater luminosities. Thus, in Fig. 7a, luminosity increases upward. Knowledge of M_V requires measurements of both the observed flux and the distance to the star.

The abscissa of Fig. 7a gives color magnitude, $B–V$, the difference of the apparent B (blue) and V (visual or yellow) magnitudes, instead of temperature. This is a logarithmic measure of the ratio of fluxes in the B and V bands; hence, it specifies a color. Bluer stars

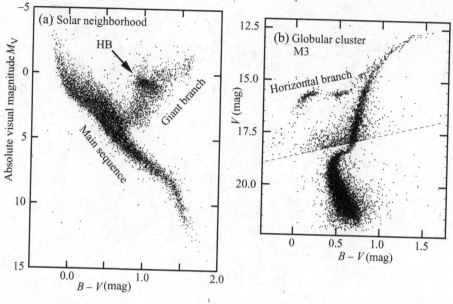

Figure 4.7. Hertzsprung–Russell color-magnitude diagrams for two very different samples of stars. (a) Diagram of 16 631 stars of various ages and various distances that are within about 330 LY of the sun, from the *V*-band and parallax distance measures by the Hipparcos satellite and ground-based color measurements. Horizontal-branch (HB) stars overlie the giant branch. (b) Color-magnitude diagram from ground-based studies of ~12 200 stars in globular cluster M3. The stars in the cluster are at a common distance of ~32 000 LY, and thus apparent magnitudes suffice for the ordinate of this H-R diagram. The stars are mostly of the common age of ~12 Gyr. The data come in two distinct samples. A deep photographic sample yields most of the stars in the diagram both above and below the diagonal line. Short exposures with a charge-coupled device (CCD) camera provide fluxes for the rarer brighter stars plotted only above the dashed line representing B = 18.6. The scatter just above the diagonal line is mostly due to increased uncertainty at the fainter end of the CCD sample. The distance modulus of M3 is 14.93. Thus, $M_V = 0$ in (a) corresponds to $V = 14.93$ in (b). Note the absence of bright main-sequence stars in M3 and also the well-defined horizontal branch. [(a) ESA SP-1200, Hipparcos catalog (1997) in J. Kovalevsky, *ARAA* **36**, 121 (1998); (b) F. Ferraro *et al.*, *A&A* **320**, 757 (1997); photographic data from R. Buonanno *et al.*, *A&A* **290**, 69 (1994)]

have larger ratios of blue to visual fluxes and hence smaller values of *B–V*. Because blue stars are hotter than yellow stars, the hotter stars are to the left. The abscissa is therefore a measure of log *T* increasing to the left, as is standard for H-R diagrams.

A plot with these quantities, M_V and *B–V*, on the axes is called a *color-magnitude diagram* (CMD). See AM, Chapter 9 for more on color magnitudes, but note that Eq. (9.11) therein should have an additive constant on the right side to account for the different zero-magnitude fluxes of the two bands.

Parallax measures carried out by the Hipparcos satellite (1989–1993) yielded precise distances to stars in the solar neighborhood as well as fluxes in the form of *V* magnitudes. Together, these provided absolute visual magnitudes of the ~17 000 stars plotted in Fig. 7a. These stars range in distance out to about 330 LY from the sun and have various masses and ages. The *B–V* colors were obtained from ground-based measurements.

The two quantities represented in an H-R diagram, luminosity and temperature, are fundamental to the star itself. The restricted areas occupied by the stars in Fig. 7a are clearly an indicator of the physics constraining their makeup. Initially, the challenge was to deduce the nature of that physics. Now, we are well aware of it and thus know the masses and evolutionary stages of stars in the several regions of the diagram. An H-R plot of a new sample of stars can consequently reveal much about the stellar content and evolutionary state of the sample.

The most common stars are the main-sequence type that burn hydrogen in their cores. In the H-R diagram (Fig. 7a), they fall along the diagonal band running from upper left to lower right. The luminous hot stars at the upper left are the most massive. The low-luminosity cool stars to the lower right are the least massive.

Effective temperature and radius

The luminosity L at the outer surface of a star of radius R may be written as

$$L \equiv 4\pi R^2 \sigma T_{eff}^4, \tag{4.16}$$

where T_{eff} is the effective temperature of the star. This is the temperature that would give the actual *bolometric* luminosity if the surface radiated perfectly as a blackbody at σT^4 (6.18); T_{eff} is thus defined by (16). ("Bolometric" is defined in Footnote c to Table 2.)

For the sun, $T_{eff} \approx 5800$ K, which is somewhat lower than the actual temperature of $T \approx 6500$ K in the photosphere (at optical depth $\tau = 1$). The overall spectrum approaches that of a blackbody, but absorption lines reduce the total flux below that expected for a blackbody; hence, $T_{eff} < T$. This ad hoc relation (16) may be applied to any star, not solely those of the main sequence.

If the temperature plotted on the H-R diagram of Fig. 7a is T_{eff} (or its equivalent), each and every coordinate on the plot with effective coordinates L and T_{eff} will yield a stellar radius R through (16). For example, stars of low temperature and high luminosity would have large radii and would be plotted in the upper right of the H-R diagram.

Such stars do exist; they are called *giants* as distinct from main-sequence stars. They lie in the *giant branch* of Fig. 7a and are discussed below, Similarly, stars of low luminosity and high temperature would have small radii and would lie in the lower left region. These "compact" stars also exist; they are known as white dwarfs. A few of the sample in Fig. 7a lie in this region (lower left).

Figure 7b shows an entirely different sample of stars, those in the globular cluster M3. The stars in a cluster are mostly of a common age but have different masses and thus different evolutionary histories. We discuss them below in the context of evolution.

Giants and supergiants

A giant star with the same temperature (stellar type) as a main-sequence star will have less surface gravitational acceleration, $g_s = GM/R^2$, because of its large size. This results in less gas pressure and density. The spectral features are thus modified by two effects. First, the atoms will be in higher ionization states according to the Saha equation (15); at a fixed temperature, a lower n_e gives a larger ratio, n_{r+1}/n_r, and the spectrum will reflect this.

Second, the lower pressure leads to less collisional broadening of the spectral lines (AM, Chapter 11).

The spectra of stars can thus be used to categorize "normal" stars into five luminosity classes: main sequence (V), subgiants (IV), giants (III), bright giants (II), and supergiants (I). The radii and masses of some giants and supergiants are given in Table 2.

Evolution of single stars

Knowledge of the nature of pre- and post-main-sequence stars is obtained through modeling that sometimes must take into account nonequilibrium situations and extreme variations of chemical composition within the star. We present here a very brief look at the evolution of a star like the sun. We then see the effects of evolution on clusters of stars in and near the Galaxy.

Solar evolution

Stars form from the gases of the interstellar medium. They must give up energy and angular momentum and also overcome magnetic pressures to succeed in their collapse to densities at which nuclear burning can start. On an H-R diagram, *protostars* approach the main sequence from the upper right, descending along a vertical *Hayashi track* at $T \approx 4000$ K ($\log T = 3.6$ in Fig. 8a). The photospheric temperature of the descending stars is that found at the depth in the stellar atmosphere where hydrogen becomes ionized, at ~4000 K. At lesser depths, the hydrogen is neutral and largely transparent. An observer thus "sees" the 4000-K "surface".

Stars do not exist to the right of the Hayashi track because, with the implied cool surfaces, they can not achieve hydrostatic equilibrium. Thus, the protostars must have a relatively high temperature, which, with their large sizes, indicates a large luminosity. This luminosity arises mostly from the release of gravitational energy by the shrinking star.

The motion on the H-R diagram slows as the star nears the main sequence. The star then moves to the left, settling on the main sequence at the position appropriate to its mass. It then resides on the main sequence for periods that range from several million years for the most massive stars to more than 10 billion years for the least massive.

The post-main-sequence evolution of the sun is shown as a track in Fig. 8a. The light dashed lines indicate rather fast evolution. Stars pass through these regions so rapidly it is unlikely that any would be found there in our observing epoch. In contrast, one expects to find stars in regions of slower evolution (light solid lines).

Stars are initially on the *zero-age main sequence* (ZAMS) but will move above it by a small amount as the hydrogen in the core is being depleted by the nuclear reactions leading to helium. The solar track in Fig. 8a shows this small upward movement.

When the hydrogen at the center of the sun is exhausted, its residue is an inert (nonburning) core of helium. Hydrogen burning continues in a shell around the core, which gradually increases the mass of the inert helium core. In this process, the core shrinks and becomes more compact. This moves the burning shell to lower potentials, causing its temperature to increase.

The burning rate thus increases markedly, and the star responds by expanding. The expansion is so great that, despite the increasing luminosity, the energy emerging per square meter

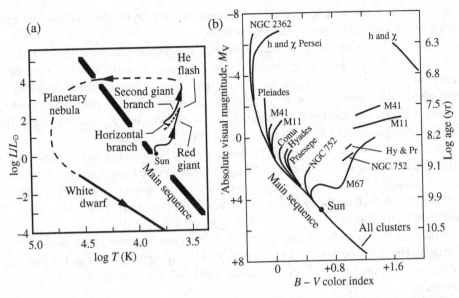

Figure 4.8. (a) Track of the sun on a Hertzsprung–Russell (H-R) diagram showing its approximate evolution until it becomes a white dwarf. This is typical of the evolution of stars with masses $\lesssim 8\ M_\odot$. The main sequence (heavy line) is the location of stars burning hydrogen at their core. High-mass stars are at the upper end, and low-mass stars at the lower end. (b) Color-magnitude diagram (H-R diagram) for ten open clusters of a variety of ages. The stars in each cluster populate the diagram along the lines shown down to the lower right, where the longest-lived (and lowest mass) stars reside. Young clusters still have short-lived, high-mass stars on or near the upper part of the main sequence. In somewhat older clusters, these stars have evolved off the main sequence to their giant phase or on to their final compact states. White dwarfs found in the older clusters are not shown. The cluster M67 is quite old, 4.0 billion years according to a recent study that postdates the plot. Nevertheless, its constituent stars of solar mass have not yet evolved off the main sequence. The gaps in the lines are regions of rapid evolution, pulsations, or both. The right axis gives the ages of stars just leaving the main sequence. [(a) After F. Shu, *The Physical Universe*, University Science Books, 1982, p. 152); (b) After A. Sandage, *ApJ* **125**, 435 (1957)]

of the surface, roughly σT^4 (W/m^2), decreases; hence, the surface temperature decreases. On the H-R diagram, the star moves upward to the right into the giant region. It becomes larger and redder and is thus called, appropriately, a *red giant*. The most massive stars will become highly luminous *supergiants*.

Eventually, the shrinking helium core becomes sufficiently dense and hot that helium begins to undergo nuclear interactions to yield heavier elements (carbon, oxygen, and neon) with the associated release of nuclear energy. The energy release further increases the temperature, which in turn increases the burning rate.

The increase in burning rate would normally be damped by an increase of pressure that would expand the star. This cools the center, and thus stabilizes the burning, as discussed in Section 2.6 under "Stable equilibrium. However, for stars of mass $\lesssim 2.2\ M_\odot$, the helium core will be degenerate. Because the degenerate equation of state $P \propto \rho^{5/3}$ (3.64) is not sensitive to temperature, this addition of thermal energy does not increase the pressure. The increasing temperature consequently accelerates the helium burning without damping. This positive feedback is known as the *helium flash*.

After $\sim 10^3$ yr of runaway He burning, the energy release lifts the core out of degeneracy, and stable helium-core burning begins on the *horizontal branch*. This is reminiscent of the core burning on the main sequence; the star moves very little on the H-R diagram during this phase. Hydrogen-shell burning continues outside the helium core. Stars more massive than 2.2 $M_\odot$ do not develop degenerate He cores, and so stable He burning begins without a "flash." The products of helium burning are carbon and oxygen.

Stars can eject large and variable amounts of mass by means of *stellar winds* during their giant phases. Thus, stars beginning with a given mass on the main sequence will have a range of masses during their helium-core-burning stage. Consequently, they will spread out along a horizontal branch on the H-R diagram. They will have comparable luminosities but quite different temperatures owing to the differing sizes of their envelopes.

The stars of the globular cluster M3 (Fig. 7b) clearly show the extended range of the horizontal branch. The heavily populated region overlaying the giant branch in Fig. 7a consists of such stars in the solar vicinity. Figure 8a shows only the postulated position of the sun on this branch.

Eventually, the helium in the core becomes depleted; the core is then inert carbon and oxygen. Helium burning continues in a spherical shell around the core. This shell is surrounded by an overlayer of helium with hydrogen shell burning still taking place at its outer edge. For stars of mass less than ~ 8 $M_\odot$, this is the end of nuclear burning.

The star will eventually eject most of its diffuse envelope, giving rise to the beautiful *planetary nebula* in which the energetic photons from the core fluoresce the ejected gases. The core shrinks until degeneracy pressure stabilizes it. It will then become a *white dwarf*, which cools at approximately constant radius following the straight-line power-law track seen in Fig. 8a and dictated (to first order) by (16).

Massive stars

A star greater than about 8 $M_\odot$ could provide sufficient pressure and temperature to commence the nuclear burning of its carbon-oxygen core. This and subsequent burning eventually lead to an iron core. As the iron core grows and the central pressure increases, no further burning can take place because iron is the most stable nucleus. Neither fission nor fusion will lead to less mass and hence energy release.

Gravity thus continues to shrink the core, which rises in temperature as demanded by the virial theorem. Eventually, the particle and photon energies become so great that the iron nuclei begin to break up into helium nuclei. These are endothermic reactions; they absorb energy. The product nuclei are more massive than the input nuclei, and so kinetic energy is removed from the core material according to $E = mc^2$. This reduces the supporting pressure, which leads to more shrinkage, higher kinetic energies, and hence more such reactions.

In this runaway process, the core collapses in a few seconds down to a neutron star or possibly a black hole with parts of the envelope being ejected into space. This is a *supernova*. Elements heavier than iron are created in the outgoing shock wave. These together with elements previously created in the stellar interior are ejected into space to become part of the interstellar medium from which later generations of stars are formed.

The cores of more massive stars (greater than $\sim$25 $M_\odot$ on the main sequence) may well collapse directly to a black hole, a *collapsar*. *Gamma-ray bursts* (GRBs) are likely emitted from such events (see the following section); however, some such stars may lose so much mass via stellar winds before collapse that they become neutron stars. The study of the galaxies containing supernova remnants and GRBs gives insight into the progenitors of the neutron stars and black holes.

Computer simulations of the collapse process add insight into the development of a supernova, but a proper full calculation in three dimensions would require computer power far exceeding that of today's supercomputers. Such calculations should take into account stellar rotation, mass loss through stellar winds, and magnetic fields because each of these can materially affect the characteristics of the collapse and associated explosion. The structure would need to be calculated repetitively on millisecond time scales. Calculations with simplifying assumptions are being carried out today, but much remains to be done.

Gamma-ray bursts

In the collapse to a black hole, a large fraction of the envelope may be accreted into the black hole. In the process, the angular momentum of the matter could create an accretion disk with matter and radiation being ejected explosively along the angular momentum axis. The ramification at the earth is a flash of gamma rays lasting from a few seconds to a few minutes.

Gamma-ray bursts were discovered in 1967. They occur about once per day from seemingly random positions on the sky. Their cosmological origin in distant galaxies was not known until accurate celestial positions became available in 1997 with the Italian–Dutch Beppo-SAX satellite. From 2003, the international HETE mission and later the international Swift mission convincingly demonstrated that the longer-duration GRBs are sometimes associated with supernova explosions. The shorter GRBs could be associated with the coalescence of two neutron stars in a binary orbit into a black hole.

Gamma-ray bursts have been detected out to redshift distances exceeding $z = 6$ (i.e., when the universe was less than 1/7 its current scale size). They are powerful probes of their host galaxies at early times and also of the intergalactic medium through which the radiation travels en route to the earth. See AM, Chapter 6 for a description of the BATSE space experiment for detecting GRBs.

One can show that $1 + z = R_{ob}/R_{em}$, where R_{ob} and R_{em} are scale factors for the universe size at the times of emission and observation, respectively, of a photon signal; see Prob. 6.31.

Globular clusters

Globular clusters are relatively tightly bound clusters of 10^5 to 10^6 stars that date back to about the time the Galaxy formed. The stars in globular clusters are mostly old stars. Most of the gas from which new stars might form has been exhausted by early star formation or has escaped the gravitational potential well of the cluster. Thus, little star formation is now taking place. The sample of stars in a globular cluster are therefore mostly of the same age and have evolved more or less undisturbed since the cluster was formed. Important exceptions are the effects of stellar collisions in the dense core of a cluster.

A notable consequence of stellar evolution is seen in the H-R diagram of the globular cluster M3 (Fig. 7b). The high-mass stars on the upper portion of the main sequence are missing;

compare with Fig. 7a. Such stars are highly luminous and soon exhaust their nuclear fuel. Their lifetimes on the main sequence are thus very short (a few million years) compared with those of stars like the sun (about 10 billion years).

At our epoch, when this H-R "snapshot" of M3 was taken, the most massive stars had moved on to become giants and then eventually to collapse to nonluminous neutron stars and black holes or low-luminosity white dwarfs. Thus, most will have disappeared from the diagram.

Large numbers of stars populate the giant and horizontal branches of M3 (Fig. 7b). These are stars that were initially somewhat more massive than the sun. Not long before this snapshot was taken, they were on the main sequence at points slightly above the "turnoff" to the giant branch. Stars with $\sim 1.0\,M_\odot$ in M3 are still on the main sequence.

Open clusters

Stars that have formed in our Galaxy in more recent times have been produced in clusters because they originated from the fragmentation of interstellar clouds; see Section 2.2. Typically, these clusters are not gravitationally bound, and so they are dispersing and eventually will no longer be recognizable as clusters. The clusters visible today have a variety of ages, but all the stars in a given cluster will have about the same age. The stars in a particular open cluster will have a variety of masses and will therefore evolve at different rates just as we have seen in globular clusters. Open clusters are associations of stars that are younger than globular clusters.

The regions on an H-R diagram occupied by individual stars in several open clusters are shown in Fig. 8b as lines that indicate the mean values of M_V at each $B–V$. The diagram is a snapshot, taken at the present time, of the stellar content of each of these clusters, some of which are quite young whereas others are quite old.

As we have seen, massive stars evolve off the main sequence much faster than do less massive stars; the durations on the main sequence are given on the right axis of Fig. 8b. The youngest clusters, such as NGC 2362, therefore still have all their stars on or near the main sequence. In contrast, the massive stars in older clusters (e.g., M41) have left the main sequence. Some are found as giants or horizontal-branch stars to the right of the H-R diagram. The most massive stars will already have collapsed to compact stars.

The turnoff points in the figure thus indicate the cluster ages. The lower the turnoff, the greater the age. M67 is the oldest with a turnoff point not much above the sun. Given more time, stars of solar mass in M67 will move off the main sequence to become red giants. If, for example, one finds a neutron star or a black hole in an open cluster, its progenitor must already have evolved off the main sequence. The progenitor must then have been more massive than the mass at the turnoff point.

Variable stars

The notable gap in the horizontal branch of Fig. 7b is the location of variable stars, which are stars that oscillate in radius and luminosity. These are the RR Lyrae stars with periods of 0.2 to 0.9 d and cepheid variables with periods 1–100 d. (A criterion for pulsation is developed in Section 2.8.) The period of such a star is coupled to its mean luminosity. These variable stars can thus be used as standard candles in determining distances (AM, Chapter 9). This

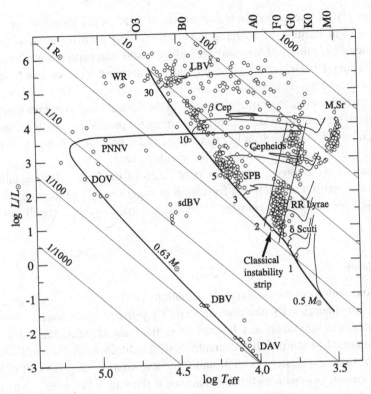

Figure 4.9. Examples of various types of pulsating variable stars plotted as small circles on the Hertzsprung–Russell diagram. The dark line to the right is the main sequence with evolutionary tracks branching off to the right for different stellar masses. The ultimate evolutionary track of a star that ends its life as a compact star of 0.63 $M_\odot$ is shown. It moves leftward through the planetary nebulae nuclei variables (PNNV) and then downward as a cooling white dwarf, passing through regions of pulsational instability sequentially classified as DOV, DBV, and DAV (DAV = Dwarf + type/temperature A + Variable). Other types of intrinsic variables are shown: β Cephei stars, Mira (M), Semiregular (Sr), luminous blue (LBV), Wolf–Rayet (WR), slowly pulsating B stars (SPB), and subdwarf B stars (sdBV). The *classical instability strip* is shown as two parallel lines encompassing Cepheid, RR Lyrae, and δ Scuti variables; if extended, it intersects the pulsating DAV stars. The thin lines represent loci of constant radius. [Provided by A. Gautschy; see A. Gautschy & H. Saio, *ARAA* **33**, 75 (1995)]

gap, known as the *Hertzsprung gap*, also contributes to the breaks in the lines of Fig. 8b (e.g., M11 and M41).

Examples of fourteen classes of periodically pulsating variables are plotted on an H-R diagram in Fig. 9. Because measurements are becoming more sensitive, even more types are being found. The clustering, clearly distinct from the general populations of Fig. 7, illustrates that substantial pulsations are obtained only under certain conditions. The thermodynamic condition worked out in Section 2.8 underlies the *classical instability strip* indicated in Fig. 9, if not other groupings. Clearly, pulsational instability occurs at many different stages of stellar evolution.

Many variable stars are not periodic pulsators but rather exhibit marked aperiodic variability. *Cataclysmic variables* are binary systems containing a white dwarf. Some exhibit episodic accretion, giving rise to occasional outbursts of light called *dwarf novae*. Others undergo sudden nuclear burning of accreted material, creating a *nova* outburst that may repeat after decades or centuries. *Flare stars* release coronal magnetic energy; pre-main-sequence stars (*T Tauri, FU Orionis*) accrete and eject matter; and massive *Luminous Blue Variables* exhibit episodic mass ejections; see after (2.79).

We have listed stars that emit most of their energy in the optical band. Systems containing compact objects provide radiation via much more energetic processes owing to the large gravitational potentials. These often yield pronounced radio and x-ray fluxes that can be highly variable on time scales ranging from milliseconds to years.

This brief outline of evolution neglects important facets of the field – most notably, the processes of stellar formation and element formation in supernovae. The major changes in stellar lives brought about by mass transfer in accreting binary systems are discussed in Section 5.

Scaling laws

Dimensional approximations of the fundamental equations can be used to gain some insight into the magnitudes of quantities that characterize stars. They give rough estimates of density, pressure, temperature, and luminosity as a function of stellar mass M and stellar radius R but do not provide the internal variation of these quantities with radius r, as would a full solution.

The variations of these expressions with M and R are known as scaling laws. If properly qualified, they are known as *homology* transformations. If the internal structure of a certain star is known, it is possible to use the transformations to scale reliably from it to another of similar structure but of somewhat different M and R.

The radius R can be eliminated from the scaling relations to obtain other scaling laws that depend solely on stellar mass M for a fixed chemical composition. This is in accord with the fact that the entire character of a newly formed star derives, in the basic case, solely from its total mass and chemical composition.

Matter density

Consider the matter distribution equation (2) for a star of mass M and radius R. For most of the stellar volume, the variable r in (2) has an order of magnitude value of R, and so we will approximate it as such. Because $\mathfrak{M}(r)$ ranges from 0 to M as r ranges from 0 to R, the derivative $d\mathfrak{M}/dr$ may be approximated as M/R. With these approximations, we obtain from (2) a characteristic density for the star,

$$\rho \approx \frac{M}{R^3}, \qquad \text{(Characteristic density)} \quad (4.17)$$

which is a factor of ~ 4 greater than the actual average mass density of the star. We note that the central density of the sun is ~ 110 times greater than the average density (Table 1) and about 25 times M/R^3.

Pressure

A characteristic internal pressure of a star of mass M and radius R follows in a similar manner from the equation of hydrostatic equilibrium (1). At the surface of the star set $r = R$ and

$\cdot P = 0$, and, at the center, set $r = 0$ and $P = P_c$. We can thus approximate dP/dr as $-P_c/R$. Further, we can make the order-of-magnitude approximations $\mathfrak{M} \rightarrow M$, $r \rightarrow R$, and from (17), $\rho \rightarrow M/R^3$. This yields, from (1), a crude approximation to the central pressure:

$$P_c \approx \frac{GM^2}{R^4} \propto \frac{M^2}{R^4}. \qquad \text{(Central pressure)} \qquad (4.18)$$

The middle term, GM^2/R^4, gives the magnitude, and the last, M^2/R^2, is the scaling law with which one can compare stars. For the sun, the former not surprisingly yields a value a factor ~ 20 less than the actual central pressure (Table 1) given the high concentration of mass toward the center that our approximation for density did not take into account.

An alternative estimate of the central pressure P_c is to assume uniform mass density and to integrate the equation of hydrostatic equilibrium (1) from the surface to the center (Prob. 31). The result is less than (18) by a factor of ~ 8 and a factor of ~ 160 less than the actual solar value owing to our underestimate of the central density.

Temperature

At the center of stars like our sun, the nondegenerate equation of state, $P = (\rho/m_{av})kT$ (14), applies. This yields the following estimate of the central temperature if one adopts (17) for the central density and (18) for the central pressure:

$$T_c \approx \frac{Gm_{av}}{k} \frac{M}{R} \propto \frac{M}{R}. \qquad \text{(Central temperature)} \qquad (4.19)$$

For solar quantities and a hydrogen plasma with $m_{av} \approx m_p/2$, this yields a value which, fortuitously, is only about 30% less than the actual value, $T_{c,\odot} = 1.6 \times 10^7$ K (Table 1). Again, we have given both the approximate magnitude and the scaling factor.

Luminosity

An estimate of luminosity in terms of the mass and radius of the star can be obtained in a similar manner from the luminosity distribution (3). Approximate the derivative as L/R, use the parameterized version of the energy-generation function (2.66), and make use of the scaling above for T (19) and ρ (17) to obtain (Prob. 36a)

$$L \approx \epsilon_0 X^2 \left(\frac{Gm_{av}}{k}\right)^\beta \frac{1}{10^{7\beta+5}} \frac{M^{\beta+2}}{R^{\beta+3}} \propto \frac{M^{\beta+2}}{R^{\beta+3}}. \qquad \text{(Luminosity)} \qquad (4.20)$$

For $m_{av} = m_p/2$, the factor in parentheses is 4.04×10^{-15} (SI units).

This expression is only an order-of-magnitude approximation. It is highly sensitive to the value of β. Additionally, both β and ϵ_0 are themselves sensitive to temperature. Furthermore, we have again made whole-star approximations that do not take into account the increased density at the star center where the nuclear reactions take place.

An estimate of the solar luminosity can be obtained from (20). One would use the solar mass $M_\odot$ and radius $R_\odot$, the coefficient $\epsilon_0 = 2 \times 10^{-3}$ W/kg given after (2.66), the index $\beta = 4$ for the pp process, and the hydrogen mass fraction at the center of the sun, $X = 0.36$. In addition, it is necessary to calculate the average particle mass m_{av}, counting all nuclei and free electrons. For the sun's center, complete ionization can be assumed. For m_{av}, one also needs the mass fractions for helium and heavies, Y and Z. (The quantities X, Y, and Z are the fractions of the mass in hydrogen, helium, and heavies, respectively; see also footnote c to Table 1.

The result of this evaluation (Prob. 36) is a luminosity within 10% of the actual solar luminosity. This is much too close given the extent of our approximations – especially given the strong dependence on β. Nevertheless, it does suggest that the scaling (20) would give us a reasonable luminosity estimate for a star similar to the sun but with somewhat different mass and radius.

Mass dependence

The luminosity L at the outer surface of a star of radius R may be written as $L = 4\pi R^2 \sigma T_{\text{eff}}^4$, as we have seen (16). The effective temperature is approximately the photosphere temperature $T(R)$, which, it can be argued, may be scaled from star to star in the same manner as the central temperature (19) – namely as M/R. Thus, at the surface,

$$L \propto R^2 T^4 \propto R^2 \left(\frac{M}{R}\right)^4 = \frac{M^4}{R^2}. \tag{4.21}$$

This is a second independent equation for the scaling of the luminosity L with M and R. The first is (20), which scales as $M^{\beta+2}/R^{\beta+3}$.

Eliminate R from these two equations to find the scaling solely with mass M:

$$\quad\quad L \propto M^{(2\beta+8)/(\beta+1)}. \qquad\qquad\text{(Luminosity: } L(M)) \tag{4.22}$$

Use the same two equations, (20) and (21), and eliminate L to obtain the scaling of stellar radius as a function of M:

$$\quad\quad R \propto M^{(\beta-2)/(\beta+1)}. \qquad\qquad\text{(Radius: } R(M)) \tag{4.23}$$

These expressions are evaluated in Table 3 for the values $\beta = 4$ and $\beta = 15$ for the pp and CNO reactions, respectively. The stellar radius increases with stellar mass, but less than linearly, to the $\sim$2/5 and $\sim$13/16 power for the two processes. The luminosity is a very strong function of mass to the $\sim$3 and $\sim$2.4 power for the two processes. A small difference in mass makes a large difference in luminosity. These exponents are roughly in accord with the ranges of parameters mentioned above in this section (just under "Main-sequence stars").

The dependence of ρ, P, and T on mass M may be obtained by using (23) to eliminate R from (17), (18), and (19). The resultant scaling laws are also given in Table 3. It is noteworthy that the mass density ρ decreases with increasing mass, albeit very slowly for the pp process. Massive stars have *less* mass density than do stars of lesser mass! The pressure, which is equivalent to energy density, increases slowly for the pp process but decreases as $M^{-5/4}$ for the CNO process.

One might wonder why the more massive stars burn so brightly if neither the mass density nor the energy density increases dramatically with stellar mass. The answer lies in the temperature dependence, which increases modestly with mass to the 3/5 and 3/16 power for the two processes. A small increase in temperature greatly increases the ability of the nuclei to penetrate the Coulomb barrier and thus to undergo energy-producing nuclear interactions.

Table 4.3. *Scaling laws for stars*

Scaling[a]	pp chain ($\beta = 4$)	CNO cycle ($\beta = 15$)
$L \propto M^{(\beta+2)}/R^{(\beta+3)} \rightarrow L \propto M^{(2\beta+8)/(\beta+1)}$	$L \propto M^{16/5}$	$L \propto M^{19/8}$
$R \propto M^{(\beta-2)/(\beta+1)}$	$R \propto M^{2/5}$	$R \propto M^{13/16}$
$\rho \propto M/R^3 \propto M^{(7-2\beta)/(\beta+1)}$	$\rho \propto M^{-1/5}$	$\rho \propto M^{-23/16}$
$P \propto M^2/R^4 \propto M^{(10-2\beta)/(\beta+1)}$	$P \propto M^{2/5}$	$P \propto M^{-5/4}$
$T \propto M/R \propto M^{3/(\beta+1)}$	$T \propto M^{3/5}$	$T \propto M^{3/16}$
$L \propto T_{\text{eff}}^{(2\beta+8)/3}$	$L \propto T_{\text{eff}}^{16/3}$	$L \propto T_{\text{eff}}^{38/3}$

[a] Scaling as functions of M and R from expressions in the text and as a function of M alone from (23). The expression for $L(T)$ is from the elimination of M from the expressions $L(M)$ and $T(M)$.

H-R diagram comparison

It is interesting to obtain a scaling law for L as a function of T, the two parameters plotted on the H-R diagram (Prob. 37). Eliminate M from the expressions for $L(M)$ and $T(M)$ (Table 3) and apply the result at the photosphere, where the temperature is $\sim T_{\text{eff}}$ and $L = \mathcal{L}(R)$:

$$L \propto T_{\text{eff}}^{(2\beta+8)/3}. \qquad \text{(Luminosity and surface temperature)} \qquad (4.24)$$

This too is evaluated for the appropriate values of β in Table 3. The luminosity is found to be a steep function of the surface temperature to the powers ~ 5 and ~ 13 for the two processes. One finds (very) rough agreement with the shape of the main sequence, which rises to the power ~ 7 over much of its extent. This exercise illustrates how, in principle, the physical laws can lead to the H-R diagram, but it also makes clear the limitations of simple scaling. One obvious shortcoming of our scaling is that we did not take radiation pressure into account.

Homology transformations

The scaling relations given in Table 3 can be shown (not here) to be strictly correct at a given fractional radius $x = r/R$ if the stars under comparison are *homologous*. Stars are homologous if they have similar structures as a function of x. For example, two stars are homologous if they have the same constant ratio of densities as a function of the relative radius, $x = r/R$, as shown in Fig. 10a,b. This implies that the fractional density $\rho^* \equiv \rho(x)/\rho_c$ is the same function for the two stars (Fig. 10c). Homologous stars also have similar radial variations of pressure, temperature, luminosity, and so forth.

Our scaling laws apply to such stars if the parameters in the left column of Table 3 are taken to be the physical quantities at a given x – that is, $L(x)$, $\rho(x)$, $P(x)$, and $T(x)$. They are then called *homology transformations*. As an example, the density of star 1 at radius $R_1/4$ will scale exactly as M/R^3 to the density of star 2 at radius $R_2/4$. That is, star 2 will thus be exactly $(M_2/M_1)(R_2/R_1)^{-3}$ more dense at $R_2/4$ than star 1 is at $R_1/4$. Similar scaling applies to $L(x)$, $P(x)$, and $T(x)$. The scaling with mass alone again has similar validity for equivalent radial positions in homologous stars.

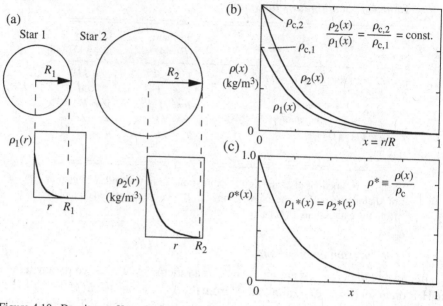

Figure 4.10. Density profiles of two homologous stars. The densities are at a fixed ratio to each other as a function of the fractional radius $x = r/R$. (a) Densities ρ versus radius r. (b) Densities ρ versus the fractional radius $x = r/R$. The ratio of densities at all x equals the ratio of the central densities. (c) Density relative to ρ_c, the central density, $\rho^* = \rho(x)/\rho_c$, which is the same for the two stars.

The homology relations may be used to extrapolate the internal structure from one star to another at the same position x without carrying out additional time-consuming calculations. This was highly useful before the days of electronic computers. Even so, the transformations were of limited utility. They are valid only if the structures are approximately homologous. In practice, they are useful only over quite limited ranges of masses (say over one stellar type such as from F5 V to G5 V). Nevertheless, the scaling laws of Table 3 remain quite helpful for qualitative discussions.

4.4   Compact stars

After its long history of evolution as a gaseous star, the final state of a star may be as a white dwarf, neutron star, or black hole. These objects are remarkable for their extreme densities and their quantum and gravitational effects. The existence of white dwarfs and neutron stars is indisputable from several lines of evidence.

Evidence for the existence of black holes does not reach the same level of persuasion; nevertheless, the circumstantial evidence for their existence is very strong. Most astrophysicists accept their existence with a very high level of confidence.

Here we present and derive some structural parameters pertinent to each type of object.

White dwarfs

The EOS of nonrelativistic degenerate gases (3.64) leads to two interesting phenomena pertaining to white dwarfs. The first is that degeneracy pressure of electrons can support a star against collapse even if its nuclear fuel has been exhausted. White dwarf stars are so

4.4 Compact stars

supported. The second is that, if a white dwarf is more massive than $\sim 1.4\,M_\odot$, the electrons become relativistic and the EOS becomes softer. In this situation, degeneracy pressure is no longer able to support the star. White dwarfs more massive than $\sim 1.4\,M_\odot$, therefore, are not expected to exist, and they are not found observationally. We now develop these ideas in a semiquantitative fashion.

Mass-radius relation

A white dwarf star is the collapsed final state of a star such as our sun. Such a star might have a radius comparable to that of the earth ($R \approx 0.01\,R_\odot$) and $M \approx 1.0\,M_\odot$ with an average density, therefore, that is $\sim 10^6$ times that of the sun. The extreme compression of the star forces the electrons into degeneracy (Section 3.6). At some degree of compression, the nonrelativistic degeneracy pressure, $P \propto \rho^{5/3}$ (3.64), becomes sufficient to withstand the inward pull of gravity. The star then stabilizes in its white dwarf state.

The approximate radius at which this stabilization occurs can be obtained from the EOS for nonrelativistic degenerate matter (3.64) and the equation of hydrostatic equilibrium (1). The former is the pressure provided by electrons at a certain matter density ρ, whereas the latter provides an estimate of the central pressure $P_c \approx GM^2/R^4$ (18) that is essential if the star is to be stabilized against gravity.

Equate these two pressures, (3.64) and (18), approximate ρ as M/R^3, and use $\mu_e = 2$ (for a helium core) to find (Prob. 41)

$$\frac{R}{R_\odot} \approx 0.01 \left(\frac{M}{M_\odot}\right)^{-1/3}. \qquad \text{(Radius of white dwarf star)} \qquad (4.25)$$

A white dwarf of one solar mass is thus expected to have a radius about 1% that of the sun, and this is in accord with observations. White dwarfs have quite a significant range of measured radii and masses, 0.004–0.03 $R_\odot$ and 0.1–1.4 $M_\odot$. Note from (25) that the stellar radius decreases with increasing mass. In contrast, on the main sequence, radius increases with mass. The relation (25) may be confirmed by constructing white dwarfs from first principles (Prob. 43).

Stability

Our equations also illustrate that the pressure balance is a stable equilibrium. The supporting degeneracy pressure (3.64) varies as $(\rho^{5/3}) \approx (M/R^3)^{5/3} \propto R^{-5}$, whereas the pressure required to offset gravity, P_c, varies as R^{-4} (18). If the star is perturbed to a smaller size, the supporting pressure increases faster than the required pressure, and the net pressure difference will return the star to its equilibrium size.

Sirius B

The first white dwarf known was the faint companion to Sirius called Sirius B. It was first observed in 1862. In 1915, measurements of its color yielded an effective temperature $T_{\text{eff}} \approx 27\,000$ K. The measured flux and distance gave a luminosity, which, in turn, yielded the radius R through the defining relation for T_{eff} (16); it was about 0.01 $R_\odot$.

Concurrent measures of the binary motion of Sirius A/B yielded a mass of about 1 $M_\odot$ for Sirius B. This implied an incredibly dense star for which there was no known support mechanism. Its physical nature was not understood until the development of the quantum-statistical theory of electron gases by Fermi and Dirac in the mid-1920s.

Chandrasekhar mass limit

If the white dwarf accretes matter from a binary companion, its mass will gradually increase. This will cause it to decrease slowly in size according to (25). This decrease in physical x, y, z volume will eventually drive the degenerate electrons at the star's center to relativistic velocities. At this point, the EOS will approach the limiting EOS for totally relativistic electrons (3.69), namely, $P \propto \rho^{4/3} \approx M^{4/3}/R^4$.

In this limit, it is immediately apparent that stable equilibrium has been lost. Consider a perturbation to a more dense state with the same mass and lesser radius. The pressure required to offset gravity (18) increases as $M^2/R^4 \propto R^{-4}$, and the supporting degeneracy pressure just given also varies as R^{-4}. Both increase by the same factor, and so there is no net restoring source. See the discussion of this state in the context of the virial theorem in "Stars", just after (2.26).

The perturbation can therefore continue on to even smaller sizes. The collapse is stopped only if neutron degeneracy and nuclear forces restabilize it. For stellar cores of $\lesssim 3\,M_\odot$, it may become a neutron star with radius 10 km or $\sim 1/700$ that of our typical white dwarf. More massive cores could result in a black hole. If the collapse were initiated by gradual accretion onto the white dwarf, it would probably destroy itself owing to explosive nuclear burning. This leads to a rather characteristic supernova known as *supernova type Ia*. These supernovae are used as standard candles in cosmological studies.

The mass at which the collapse would occur, namely $\sim 1.4\,M_\odot$, is known as the Chandrasekhar mass limit. As the star approaches this limit, the EOS at the star center evolves from the nonrelativistic $\rho^{5/3}$ toward the relativistic $\rho^{4/3}$ with exponent moving toward 4/3. Until the star reaches this value, there is a net restoring force that prevents collapse. Just as the star reaches 4/3, or a value so close that a perturbation forces it over 4/3, collapse commences.

The mass at which this occurs can be estimated by equating the relativistic degeneracy pressure $P \propto \rho^{4/3}$ with all its coefficients (3.69) to the required support pressure $P_c \approx GM^2/R^4$ (18). This equality is valid at the moment before collapse. As demonstrated, the R^{-4} dependence drops out. Again, approximate the density as M/R^3, set $\mu_e = 2$, and solve for M. The result is about $0.3\,M_\odot$. Alternatively, one can use the somewhat lower central pressure obtained from a constant density model (Prob. 31) to find that the collapse mass is about $8\,M_\odot$. Using the intermediate central pressure from a linearly decreasing density model (Prob. 32) yields $\sim 1.3\,M_\odot$. These three results are worked out in Prob. 41.

Our approximations show that the collapse mass is on the order of one solar mass. A proper calculation results in

$$M_{\mathrm{Ch}} = 1.46 \left(\frac{2}{\mu_e}\right)^2 M_\odot \underset{\mu_e = 2}{\rightarrow} 1.46\,M_\odot, \qquad \text{(Chandrasekhar mass limit)} \qquad (4.26)$$

where μ_e is approximately the number of nucleons per electron (3.62). The maximum mass of a white dwarf is likely to be somewhat less than this. A star with an iron core, $\mu_e = 56/26$, gives $1.26\,M_\odot$. The conversion of protons to neutrons in the dense core will further increase μ_e and modestly lower the mass. Furthermore, a general relativistic instability can further lower the maximum mass for helium and carbon white dwarfs.

The result (26) is named after the young Indian physicist who first calculated it in 1932; he was awarded the Nobel prize for this work in 1983.

Neutron stars

A collapsing white dwarf must stabilize itself as a neutron star if it is to avoid gravitational collapse into a black hole. A neutron star is constituted of nuclear matter, largely in a superfluid of neutrons. It probably has more exotic matter at its center – possibly *quark matter*. It is an oversized liquid atomic nucleus supported from collapse by nuclear forces and to a smaller extent by neutron degeneracy pressure.

The neutrons come about because of the high energy of the electrons at the top of the degenerate Fermi sea. The electrons interact with nuclei and convert protons to neutrons, thus making the nuclei neutron rich. The neutrons can not easily decay back to protons because the states into which the ejected electrons would go are already occupied as a result of the electron degeneracy.

Finally, if the nucleus is very neutron rich, the neutrons are no longer bound to the nuclei and they float off, becoming free neutrons. The density at which the neutrons become independent of nuclei is called the *neutron drip point*. Eventually only a few free protons will be present in a sea of neutrons.

Radius of a neutron star

We find in Section 1.6 that the mass of a neutron star is typically determined from orbital motions of its companion in a binary system. Here we discuss the other basic parameter of the star, its radius.

The radius of a neutron star can be estimated (*i*) from the density of nuclear matter, or (*ii*) from neutron degeneracy pressure (Prob. 44c). The two methods give comparable answers. Because the dominant force involved is nuclear, the former is more relevant. Here we present a simple scaling from an atomic nucleus.

Assume that the density of the neutron star is comparable to or greater than that of nuclear matter and that its mass is of order 1 $M_\odot$. More precisely, we choose to adopt a density five times that of nuclear matter and a mass of 1.4 $M_\odot$, which are values closer to the current nominal ones. The density ρ_n of a single neutron of radius r_n and mass m_n may be adopted as characteristic of nuclear matter,

$$\rho_n = \frac{m_n}{\frac{4}{3}\pi r_n^3} = 1.2 \times 10^{17}\,\text{kg/m}^3, \qquad \text{(Density of a neutron)} \qquad (4.27)$$

where we use $m_n = 1.7 \times 10^{-27}$ kg and $r_n = 1.5 \times 10^{-15}$ m.

Now the radius, mass, and average density of the neutron star are related as follows if we assume a constant density throughout the star:

$$M_{ns} = (4/3)\pi R_{ns}^3 \rho_{ns}. \qquad (4.28)$$

Adopt $\rho_{ns} = 5\rho_n = 6.0 \times 10^{17}$ kg/m^3 and $M_{ns} = 1.4\,M_\odot$, where $M_\odot = 2.0 \times 10^{30}$ kg, to obtain the approximate radius of a neutron star:

$$\Rightarrow \qquad R_{ns} \approx \left(\frac{3M_{ns}}{4\pi\rho_{ns}}\right)^{1/3} = 10.3 \times 10^3\,\text{m} \approx 10\,\text{km}. \qquad (4.29)$$

Such a star is remarkably compact. It is only about the size of Manhattan island, about 10^{-5} times the radius of the sun, and almost 10^{15} times more dense!

This 10-km radius is the nominal size often attributed to a neutron star. As with normal stars, the size is expected to be a function of the mass of the neutron star. The mass we

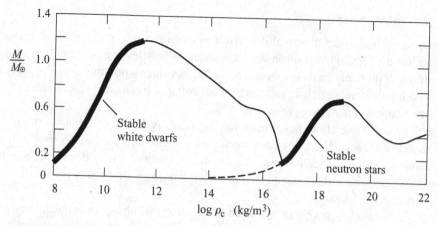

Figure 4.11. Mass of compact stars as a function of central density ρ_c for equations of state known as HW (1958) and OV (1939). The rising portions of the curve (heavy lines) are the regions where stable white dwarfs and neutron stars could exist. For these equations of state the maximum masses for white dwarfs and neutron stars would be 1.2 and 0.7 $M_\odot$, respectively. Nuclear forces, not included, would raise the maximum mass of a neutron star to about 2 M. [From S. Shapiro and S. Teukolsky, *Black Holes, White Dwarfs, and Neutron Stars*, Wiley Interscience, 1983, p. 244]

chose, 1.4 $M_\odot$, is consistent with the values currently measured through studies of binary systems that include neutron stars. The 10-km size is consistent with theoretical models of neutron stars based on plausible (nuclear) equations of state. Sizes of ∼10 km have also been measured through the luminosity-temperature dependence ($L = 4\pi R^2 \sigma T^4$) of *x-ray bursts*. These are brief (∼1 min) outbursts of x rays that arise from rapid nuclear burning of accreted material on the surface of a neutron star.

Equations of state and structure

A correct calculation of the structure of a neutron star, and its variation with mass, must take into account both the degeneracy and nuclear forces. The nuclear physics is somewhat uncertain at the high densities of the interior, and so the EOS is also somewhat uncertain. Various equations of state have been put forward. When one of them is used to calculate a neutron-star model for a given stellar mass, a stellar radius R and central density ρ_c are obtained.

A relation between stellar mass M and central density ρ_c for white dwarfs and neutron stars is plotted in Fig. 11. The latter is for a pure ideal neutron gas; nuclear forces are not included. A star that exists on the rising portion of a curve is stable. The addition of a bit of mass results in an increase in central density, which, from the EOS, provides an increase in the pressure necessary to resist the greater inward gravitational pull. If, on the other hand, the central density were to decrease with the addition of mass, there would be no additional pressure to oppose the greater gravitational pull. A star on the falling part of the curve would thus not be stable against such perturbations; it could not exist.

The structure of a 1.4-$M_\odot$ neutron star for a moderate (i.e., not extremely soft or hard) EOS could be as shown in Fig. 12. The several levels in the star are

(*i*)   a gaseous atmosphere less than 0.1 m thick (not shown);
(*ii*) a thin outer, electrically conductive crust containing nuclei in a rigid lattice together with relativistic degenerate electrons about 300 m thick;

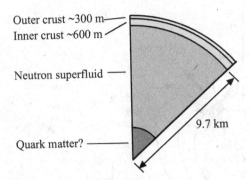

Figure 4.12. Neutron star structure for a 1.4-$M_\odot$ star with the "TMI" equation of state. [After D. Pines, *J. Phys. Colloq.* **41**, C2/111 (1980); S. Shapiro and S. Teukolsky, *ibid.*, p. 251]

(*iii*)   an inner crust about 600 m thick consisting of a lattice of neutron-rich nuclei, a relativistic electron gas, and (probably superfluid) neutrons;

(*iv*)   a region of neutron liquid (a superfluid), which may also contain some electrons and superfluid protons; and

(*v*)   a core of uncertain nature – possibly some exotic material such as *quark matter*.

A superfluid has several interesting properties, one of which is that it does not rotate as a whole but rather forms cylindrical, regularly spaced, quantized *vortices* of rotating fluid aligned parallel to the rotation axis. Each of these rotates independently, and their collective angular momentum mimics a normally rotating fluid. Each vortex core may be only $\sim 10^{-14}$ m in radius, and the several vortices may be spaced roughly 0.1 mm apart. The density of vortices (vortices per square meter) determines the magnitude of the angular momentum of the superfluid. If the rotation of the fluid slows, the vortices move apart. This reduces their spatial density and hence the fluid's angular momentum.

The vortices in the crust of a neutron star, not necessarily coupled to those in the superfluid core, are "pinned" to the nuclei in the crust. Also, the ends of the tubes of magnetic flux expected in the superconducting proton fluid in region (*iv*) are anchored to the field lines in the highly conductive rigid crust. These couplings between the crust and core of a neutron star affect the way it responds to a changing spin rate or to externally applied torques.

Isolated neutron stars (radio pulsars) exhibit occasional sudden spinups, sometimes called *starquakes*, which are probably due to readjustments of the crust and vortices as the spin rate decreases. In accreting binary systems, time-varying torques are applied to the crust by the accretion flow, and the spin-rate response depends on the degree of coupling of the crust and core. Observations of spin-rate changes through pulse-timing (Fig. 1.13) thus provide information about the internal structure of neutron stars.

Evidence for neutron stars

The discovery of isolated *radio pulsars* in 1967, and in particular the pulsar in the Crab nebula the next year, demonstrated conclusively that radio pulsars are spinning neutron stars. The high spin rate (30 Hz) of the Crab pulsar excludes a spinning white dwarf, which would break up at that spin rate as a result of the large centrifugal forces. The spin rate was also found to be decreasing slowly at just the rate that would be expected if the luminosity of the Crab nebula, $\sim 10^5 \, L_\odot$, were powered by the rotational energy of a compact star with the

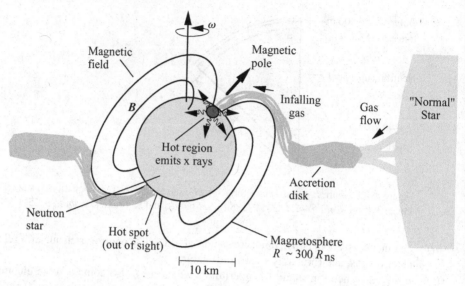

Figure 4.13. High-mass x-ray binary with x-ray pulsar. Gaseous matter accretes from the large normal star, size $\gtrsim 10^9$ m, to the compact neutron star, size $\sim 10^4$ m. The gas accumulates in an accretion disk and eventually is guided to the magnetic pole of the neutron star by the strong magnetic field. The hot region on the star is seen as a pulsing source as it comes into and out of sight while the neutron star rotates.

moment of inertia of a neutron star, $\sim M_{ns}R_{ns}^2$. In the simplest picture, a spinning neutron star with off-axis magnetic field radiates magnetic dipole radiation into the nebula. This leads to acceleration of electrons in the nebula, probably via shocks, and these electrons radiate the observed photons. The energy source of this radiation had been a major puzzle of astronomy since the 1950s.

The current view is that the rotating neutron star has a strong magnetic field, which is conducive to acceleration of electrons and to beaming of radiation in the direction along the magnetic axis – most likely by curvature radiation (Section 8.6). The magnetic pole is offset from the spin axis, as is the case for the earth. The radio beam thus sweeps around the sky like a lighthouse beam (Fig. 8.12), and observers on earth consequently detect brief, regularly spaced pulses of radio emission.

The discovery of *pulsing accreting x-ray-emitting binary systems* about 4 years later, in 1971, was humankind's second view of neutron stars. In this case, a spinning neutron star is in a close binary orbit with a normal star that, because of its proximity, accretes gas into the potential well of the neutron star (Fig. 13). The gas organizes itself into an accretion disk and drifts toward smaller radii as it gives up bulk-motion energy through dissipation. The inner disk, which is deep in the potential well, is so hot that it emits x rays.

If the neutron star has a sufficiently strong magnetic field (e.g., $\sim 10^8$ T), the gas will be guided by the magnetic field onto localized regions of the star, which become x-ray hot and perhaps the dominant source of x rays. These hot spots come into and out of view as the neutron star rotates, thus giving rise to pulses of x rays. Only neutron stars could have the deep gravitational potential wells and strong offset magnetic fields sufficient to bring about the observed intense x-ray pulsing.

These x-ray sources are now categorized broadly into *high-mass x-ray binaries* (HMXBs) and *low-mass x-ray binaries* (LMXBs) as determined by whether the normal companion is substantially more massive than the $\sim$1.4-$M_\odot$ neutron star or less than (or comparable to) it. The two types have quite different observable properties in several respects; for example, the HMXBs are often x-ray pulsars, as in Fig. 13, and the LMXBs usually are not.

Isolated, pulsing neutron stars with exceptionally high magnetic fields ($\sim 10^{10}$ to 10^{11} T) are known as *magnetars, anomalous x-ray pulsars,* or, in some cases, *soft gamma-ray repeaters* in accordance with the particular phenomena exhibited. Their x-ray luminosity and occasional huge outbursts most likely derive from the magnetic field energy contained in these stars.

Maximum mass

A neutron star can not be arbitrarily massive according to current thinking. If, at some sufficiently large mass, the nuclear and degeneracy pressures could no longer withstand the inward pull of gravity, the ensuing collapse would inevitably carry the entire star to within the Schwarzschild radius to become a black hole. There is no known force that could withstand gravitational forces of this magnitude.

An elementary upper limit to the possible mass of a neutron star follows directly from the concept of the Schwarzschild radius, $R_S = 2GM/c^2$, the event horizon of a nonrotating black hole; see (36) below. Consider a hypothetical neutron star of average density ρ_{av} so that its radius is related to its mass as $M = 4\pi R^3 \rho_{av}/3$, or,

$$R \approx \left(\frac{M}{4\rho_{av}} \right)^{1/3}. \tag{4.30}$$

As mass is added to a neutron star, the neutron-star radius R increases as $M^{1/3}$ for constant ρ_{av} and more slowly (or even decreases) if ρ_{av} increases. In contrast, the Schwarzschild radius (36) increases linearly with M. It thus eventually becomes greater than the neutron-star radius. At this point, the neutron star becomes, by definition, a black hole. The requirement for this is thus that $R_S > R$, or

$$\frac{2GM}{c^2} > \left(\frac{M}{4\rho_{av}} \right)^{1/3}. \qquad \text{(Condition for neutron-star collapse)} \tag{4.31}$$

Solve for M:

$$M > \frac{c^3}{2^{5/2} \rho_{av}^{1/2} G^{3/2}}. \tag{4.32}$$

For the typical neutron star of density $\rho_{av} = 5 \times 10^{17}$ kg/m^3, one obtains $M > 6\,M_\odot$, or according to a better calculation,

➡ $$M \gtrsim 5\,M_\odot. \qquad \text{(Mass limit for neutron-star collapse)} \tag{4.33}$$

This argument, which is based solely on the general relativistic notion of an event horizon, indicates that a neutron star with the specified average density could not exist as a visible object with a mass greater than $\sim 5\,M_\odot$. No matter how resistant the star is to collapse (with a very stiff EOS), the entire star would lie within the event horizon and hence would be invisible to external observers.

This calculation can be repeated with a more appropriate mass-radius relation, $R \propto M^{-1/3}$, to find a somewhat lower limit, $3.6 M_\odot$ (Prob. 44g). In this case, as mass is added, the star surface descends as the event horizon expands toward it, and so less mass is needed. One can also approximate that the support pressure is due solely to degeneracy of neutrons to find a limit of $5.0 M_\odot$ (Prob. 44f).

This mass limit is further reduced to about $3 M_\odot$ if one requires the speed of sound $(\mathrm{d}P/\mathrm{d}\rho)^{1/2}$ in the neutron star to be less than the speed of light,

$$\left(\frac{\mathrm{d}P}{\mathrm{d}\rho}\right)^{1/2} \leq c, \qquad \text{(Causality condition)} \qquad (4.34)$$

which is the *causality condition*. Further, calculations of limits based on realistic equations of state that incorporate the effects of general relativity (GR) yield mass limits in the range 1.5–$2.7 M_\odot$. In all of these cases, the analysis shows that the star can not withstand the pull of gravity and that it must collapse inward if it exceeds the calculated mass limit. The inevitable result of adding mass to a neutron star would seem to be a black hole.

Black holes

The compact object in some x-ray–emitting binary systems is most likely a *stellar black hole* of mass $\sim 10 M_\odot$. The active galactic nucleus in the center of some galaxies is believed to be powered by a *massive black hole* with mass of 10^6–$10^8 M_\odot$. In both types, the black hole is evident to us because of the substantial gaseous matter being accreted into it. The highly ionized plasma reaches x-ray temperatures before entering the black hole and thus can radiate to a distant observer. In the stellar case, it is gas from the normal gaseous binary companion, and in the galactic case most likely gas from disrupted stars near the galactic nucleus.

A black hole is, in one sense, simply an extension of the neutron-star concept; it is an object even more compact with an even deeper potential well. A photon emitted trying to escape would find itself pulled back, just as an underpowered rocket is pulled right back to the earth. An observer sees only the environs of the black hole. The challenge today is to find characteristics in the radiation that distinguish black holes from neutron stars.

Event horizon (Schwarzschild radius)

It is an effect of Einstein's general theory of relativity that photons will lose energy as they climb out of a gravitational potential well; the photons will be redshifted. Let a photon be emitted with frequency ν_r from a radial position r measured from the position of a gravitational mass M. The frequency ν observed at an infinite distance is related to ν_r as

$$\frac{\nu}{\nu_r} = \left(1 - \frac{2GM}{c^2 r}\right)^{1/2}. \qquad \text{(Gravitational red shift; } J = 0) \qquad (4.35)$$

If the photon starts out at a very large radius, $r \gg 2GM/c^2$, negligible frequency shift occurs ($\nu \approx \nu_r$). If the photon starts deeper in the well (e.g., at $r = 4GM/c^2$, its frequency undergoes a significant red shift as it climbs out of the well ($\nu = \nu_r/\sqrt{2}$). Because the energy of the photon is $E = h\nu$, this corresponds to a large decrease in its energy.

If the photon starts even deeper in the well, at $r = 2GM/c^2$, the frequency at infinity is redshifted to zero, $\nu = 0$. It takes the entire energy of the photon to escape the gravitational pull of M from this radius. This radius defines the *event horizon* or *Schwarzschild radius*,

$$\Rightarrow \quad R_S = \frac{2GM}{c^2}. \qquad \text{(Schwarzschild radius or event horizon)} \qquad (4.36)$$

There is no way to get information to the outside world from positions closer to the star than the Schwarzschild radius. If a mass M is contained within the radius R_S, it can not be seen by an outside observer. Any mass, in principle, can be a black hole if it is small enough. A 1-$M_\odot$ object has $R_S = 3.0$ km. This is barely smaller than our nominal neutron star of 10 km; if a 1.4-$M_\odot$ neutron star were 40% its nominal size, it would be a black hole and photons from it could not reach exterior observers.

Note the linearity of M and R_S. A mass of $\sim 6\,M_\odot$ of radius less than 18 km would be a black hole. A cubic kilometer of water (10^{12} kg $= 0.5 \times 10^{-18} M_\odot$) would be a black hole if it could be compressed to a sphere of radius 1.5×10^{-15} m, the radius of a proton.

Angular momentum

The matter entering a black hole can have a variety of attributes, but the only ones that survive entry into the black hole are the mass M, the charge Q, and the angular momentum J. It is also possible that, for $Q \neq 0$, $J \neq 0$, the black hole could exhibit an axial magnetic field, but electrostatic forces would quickly neutralize any net charge giving rise to it. All other attributes (e.g., distributions of shape) are radiated away in the form of electromagnetic and gravitational radiation. The remaining attributes, M, Q, and J, can in principle be measured with test masses, charges, and gyroscopes at distances well outside the event horizon according to GR.

A black hole with $Q = J = 0$ is called a *Schwarzschild black hole*. A *Kerr black hole* has nonzero values of angular momentum J, but no charge. Angular momentum is invariably present in collapsing systems and compact objects, and so black holes are usually expected to have $J \neq 0$. Most astrophysical systems are electrically neutral; thus, $Q = 0$ should be appropriate for most astronomical black holes.

The event horizon of a Kerr black hole with angular momentum coefficient $j \equiv J/J_{max}$ can be shown from GR to be at radius

$$R_h = \frac{GM}{c^2}\left[1 + (1 - j^2)^{1/2}\right]. \qquad \text{(Event horizon; } j \equiv J/J_{max}) \qquad (4.37)$$

A Schwarzschild black hole ($j = 0$) has an event horizon at $R_h = R_S = 2GM/c^2$ consistent with (36). In contrast, a *maximally rotating* Kerr black hole ($j = \pm 1$) has an event horizon at $R_h = GM/c^2$, or $1/2\,R_S$. These values are tabulated in Table 4 in units of GM/c^2.

The maximum angular momentum that a black hole can have may be obtained from Newtonian considerations. Consider a mass M at radius R orbiting a central point at speed v. The Newtonian angular momentum is $J = MvR$. To obtain the maximum angular momentum, consider M to be the entire mass of the black hole, the speed $v = c$, and the orbital radius that of the Kerr event horizon, $R = GM/c^2$, as follows:

$$J_{max} = Mc\frac{GM}{c^2} = \frac{GM^2}{c}. \qquad \text{(Maximum angular momentum for BH)} \qquad (4.38)$$

This is the correct GR result.

Table 4.4. *Radii of event horizon R_h and innermost stable orbit R_{iso}*

$j = J/J_{max}$[a]	$R_h(GM/c^2)$	$R_{iso}(GM/c^2)$	
-1[b]	1	9	retrograde
0[c]	2	6	no spin
$+1$[b]	1	1	prograde

[a] $J_{max} = GM^2/c$
[b] Kerr black hole (maximally rotating)
[c] Schwarzschild black hole (nonrotating)

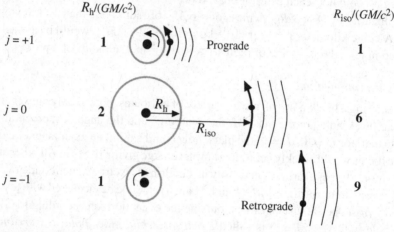

Figure 4.14. Radii of event horizons R_h (boldface numerals to left) and innermost stable orbits R_{iso} (boldface numerals to right) for spin zero and maximal spin compact objects ($j \equiv J/J_{max}$) in units of GM/c^2. The substantial differences in R_{iso} may permit astrophysicists to determine the angular momentum of a black hole.

Innermost stable orbit

The orbit of a test particle about a black hole results in capture if it comes too close to the hole; in such cases, the orbit is unstable. In contrast, a Newtonian orbit is stable at any radius unless the particle collides with the central object or otherwise suffers energy loss or gain. In GR, the stable orbit closest to the black hole is circular and is called the *innermost stable orbit* or the *marginally stable orbit*. Its radius is designated R_{iso}. This radius depends on both the mass and the angular momentum of the black hole.

The expression for R_{iso} is rather involved, and so we present in Table 4 and Fig. 14 only the results for Schwarzschild ($j = 0$) and maximally rotating Kerr black holes ($j = \pm 1$) in units of GM/c^2. The positive value of j indicates the test particle rotates in the prograde sense (i.e., its angular momentum is in the same direction as the spin of the central mass).

For a Schwarzschild black hole ($j = 0$), the innermost stable orbit is three times farther out than the event horizon; that is, $R_{iso} = 3R_h$. For a maximally rotating black hole with prograde orbit, it is coincident with the event horizon, $R_{iso} = R_h$, which itself is smaller by a factor of two than for the Schwarzschild case. For retrograde motion, the innermost stable orbit is quite far out, $R_{iso} = 9R_h$. (For neutron stars, see Prob. 47.)

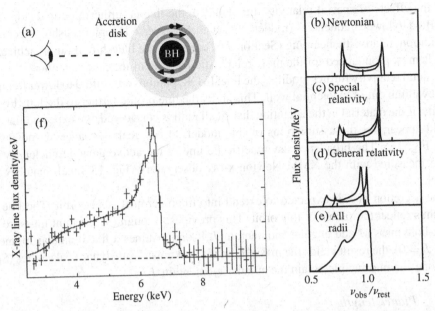

Figure 4.15. X-ray iron line profile. (a) Accretion disk around a black hole with two annular elements with observer in disk plane. (b)–(e) Line profiles for the two annular elements, including (b)
Newtonian Doppler shifts, (c) beaming and transverse Doppler of special relativity, (d) the gravitational redshift of general relativity, and finally (e) the contributions of all radial bands. (f) Profile in
MCG 6-30-15 from the XMM-Newton x-ray observatory with model (solid line). [(a–e) A. Fabian
et al., *PASP* **112**, 1145 (2000); (f) A. Fabian *et al.*, *MNRAS* **335**, L1 (2002)]

We thus see that, in GR, the angular momentum of a massive object renders the surrounding
space azimuthally asymmetric. It has a twist or torsion. This is reflected in the different
behaviors of particles in prograde and retrograde orbits outside the event horizon.

One can hypothesize that an accretion disk will typically extend down to the innermost
stable orbit about a black hole. The material orbiting closest to the black hole would be deep
into the potential well and thus would have the highest temperature with value representative
of R_{iso}; recall the virial theorem (Section 2.4). The observed spectrum would thus reflect this
temperature, and this could, in principle, give one a handle on the angular momentum of the
black hole.

Broad, distorted iron line

Spectral lines can be radically broadened and distorted by the effects of gravity near a compact
object. The prominent Kα iron line at 6.4 keV that appears in x-ray spectra of compact objects
is a prime example.

Consider the azimuthally symmetric accretion disk of Fig. 15a. If all the elements in
a circular annulus (darkened) are emitting a narrow Fe line, the Newtonian Doppler shift
will yield the symmetric spectral shape of Fig. 15b. The horns at either extreme represent
the material directly receding or approaching the observer; it has the greatest line-of-sight
velocity. The two curves represent the two annuli of Fig. 15a; the inner one yields the broader
line because the inner material rotates with greater speed according to Kepler's third law.

If the effects of special relativity are included, the line is further modified (Fig. 15c). It is shifted to lower frequencies (reddened); see the transverse Doppler redshift (Section 7.4). In addition, relativistic beaming (Section 7.6) enhances the blueshifted (approaching) horn. This is most pronounced for the more relativistic material of the inner annulus.

General relativity further modifies the line. In GR, photons are shifted to lower frequencies as they climb out of a potential well. This shifts the line profile farther to the left (Fig. 15d). Finally, if the material in the accretion disk at all radii is considered, the expected line profile would appear, for the assumptions of this model, to be greatly broadened and distorted, as in Fig. 15e. The broadened asymmetric Fe line of the active galactic nucleus MCG 6-30-15 measured with the XMM-Newton x-ray observatory (Fig. 15f) well matches such a model.

The accretion disk is expected to extend into the innermost stable orbit. The line shape will thus reflect the radius of this orbit. This provides, through (37), a joint measure of the black-hole mass M and angular momentum J. If one assumes a fixed angular momentum, (e.g., $J = 0$), the result yields the mass M, or, if M is known from dynamical studies (Section 1.6), one can in principle obtain the angular momentum J.

Planck length

What happens to an object that falls into a nonrotating black hole? After crossing the event horizon, the theory indicates the object will continue to fall inward, reaching a central point, where it will encounter, in a very short time, the *Schwarzschild singularity*. The singularity is a breakdown of the equations of GR at the center of a black hole. Because the theory does not include quantum effects, this singularity may not be a physical reality.

Let us now estimate the size scale at which quantum effects become important. This occurs when the radius of the event horizon, $\sim GM/c^2$, for a given mass is so small that it matches the *Compton wavelength* $\hbar/mc$ of that mass; that is,

$$\frac{Gm}{c^2} = \frac{\hbar}{mc}, \tag{4.39}$$

where $\hbar \equiv h/2\pi$. The Compton wavelength is the length implied by the uncertainty principle $\Delta x \, \Delta p \gtrsim \hbar$ when Δp is set to mc. In its alternate definition, h/mc, it is the wavelength shift suffered by a photon that is Compton scattered at $90°$ by an initially stationary electron (Section 9.2).

Solve (39) for $m = m_P$, the *Planck mass*:

$$m_P = \left(\frac{\hbar c}{G}\right)^{1/2} = 2.2 \times 10^{-8} \text{ kg}. \qquad \text{(Planck mass)} \tag{4.40}$$

This is the mass for which the Schwarzschild radius equals the Compton wavelength. It is not necessarily a physical mass. The Compton wavelength for the mass m_P is known as the *Planck length*,

$$\ell_P = \frac{\hbar}{m_P c} = \left(\frac{\hbar G}{c^3}\right)^{1/2} = 1.6 \times 10^{-35} \text{ m}. \qquad \text{(Planck length)} \tag{4.41}$$

The region where quantum effects must enter is small indeed.

The implied mass density in this region is huge:

$$\rho_P = \frac{m_P}{\ell_P^3} = \frac{c^5}{\hbar G^2} = 5.2 \times 10^{96}\ \text{kg/m}^3. \qquad \text{(Planck mass density)} \qquad (4.42)$$

Because the energy and (total) mass are related as $E = mc^2$, the mass density (42) corresponds to an *energy* density of

$$u_P = \rho_P c^2 = \frac{c^7}{\hbar G^2} = 4.6 \times 10^{113}\ \text{J/m}^3. \qquad \text{(Planck energy density)} \qquad (4.43)$$

Finally, it is useful to note the time it takes for a light signal to travel the Planck length, which is known as the *Planck time*,

$$t_P = \frac{\ell_P}{c} = \left(\frac{\hbar G}{c^5}\right)^{1/2} = 5.4 \times 10^{-44}\ \text{s}. \qquad \text{(Planck time)} \qquad (4.44)$$

The opposite sides of the spherical event horizon can not communicate with each other in times less than this; otherwise, causality would be violated. There is no way to discuss times shorter than the Planck time sensibly or distances smaller than the Planck length without a theory of quantum gravity.

Particle acceleration

A black hole with angular momentum can, in principle, give energy to particles that interact with it. If a particle approaches the hole with energy E_1 in just the proper orbit, and if it can be instructed to split in two at the proper position near the black hole, one piece will go into an orbit with *negative*(!) energy E_2 and be captured by the hole, whereas the other will escape out to infinity with energy E_3 greater than E_1, as required to conserve energy: $E_1 = E_2 + E_3$. One could imagine extracting huge amounts of energy from a black hole with this mechanism, but unfortunately this may not be physically realizable.

Evaporation

Another surprising (theoretical) characteristic of black holes, whether or not rotating, is that they emit a thermal spectrum of particles if quantum effects are taken into account. This is known as the *Hawking process*. The energy radiated arises from "virtual" pairs of particles (e.g., $e^+ e^-$) that are continually being created and annihilated in the fluctuations of the vacuum. The strong gravity near a black hole can separate the charges, and so they become real particles. One is captured by the black hole, and the other escapes (tunnels) to infinity. In optically thick conditions, the emerging particle kinetic energies are converted to radiation with a blackbody spectrum.

The temperature of this radiation is, according to the theory,

$$T = \frac{\hbar c^3}{8\pi k G M} = 6.1 \times 10^{-8} \left(\frac{M_\odot}{M}\right)\ \text{K}, \qquad \text{(Hawking radiation)} \qquad (4.45)$$

which is inversely proportional to the mass M of the black hole. The temperature is only $\sim 10^{-7}$ K for a one-solar-mass black hole, but it can become substantial for small masses. At $M = 5 \times 10^{-19} M_\odot = 10^{12}$ kg, for example, one has $T \approx 10^{11}$ K, which corresponds to an average photon energy of ~ 10 MeV. This value is equivalent to ~ 20 times the 0.5-MeV rest energy of an electron.

The luminosity of the black hole follows if one takes the emitting area to be spherical with the Schwarzschild radius. Because $R_S \propto M$ (36) and $T \propto M^{-1}$ (45), the total power (luminosity) radiated is

$$L = 4\pi R_S^2 \, \sigma T^4 \propto M^2 \times M^{-4} \propto M^{-2}. \tag{4.46}$$

The luminosity increases rapidly as mass decreases.

The approximate (characteristic) time τ_c to radiate away the entire energy content Mc^2 of the black hole at a given luminosity is $\sim Mc^2/L \propto M^3$. A relation for τ_c as a function of the initial mass M_i follows from (36), (45), and (46) (Prob. 48). A proper calculation yields

$$\tau_c = 1.4 \times 10^{10} \left(\frac{M}{5 \times 10^{11}\,\text{kg}} \right)^3 \text{ yr.} \quad \text{(Characteristic evaporation time)} \tag{4.47}$$

The characteristic time shortens drastically as the mass decreases. The evaporation therefore accelerates and the black hole actually *evaporates* away to nothing. In its last 0.1 s, or perhaps earlier when it reaches temperature $\sim 10^{12}$ K, the photon energies become sufficient to create pi mesons; the black hole explodes in a burst of particles and gamma rays. Such bursts could, in principle, be detected with radio or gamma-ray instruments of sufficient sensitivity.

Equation (47) indicates that a black hole of mass $\sim 5 \times 10^{11}$ kg would radiate its energy away in $\sim 10^{10}$ yr, the age of the universe. This is the mass of half a cubic kilometer of water that has a Schwarzschild radius of $\sim 10^{-15}$ m, which is the size of the proton; it is a "mini" black hole of mass only $\sim 10^{-19}\ M_\odot$.

It has been suggested that many of these "mini" black holes could have been created in the hot dense phases of the early universe and that these would be evaporating and exploding throughout the life of the universe. The less massive ones would long since have exploded, whereas the ones of 5×10^{11} kg would just now be ending their lives.

It is fascinating to ponder the possibility that these low-mass black holes could return their entire energy content to the accessible universe just as they once removed energy content from it when they were formed. Indeed, in their hot radiative phases, they could be called *white holes*.

The existence of these "mini" black holes is highly speculative; they may never have formed in the early universe. There is, at present, no observational evidence for their existence. The gamma-ray bursts that are the focus of much current research (Section 3 above under "Gamma-ray bursts") do not have the expected characteristics – most notably, uniformity from burst to burst.

Existence of black holes

Black hole existence gained credibility with the discovery that the x-ray binary Cygnus X-1 has a compact companion with mass in excess of that plausible for a neutron star (Section 1.6 under "Examples"). Some two dozen additional compelling examples are now known. The spectral and temporal study of radiation from accreting stellar black holes provides much of our knowledge of them (e.g., with the *Rossi X-ray Timing Explorer satellite* launched in 1995).

The case for a black hole's being the power source in active galactic nuclei is quite strong. The high luminosities argue for high masses; see the discussion of the Eddington limit

(Section 2.7). Rapid temporal variability (hours and days) limits source regions to light travel distances unless beaming is involved. Masses of the central object are also obtained from spectral line broadening arising from orbiting gases and by tracking nearby orbiting bodies with high-resolution imaging (Section 1.7). The large masses so obtained are constrained to volumes so small that alternative energy sources such as dense star clusters are not plausible.

Another argument favoring the existence of black holes is that, in GR, pressure is an additional source of gravity, the force that attracts matter to other matter. Consider an accreting neutron star or the growing inert core of a giant star. As the mass increases, the central pressure grows so as to maintain hydrostatic equilibrium, but it also creates more gravity that produces another inward force! An increase of compactness must therefore eventually lead to collapse if this aspect of GR is correct. The necessary result would be continued collapse to a singularity. This takes place inside the event horizon, and so astronomers can not observe it.

We caution that these arguments do not totally ensure the existence of black holes. These applications make use of GR in the "strong-field" regime of the theory, which has been tested only in the "weak-field" limit. Thus, the discovery of compact objects of mass $\gtrsim 3$–$5\ M_\odot$ does not necessarily mean that black holes have been discovered. However, we know of no other kind of object, or type of physics, that is so in accord with the observations.

4.5 Binary evolution

The two stars in a binary system can directly affect each other's evolution through accretion of matter from one to the other, tidal forces, and gravitational radiation, among other factors. In their collapse and formation from the interstellar medium, binaries can be formed with a wide range of mass combinations, energy, and angular momenta. In subsequent evolutionary stages, there can be losses of mass and angular momentum from the system in highly uncertain amounts. The evolution of a particular system may thus take a variety of paths.

Nevertheless, observations of various types of binaries (normal stars, x-ray emitters, radio pulsars, etc.) together with theoretical modeling have enabled astrophysicists to deduce much about the evolutionary paths that do take place. For stars of sufficient mass, the inexorable action of gravity will, in time, carry the binary to a final state of one or two compact objects.

We do not attempt here to give a comprehensive view of all the possibilities. Rather, we consider some of the basic principles that govern the evolution of binaries and then present two characteristic examples of the multistage evolution a binary system might undergo.

Time scales

The processes that govern binary evolution are rooted in the evolution of single stars discussed in Section 3. A star in a binary will evolve as a single star until and unless it is perturbed by its partner.

We learned, for example, that a more massive star will evolve more rapidly than a low-mass star. A massive companion will thus enter the giant phase before its lower-mass companion, thereby losing mass to stellar winds or by overflow (accretion) of its gaseous matter onto the lower-mass companion. The increased mass of the latter star, if substantial, will lead to an increased rate of evolution, and it will enter its giant phase earlier than it otherwise would have.

On the other hand, mass lost by the giant star would be primarily from its diffuse envelope, and so the evolution of its helium core would be largely unaffected. It would proceed to its final compact state on its own time scale – possibly via a supernova.

Binaries containing low-mass stars ($\lesssim 0.7\ M_\odot$) live as long as or longer than the age of the universe. Thus, binaries containing a low-mass star may persist for a very long time compared with those with higher mass stars.

Gravitational radiation

Two masses orbiting one another will radiate gravitational waves according to GR because the mass distribution has a time-varying quadrupole moment. The energy loss due to this has been dramatically demonstrated through the observation of the orbital decay of a two-neutron-star binary system, the "binary pulsar" PSR B1913+16, which is also known as the Hulse–Taylor pulsar (AM, Chapter 12).

This energy loss is the analog of accelerating charges radiating electromagnetic waves, though the latter can emit dipole radiation whereas gravitational matter can not. Quadrupole radiation is the lowest order possible for gravitational radiation because there is only one sign of mass. In both cases the emitted waves travel at the speed of light, though gravitational waves have not yet been directly detected.

Energy loss rate

The rate of energy change through gravitational radiation of a binary system with masses m_1 and m_2 in circular orbits with star separation s is, from GR,

$$\frac{dE}{dt} = -\frac{32}{5}\frac{G^4}{c^5}\frac{M_T^3\mu^2}{s^5} \qquad \text{(Gravitational energy loss;}$$

$$\xrightarrow[m_1=m_2\equiv m]{} -\frac{64}{5}\frac{G^4 m^5}{c^5 s^5}, \qquad \text{circular orbits; W)} \qquad (4.48)$$

where $\mu = m_1 m_2 / M_T$ is the reduced mass (1.59), $M_T = m_1 + m_2$ the total mass, and s the star separation. Note that large masses and small separations greatly increase the rate of energy loss. Energy loss in a binary shrinks the orbit with an associated period decrease.

Recall that, in general, each star in a binary sweeps out an ellipse with focus at the barycenter (BC) (Fig. 1.10b) and that the relative orbit (referenced to m_1) also sweeps out an ellipse (Fig. 1.10c). We adopt the notation of Chapter 1 wherein the relative vector separation of the two stars is $\boldsymbol{s} = \boldsymbol{r}_2 - \boldsymbol{r}_1$ and the semimajor axis of the ellipse swept out by $\boldsymbol{s}$ is a_s. For a circular orbit, $a_s = s$, where s is the magnitude of $\boldsymbol{s}$.

The total energy (potential plus kinetic) of a binary with elliptical orbit and unequal masses is, from (1.78),

$$E_t = -\frac{GM_T\mu}{2a_s} \xrightarrow[\substack{m_1=m_2\equiv m \\ a_s=s}]{} -\frac{Gm^2}{2s}. \qquad \text{(Total energy; circular orbit; J)} \qquad (4.49)$$

The limiting value for equal masses and circular orbit is also given.

The time it takes to radiate away a substantial portion of this energy, the characteristic time τ, follows from (48) and (49). For two neutron stars of $1.4M_\odot$ in circular orbits,

$$\tau = \frac{E_t}{dE/dt} = \frac{5}{128} \frac{c^5}{G^3} \frac{s^4}{m^3}$$

$$\underset{m=1.4\,M_\odot}{\rightarrow} 1.5 \times 10^{-20} s^4,$$

(4.50)

where τ and s are, in SI units, seconds and meters, respectively. This is the time it would take the energy to decrease to twice its initial (negative) value if, hypothetically, dE/dt were held fixed at its initial value. This energy decrease would cause the orbit to shrink a factor of two; see (49). The actual time for this shrinkage is less than τ because the rate of energy loss increases as the orbit shrinks.

If the time constant τ is set equal to the age of the universe, $\sim 1 \times 10^{10}$ yr, we find from (50) that $s \approx 3\,R_\odot$. This is about the spacing of the two neutron stars of PSR B1913+16. The integrated effect of the pulsar's orbital speedup due to the minuscule shrinkage of its orbit is detectable in a few years because of the precision of pulse-timing measurements (Prob. 51). Orbits of substantially greater spacing will be negligibly affected by gravitational radiation. Note that $3\,R_\odot \approx 200\,000$ neutron-star radii ($1.0\,R_{ns} = 10$ km).

A compact star can come quite close to its partner because of its small size. If the stars had one-half the just quoted spacing (i.e., $s = 100\,000\,R_{ns}$), the decay time would be reduced a factor of 16 to $<10^9$ years, which is well under the age of the solar system, $\sim 10^{10}$ yr. At one-tenth the spacing ($s = 20\,000\,R_{ns}$), the characteristic time would be $\sim 10^6$ yr, which is comparable to the times the more massive stars are on the main sequence.

Gravitational radiation can thus bring compact stars in close binaries substantially toward one another on evolutionary time scales. Compact stars in accreting low-mass binaries may be kept in "contact" by gravitational radiation and also by magnetic braking (see just below), counteracting the tendency of accretion to drive them apart (60).

Final chirp

An extreme case is two neutron stars separated by only $10\,R_{ns} = 10^5$ m. We find from (50) that $\tau = 1.5$ s. The neutron stars will spiral into each other with a ~ 1-s "chirp" of gravitational radiation so intense that such events in nearby galaxies should be detectable with gravitational wave detectors now coming on line (AM, Chapter 12). The chirp signal has a rapidly increasing amplitude and frequency, and it suddenly terminates at about 1 kHz when the two stars merge.

Tidal interaction

Stars in a binary system will interact tidally. Within a few stellar radii, the forces can be very strong. Such interactions result in a loss of energy that leads to *circularization* of an elliptical orbit and also to *synchronization* of the spin periods of the stars and the orbital period.

The circularization and synchronization are due to the gravitational effect of tidal bulges raised on one or both stars. When the orbit is circular and both stars are synchronized to the orbit, there is no more energy loss due to tidal dissipation. The system can then rotate with stable period and separation until some other effect perturbs it – perhaps the evolution of one of the stars.

The earth-moon system is a familiar example. The earth's oceanic tidal bulge is dragged ahead of the moon by the rapidly rotating earth. The mutual force between the tidal bulge and the moon tends to raise the moon's orbit and slow the earth's spin. The end result is that the system tends toward synchronization in which the earth's rotation will eventually be synchronized with the earth-moon orbital period. The moon's spin is already synchronized with the orbit. (The earth-moon interaction is worked out in the problems of AM, Chapter 4.)

The circularization can be understood as a minimization of energy with angular momentum being conserved. If mass is not ejected from the system and external torques are negligible, the system angular momentum is expected to be conserved. Most of the angular momentum is carried in the orbital motion; spin angular momentum is usually negligible. Under conservation of orbital angular momentum, as given in (1.73), it can be shown (Prob. 54) that circularization lowers the apastron and raises the periastron.

Consider how a NASA controller might perform such a maneuver for an earth-orbiting satellite. (See our discussion of this in Section 1.3.) A slowing burn at periastron of the proper duration would lower the apastron to the desired radius of the final circular orbit. Then, an energy-increasing burn at the new apastron would raise the periastron to the desired radius. The orbit would then be circular.

Tides can have the same effect through many small momentum exchanges throughout the orbit. Assume that the star with the largest tides (like the earth in the earth-moon system) has already been synchronized; it is spinning with a period equal to that of the elliptical orbit. At periastron, the rapid orbital angular velocity of the partner exceeds that of the tidal star's spin, whereas at apastron it lags that of the spin.

The tidal hump tries to align itself with the partner but because of viscosity will be dragged somewhat off this direction by the tidal star's spin. At perigee, the star spin lags the rapid passage of the orbiting partner, and thus so does the tidal hump. The gravitational attraction between the hump and partner will thereby slow the partner somewhat. At apogee, the star spin exceeds the partner motion, and so the hump is dragged ahead of the orbiting partner. The gravitational pull thus gives energy to the partner.

These two cases indicate how small tidal impulses all around the orbit can have the same effect as the actions of the NASA controller in circularizing the orbit. In this discussion, we visualized a massive tidal star and lighter orbiting partner, but the arguments also apply to the relative orbit of two stars with arbitrary masses.

Magnetic braking

Magnetic braking serves to bring stars in a binary together. Consider a rotating star with open magnetic field lines that can guide a solar wind to large radii while forcing the wind to corotate with the star until it finally becomes free of the magnetic field. The angular momentum thus imparted to the wind is lost to the star.

As an analog, visualize a student on a free-to-spin stool with a weight in each hand held close to the body. Upon stretching out his or her arms with the weights, the spin rate would decrease in accord with angular momentum conservation. If the student then lets go of the weights, applying no azimuthal force to them, they would fly off tangentially (gravity neglected), and the stool's spin would remain at its reduced rate with lesser angular momentum.

Consider a star in a binary in which the spin is synchronized with the orbit. Let one of the stars undergo mass loss with magnetic braking. Further assume that the ejected wind is lost to the entire system. As the star loses angular momentum, its spin would tend to slow and cause it to fall out of synchronization, but then the tidal forces would try to bring it back into synchronization. In this manner, the angular momentum loss (and the associated energy loss) would be shared with the entire system. The binary orbit would thus shrink with a shortening of the period.

Magnetic braking can serve to bring a low-mass star and its neutron-star companion closer together. This keeps the two stars in contact with each other, thus perpetuating accretion and x-ray emission; otherwise, as already noted, they would tend to separate (see (60) below).

Effective equipotentials

A test particle in the presence of, but outside the atmospheres of, two stars in a binary system will feel the gravitational pull of each star. In the two stars' rotating frame of reference, the test particle will also be acted on by the "fictitious" centrifugal and Coriolis forces. If the test particle is at rest in this frame, the Coriolis force is zero. If the stars are in circular orbits and are rotating synchronously with the orbit, they and their atmospheres will appear stationary in the rotating frame, and the forces in the rotating frame will not vary with time.

Accordingly, one can construct *effective equipotentials* that describe the forces on any given (stationary) particle in the rotating frame of reference. These forces are the gravitational forces of the two stars, assumed to be point masses, and the centrifugal force due to the frame rotation. The equipotentials are surfaces in three-dimensional space, and the direction of the net force on a test particle (neglecting pressure-gradient forces) at any point is opposed to the gradient of the potential.

A cross section through the two stars (Fig. 16a) in the orbital plane shows sample equipotentials as curved lines for two stars of masses 15.0 and 7.0 $M_\odot$ and separation $100\,R_\odot$. The lines are contours of constant potential analogous to the altitude contours on a hiker's map. A three-dimensional plot of the effective potential in the orbital plane of a similar system is shown in Fig. 16b.

At large distances, the centrifugal force, $F_c = m\omega^2 r$, dominates and increases with radius, thus yielding decreasing potentials that become increasingly steep and circular with radius. Close to the stars, the inward gravitational force, $F_g \propto r^{-2}$ dominates, and so the equipotentials there (not shown) are also nearly circular and increasingly steep toward the star.

Roche lobes

In the intermediate regions, the potentials of Fig. 16 exhibit five maxima or saddle points. These are positions of unstable equilibrium known as Lagrangian points. At each, the net force is zero. The one with the lowest potential (L1) lies on the line connecting the two stars. The figure-eight equipotential contour that passes through it defines the *Roche lobes* of the two stars. In three dimensions, the lobes are shaped like two end-to-end pears; see Fig. 17 below. At sequentially higher potentials, one finds Lagrangian points L2 beyond the less massive star (m_2), L3 beyond m_1, and L4 and L5 off to the sides at equal potentials. The two stars and L4 (or L5) form an equilateral triangle.

Each Lagrangian point is nominally a position of unstable equilibrium. The NASA satellites SOHO and WMAP have been stationed at the earth-sun L1 and L2 points, respectively, the first for continuous viewing of the sun and the latter for continuous viewing of the dark sky. At these locations the satellites require occasional orbit adjustments to keep them on station.

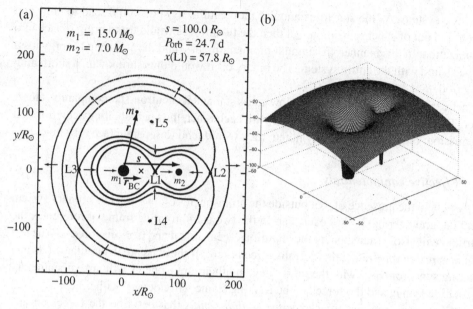

Figure 4.16. (a) Selected effective equipotentials ("pseudopotentials") for a binary with the indicated parameters shown in cross section at the orbital plane. The orbital angular momentum vector is out of the page, and the barycenter (BC) is indicated with a cross. The contours indicate levels of constant pseudopotential. The Lagrangian points L1 to L5 locate maxima or saddle points that are sequentially at higher and higher potentials. The potentials fall away to lower values at large distances owing to the centrifugal force and toward the two masses owing to the gravitational forces. Force directions, but not magnitudes, are shown with small arrows. The vector r to a test mass m is indicated. The Roche lobes (figure-eight contour) are at the potential of L1. The star with the smaller mass has the smaller Roche lobe. (b) Pseudopotential in the orbital plane of a binary with unequal masses plotted against arbitrary spatial coordinates. [(a) Adapted from T. Tauris and E. van den Heuvel, in *Compact Stellar X-ray Sources*, eds. W. Lewin and M. van der Klis, Cambridge 2006, Fig. 16.9, with permission; (b) Courtesy S. Rappaport.]

In contrast, the forces at and near L4 and L5 will maintain a test mass (or asteroid) close to the Lagrangian point without orbit adjustments if the two orbiting masses, m_1 and m_2 in Fig. 16a, have a mass ratio greater than $(\sqrt{27} + \sqrt{23})/(\sqrt{27} - \sqrt{23}) = 24.96$. (Motion of the test mass brings the Coriolis force into play.) The *Trojan asteroids* are located at the L4 and L5 points of the sun–Jupiter system. The asteroids precede and follow Jupiter by 60° in its orbit about the sun.

Lagrangian-point positions

It is highly useful to know the size (radius) of the Roche lobe of a star. If a star fills its lobe, matter will spill over to the other star via the L1 Lagrangian point. We quote here the radius of the lobe of star m_1. It is a fit to numerical integrations and represents the radius of a spherical volume that will equal the volume of the pear-shaped Roche lobe. It is a function only of the ratio of the masses, $q \equiv m_1/m_2$, and gives the "radius" of the lobe encompassing m_1 (see P. Eggleton, *ApJ* **268**, 368 (1983)):

$$R_{\mathrm{L},m_1} = \frac{0.49\,q^{2/3}}{0.6\,q^{2/3} + \ln(1 + q^{1/3})}\,s. \qquad \text{(Radius of Roche lobe of } m_1) \qquad (4.51)$$

If the star radius grows to this effective radius, accretion will begin.

The positions of Lagrangian points L1, L2, and L3 may be found by searching for positions of zero net force along the line passing through the two stars. The net force on a test mass m at any point in the equatorial plane is

➡ $$F = +m\omega^2(r - r_{BC}) - \frac{Gm_1 m}{r^3} r - \frac{Gm_2 m}{|r - r_2|^3}(r - r_2), \qquad \text{(Force; N)} \qquad (4.52)$$

where the origin is taken to be at m_1, the more massive object, and where the vector r locates the test mass m, r_{BC} locates the barycenter (BC), and $r_2 = s$ locates m_2 (Fig. 16a). This expression, as written, is valid at any position in the equatorial plane.

For positions along the line connecting the two masses (designated the x-axis), this expression becomes a one-dimensional equation in $u \equiv x/s$. Following (52), let x, $x_2 = s$ and $x_{BC} = (m_2/M_T)s$ be, respectively, the positions of the test mass, the mass m_2, and the BC as measured from the origin at m_1. Use Kepler's third law (1.75) to rewrite ω in terms of the total mass and separation s as

$$\omega^2 = \frac{GM_T}{s^3}, \qquad (4.53)$$

and finally set $F_x = 0$. After some rearrangement, one obtains an equation in $u \equiv x/s$,

$$0 = \left(u - \frac{m_2}{M_T}\right) - \frac{m_1}{M_T}\frac{u}{|u^3|} - \frac{m_2}{M_T}\frac{u-1}{|u-1|^3}, \qquad (4.54)$$

where u can take on values $-\infty < u < \infty$, which is satisfied by the positions u_i of the three Lagrangian points. For given values of m_1 and m_2, one can solve this for the Lagrangian positions u_i graphically or on a calculator by trial and error (Prob. 52).

Accretion

Main-sequence stars in a binary are typically much smaller than their Roche lobes, as in Fig. 16. Such a system is referred to as a *detached binary*. As one of the stars evolves into the giant phase, it can fill its Roche lobe and matter will spill over toward the companion star through the L1 point. At this stage, it is known as a *semidetached binary*.

Top and side views of an overflowing Roche lobe with a compact accretor are shown in Fig. 17. As the flowing material leaves the L1 point (top view), the Coriolis force, $F_{cor} = -2m\omega \times v'$, bends it to the right (Fig. 17a), but the increasingly strong gravity then bends it to the left into the accretion disk, which it strikes, possibly creating a "hot spot" in the accretion disk. The arrival direction of the accretion stream leads to disk rotation in the same direction as the binary itself (counterclockwise) just as a low-pressure region in the northern hemisphere of the earth leads to counterclockwise circulating winds.

The masses of the two stars may be changed substantially by the accretion. This leads to changes in the star separation and orbital period. In general, added mass will cause a star to evolve more rapidly; thus, it may enter its giant phase and fill its Roche lobe earlier than it would have otherwise (as noted above under "Time scales"). Under certain circumstances, mass can be lost to the entire system via overflow at the L2 point, via stellar winds, or via mass ejection during a *common envelope* phase. The latter occurs when the stars are so close that one is orbiting within the gaseous body (envelope) of the other. Such a system is sometimes known as a *contact* binary.

Accretion onto a compact object yields x-ray emission because the compact star has a very deep potential well. The relation between the mass accretion rate and the x-ray luminosity

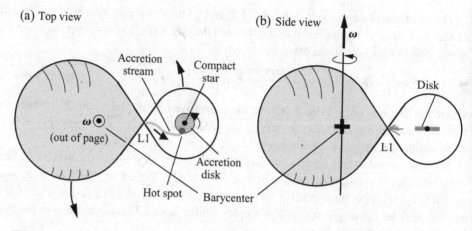

Figure 4.17. Overflowing Roche lobe. The matter from the filled lobe of the more massive star (left) exits through the L1 point and forms a counterclockwise rotating accretion disk surrounding the compact object. The angular velocity vector of the binary system is indicated.

was presented in Section 2.7 in the context of the limiting Eddington luminosity; see (2.80). See also Fig. 13 and our discussion in Section 4 of the evidence for neutron stars.

Star separation

Let us now examine how the star separation in a binary will be modified by the transfer of mass. Assume that the stars are in circular orbits – a not unlikely situation given tidal dissipation for stars in close proximity. Further, for simplicity, let both mass and angular momentum be conserved as matter is transferred (i.e., *conservative mass transfer*). This is true only in limited situations, but it gives us some insight into the processes.

The angular momentum magnitude, $J = J_1 + J_2$, of a two-body system (i.e., arbitrary m_1 and m_2) in elliptical orbits is given in (1.73) as a function of the total mass $M_T = m_1 + m_2$ and the relative semimajor and semiminor axes a_s and b_s (Fig. 1.10c). It is further assumed in the derivation of J that there is negligible (spin) angular momentum in the stars themselves. We quote (1.73) here,

$$J = \left(\frac{GM_T}{a_s}\right)^{1/2} \mu b_s,$$
(4.55)

where the reduced mass μ is defined (1.59) by

$$\mu \equiv \frac{m_1 m_2}{m_1 + m_2}.$$
(4.56)

The parameter J is the sum of the angular momenta of the two masses in the frame of their barycenter; see (1.72). The parameters a_s and b_s, though, pertain to the ellipse swept out by the relative radius vector $s = r_2 - r_1$.

For our circular orbit, we can set $a_s = b_s = s$, which is the constant distance between the two stars. (A circular binary orbit is illustrated in Fig. 1.6.) The angular momentum quoted thus may be written as

$$J = \left(\frac{G}{M_T}\right)^{1/2} m_1 m_2 s^{1/2}.$$
(4.57)

For a constant angular momentum and total mass, we have

$$s \propto \frac{1}{(m_1 m_2)^2}. \tag{4.58}$$

For the sum of the masses being constant (no mass lost), one can quickly confirm by trial and error that s is at a minimum when $m_1 = m_2$. (Try $m_1 + m_1 = 10$ to find that $5 \times 5 = 25$, $6 \times 4 = 24$, and so on.) Thus, mass transfer that tends to equalize the masses brings the stars together and vice versa.

This result may be obtained formally as follows. Take the differential of J (57) with variables m_1, m_2, and s and divide the result by J to obtain

$$\frac{dJ}{J} = \frac{dm_1}{m_1} + \frac{dm_2}{m_2} + \frac{1}{2}\frac{ds}{s}. \tag{4.59}$$

The fractional mass terms represent small changes in the masses of the two stars. For the case in which a small quantity of mass is moved from star 1 to star 2 without mass loss, one has $dm_1 = -dm_2$. In this case, $dm_2 > 0$; the mass of star 2 is increasing. Conservation of angular momentum gives $dJ = 0$. The fractional change in spacing is thus

$$\text{➡} \qquad \frac{ds}{s} = 2\frac{dm_2}{m_2}\left(\frac{m_2}{m_1} - 1\right). \qquad \text{(Fractional spacing change;} \atop m_1 \text{ is the donor star)} \tag{4.60}$$

It is the sign of ds that interests us. For $dm_2 > 0$, the factor in parentheses dictates the sign of ds.

Consider the case in which donor star m_1 is more massive than the accretor m_2 (e.g., Fig. 17). The parenthetical term is negative, and so $ds < 0$ and the two stars approach one another. This is in accord with our statement regarding (58) that equalizing the masses brings the stars together. If the donor is less massive than the accretor, the separation increases.

We thus summarize a fundamental result of conservative mass transfer. *If a low-mass star accretes onto a higher-mass star, the stars will tend to move apart. If a high-mass star accretes onto a lower-mass star, they will tend to move together.* This is valid if angular momentum and mass are conserved, but it could still be true if only modest amounts of matter and angular momentum are lost to the system.

We further note from (60) that the degree of separation change for a given amount of mass transfer is large for mass ratios, m_2/m_1, that are large or small relative to unity. For mass ratios near unity, the separation change is minimal.

The changing separation can have a major effect on the evolution. For example, a massive giant star accreting onto a 1.4-$M_\odot$ neutron star will bring the stars together, leading to a reduced Roche lobe size, enhanced accretion, and hence further shrinkage of the orbit. This positive feedback yields an unstable system that quickly becomes a common-envelope system wherein the neutron star is orbiting inside the envelope of the giant. The resultant turbulence can eject the entire envelope from the system.

The shrinkage of the orbit in this case need not be unstable if the donor's size decreases in response to its reduced mass sufficiently to offset the effect of the decreasing Roche lobe radius. For example, intermediate mass stars have an immediate adiabatic response to mass loss; they will shrink. This tends to stop the accretion (negative feedback). Thereafter the donor star will readjust itself to thermal equilibrium by expanding on the longer thermal time scale (Section 2.5). This would tend to restart accretion.

This all happens continuously, and so the result can be steady controlled mass transfer called *thermal time-scale mass transfer*. This type of transfer is possible if the closing rate is not too great – that is, if the mass ratio is not too extreme; see (60). For example, a 2-$M_\odot$ star can accrete stably onto a 1.4-$M_\odot$ neutron star. In contrast, a 15-$M_\odot$ star can not.

Period change

Kepler's third law (Kepler III) (1.75) tells us that, given the two star masses, the binary period defines the semimajor axis,

$$GM_T P^2 = 4\pi^2 a_s^3, \tag{4.61}$$

which for a circular orbit becomes

$$GM_T P^2 = 4\pi^2 s^3, \tag{4.62}$$

where s is the separation of the two stars. We thus see that the accretion of Fig. 17 results in a shorter and shorter period as the accretion drives two stars toward one another. If the accretion continues until and after the masses become equal, the period and separation will begin to increase again.

Stellar winds

Stars in their giant phases are known to eject substantial and uncertain fractions of their mass via stellar winds (see "Evolution of single stars" in Section 3 above). Stars in binary systems are no exception to this. A stellar wind will carry off angular momentum with the mass it ejects. Evolutionary calculations thus become quite dependent on the exact nature of the outflows. A given initial state (m_1, m_2, a_s, e) can have a variety of outcomes, which may be described with distributions derived from Monte Carlo simulations.

Pulsar wind and x-ray irradiation

A radio pulsar in a binary system can irradiate its partner with a *pulsar wind* of photons and relativistic particles that can expel material from the partner. Systems doing this are known and include PSR B1957+20 ($P_{orb} = 0.38$ d; $m_{partner} = 0.02\ M_\odot$). These systems have been called "black widow" pulsars because they are destroying their partners.

The final result could in principle be an isolated millisecond pulsar. It is more likely, however, that most millisecond radio pulsars we know were removed from the binary systems (in which they were spun up to millisecond periods) by near collisions with other stars in the crowded regions of globular clusters.

Another mechanism whereby a neutron star can affect its partner is *x-ray irradiation*. The intense x-ray flux from the neutron star in an accreting x-ray binary will irradiate and heat the atmosphere of the donor star. This could be a significant effect in low-mass x-ray binaries with their small star separations and long accretion phases (see discussion of Fig. 20 below). The heated atmosphere traps energy, causing it to expand, and this increases mass loss via Roche lobe overflow. The importance of this effect has not yet been exhibited observationally.

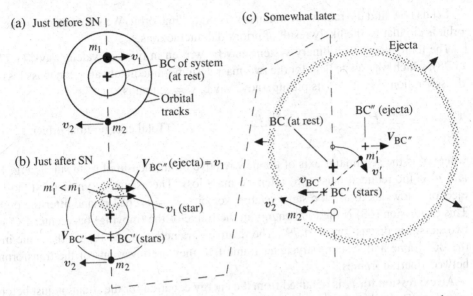

Figure 4.18. The effect of an isotropic supernova mass-ejection event in a binary system shown in the barycenter (BC) frame of reference of the entire system. (a) Two stars in circular orbits just before the ejection. (b) The stars and shell of ejected matter just after the ejection from star 1. The relative magnitudes of the masses are reversed (in this example), but the star velocities have not changed. The barycenter of the ejecta (BC″) has the same velocity v_1 (to the right) as its parent star, and the barycenter of the new two-star system (BC′) has jumped down and is moving to the left. (c) The spherical shell of ejecta has expanded beyond the two stars and thus exerts no gravitational effect on them. The two stars orbit their new BC′ in elliptical orbits in the moving frame of reference. If more than one-half the original system mass is ejected, the stars will no longer be bound to each other.

Sudden mass loss

The core of a formerly massive star ($\gtrsim 8\,M_\odot$) can undergo a supernova collapse to a neutron star or black hole with the "sudden" ejection of the stellar envelope. If the star is one partner of a binary system, the system will be highly perturbed. Here we examine the consequences to the orbit for the special, but common, case of an initially circular orbit.

Semimajor axis and period

The mass ejection can be asymmetric and thus impart momentum to the neutron star. In fact, it may unbind the binary. Measured velocities of neutron-star pulsars are quite high (of order 100 km/s to more than 1000 km/s), indicating they must receive substantial kicks at their birth in a supernova event. Nevertheless, it is instructive to make the simplifying assumption that the ejection does not change the velocity of the exploding star.

This would be the case if the matter were ejected as a spherical shell with barycenter (BC″) moving with the initial velocity of its parent star (Fig. 18). The shell is also presumed to expand beyond the two stars before the stars have moved significantly in their new orbits. Thereafter, it has no gravitational effect on the binary; remember Gauss's law. The effect on the orbit is solely that of the sudden mass decrease of one star. The result is either an elliptical orbit or an unbound system.

Let us now find the relative semimajor axis of the final orbit. We state again that the initial orbit is circular as the likely result of prior tidal interactions.

The total energy of a binary system may be written in two equivalent ways, (1.77) and (1.78), which we equate here for the two-mass system immediately after the mass loss when the separation s' is still at its presupernova value, $s' = s$; that is,

$$-\frac{GM_T' \mu'}{2a_s'} = \frac{1}{2}\mu' v_s'^2 - \frac{GM_T' \mu'}{s}, \qquad \text{(Total energy; final orbit)} \qquad (4.63)$$

where a_s' is the semimajor axis of the new relative orbit (measured from star 1; Fig. 1.10c) and μ' is the reduced mass (56) after the mass loss. The assumed symmetric ("no kick") ejection allows us to use the same relative velocity, $v_s' = v_s$, before and after the outburst. This expression (63) is the total energy in the frame of the (moving) barycenter BC' of the two stars; see discussion of (1.79). The relative parameters, s and v_s, are the same in both the BC' frame and system barycenter frame BC; they are invariants of the transformation between inertial frames.

An expression for v_s is obtained from the energy equation for the situation just before the mass loss. Following (63), we have

$$-\frac{GM_T \mu}{2s} = \frac{1}{2}\mu v_s^2 - \frac{GM_T \mu}{s}, \qquad \text{(Initial circular orbit)} \qquad (4.64)$$

where s has been substituted for a_s because the initial orbit is circular; $a_s = s$. This immediately gives the relation

$$v_s^2 = \frac{GM_T}{s}. \qquad (4.65)$$

This can be substituted into (63) and the result solved for a_s',

➡ $$a_s' = \frac{M_T'}{2M_T' - M_T}s = \frac{f}{2f - 1}s; \qquad f \equiv \frac{M_T'}{M_T}, \qquad (4.66)$$

where we define f to be the fractional mass remaining after the supernova. This result was found in a different context ($m_1 \gg m_2$) in Prob. 1.46a. If the system loses 20% of its mass ($f = 0.8$), this would yield $a_s' = 4s/3$. The mass loss causes the semimajor axis to increase.

In the prior circular orbit, the velocity v is normal to the radius vector s, and it must also be so just after the mass loss. This condition holds in an elliptical orbit only at periastron and apastron. Thus the relative star position at ejection must be either at periastron or apastron of the final orbit. If the fraction f lies between $1/2$ and 1, we see from (66) that $a_s' > s$. Hence, at the time of the ejection, the relative star position must be at periastron of the postexplosion orbit (Fig. 18). (If $f \leq 1/2$, the orbit becomes unbound; see (73) below.)

A numerical value of a_s' requires knowledge of s as well as of f. The former is obtained in terms of the period of the initial circular orbit directly from Kepler III (62):

$$s = \left(\frac{GM_T P^2}{4\pi^2}\right)^{1/3}. \qquad (4.67)$$

The period P' of the final orbit follows from a'_s Kepler III (61),

$$P' = 2\pi(GM'_T)^{-1/2}(a'_s)^{3/2}, \qquad (4.68)$$

where a'_s is given by (66).

Eccentricity

The eccentricity of an ellipse is related to the semimajor axis and periastron distance according to (1.5) as $r_{min} = a(1 - e)$. For the final orbit in our case we have $r_{min} = s$, the initial separation, and thus

$$s = a'_s(1 - e'). \qquad (4.69)$$

Solve for e' and invoke (66):

$$\Rightarrow \qquad e' = 1 - \frac{s}{a'_s} = \frac{M_T - M'_T}{M'_T} = \frac{1-f}{f}. \qquad (4.70)$$

This is the eccentricity of the final orbit. For the range $0.5 < f < 1.0$, the eccentricity ranges from 1.0 (linelike) to 0.0 (a circle).

If, again, 20% of the system mass is lost ($f = 0.8$), we find $e' = 0.25$. We also find from (1.12) and (66) that the semiminor axis is

$$b'_s = a'_s(1 - e'^2)^{1/2} = \frac{4}{3}s\left(1 - \frac{1}{16}\right)^{1/2} = 1.29\,s, \qquad (4.71)$$

which is also greater than the original separation. The mass loss makes the circular orbit slightly elliptical as well as larger.

Unbinding of the orbit

If the mass loss is too great, the orbit will become unbound. This is apparent from an examination of the expressions for a'_s (66) and e' (70). If f is reduced to $1/2$, the semimajor axis becomes infinite and the eccentricity unity (linelike). This is a clear indication that if one-half or more of the system mass is ejected, the binary becomes unbound.

This result is readily obtained directly as follows. The orbit will become unbound when the total energy is zero or greater. Write the total energy equation (63), invoke (65) for v_s, and set the left side to zero,

$$0 = \frac{1}{2}\mu'\frac{GM_T}{s} - \frac{GM'_T\,\mu'}{s}, \qquad (4.72)$$

to obtain

$$\Rightarrow \qquad M'_T = \frac{M_T}{2}, \qquad \text{(Mass for just unbinding)} \qquad (4.73)$$

where M_T and M'_T are, respectively, the two-star system mass before and after the mass ejection event. This is the expected result. Keep in mind that this limit applies strictly only to our ideal case of an initially circular orbit and a spherical outburst in the frame of the exploding star.

This limit directly affects evolution scenarios in that, for the binary to survive the mass loss entailed in the collapse to a 1.4-$M_\odot$ neutron star, the precursor star must have had a low presupernova mass. Otherwise, the expelled mass could amount to more than one-half of the system mass, which would unbind the binary. This pertains to systems that are now x-ray binaries such as shown in Fig. 13. Mass loss from a binary system can also take place via stellar winds, accretion overflow, and ejection during a common-envelope phase.

We state again that the assumptions in the calculation of the mass limit (73) are highly idealistic. A proper calculation must include kick velocities imparted to the neutron star during the supernova outburst. The direction, magnitude, and timing of these kicks can not be predicted, and so this restricts us to statistical predictions.

Evolutionary scenarios

Here we present the possible evolutionary sequences (scenarios) for two binary systems from a recent review paper. One has two rather massive components, 14.4 and 8.0 $M_\odot$, and evolves to a high-mass x-ray binary (HMXB) and then to a two neutron-star system, one of which is a radio pulsar. The other system of 15.0 and 1.6 $M_\odot$ has only one star sufficiently massive to become a neutron star. It evolves into a low-mass x-ray binary (LMXB) and then to a binary containing a neutron star (a millisecond radio pulsar) and a white dwarf.

In each case, the sequence is a plausible one; the x-ray and radio states match known sources. Keep in mind that the details of some phases are quite uncertain (e.g., the in-spiral phase). Recall that HMXB and LMXB systems are neutron-star x-ray emitters distinguished by the mass of the donor companion (see "Evidence for neutron stars" in Section 4 above). In HMXB systems, the donor is much more massive than the nominal 1.4-$M_\odot$ neutron star mass, and in LMXB systems, it has a comparable or lesser mass.

High-mass x-ray binary

The HMXB sequence is illustrated in Fig. 19. The ten stages of the evolution can be understood, more or less, in terms of the various processes described in the preceding sections. For each stage, the component masses and the period determine the relative semimajor axis a'_s through Kepler III (61). The given age is the time the system arrives at the current stage counting from the arrival time of the more massive star on the (zero-age) main sequence. The total mass is sometimes (mostly) conserved, but other times it is not. The scenario includes two supernova events, each of which yields a neutron star.

The scenario starts with two main-sequence stars of masses 14.4 and 8.0 $M_\odot$ in a circular orbit with a 100-d orbital period, as illustrated in the first stage (A) of Fig. 19. After a relatively long wait of 13.3 Myr, the more massive star moves off the main sequence and becomes a giant star and overflows its Roche lobe (B). A large amount of mass is transferred during the overflow, and most of it remains in the system, as indicated by the masses of the helium 3.5-$M_\odot$ star remnant and normal star of stage (C).

The helium star remnant (C) then undergoes a core-collapse supernova event (D), giving rise to a wide orbit (3390 $R_\odot$) with high eccentricity (E). Another long wait of ~10 Myr ensues while the 16.5-$M_\odot$ (formerly 8-$M_\odot$) normal star burns hydrogen before entering its giant phase. As it enters its giant phase, tidal forces come into play, decreasing the semimajor axis so that the orbit is close to circular (Prob. 54). Also, stellar winds carry off 1.5 $M_\odot$ of its envelope.

When the giant nearly fills its Roche lobe (F), the stellar wind ejects large quantities of mass, some of which accretes onto the neutron star, giving rise to x-ray emission. The

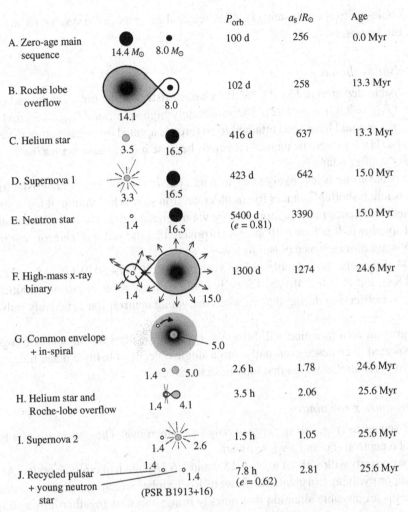

	P_{orb}	$a_s/R_\odot$	Age
A. Zero-age main sequence — $14.4\,M_\odot$ $8.0\,M_\odot$	100 d	256	0.0 Myr
B. Roche lobe overflow — 14.1 8.0	102 d	258	13.3 Myr
C. Helium star — 3.5 16.5	416 d	637	13.3 Myr
D. Supernova 1 — 3.3 16.5	423 d	642	15.0 Myr
E. Neutron star — 1.4 16.5	5400 d ($e = 0.81$)	3390	15.0 Myr
F. High-mass x-ray binary — 1.4 15.0	1300 d	1274	24.6 Myr
G. Common envelope + in-spiral — 5.0 1.4 5.0	2.6 h	1.78	24.6 Myr
H. Helium star and Roche-lobe overflow — 1.4 4.1	3.5 h	2.06	25.6 Myr
I. Supernova 2 — 1.4 2.6	1.5 h	1.05	25.6 Myr
J. Recycled pulsar + young neutron star — 1.4 1.4 (PSR B1913+16)	7.8 h ($e = 0.62$)	2.81	25.6 Myr

Figure 4.19. Evolutionary scenario that leads to an HMXB system and then to a millisecond radio pulsar binary consisting of two neutron stars. The final parameters given here are for the 59-ms radio-emitting "binary pulsar" PSR B1913+16. The indicated values (masses, period, semimajor axis of relative orbit, and age) apply to the beginning of each stage. See text for additional explanation. [Adapted from T. Tauris and E. van den Heuvel, *ibid.*, Fig. 16.12, with permission]

Roche lobe is not filled, and thus Roche-lobe overflow does not occur – at least not initially. Continuing evolution further expands the star. The resulting Roche-lobe overflow is unstable because of the large mass ratio (60). The neutron star is soon enveloped by the atmosphere of its partner. Rapid in-spiral follows with large mass ejections so that only the $5\text{-}M_\odot$ helium core of the normal star remains in a detached binary (G). This all happens very rapidly (in about 10^4 yr).

The detached binary remains in this state for $\sim$1.0 Myr until the evolution of the helium star leads to mass loss via winds and then to Roche lobe overflow with, possibly, x-ray emission (H). After additional mass loss, the helium star undergoes a supernova explosion (I) to yield a double-neutron star system (J). After residual gases leave the system, radio pulsing, if present, can be observed.

Pulsar evolution

This final system depicted in Fig. 19 has the characteristics of the binary radio pulsar PSR B1913+16 ($P_{orb} = 7.8$ h, $e = 0.62$). The moderately high spin rate ($P_{spin} = 59$ ms) of the "recycled" pulsar could have been attained from torques applied by the accreting material in stages (F) and (H). It gained the name, "recycled" because it could have been a (radio) pulsar when it was "young" (stage E).

A young neutron star is less likely to be seen as a pulsar because its characteristic lifetime as a pulsar is much shorter than it is for an older neutron star. When young, it has a stronger magnetic field and hence radiates more energy via magnetic dipole radiation. The result is a more rapid spindown. It is thus perhaps not surprising that the younger neutron star in PSR B1913+16 is not observed as a pulsar today.

The 2004 discovery of a double-pulsar system PSR J0737−3039 ($P_{spin,A} = 22.7$ ms, $P_{spin,B} = 2.8$ s, and $P_{orb} = 2.4$h) dramatically confirms these ideas of pulsar evolution. We fortuitously are observing during the era when the young neutron star is actively pulsing at 2.8 s.

Finally, gravitational radiation will bring the two neutron stars of our scenario together. They are expected to coalesce eventually into a single object (probably a black hole) with a ~1-s burst of gravitational waves that may be detectable.

Low-mass x-ray binary

The LMXB sequence (Fig. 20) undergoes only one supernova. The end result is a binary consisting of a neutron star and a white dwarf.

This scenario starts with stars of masses 15.0 and 1.6 $M_\odot$ again in circular orbits (A). The 15-$M_\odot$ star, upon evolving to a giant, overflows its Roche lobe, accreting onto the less massive star (B). This is an unstable situation that quickly brings the stars together into a common envelope and a rapid spiral-in (C). The entire envelope of the giant is ejected, leaving a tight binary consisting of a helium star and a normal star (D).

The 1.6-$M_\odot$ star is presumed to have survived all this with essentially its original mass; it is plausible that only negligible amounts of mass were acquired or lost in stages (B) and (C). The He star then undergoes a core-collapse supernova (E), leaving a neutron star and the 1.6-$M_\odot$ star in a somewhat wider and eccentric orbit (F). After a very long time (2.24 Gyr), the normal star finally starts evolving onto the giant branch and overflows its Roche lobe onto the neutron star. It is now an LMXB (G).

This continues for a long time, 400 Myr, until the envelope is exhausted. It is during this phase that the accretion spins up the neutron star to periods as rapid as 2 ms. At first, the mass transfer tends to equalize the masses and causes the stars to move together, but then it increases the mass disparity, and so the stars move apart. They are kept in contact by the continuing expansion of the evolving donor star. The envelope is eventually totally expended when the separation reaches $27 R_\odot$. The final result is a detached binary consisting of a neutron star and low-mass white dwarf.

As the gaseous material clears from around the neutron star, it becomes a visible radio millisecond pulsar with a companion white dwarf (H) similar to the 5.3-ms PSR B1855+09

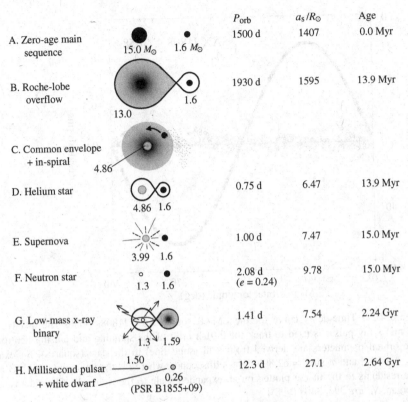

		P_{orb}	$a_s/R_\odot$	Age
A. Zero-age main sequence	15.0 $M_\odot$ 1.6 $M_\odot$	1500 d	1407	0.0 Myr
B. Roche-lobe overflow	13.0 1.6	1930 d	1595	13.9 Myr
C. Common envelope + in-spiral	4.86			
D. Helium star	4.86 1.6	0.75 d	6.47	13.9 Myr
E. Supernova	3.99 1.6	1.00 d	7.47	15.0 Myr
F. Neutron star	1.3 1.6	2.08 d ($e = 0.24$)	9.78	15.0 Myr
G. Low-mass x-ray binary	1.3 1.59	1.41 d	7.54	2.24 Gyr
H. Millisecond pulsar + white dwarf	1.50 0.26 (PSR B1855+09)	12.3 d	27.1	2.64 Gyr

Figure 4.20. Evolutionary scenario that leads to an LMXB system and then to a 5-ms binary radio pulsar (neutron star) with a white dwarf companion. See caption to previous figure. [Adapted from T. Tauris and E. van den Heuvel, *ibid.*, Fig. 16.12 with permission]

system. At a separation of 27.1 $R_\odot$ and with a rather low-mass partner, the characteristic time τ for gravitational decay, from (48) and (49), is $\sim 1 \times 10^4$ Hubble times. For all practical purposes, the stars never merge.

Neutron-star spinup

The mechanism that spins up neutron stars to the millisecond periods exhibited by some radio pulsars was long suspected to be the torque applied to the neutron star by the accreting gases. The gases circulate in the accretion disk, gradually descending as they give up energy. At the inner edge of the accretion disk, they hook onto the outer edge of the magnetosphere, which corotates with the star, as illustrated in Fig. 13.

The torque applied will spin up the neutron star if the orbital speed at the inner edge of the accretion disk exceeds the equatorial speed of the outer edge of the magnetosphere. This is most likely if the magnetosphere is small so that the accretion disk can extend down toward the star where the Keplerian velocities are greater. A small magnetosphere indicates a weaker magnetic field at the stellar surface.

It is well accepted, in fact, that the (isolated) millisecond radio pulsars have weak magnetic fields because their spindown rates are very small. The spindown rate, in a simplistic view, is caused by energy loss in the form of *magnetic dipole radiation*; thus, these pulsars are believed to have low magnetic dipoles and hence weak magnetic fields. Moreover, there

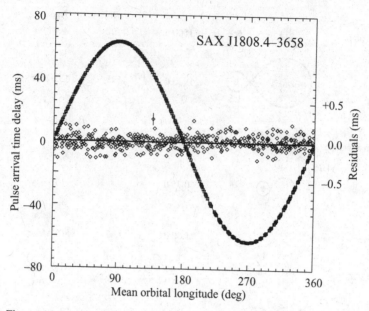

Figure 4.21. Time-delay curve for the LMXB source SAX J1808−3658. The 401-Hz ($P = 2.5$ ms) x-ray pulsar is used to track the 2.01-h orbit of the spinning and pulsing neutron star. The orbital parameters are derived from a fit (solid line) to the data (squares); for example, $e < 4 \times 10^{-4}$ and $a_x \sin i = 62.809$ light-milliseconds, where i is the unknown orbit inclination. The residuals to the fit are plotted on an expanded scale to the right. [D. Chakrabarty and E. Morgan, *Nature* **394**, 346 (1998)]

is evidence that the neutron stars in LMXB systems also have weak fields (e.g., most do not exhibit x-ray pulsing). All in all, it has been reasonable to believe that the rapid spin rates of these pulsars originated during the x-ray–emitting phase of LMXB systems, but the observational evidence had been lacking.

Pulsing x-ray emission with a 2.5-ms period from an accreting binary system was first found in 1998 in data from an LMXB source (SAX J1808−3658) by observers using the Rossi X-ray Timing Explorer (RXTE), an orbiting x-ray observatory. Figure 21 shows the variation in the 2.5-ms pulse period caused by the changing time delays arising from the orbital motion of the neutron star in its binary system. The excellent sinusoidal fit demonstrates the stability of the 2.5-ms period, which must arise from a spinning compact star's large moment of inertia. Hence, the 2.5-ms pulse period must be the spin period of the accreting neutron star.

One thus actually "sees" the spunup neutron star while it is still in the accreting phase. This solidly confirms this aspect of binary evolution – namely, that millisecond pulsars are spun up by accretion torques in x-ray binaries of the low-mass variety. There are now about eight known examples of this type of system.

Problems

4.2   Equations of stellar structure

Problem 4.21. Derive or justify from first principles, without reference to the text (insofar as possible), three of the fundamental equations of stellar structure, namely, (a) hydrostatic equilibrium (1), (b) mass distribution (2), and (c) luminosity distribution (3). In each case, comment on the underlying assumption (e.g., conservation laws, etc.).

Problem 4.22. Derive, approximately, from first principles the equation of radiation transport (7). Follow the reasoning and suggested approximations in the text. By what factor does your answer differ from (7)? [Ans. 4/3]

Problem 4.23. Derive from first principles the equation of convective transport (13) beginning with (9). Follow the suggestions in the text.

Problem 4.24. (a) Use approximate "whole-sun" values to deduce a rough value for the opacity κ within the sun from the radiation transport equation (7). Adopt a "whole-sun" temperature $T \approx 5 \times 10^6$ K, central temperature $T_c = 1.6 \times 10^7$ K, characteristic radius $R_\odot$, density $\rho = \rho_{av} \approx M_\odot/R_\odot^3$, $dT/dr \approx T_c/R_\odot$, and luminosity $L = L_\odot$. See Table 1 for solar values. (b) Argue that the proper expression for the opacity κ (m²/kg) in terms of the (Thomson) cross section is $\kappa = (\sigma_T n_e)/\rho$ (refer to AM, Chapter 10). Calculate the opacity (solar average) from this expression under the assumption that it arises from photon-electron interactions with the Thomson cross section, $\sigma_T = 6.7 \times 10^{-29}$ m². Assume a totally ionized hydrogen plasma; note that the expression for κ can be simplified in this case. Compare the opacities in parts (a) and (b); they should agree within two orders of magnitude, demonstrating the (very) rough validity of our approximations. [Ans. $\sim$2 m²/kg; $\sim$0.04 m²/kg]

Problem 4.25. (a) Use the equation of state for an ideal gas, $P = \rho kT/m$, and the adiabatic condition, $PV^\gamma = $ constant, to determine how the mass density ρ and temperature T depend on pressure in an adiabatically expanding monatomic gas. Give your answer in terms of the pressure P and the ratio of specific heats, γ, and for $\gamma = 5/3$. (b) Under convective conditions, by what fraction does the temperature in a star decrease over the radial distance in which the pressure drops by a factor of two? Assume a monatomic gas. How much is the mass density reduced over the same distance? [Ans. $P^{0.6}$, $P^{0.4}$; $\sim$0.75, $\sim$0.65]

4.3 Modeling and evolution

Problem 4.31. (a) Find the central pressure, $P_c(M, R)$, of a star as a function of its mass M and radius R and other physical constants under the simplifying assumption of constant density $\rho_0 = \rho_{av}$ throughout the star. Hints: express $\mathfrak{M}(r)$ in terms of ρ_0 and r and integrate the equation of hydrostatic equilibrium (1) from the solar surface to its center. Substitute solar values into your expression to obtain a numerical value. (b) Apply the EOS for a Maxwell–Boltzmann gas to find the central temperature $T_c(M, R)$. Again, find the solar value. (c) Compare your values of P_c and T_c with those in Table 1. Explain the differences. (d) Consider a degenerate white dwarf of one solar mass and radius $10^{-2}R_\odot$. How would its central pressure compare with that of the sun? What can you say about its central temperature?

$$\left[\text{Ans. } P_c = \frac{3}{8\pi}\frac{GM^2}{R^4} \approx 10^{14} \text{ Pa}; T_c = \frac{G m_{av}}{2} \frac{M}{k} \frac{M}{R} \approx 10^7 \text{ K}; -; \sim 10^8 P_{c,\odot}\right]$$

Problem 4.32. (a) Find the included mass function $\mathfrak{M}(r)$ and the total mass M of a spherical star of radius R whose mass density (kg/m³) varies with radius r as $\rho(r) = \rho_c[1 - (r/R)]$. (b) Find the central pressure P_c of this star in terms of M and R. (c) Find the central temperature T_c in terms of M and R. (d) Compare your results for (b) and (c) with the answers to Problem 31 and comment. Hint: use (2), (1), and (14) for parts (a)–(c). [Ans. $M = \pi\rho_c R^3/3$; $P_c = (5/4\pi)GM^2/R^4$; $\sim$0.4$(Gm_{av}/k)(M/R)$; $-$]

Problem 4.33. (a) What is the ratio of photon energy density u_ν to particle (electrons and protons) kinetic energy density u_p at the center of the sun, where the temperature is 1.6×10^7K and the particle mass density is 1.5×10^5 kg/m^3? Assume a blackbody photon spectrum and a totally ionized hydrogen plasma that obeys the Maxwell–Boltzmann distribution. Hint: use the blackbody relation for radiation energy density (6.25) and recall the average particle energy in an M-B gas. (b) What is the ratio of photon number density n_ν to particle number density n_p in this region? Use (6.31). (c) What do these two values tell you about the average energy per particle for the two constituents? Demonstrate that this is in accord with the expected average energies for particles and photons in thermal equilibrium. Use (6.32). ([Ans. $\sim 1/1200$; $\sim 1/2200$; $-$]

Problem 4.34. (a) Obtain the radial variation of pressure, $P(r)$, for a model star in which the density ρ_0 is constant throughout by integrating the equation of hydrostatic equilibrium. (This is an elaboration of Prob. 31; see hints therein.) Express $P(r)$ in terms of r, R, and M and also in terms of r, R, and the pressure at the center of the star, P_c. (b) Find the temperature variation $T(r)$ for the ideal gas equation of state as a function of m_{av}, M, R, and r and also in terms of T_c, r, and R. Note that $P(r)$ and $T(r)$ have identical functional forms in this simple model. (c) The energy-generation rate (W/kg) inside stars can have the form (2.66), $\epsilon = \epsilon_0 \rho T^\beta$, where $\epsilon_0 = $ constant, $\beta \approx 4$ for the pp chain, and $\beta \approx 15$ for the CNO chain. Find the radial variation of ϵ by using the function $T(r)$ from (b). Express your answer in terms of ϵ_c, the energy-generation rate at the center of the star. (d) For both the pp chain and the CNO cycle, find the fractional radial distance r/R at which the energy-generation rate has fallen to 10% of the central value (i.e., the radius at which $\epsilon/\epsilon_c = 0.1$). What does this tell you about the relative difference in the nature of the energy generation between the two cycles? Qualitatively and briefly, how would this result be modified for real stars in which $\rho \neq $ constant? [Ans. $P_c(1 - (r^2/R^2))$; $T_c(1 - (r^2/R^2))$; $\epsilon_c(1 - (r^2/R^2))^\beta$; ~ 0.7, ~ 0.4]

Problem 4.35. (a) Use the profiles for $T(r)$ and $\rho(r)$ in Fig. 5c,d to determine the pressure function $P(r)$ for the sun and plot your values on a similar graph with arbitrary pressure scale. Assume the gas consists of ionized hydrogen and that radiation pressure is not important. (b) Calculate the energy-generation rate $\epsilon \propto \rho T^4$ as a function of radius and plot it on the same graph. Adjust the scale of ϵ to have the same value as P at $r = 0$. Comment on your plots.

Problem 4.36. (a) Construct the approximate expression for luminosity $L(M, R)$ (20) from the differential equation (3) and the scaling expressions for ρ (17), T (19), and ϵ_{pp} (2.66), as indicated in the text. (b) Find an approximate expression for the average particle mass m_{av} in terms of the elemental mass fractions, X, Y, and Z. Assume complete ionization and count all nuclei and free electrons in your average. (X, Y, and Z are the fractions of the total mass in hydrogen, helium, and heavies, respectively, in a gas sample, e.g., in unit volume.) Assume that the Z elements consist only of carbon, that the mass of each element is a multiple of the atomic mass unit m_{amu}, that electron masses are negligible, and that all elements are completely ionized. Hint: what is the number density of helium atoms in terms of ρ, Y, and m_{amu}? Evaluate your expression for the nominal solar abundances $X = 0.71$, $Y = 0.27$, and $Z = 0.02$ as well as for those at the center of the sun, $X = 0.36$, $Y = 0.62$, and $Z = 0.02$. (c) Find the approximate luminosity of the sun from (20). Use the solar mass and radius, your value of m_{av}, and X for the center of the sun and assume the pp process dominates ($\beta \approx 4$). Compare with the actual luminosity. [Ans. $-$; $m_{av} \approx m_{amu}(2X + (3/4)Y + (7/12)Z)^{-1}$, ~ 0.8 (sun center); $\sim 1 L_\odot$]

Problem 4.37. (a) Use equations (20) and (21) to obtain the scaling relations $L(M)$ and $R(M)$. From this and the equation of state, find $T(M)$ and finally $L(T)$ (24). (Follow the text as needed.) What would be the slope on a $\log L$ versus $\log T$ plot (a Hertzsprung–Russell diagram with temperature axis increasing to right) for the pp chain ($\beta \approx 4$) and for the CNO chain ($\beta \approx 15$)? (b) Compare these with the slope defined by the positions on the same plot of a G0 V and a K0 V star. Repeat for an O5 V and a B0 V star. Use $L = 4\pi R^2 \sigma T^4$ (16) and the data for $R/R_\odot$ and $T_{\rm eff}$ given in Table 2 to find the ratio of *bolometric* luminosities $L/L_\odot$; use $T_{\rm eff,\odot} = 5777$ K. Comment on the comparison. [Ans. ~5, ~13; ~7, ~8]

4.4 Compact stars

Problem 4.41. (a) Find the approximate radius-mass relation (25) for a white dwarf star supported by nonrelativistic degeneracy pressure. Follow suggestions in the text. (b) Find three approximations of the Chandrasekhar mass limit in the manner suggested in the text immediately preceding (26).

Problem 4.42. After its formation, a white dwarf will remain approximately at a constant radius. (a) Find, under this assumption, the slope of the straight-line track followed by a white dwarf as it cools on a $\log L - \log T_{\rm eff}$ plot (Hertzsprung–Russell diagram) if it obeys the luminosity relation (16). Compare with, and comment on, the slopes of the white dwarf tracks in Figs. 8a and 9. (b) Find the relative radii $R/R_\odot$ implied by (16) for the white dwarfs in the two figures in each case when the luminosity is $10^{-2} L_\odot$. [Ans. ~4; ~0.009, ~0.013]

Problem 4.43. (Build a white dwarf with a computer or calculator.) (a) Use the nonrelativistic EOS with its numerical coefficient (3.74), the condition of hydrostatic equilibrium (1), and the mass distribution formula (2) to construct a white dwarf. Let the central mass density be $\rho_c = 1.0 \times 10^9$ kg/m^3 and adopt $\mu_e = 2$. Proceed by constructing the star layer by layer starting at the center. Use the EOS to find the central pressure P_c and save it. Choose a small radial interval Δr (e.g., 10^4 m), add a sphere of matter of radius Δr at density ρ_c, save the value of its mass $\mathfrak{M}(r)$, find the (small negative) pressure change dP across this radial interval from the equation of hydrostatic equilibrium, calculate and save the new (lesser) pressure P, calculate the new density from the EOS, increase the radius r by Δr, add a shell of mass $d\mathfrak{M}(r)$ of thickness Δr at the new density and radius as given by the mass distribution equation, find and save the new total mass, and then loop back to the calculation of dP. Continue until the density decreases to $\sim 10^{-4} \rho_c$. Stop the process every half decade of density change to record the mass and radius. What is the final radius and mass in solar units? If you are ambitious, you could try other central densities to construct an entire family of white dwarfs with central densities 10^8 to 10^{10} kg/m^3, which would yield plots of M versus ρ_c and R versus M. (b) A more accurate result would make use of an EOS, namely $P(P_N, P_R)$, that matches both the low-density (nonrelativistic) and high-density (relativistic) limits, $P_N = k_N \rho^{5/3}$ and $P_R = k_R \rho^{4/3}$, respectively. Find $P(P_N, P_R)$, a simple function of P_N and P_R, that satisfies these limits, and then rewrite it in terms of k_N, ρ, and ρ_0, the latter being the density at which $P_N = P_R$. Hint: sketch a $\log P - \log \rho$ plot of the two functions. (c) Optional: construct another white dwarf with this combined EOS. [Ans. ~0.016$R_\odot$, ~0.5$M_\odot$; $P \doteq k_N \rho^{5/3}(1 + (\rho/\rho_0)^{2/3})^{-1/2}$; –]

Problem 4.44. For a neutron star of $M = 1.4 M_\odot$ and $R = 10^4$ m, adopt the linearly-decreasing-density model of Prob. 32. Let the star consist solely of free neutrons, which, in degeneracy, behave as electrons. Assume the star is supported solely by completely degenerate, nonrelativistic neutron pressure. Find the following: (a) the approximate central pressure, (b) the neutron number density n_n and mass density ρ_c at the center, (c) the mass-radius relation and neutron-star radius, (d) the Fermi momentum of neutrons in the center of the star in energy units, $p_F c$, with comparison to $m_n c^2 = 940$ MeV, (e) the kinetic energy E_k and speed factor $\beta = v/c$ of the most energetic neutrons at the center of the star calculated relativistically (Chapter 7) with comment, (f) an approximate maximum neutron-star mass limited by the (degenerate) neutrons' becoming relativistic, and (g) an approximate upper mass limit by requiring the Schwarzschild radius to equal the neutron-star radius obtained in part (c). The neutron mass is 1.675×10^{-27} kg. [Ans. $\sim 10^{34}$ N/m^2; $\sim 10^{45}$ m^{-3}, $\sim 10^{18}$ kg/m^3; ~ 15 km; ~ 700 MeV; ~ 200 MeV, ~ 0.6; $\sim 5\, M_\odot$; $\sim 4\, M_\odot$]

Problem 4.45. What is the spin rotation rate (Hz) of a spherical neutron star ($M = 1.4\, M_\odot$, $R = 10^4$ m) when centrifugal force equals gravity at the equator? Explain why the neutron star will begin to break up at rotation rates just above this no matter how strong the nuclear binding forces are. Hint: consider the surface area and volume of an element of mass on the equator. [Ans. ~ 2 kHz]

Problem 4.46. Find, from (35), the fractional gravitational wavelength shift $\Delta\lambda/\lambda$ for photons arriving at a distant observer from the surface (or indicated radius) of each of the following objects: earth, sun, white dwarf of mass $1\, M_\odot$ (and radius corresponding thereto), a neutron star of $1.4\, M_\odot$ and radius 10 km, and nonrotating black holes of $10\, M_\odot$ and of $10^7\, M_\odot$ (innermost stable orbit). For each object, give the wavelength shift $\Delta\lambda$ for the Hα line at $\lambda = 656$ nm and find the Schwarzschild radius. Define $\Delta\lambda/\lambda \equiv (\lambda - \lambda_0)/\lambda_0$, where λ_0 is the wavelength at the surface and λ the wavelength at infinity. Tabulate your inputs and results and comment on features of interest. [Ans. $\Delta\lambda/\lambda$: $\sim 10^{-9}$, $\sim 10^{-6}$, $\sim 10^{-4}$, ~ 0.3, ~ 0.2, ~ 0.2]

Problem 4.47. (a) A neutron star of $1.4\, M_\odot$ and 10-km radius is not rotating significantly. Accreting gaseous material forms an accretion disk about it. In the absence of a magnetic field, what is the expected inner radius of the accretion disk (i.e., the innermost stable orbit of GR)? Does it reach the neutron-star surface? (b) If this neutron star rotates at 100 Hz, what fraction of the maximum angular momentum GM^2/c does it have? As approximations, use the Newtonian expression for angular momentum and assume a constant-density sphere (moment of inertia $0.4\, MR^2$). What is the effect of this spin on the event horizon? Would you expect a substantial change to the innermost stable orbit at this spin rate? (c) Repeat (b) for rotation at 10^3 Hz. (d) What spin frequency corresponds to maximum angular momentum, GM^2/c, with our approximations? [Ans. ~ 10 km; ~ 0.04; ~ 0.4; $\sim 2 \times 10^3$ Hz]

Problem 4.48. Consider an evaporating black hole with an initial mass M_i. (a) Use the expressions of R_S (36) and T (45) to find the numerical proportionality factor a in the evaporation luminosity, $L = aM^{-2}$ (46). According to this, what is the luminosity of an evaporating black hole of mass 5×10^{11} kg? (b) Write the differential expression that relates the change of mass dM to the luminosity L and the time dt in which the incremental mass loss takes place. How does the mass loss rate, dM/dt, depend on mass $M(t)$? Comment. (c) Integrate the differential equation to obtain an expression for the remaining mass $M(t)$ of a black hole as a function of time t, M_i, a, and c^2. Let $t = 0$ when $M(t) = M_i$, (d) From this, find the mass of a black hole that would evaporate to zero mass in the age of the universe, $\sim 1.4 \times 10^{10}$ yr. By what factor does this mass differ from the mass given (for this evaporation time) in (47)? [Ans. $\sim 10^9$ W; $\propto M^{-2}$; –; $\sim 2 \times 10^{11}$ kg]

Problem 4.49. (a) According to (47), what is the mass of a black hole that will completely evaporate within ~ 0.1 s? What would its temperature be? (b) In the subsequent decay to zero mass, how much total energy is released? How does this energy compare with the rest energy of the sun? (c) Assume that 10% of this energy is released isotropically as 100 MeV photons (gamma rays). Up to what distance could you detect such an outburst with a gamma-ray detector of area 1 m^2 if five photons detected in 0.1 s constitutes a reliable detection? (d) If the black hole explodes prematurely when it reaches $T = 10^{12}$ K, what is the maximum distance for detection under similar assumptions? [Ans. $\sim 10^5$ kg, $\sim 10^{18}$ K; $\sim 10^{22}$ J; ~ 0.2 LY; ~ 100 LY]

4.5 Binary evolution

Problem 4.51. Consider a binary pulsar containing two neutron stars, each of mass $m = 1.4 M_\odot$, in *circular* orbits of the same period as the PSR B1913+16 (H-T) binary pulsar, $P = 7.75$ h. The stellar separation, $s = a_s$, follows from Kepler III (1.75). The system loses energy to gravitational radiation, and pulse-timing measurements track the location of the pulsar in the shrinking orbit as a function of time. (a) From the rate of energy loss to gravitational waves (48), estimate the advance of the time of periastron in 25 years. Make simplifying assumptions as needed (e.g., that the rate of energy loss does not change appreciably in 25 years). Hints: find a relation between ΔP and Δs and also one between ΔE and Δs to determine the period change after one-half the 25 yr, and use this as an average value for the entire 25 yr. (A fun aside: how much does the separation decrease in 12.5 yr?) Compare with the 26 s actually measured for the H-T pulsar, which has eccentricity $e = 0.617$. How would you explain the difference? (b) Recalculate the phase advance for a circular orbit that is at the closer periastron distance of the H-T pulsar, $s' = 0.383s$. Again compare with the H-T pulsar phase advance and comment. [Ans. ~ -2 s; ~ -25 s]

Problem 4.52. Find the locations, $u = x/s$, relative to the more massive star, m_1, of the Lagrangian points (a) L1, (b) L2, and (c) L3 in a binary system of circular orbit with separation s, and unequal masses ($m_1 = 3m_2$), where x is measured from m_1 along the line connecting the stars. Proceed by confirming the correctness of (52) and (54) and then by solving for the points by trial and error on your programming calculator. Make a sketch of the force component F_x versus $u \equiv x/s$ to guide you to approximate solutions. Compare your result for (a) with the "Roche lobe radius" given by (51) and comment. [Ans. $\sim +0.6$; $\sim +1.5$; ~ -0.9]

Problem 4.53. Consider an accreting HMXB with a circular orbit, a donor of 15.0 $M_\odot$ filling its Roche lobe, and a recipient neutron star of mass $M_{ns} = 1.4 M_\odot$ and radius $R_{ns} = 10^4$ m. Comment on how the changing star separation would affect the accretion. If two-thirds of the donor mass were to accrete onto the neutron star (as in stage F of Fig. 19) in a continuous stream over $\sim 10^4$ yr, what is the expected luminosity (primarily x rays)? How does it compare with the Eddington luminosity (2.75)? Where do you think the 10 $M_\odot$ of donated gas actually would end up?

Problem 4.54. Consider the large reduction of the semimajor axis of the relative orbit between stages E and F of Fig. 19 from $a_s = 3390 R_\odot$ with $e = 0.81$ to $a'_s = 1274 R_\odot$. This is due to tidal interaction. Assume that the tidal interactions do not change the angular momentum of the two-star system and that the stellar masses remain fixed. Neglect the spin angular momenta of the stars; in practice it is usually negligible. (a) Find an expression for, and the value of, the final stage F eccentricity e' of the orbit. Compare the shape to that of a circle.

That is, what is the ratio of the semiminor to semimajor axes? (b) If the interaction should continue until the orbit is circularized, what would be the radius s' of the relative circular orbit in terms of a_s and e? How does s' compare with the initial periastron distance s_p? ·[Ans. $e' \approx 0.3$, ~ 0.95; $s' = a_s(1 - e^2) = s_p(1 + e)$]

Problem 4.55. Consider the evolution scenario in Fig. 19. For each stage calculate and tabulate the total angular momentum as given in (55). Comment on the degree to which angular momentum is or is not conserved in each stage and by what mechanism(s) it may have been lost. Similarly, tabulate the mass lost from the system in each stage and comment on significant expulsions. Assume eccentricity $e = 0$ except for stages E, F, and J; see Fig. 19 and the answer to Prob. 54 for these eccentricities. Suggested column headings are m_1, m_2, a_s, e, J, Δm. Write your comments in separate short paragraphs. Note that J can be written in terms of e rather than b_s with the aid of (1.12).

Problem 4.56. There are three supernova events in Figs. 19 and 20. Assume the orbits just before the supernovae were circular with the masses and semimajor axis a_s given next to the supernova sketch in the figure. In each case, what would be the semimajor axis and eccentricity just after the supernova for the indicated mass loss and if the outbursts gave no "kick" to the remnant neutron star as assumed in our development (e.g., in Fig. 18)? Compare your answers with those given in the figures and comment.

5

Thermal bremsstrahlung radiation

What we learn in this chapter

A **hot plasma** of ionized atoms emits **thermal bremsstrahlung radiation** through the **Coulomb collisions** of the electrons and ions. The electrons experience **large accelerations** in the collisions and thus efficiently radiate photons, which escape the plasma if it is **optically thin**. The energy Q radiated in a single collision is obtained from **Larmor's formula**. The characteristic frequency of the emitted radiation is estimated from the duration of the collision, which, in turn, depends on the **electron speed** and its **impact parameter** (projected distance of closest approach to the ion). Multiplication of Q by the electron flux and **ion density** and integration over the range of speeds in the **Maxwell–Boltzmann distribution** yield the **volume emissivity** $j_v(\nu)$ (W m^{-3} Hz^{-1}), the power emitted from unit volume into unit frequency interval at frequency ν as a function of frequency. It is proportional to the product of the electron and ion densities and is approximately **exponential** with frequency. A slowly varying **Gaunt factor** modifies the spectral shape somewhat. Most of the power is emitted at frequencies near that specified by $h\nu \approx kT$.

Integration of the volume emissivity over all frequencies and over the volume of a plasma cloud results in the **luminosity of the cloud**. By integrating over the line of sight through a plasma cloud, one obtains, the **specific intensity** I (W m^{-2} Hz^{-1} sr^{-1}), which is directly measurable. The specific intensity is proportional to the **emission measure** (EM), which is the line-of-sight integral of the product of the electron and ion densities. Integration of the specific intensity over the solid angle of a source yields the **spectral flux density** S (Wm^{-2} Hz^{-1}).

Measurement of the spectrum can provide **two basic parameters** of the plasma cloud, its temperature and its emission measure, **without knowledge of its distance**. H II regions that are kept ionized by newly formed stars are copious emitters of thermal bremsstrahlung radiation such as those in the **W3 complex** of radio emission. **Clusters of galaxies** commonly contain a plasma that has been heated to **x-ray temperatures**. In both cases, the radiation detected at the earth reveals the nature of the astronomical plasmas. The x-ray spectra from astrophysical plasmas are rich in **spectral lines**. Here we develop the **continuum spectrum** from first principles.

5.1   Introduction

Coulomb collisions between electrons and ions in a hot ionized gas (plasma) give rise to photons because electrons are accelerated by the Coulomb forces and thereby emit radiation, thus losing energy. The German word *bremsstrahlung* means "braking radiation." These near collisions are *free-free* transitions because they are transitions from one free (unbound) state of the atom to another such state. If the gas is in thermal equilibrium, the velocities of the ions and electrons will obey the Maxwell–Boltzmann distribution (Section 3.2).

One can further assume that the gas is *optically thin* – that is, the emitted photons escape the plasma without further absorption or interaction. This and the assumptions of nonrelativistic particle speeds and small interaction energy losses make possible a relatively straightforward classical derivation of the continuum spectrum of the emitted photons. To be precise, one would call this radiation "thermal bremsstrahlung radiation from an optically thin, nonrelativistic plasma."

Our derivation here will yield the volume emissivity j_v as a function of frequency ν and temperature T. For an ionized hydrogen gas with free electron density n_e throughout its volume, the result is

$$j_v(\nu, T) \propto g(\nu, T)\, n_e^2 T^{-1/2} e^{-h\nu/kT}, \qquad \text{(Volume emissivity;}$$
$$\text{hydrogen plasma; W m}^{-3}\text{ Hz}^{-1}\text{)} \qquad (5.1)$$

where h and k are the Planck and Boltzmann constants, respectively. In AM, we suppressed the subscript in j_v, but here we keep it to distinguish it from the integrated (over frequency) volume emissivity j (W m^{-3}).

The volume emissivity j_v is the power emitted from unit volume of the plasma at some frequency in unit frequency interval. The *Gaunt factor* $g(\nu, T)$ is a slowly varying (almost constant) function of frequency that modifies the shape of the spectrum somewhat. If it is treated as a constant, the spectrum emitted by a plasma at some temperature T becomes a simple exponential. This reflects the exponential character of the Maxwell–Boltzmann distribution of emitting particle speeds.

The observed specific intensity $I(\nu, T)$ (W m^{-2} Hz^{-1} sr^{-1}) in a view direction that intercepts the plasma may be derived from the volume emissivity $j_v(\nu, T)$ according to the relation $4\pi I = j_{v,\text{av}}\Lambda$ (see (42) below), where $j_{v,\text{av}}$ is the average volume emissivity along the line of sight and Λ is the line-of-sight depth or thickness of the plasma. The conversion from j_v to I is purely geometrical. Thus, the variation with ν and T is the same for both functions, and the specific intensity is

$$I(\nu, T) \propto g(\nu, T)\, n_e^2 T^{-1/2} e^{-h\nu/kT}\, \Lambda. \qquad \text{(Specific intensity; hydrogen}$$
$$\text{plasma; W m}^{-2}\text{ Hz}^{-1}\text{ sr}^{-1}\text{)} \qquad (5.2)$$

The objective of this chapter is to derive $j_v(\nu, T)$ from fundamental principles and thereby to find $I(\nu, T)$. The exponential version (without the Gaunt factor) will be obtained in a *semiquantitative classical derivation*. Such a derivation makes use of approximations while keeping track of the physics.

We use several quantities that describe energy content and flow of photons (e.g., the specific intensity I, the volume emissivity j_v, and spectral flux density S). These quantities and their units are summarized in the appendix, Table A4. Their basic characteristics are developed in AM, Chapter 8.

5.2   Hot plasma

A plasma of electrons and ions will exist if collisions between atoms (or between electrons and ions) are sufficiently energetic and frequent to keep the gas ionized through the ejection of atomic electrons. In other words, the gas must have a sufficiently high temperature T. For a monatomic gas in thermal equilibrium and therefore with the Maxwell–Boltzmann distribution of speeds, the temperature specifies the average translational kinetic energy $(mv^2/2)_{av}$ of the atoms (3.35):

$$\frac{3}{2}kT = \left\langle \frac{1}{2}mv^2 \right\rangle_{av}, \qquad \text{(Defines temperature)} \qquad (5.3)$$

where m and v are, respectively, the mass and thermal speed of an individual atom. This relation also applies separately to the electrons and ions in a plasma. If the two species are in thermal equilibrium, their average kinetic energies will be equal.

Consider a plasma consisting only of ionized hydrogen (i.e., protons and electrons). If the kinetic energies are in excess of 13.6 eV, one might expect the gas to be mostly ionized. According to (3), this corresponds to a temperature $T \gtrsim 10^5$ K. In fact, the required temperature also depends on the particle densities because a collision between an electron and a proton can lead to their *recombination* into a neutral atom – possibly only momentarily.

The fraction of atoms ionized at a given instant is called the *degree of ionization*, which is specified by a complicated function of the physical conditions known as the *Saha equation* (4.15). It turns out that, for hydrogen plasmas at the low densities encountered in astrophysics, collisional recombination is small, and so hydrogen gases become almost totally ionized at temperatures as low as $\sim$3000 K.

In the present derivation, the plasma is assumed to be completely ionized and to consist of electrons of charge $-e$ and ions of charge $+Ze$, where Z is the atomic number. The simplest plasma is a hydrogen plasma ($Z = 1$) of electrons and protons. We take the plasma to be in thermal equilibrium; thus, the average kinetic energy of the ions is equal to that of the electrons. Because the electron mass is much less than the proton mass (1/1836), the electrons in a hydrogen plasma move about 40 times faster than the protons. This speed difference is even more pronounced in the presence of heavier ions. We therefore consider the ions to be stationary with fast-moving electrons being accelerated by their electric fields (Fig. 1).

In general, the collisions are a quantum phenomenon. The electromagnetic force itself is an exchange of virtual photons of which a few escape to become observable. For slowly moving particles and relatively near collisions, the number of virtual photons involved is large, and classical approximations are appropriate. The nonrelativistic speeds ($v \ll c$) in our plasma are in accord with this.

The classical approximation also requires that the electrons emit photons with energies $h\nu$ substantially less than their own kinetic energies. If the latter condition is not met, most of the electron's energy could be radiated away with only a few quanta, and quantum effects would be important. We further assume that the energy loss of an electron, integrated over an entire collision, is only a small part of the electron kinetic energy. Despite these restrictions,

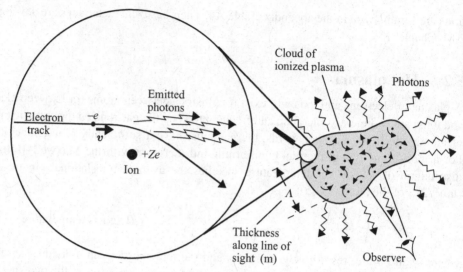

Figure 5.1. Cloud of plasma (ionized gas) giving rise to photons owing to the near collisions of the electrons and ions. The electrons are accelerated and thus emit radiation in the form of photons. The line-of-sight thickness of the cloud is Λ.

the classical approximation yields a result – namely a distribution of photon energies – that is valid over a wide range of conditions.

The free-free transition radiation derived here yields a continuum spectrum (i.e., without spectral lines). In practice, the continuously ionizing and recombining atoms undergo many bound-bound transitions, which yield spectral lines. Ionized plasmas are thus rich in spectral-line emission. Transitions in plasmas can be detected in the radio band between the closely spaced, very high levels of hydrogen; in the optical band as, for example, the hydrogen Balmer spectral lines; and in the x-ray band as spectral lines of heavy elements.

We take the density of electrons to be sufficiently low that the photons escape the plasma with negligible probability of interaction; that is, the optical depth τ for photon-electron inter-action is small, $\tau \ll 1$ (see AM, Chapter 10 for more on τ). The validity of this assumption in a given case depends on the frequency of the radiation. The plasma in a typical emission nebula (e.g., the Orion nebula) is transparent to high-frequency radio photons and opaque to lower-frequency radio waves. Our calculation here is no longer valid in the latter case. Rather, the optically thick (blackbody) radiation presented in Chapter 6 would apply.

The classical semiquantitative derivation in this chapter consists of finding the following quantities in order:

- (*i*)   the radiative energy emitted by a *single* electron during its near collision with an ion;
- (*ii*)   the frequency of the emitted radiation in terms of the electron speed and impact parameter;
- (*iii*) the power emitted at frequency ν into $d\nu$ by *all* the electrons of different speeds (the Maxwell–Boltzmann distribution) that collide with a single proton;
- (*iv*)   the emitted power at frequency ν in $d\nu$ from all collisions in a 1-m^3 volume, which is the volume emissivity $j_\nu(\nu, T)$;
- (*v*)   the integrated (over frequency) volume emissivity $j(T)$; and
- (*vi*)   the specific intensity $I(\nu, T)$ in terms of the line-of-sight thickness of the plasma.

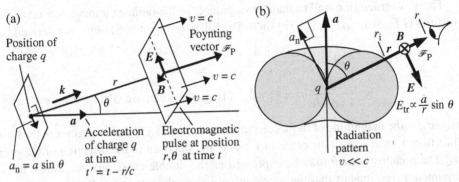

Figure 5.2. (a) Electric and magnetic field vectors, E and B, at the position r, θ at time t that arise from the horizontal acceleration a of a positive charge at the earlier time, $t' = t - r/c$. The two planes shown are normal to the propagation vector $k = (2\pi/\lambda)\hat{k}$, where the hat indicates a unit vector. In the plane-wave approximation (large r), the E and B fields are constant over the right-hand plane. (b) Dipole radiation pattern for a vertical acceleration a of a positive charge. The radial distance r_i along the line of sight from the origin to the intercept of the circle represents the relative magnitude of the transverse electric vector. At a fixed observer distance r, the magnitude varies as $\sin \theta$; see (4). The radiated power is proportional to E^2. The radiated B vector in (b) points into the paper.

5.3 Single electron-ion collision

Consider the collision of a single electron with a single ion. As noted just above, the velocities of the ions are on average much less than those of the electrons because both components are in thermal equilibrium. Furthermore, in a given collision, the acceleration of the ion is much less than that of the electron owing to momentum conservation or Newton's third law. Thus, we can consider the electron to be traveling through a region of static electric fields and neglect radiation from accelerating ions. We begin with a brief review of radiation in a vacuum; for more see Section 11.2.

Radiation basics

Radiated electric vector

The instantaneous pattern of electric-field lines of a charge q moving in a straight line at speed $v \ll c$ is isotropic (Fig. 7.3a). If this nonrelativistic charge undergoes acceleration, the field lines become distorted, and these distortions propagate outward at speed c to form a propagating electromagnetic wave that consists, at large distances, of transverse electric and magnetic field vectors (Fig. 2a).

The propagating electric vector E due to a previous instantaneous acceleration lies in the plane defined by the acceleration vector a and the propagation direction k (Fig. 2a). For a positive charge, the E vector is directed opposite to, and has amplitude proportional to, the projection a_n of a on the plane normal to the line of sight. In addition, its amplitude is proportional to charge q and varies with distance as $1/r$. The energy flux is proportional to E^2 and will thus decrease as $1/r^2$, as required by energy conservation. For a negative charge, the E vector is reversed; it is in the direction of a_n.

These considerations tell us that the magnitude of the radiated transverse electric field in a vacuum is $E_{tr} \propto q a_n / r = (qa/r)\sin\theta$. The full expression in SI units, not derived here, is

$$\boldsymbol{E}(\boldsymbol{r}, t) = E_{tr}\hat{\boldsymbol{n}} = \frac{qa(t')\sin\theta}{4\pi\varepsilon_0 c^2 r}\hat{\boldsymbol{n}}$$

$$= \frac{q}{4\pi\varepsilon_0 c^2 r}(\boldsymbol{a} \times \hat{\boldsymbol{k}}) \times \hat{\boldsymbol{k}}, \quad \text{(Transverse electric vector; V/m; } v \ll c) \quad (5.4)$$

where a is the magnitude of the acceleration vector $\boldsymbol{a}$ and $\hat{\boldsymbol{n}}$ is the unit vector in the transverse direction normal to $\boldsymbol{k}$ in the plane of $\boldsymbol{k}$ and $\boldsymbol{a}$ and opposed to the direction of the projection a_n. The quantity $a\sin\theta$ may be expressed as the double cross product shown in (4). The argument t' reminds us that the acceleration $\boldsymbol{a}$ took place at the earlier time $t' = t - (r/c)$, where t is the time of detection of $\boldsymbol{E}$ at distance r (Fig. 2a). This expression is valid only if the charge is moving, relative to the observer, at substantially less than the speed of light, $v \ll c$.

The angular dependence of E_{tr} is illustrated graphically in Fig. 2b. Here the acceleration is directed upward. The distance r_i from the origin to the intercept of the line of sight with the outer boundary of the doughnut-shaped pattern (a toroid) represents the relative magnitude of the field at that angle for a fixed observation distance r. The intercept distance and hence the amplitude vary as the sine of the angle from $\boldsymbol{a}$. The magnitude is zero at the pole and maximal at $\theta = 90°$. The power radiated is proportional to E^2, and so the angular distribution of power is much more strongly peaked in the equatorial directions than the toroid of Fig. 2b. This is a *dipole radiation* pattern.

A propagating transverse $\boldsymbol{E}$ vector in a vacuum is always associated with a propagating magnetic vector $\boldsymbol{B}$ of magnitude E/c (in SI units). It is at right angles to both $\boldsymbol{E}$ and $\boldsymbol{k}$ such that the direction $\boldsymbol{E} \times \boldsymbol{B}$ gives the propagation direction $\boldsymbol{k}$. In Fig. 2a, $\boldsymbol{B}$ is out of the paper, and in Fig. 2b it is into the paper.

In Fig. 2a,b, we show only an instantaneous pulse of $\boldsymbol{E}$ and $\boldsymbol{B}$ that propagates outward owing to a prior instantaneous acceleration of the charge. Subsequent vectors arriving at the observer will reflect the acceleration of the charge at later times. For example, if the charge position oscillates vertically at a fixed position, the observer will detect an oscillating, linearly polarized $\boldsymbol{E}$ field with an associated oscillating $\boldsymbol{B}$ field.

Poynting vector

At a given point in space at some time t, the energy flux density $\mathscr{F}$ (W/m^2) carried by an electromagnetic wave depends on the instantaneous values of the component vectors $\boldsymbol{E}$ and $\boldsymbol{B}$ of the wave at that position. From the energy densities (J/m^3) of electric and magnetic fields, $\varepsilon_0 E^2/2$ and $B^2/2\mu_0$ respectively, the speed of propagation c, and the relations $B = E/c$ and $c^2 = 1/\mu_0\varepsilon_0$, it follows directly that

$$\mathscr{F}_P = \frac{\boldsymbol{E} \times \boldsymbol{B}}{\mu_0}. \quad \text{(Poynting vector; W/m}^2) \quad (5.5)$$

This is known as the *Poynting vector*. As a vector, it specifies the direction of energy flow in the wave as well as the magnitude. It is usually designated with the symbol S, but here we use $\mathscr{F}_P$ to avoid confusion with spectral flux density S (W m^{-2} Hz^{-1}) and to be consistent with our use of $\mathscr{F}$ for energy flux density (Table A4 in the appendix).

The quantity $\mu_0 = 4\pi \times 10^{-7}$ T m A^{-1} is the *permeability of free space*, a constant. It and the *permittivity of the vacuum*, $\varepsilon_0 = 8.854 \times 10^{-12}$ s^4 A^2 m^{-3} kg^{-1}, are related to the speed of light as

$$c = (\mu_0 \varepsilon_0)^{-1/2} = 2.998 \times 10^8 \text{ m/s}, \tag{5.6}$$

which follows from the wave-equation solution to Maxwell's equations. The Poynting vector $\mathscr{F}_P$ is indicated in Fig. 2a,b; it points in the outward radial direction toward which the power flows.

Because E and B are perpendicular to each other in an electromagnetic wave (Fig. 2), the magnitude of the cross product (5) simplifies to the simple product of the vector magnitudes. Also, the magnitudes are proportional to each other, $B_{tr} = E_{tr}/c$, from differentiation and integration of one of Maxwell's equations (e.g., Faraday's law). Substitute B_{tr} into (5) and eliminate μ_0 with (6) to obtain the scalar amplitude of the Poynting vector:

$$\mathscr{F}_P = \varepsilon_0 c E \cdot E = \varepsilon_0 c E_{tr}^2. \qquad \text{(W/m}^2\text{)} \tag{5.7}$$

Substitute E_{tr} (4) into (7),

$$\Rightarrow \qquad \mathscr{F}_P(r, \theta, t) = \frac{q^2 \sin^2 \theta \, a^2(t')}{(4\pi)^2 \varepsilon_0 c^3 r^2}, \qquad \begin{array}{l}\text{(Magnitude of Poynting vector}\\ \text{in vacuum; } v \ll c; \text{W/m}^2\text{)}\end{array} \tag{5.8}$$

to find the magnitude of the Poynting vector in terms of the angle θ of the observer from the acceleration direction, the distance r from the accelerating charge to the observer, and the acceleration at the earlier time $t' = t - r/c$ at which the detected radiation was emitted. Again the $v \ll c$ restriction applies.

The flux $\mathscr{F}_P$ described in (8) is at a maximum along directions perpendicular to the acceleration direction (i.e., at $\theta = 90°$), as illustrated in Fig. 2b, for $v \ll c$. Also, the flux density varies inversely with radius squared, $\propto r^{-2}$, which is the usual inverse-squared law required by energy conservation.

Larmor's formula

The total power radiated by the electron into all directions is obtained by summation (integration) over a spherical surface at distance r from the charge. The element of area on the sphere is $r^2 d\Omega = r^2 \sin \theta \, d\phi \, d\theta$, where $d\Omega$ is an element of solid angle, θ is the polar angle in Fig. 2b, and ϕ is the azimuthal angle; see Fig. 3.2a. The total power radiated at some time t' is thus

$$\mathscr{R}(t) = \int_{\theta=0}^{\pi} \int_{\phi=0}^{2\pi} \mathscr{F}_P(r, \theta, t) \, r^2 \sin \theta \, d\phi \, d\theta. \tag{5.9}$$

Substitute (8) into (9) and integrate over all angles to obtain *Larmor's formula* (Prob. 32):

$$\Rightarrow \qquad \mathscr{R}(t) = \frac{1}{6\pi\varepsilon_0} \frac{q^2 a(t)^2}{c^3}. \qquad \begin{array}{l}\text{(Larmor's formula; instantaneous}\\ \text{emitted power; } v \ll c; \text{W)}\end{array} \tag{5.10}$$

Formally, this is the total power crossing a sphere at radius r at time t in terms of the earlier acceleration at time t'. Note that it is independent of distance r because the same total power

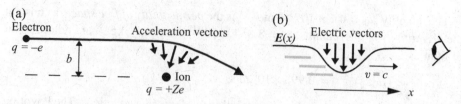

Figure 5.3. Electron trajectory showing (a) deviation of track, impact parameter b, and acceleration vectors and (b) the pulse of emitted transverse electric vectors for a negative accelerated charge. The profile $E(x)$ at a fixed time is shown as a solid line.

will cross spheres farther out at later times. We prefer, however, to think of $\mathcal{P}(t)$ simply as the power radiated instantaneously by an electron as it accelerates. Hence, in (10), we drop the prime in $a(t')$ and use simply $a(t)$.

Larmor's formula is a well known and highly useful relation that underlies much of astrophysics. It is based on the radiated electric field (4) and is valid if the radiating electron does not have a relativistic speed (i.e., $v \ll c$).

Energy radiated per collision

The instantaneous power radiated by an electron of charge $-e$ in the vicinity of an ion of charge $q = Ze$ (Fig. 3) depends on the electron's instantaneous acceleration $a(t)$ at the time t of interest (10). The acceleration depends, in turn, on the Coulomb $1/r^2$ force, which increases to a maximum at the point of closest approach to the proton and then decreases toward zero as the electron and proton separate.

The Coulomb force F between two charges q_1 and q_2 is

$$F = \frac{q_1 q_2}{4\pi\varepsilon_0 r^2}\,\hat{r}, \qquad\qquad \text{(N)} \qquad (5.11)$$

where $\hat{r}$ is a unit radius vector. The acceleration experienced by an electron of charge $-e$ and mass m at a distance r from an ion of charge Ze is

$$a = \frac{F}{m} = -\frac{1}{4\pi\varepsilon_0}\frac{Ze^2}{r^2 m}\,\hat{r}, \qquad\qquad \text{(m/s}^2\text{)} \qquad (5.12)$$

where we made use of Newton's second law ($F = ma$). Keep in mind that the vector quantities a, F, and $\hat{r}$ are all functions of time t.

The *impact parameter* b of a given collision is defined in Fig. 3 to be the projected distance of closest approach of the electron to the ion for a given encounter. If the electron loses only a small fraction of its (large) kinetic energy, the trajectory will be deflected only slightly from a straight line, and the closest approach will be approximately equal to the impact parameter b. In a plasma of randomly moving particles, the impact parameter will differ from collision to collision. For our purposes, the ion may be considered to be infinitesimally small.

The maximum acceleration of the electron during the collision is obtained from (12), where the radius is approximated with the impact parameter b:

$$a_{max} \approx \frac{1}{4\pi\varepsilon_0}\frac{Ze^2}{mb^2}. \qquad\qquad (5.13)$$

The time interval during which the acceleration is comparable to a_{max} for an electron of speed v is approximately the time it is in the immediate vicinity of the proton. If the distance traveled in the vicinity is taken to be roughly b, the interval is

$$\tau_b \approx b/v. \qquad \text{(Collision time)} \qquad (5.14)$$

In a given infinitesimal time interval dt at time t, the *energy* (joules) emitted is $\mathcal{P}(t)$dt, where $\mathcal{P}(t)$ is the instantaneous power given by Larmor's formula (10). The total energy emitted by the electron during the transit is then the sum (integral) of these contributions for the entire duration of the collision,

$$Q(b, v) = \int_{-\infty}^{+\infty} \mathcal{P}(t)\, \mathrm{d}t = \frac{1}{6\pi\varepsilon_0} \frac{e^2}{c^3} \int_{-\infty}^{+\infty} a(t)^2 \, \mathrm{d}t, \qquad \text{(J)} \qquad (5.15)$$

where the integration allows for the changing acceleration. This ideal collision between two isolated charges would last, in principle, from $t = -\infty$ to $t = +\infty$ because the $1/r^2$ force reaches to infinity, but it becomes vanishingly small at large distances. In a real plasma, the net force on an electron by an ion is strong only at small distances. At larger distances, the force goes rapidly to zero because other electrons closer to the proton yield opposing electric fields that shield (or screen) the ion (*Debye screening*).

The integration (15) may be simplified if the acceleration is taken to be constant at its maximum value while it is in the vicinity of the ion for time τ_b and to be zero before and after this period:

$$Q(b, v) \approx \frac{1}{6\pi\varepsilon_0} \frac{e^2}{c^3} a_{max}^2 \tau_b. \qquad (5.16)$$

Apply to this the acceleration (13) and the duration (14) to find

➡ $$Q(b, v) \approx \frac{1}{(4\pi\varepsilon_0)^3} \frac{2}{3} \frac{Z^2 e^6}{c^3 m^2 b^3 v}. \qquad \text{(joules/collision)} \qquad (5.17)$$

This, then, is the *total energy radiated* by a single electron of speed v as it passes an ion of charge Ze with impact parameter b. The energy loss increases strongly as b decreases. The velocity v is roughly the same before and after the collision for most collisions because only a modest part of the initial kinetic energy is given up to radiation in this classical approximation. For a given b, the radiated energy Q is less for faster electrons because they are in the vicinity of the ion for a shorter time (14).

Frequency of the emitted radiation

We have seen (Fig. 2a) that the electric vector E emitted by a *positive* accelerating charge is, at a sufficient distance at a later time, transverse to its propagation direction and opposed to the direction of the normal component of the instantaneous acceleration a_n.

In the case of our *negative* accelerating charge (Fig. 3a), the emitted electric vectors will lie in the same direction as the projected acceleration, as shown in Fig. 3b. Because the acceleration increases and decreases only once, the radiated electric vectors build up to a maximum and return toward zero only once. The radiation thus consists of a single pulse of downward-pointing E vectors rushing toward the detector (eye).

The frequency of the radiation derives directly from the duration of the central portion of the pulse as inferred by a fixed observer, $\tau_b \approx b/v$ (14). Set the angular frequency of the radiation, $\omega_{tb} = 2\pi\nu$, to be approximately the inverse of τ_b:

$$\omega_{tb} \approx \frac{1}{\tau_b} = \frac{v}{b}. \qquad \text{(rad/s)} \qquad (5.18)$$

The characteristic frequency ν of the radiation due to a single collision is thus

$$\Rightarrow \qquad \nu = \frac{\omega_{tb}}{2\pi} \approx \frac{v}{2\pi b}. \qquad \text{(Hz)} \qquad (5.19)$$

In other words, the approximate frequency for the radiation is the orbital frequency of an electron orbiting the ion at speed v and radius b.

This frequency turns out to be close to the frequency at which the most power (per unit frequency interval) is emitted. It is also approximately the maximum frequency emitted because there are no motions of the electron with shorter time scales in the problem. In contrast, there are longer time scales that follow from the slower changes in acceleration when the electron is at greater distances from the ion. These lead to emission at a broad range of lower frequencies ranging down to $\nu \approx 0$.

The power at these lower frequencies is likely to be small compared with the power in the band near the turnover frequency because of the decreasing bandwidths $\Delta\nu$. Consider a decade of frequency near the maximum, say from $0.3\nu_0$ to $3\nu_0$, which is a bandwidth of $2.7\nu_0$. The entire bandwidth below this decade to zero frequency has bandwidth $0.3\nu_0$ or nine times less. We thus make the further approximation that *all* the power for this particular collision of impact parameter b and speed v is emitted at the frequency ν given in (19).

The relation between b and the emitted frequency ν is given in (19). The smaller the impact parameter, the higher the emitted frequency. The electron experiences more of its acceleration in a shorter time during a close collision. The relation between the intervals of impact parameter db and frequency $d\nu$ is obtained by differentiation of (19):

$$b = \frac{v}{2\pi\nu} \qquad (5.20)$$

and

$$db = -\frac{v}{2\pi\nu^2} d\nu. \qquad (5.21)$$

These relations, (20) and (21), will enable us to take into account collisions at many impact parameters if a flux of electrons passes by the ion.

5.4   Thermal electrons and a single ion

Here we address the case of a single ion immersed in a sea of thermal electrons to obtain the power radiated by collisions with the ion.

Single-speed electron beam

Let the ion be immersed in a parallel beam of electrons of speed v and consider only the electrons that intersect a narrow annulus of radius b and width db surrounding the ion

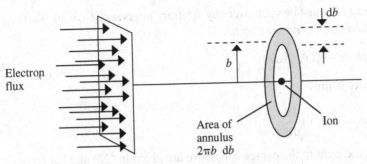

Figure 5.4. Flux of electrons approaching an ion with the annular region representing the target area at impact radius b in db. The number of electrons that pass through the annulus can be calculated from the flux of electrons and the target area.

(Fig. 4). One can then calculate the power emitted by those electrons as a function of the emitted frequency ν and the speed v.

Power from the annulus

If the density of electrons in the beam is n_e, the electron flux is $n_e v$ (electrons m^{-2} s^{-1}). The number per second that would strike an annulus of radius b and width db is just this flux times the area of the annulus, namely $n_e v\, 2\pi b\, db$. The energy emitted by these per unit time (emitted power per ion) is just this number times the energy $Q(b, v)$ (17) emitted by each electron per collision with impact parameter b and speed v:

$$\mathcal{P}_b(b, v)\, db = Q(b, v)\, n_e v\, 2\pi b\, db. \qquad \text{(W/ion in } db \text{ at } b) \qquad (5.22)$$

This is the power coming from each ion due to electrons of density n_e and speed v impinging on the ion at radius b in db.

The electrons actually arrive at the ion from all directions, and so the density in (22) should have been multiplied by $d\Omega/4\pi$. However, integration over all directions would directly yield the same result because $\int d\Omega = 4\pi$ sr.

Power per unit-frequency interval

The quantity $\mathcal{P}_b(b, v)$ is the power emitted per unit impact-parameter interval. We now convert it to power per unit-frequency interval, $\mathcal{P}_\nu(\nu, v)$, the unit used for photon spectra. The variables b and ν have a one-to-one correspondence (20) in our approximation, as do their differentials (21). Thus, $\mathcal{P}_b(b, v)$ can be integrated over some range of b to obtain the emitted power from that range, and $\mathcal{P}_\nu(\nu, v)$ can be integrated over the equivalent range in frequency. Because the ranges are equivalent, the integrated powers must be the same:

$$\int_{b_1}^{b_2} \mathcal{P}_b(b, v)\, db = -\int_{\nu_1}^{\nu_2} \mathcal{P}_\nu(\nu, v)\, d\nu. \qquad (5.23)$$

The minus sign arises from the requirement that the power be a positive quantity and from the fact that $db \propto -d\nu$ (21); an increase in ν corresponds to a decrease in b.

The equivalence (23) must be valid over any arbitrary interval of b (or ν). This can be true only if the integrands themselves are equal:

$$\mathscr{P}_b(b, v)\, db = -\mathscr{P}_\nu(\nu, v)\, d\nu.$$

(5.24)

The desired quantity is thus

$$\mathscr{P}_\nu(\nu, v) = -\mathscr{P}_b(b, v)\frac{db}{d\nu}.$$

(5.25)

The two terms on the right in the expression above are given in (22) and (21):

$$\mathscr{P}_\nu(\nu, v) = Q(b, v)\, n_e v\, 2\pi b\, \frac{v}{2\pi\nu^2}.$$

(5.26)

Finally, eliminate Q and then b with (17) and (20), respectively, to obtain, for electrons of speed v, the power emanating from each ion at frequency ν in the band $d\nu$:

➡ $$\mathscr{P}_\nu(\nu, v)\, d\nu \approx \frac{1}{(4\pi\varepsilon_0)^3}\, \frac{8\pi^2}{3}\, n_e\, \frac{Z^2 e^6}{c^3 m^2 v}\, d\nu.$$ (W/ion in $d\nu$ at ν)   (5.27)

Surprisingly, $\mathscr{P}_\nu(\nu, v)$ is independent of frequency. There will be more collisions at large b because the annulus area increases with b, but each will emit lower-frequency photons. The energy emitted by distant impacts thus provides the same emitted power per unit frequency as do the fewer, more efficient collisions at small b. We retain the argument ν in $\mathscr{P}_\nu(\nu, v)\, d\nu$ as a reminder that, in another situation, there could be a frequency dependence. The power (27) varies as v^{-1} for fixed ν due to a combination of several effects.

The classical result given in (27) can not be correct at the highest frequencies because an electron with a given kinetic energy, $mv^2/2$, can not radiate more than all of its kinetic energy. Thus, the maximum photon energy $h\nu_{max}$ that the electron can emit at a given velocity is $mv^2/2$, and therefore

$$\nu_{max} = \frac{mv^2}{2h}$$ (Maximum frequency)   (5.28)

is the cutoff frequency above which the radiated power must drop to zero.

The result (27) is approximately equal to the exact classical result except for a (frequency-dependent) factor of order unity, the *Gaunt factor*, which we discuss in Section 5. The classical result (with the Gaunt factor) is nearly correct so long as the electrons are not relativistic – that is, that the characteristic kinetic energy $\sim kT$ is substantially less than the electron's *rest energy*, mc^2. This means that the temperature of the plasma must be $T \ll 6 \times 10^9$ K.

Electrons of many speeds

Consider now the wide range of electron speeds in a real gas. The power emitted at a given frequency, ν, is a function of the electron speed v (27).

The *probability*, $\mathcal{P}(v)dv$, of an electron's having *speed* v in the interval dv, regardless of the vector direction of travel, for a nondegenerate gas is

$$\mathcal{P}(v)dv = P(v)\, 4\pi v^2 dv,$$ (Probability of speed v in dv)   (5.29)

where $4\pi v^2 dv$ is the volume of a shell in velocity space at speed v and $P(v)$ is the *Maxwell–Boltzmann distribution* of particle speeds in a gas:

$$P(v) = \left(\frac{m}{2\pi kT}\right)^{3/2} \exp\left(-\frac{mv^2}{2kT}\right). \quad \text{(Maxwell–Boltzmann distribution)} \quad (5.30)$$

Specifically, $P(v)$ is the probability of finding a particle with vector velocity v per unit 3-D velocity space, whereas the product (29) is the probability of finding speed v in dv. These expressions are those of (3.8) and (3.11), which are given in terms of momentum. The conversion from probability per unit momentum to probability per unit speed follows from $P(p)dp = P(v)dv$ and the relation $p = mv$ (Prob. 41). The conversion method is similar to that for $P_b(b, v)$ in (25).

The expression $P(v)dv$ is a *probability*; hence, it is dimensionless. Also, the sum of all probabilities is unity:

$$\int_0^\infty P(v)dv = 1. \quad (5.31)$$

The radiated power per ion, $\mathcal{P}_v(\nu, v)$, given in (27) for a flux $n_e v$ of electrons at speed v must be multiplied by the probability $P(v)dv$ (29) if it is to represent the power only from the electrons of speed v in dv. The total power summed over all speeds is an integral of this product over speed, namely,

$$\langle \mathcal{P}_v(\nu) \rangle_{\text{ion}} = \int_{v_{\min}}^\infty \mathcal{P}_v(\nu, v)\, P(v)\, 4\pi v^2 dv. \quad \text{(W ion}^{-1}\text{ Hz}^{-1}) \quad (5.32)$$

This is the power emitted at frequency ν (per unit frequency interval) from a single ion in a sea of electrons with the Maxwell–Boltzmann distribution of speeds. It is actually an average of $\mathcal{P}_v(\nu, v)$ over velocity with appropriate weighting according to the Maxwell–Boltzmann distribution, which accounts for the angle brackets. The functions in the integrand are given in (27) and (30); they would allow the integral to be evaluated.

The lower limit of the integral $v_{\min}$ is the smallest (minimum) velocity an electron can have and still emit a photon of energy $h\nu$. An electron can not give up more energy than it initially had; see discussion of (28). The speed limit follows from (28) for a fixed frequency:

$$v_{\min} = (2h\nu/m)^{1/2}. \quad (5.33)$$

This is a quantum constraint in our classical derivation.

5.5 Spectrum of emitted photons

Here we arrive at the desired spectrum by considering the radiation from a plasma that contains n_i ions per unit volume. As before, there are n_e electrons per unit volume, and they have a Maxwell–Boltzmann distribution of speeds.

Volume emissivity

The power emitted per unit volume per hertz from our plasma is known as the *volume emissivity*, $j_v(\nu)$ (W m^{-3} Hz^{-1}). It follows directly from the expressions in the previous section.

The units of volume emissivity (W m^{-3} Hz^{-1}) should be carefully noted because other references sometimes define j_ν to be (W m^{-3} Hz^{-1} sr^{-1}). The quantity used here is the power emitted in all directions, whereas the latter is the power emitted into 1 sr of solid angle. For isotropic emission, a value in the latter units is smaller by a factor of 4π.

Multiple ion targets

The integrated power given in (32) is the power emitted from collisions by electrons with a single ion in a Maxwell–Boltzmann ionized gas. Multiplication by the ion density n_i yields the volume emissivity $j_\nu(\nu)$ (W m^{-3} Hz^{-1}), which is the power emitted per unit volume at frequency ν per unit frequency interval. To obtain the power emitted in the frequency interval $d\nu$, also multiply by the bandwidth $d\nu$ (Hz):

$$j_\nu(\nu)\,d\nu = n_i \,\langle \mathcal{P}_\nu(\nu, v)\rangle_{\text{ion}}\,d\nu. \tag{5.34}$$

The average indicated by the brackets is over the electron speeds. Thus, from (32),

$$j_\nu(\nu)\,d\nu = n_i \left[\int_{v_{\min}}^{\infty} \mathcal{P}_\nu(\nu, v)\,\mathcal{P}(v)\,4\pi v^2\,dv \right] d\nu. \qquad \text{(W/m}^3 \text{ at } \nu \text{ in } d\nu) \tag{5.35}$$

Exponential spectrum

One could now substitute into (35) the expressions $\mathcal{P}_\nu(\nu, v)$, $\mathcal{P}(v)$, and $v_{\min}$ from (27), (30), and (33), respectively, and carry out the integration. Because $\mathcal{P}_\nu(\nu, v)$ is independent of frequency, the frequency dependence comes in only through $v_{\min}$. The result would incorporate the errors due to the approximations we have made; however, it would turn out to be identical to the correct result except that the numerical factor would be low by a factor of $\sim$2 and the *Gaunt factor* $g(\nu, T, Z)$ would be missing (Prob. 42). (The Gaunt factor is discussed in the next section.) For reference, we quote, instead, the correct result,

$$\Rightarrow \qquad j_\nu(\nu)\,d\nu = g(\nu, T, Z)\,\frac{1}{(4\pi\varepsilon_0)^3}\,\frac{32}{3}\left(\frac{2}{3}\,\frac{\pi^3}{m^3 k}\right)^{1/2}\frac{Z^2 e^6}{c^3}\,n_e\,n_i\,\frac{e^{-h\nu/kT}}{T^{1/2}}\,d\nu.$$

$$\text{(Volume emissivity; W/m}^3 \text{ at } \nu \text{ in } d\nu) \tag{5.36}$$

The proportionality constants in (36) may be represented as C_1:

$$j_\nu(\nu)\,d\nu = C_1 g(\nu, T, Z)\,Z^2 n_e\,n_i\,\frac{e^{-h\nu/kT}}{T^{1/2}}\,d\nu,$$

$$C_1 = 6.8 \times 10^{-51}\,\text{J m}^3\,\text{K}^{1/2}. \qquad \text{(W/m}^3 \text{ in } d\nu \text{ at } \nu) \tag{5.37}$$

This is the result anticipated in (1). It is plotted in log-log coordinates in Fig. 5 for a $T = 5 \times 10^7$ K gas with $n_e = n_i = 10^6$ m^{-3}. In a real astrophysical plasma, the relative numbers of ions with different atomic numbers Z must be taken into account. For cosmic abundances, one would have an effective Z^2 value of $\sim$1.15. We state again that the quantum effects that give rise to abundant emission spectral lines were not included in this derivation.

The horizontal axis of Fig. 5 is logarithmic and hence could extend to the left forever to smaller and smaller values and would never reach zero frequency, which occurs at log $\nu = -\infty$. At low frequencies, where $h\nu/kT \ll 1$, the exponential alone approximates unity, $\exp(-h\nu/kT) \approx 1.0$. The dashed curve in Fig. 5 is thus flat as it extends to low frequencies.

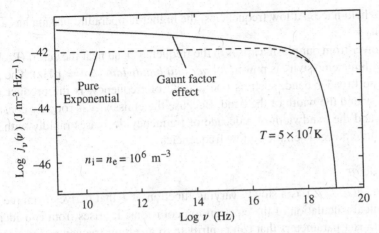

Figure 5.5. Theoretical continuum thermal bremsstrahlung spectrum. The volume emissivity (37) is plotted from radio to x-ray frequencies on a log-log plot with the Gaunt factor (38) included. The specific intensity $I(\nu, T)$ would have the same form. Note the gradual rise toward low frequencies due to the Gaunt factor. We assume a hydrogen plasma ($Z = 1$) of temperature $T = 5 \times 10^7$ K with number densities $n_{\rm i} = n_{\rm e} = 10^6 \, {\rm m}^{-3}$.

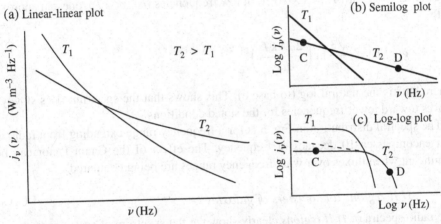

Figure 5.6. Thermal bremsstrahlung spectra (as pure exponentials) on (a) linear-linear, (b) semilog, and (c) log-log plots for two sources with the same ion and electron densities but differing temperatures, $T_2 > T_1$. The spectral shape is the same for the volume emissivity j_ν and the specific intensity I; see (42). Measurement of the specific intensities at two frequencies (e.g., at C and D) permits one to solve for the temperature T of the plasma as well as for the emission measure $\langle n_{\rm e}^2 \rangle_{\rm av} \Lambda$. The dashed line in (c) shows the effect of the Gaunt factor. [From H. Bradt, *Astronomy Methods*, Cambridge, 2004, Fig. 11.3, with permission]

The effect of the Gaunt factor is shown; it modifies the exponential response noticeably but modestly over the many decades of frequency displayed.

The curves in Fig. 6 qualitatively show the function $j_\nu(\nu, T)$, on linear, semilog, and log-log axes for two temperatures $T_2 > T_1$. The exponential term causes a rapid reduction ("cutoff") of flux at a higher frequency for T_2 than for T_1. At low frequencies, because the exponential is essentially fixed at unity, the intensity is governed by the $T^{-1/2}$ term if the other variables,

Z, n_i, and n_e, are held fixed. At low frequencies, the higher temperature plasma has a *lower* volume emissivity!

Most of the power from our plasma arises in the frequency band near the cutoff (Prob. 52). Recall that the volume emissivity is power/vol *per unit bandwidth* ($\Delta \nu = 1$ Hz). The power emitted into some broader band, such as one decade of frequency, is the product of the average emissivity and the width of the band. Because the emissivity is roughly constant at low frequencies and the bandwidth of a decade of frequency decreases rapidly with lower frequency, very little power is emitted a low frequencies.

Gaunt factor

The *Gaunt factor*, $g(\nu, T, Z)$, is a slowly varying function of ν that derives from the exact quantum mechanical calculation of the electron-ion collisions. It arises from consideration of the range of impact parameters that can contribute to a certain frequency. For example, if the impact parameter is too large, other charges in the vicinity will "screen" the electric field of the ion. Also, if the impact parameter approaches zero, quantum effects become important.

For most conditions the Gaunt factor has a numerical value of order unity. There is no single closed expression for g; it depends on the temperature and frequencies. For a hydrogen plasma ($Z = 1$) with $T > 3 \times 10^5$ K at low frequencies ($h\nu \ll kT$), one can approximate it with

$$g(\nu, T) = \frac{\sqrt{3}}{\pi} \ln \left(\frac{2.25 \, kT}{h\nu} \right), \tag{5.38}$$

where "ln" is the natural log (to base e). This shows that the spectrum rises slowly as one moves toward lower frequencies for the stated conditions.

The spectral distribution in Fig. 5 is for a frequency range extending from radio to x ray that encompasses 10 decades of frequency. The effect of the Gaunt factor can be quite significant when fluxes over wide frequency ranges are being compared.

H II regions and clusters of galaxies

The radio spectra of *H II regions* clearly show the flat spectrum of an optically thin thermal source. H II regions are star-forming regions that contain large amounts of gas and dust. The brightest of the newly formed stars in the region emit copiously in the ultraviolet and thus ionize the hydrogen gas in the region. The result is a plasma that emits the typical spectrum of thermal bremsstrahlung. An example of this is shown in Fig. 7 for two H II regions in the "W3" complex of radio emission.

The data points in Fig. 7 are the filled and open circles; the drawn lines are continuum models for the plasma and the dust that best fit the data. The huge peak is due to blackbody emission from hot dust; thus, the data points that represent the flatter and less intense free-free continuum are found only up to $\sim 10^{11}$ Hz ($\lambda \approx 3$ mm) – that is, into the microwave region. At the lowest frequencies, the plasmas become optically thick and turn over with a ν^2 spectrum typical of the low-frequency part of the blackbody spectrum (6.8).

Another example of thermal bremsstrahlung is the x radiation from hot gas interspersed between galaxies in a *cluster of galaxies*. In this case we show a purely theoretical spectrum

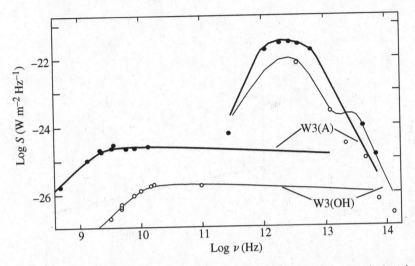

Figure 5.7. Continuum spectra (energy flux density) of two H II (star-forming) regions, W3(A) and W3(OH), in the complex of radio, infrared, and optical emission known as "W3." The data (filled and open circles) and early model fits (heavy and light lines) are shown. In each case, there is a flat thermal bremsstrahlung (radio), a low-frequency cutoff (radio), and a large peak at high frequency (infrared, 10^{12}–10^{13} Hz) due to heated, but still "cold," dust grains in the nebula. [P. Mezger and J. E. Wink, in "H II Regions & Related Topics," T. Wilson and D. Downes, Eds., Springer-Verlag, p. 415 (1975); data from E. Kruegel and P. Mezger, *A&A* **42**, 441 (1975)].

(Fig. 8) for a plasma of temperature 10^7 K that takes into account quantum effects and hence shows the expected emission lines. Comparison with real spectra from clusters of galaxies allows one to deduce the actual amounts of different elements and ionized species in the plasma as well as its temperature. It is only in the present millennium that x-ray spectra taken from satellites (e.g., Chandra and the XMM Newton) have had sufficient resolution to distinguish these narrow lines.

Integrated volume emissivity

Total power radiated

The total power radiated from unit volume is found from an integration of (37) over frequency (Prob. 53):

$$\Rightarrow \quad j(T) = \int_0^\infty j_\nu(\nu)\,d\nu = C_2\,\bar{g}(T, Z)\,Z^2\,n_e\,n_i\,T^{1/2}$$
$$C_2 = 1.44 \times 10^{-40}\ \text{W m}^3\ \text{K}^{-1/2}, \qquad\qquad (\text{W/m}^3) \quad (5.39)$$

where T is in units of Kelvin (K), and n_e and n_i, the number densities of electrons and ions, respectively, are in units of m^{-3}. The integration is carried out with $g = 1$, and a frequency-averaged Gaunt factor $\bar{g}$ is then introduced. Its value can range from 1.1 to 1.5 with 1.2 being a value that will give results accurate to ∼20%. Note that the total power increases with temperature for fixed densities, as might be expected.

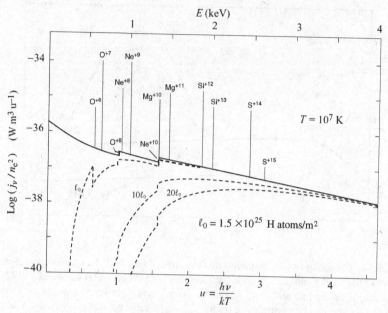

Figure 5.8. Semilog plot of theoretical calculation of the volume emissivity j_ν, divided by electron density squared, of a plasma at temperature 10^7 K with cosmic abundances of the elements as a function of $h\nu/kT$. The various atomic energy levels and ionization states are properly incorporated; strong emission lines and pronounced "edges" are the result. The dashed lines show the effect of x-ray absorption by interstellar gas. The straight-line portion of the plot falls by a factor of ~ 2.7 for each change of u by unity, as expected for the exponential e^{-u}. [From W. Tucker and R. Gould, *ApJ* **144**, 244 (1966)]

White dwarf accretion

One can use the expression (39) for $j(T)$ to deduce the equilibrium temperature of an optically thin plasma into which energy is being injected. An example is gas that accretes (Section 2.7) onto the polar region of a compact white dwarf star from a companion star. As the matter flows downward, it is accelerated by gravity to very high energies. Just above the surface, it may encounter a shock, which abruptly slows the material and raises it to a high density; the kinetic infall energy is converted into random motions (i.e., thermal energy). The material is then a hot, optically thin plasma that slowly settles to the surface of the white dwarf.

This plasma radiates away its thermal energy according to the expressions (36) and (39) above. At the same time it is continuously receiving energy from the infalling matter. In equilibrium, the energy radiated by the plasma equals that being deposited by the incoming material. The plasma temperature will come to the value required for the plasma to radiate away exactly the amount of energy it receives.

One can thus use the deposited energy as an estimate of the radiated energy. That is, if values are adopted for the accretion energy being deposited per cubic meter per second and for the densities n_e and n_i, the temperature of the plasma may be determined from (39). Conversely, measurement of the temperatures and fluxes of the emitted radiation provide quantitative information about the underlying accretion process.

If the star is highly magnetic, the infalling material is guided to the polar regions of the star by its magnetic field, and the hot plasma will be forced into a very small volume. For such magnetic systems, the plasma reaches x-ray temperatures (Prob. 51).

5.6 Measurable quantities

Here we explore the relationships between volume emissivity and two determinable quantities, the luminosity and the specific intensity.

Luminosity

The luminosity $L(T)$ as a function of temperature of an entire plasma cloud follows from $j(T)$ (39). If j is constant throughout the volume, the luminosity is simply the product of $j(T)$ and the volume V of the plasma. If not, an integration over the cloud must be carried out as follows:

$$L(T) = \int_{\substack{\text{volume} \\ \text{of source}}} j(T)\, dV. \qquad \text{(W)} \qquad (5.40)$$

Substitute into this the expression for $j(T)$ (39) and assume a hydrogen plasma ($Z = 1, n_e = n_i$),

$$\blacktriangleright \qquad L(T) = C_2\, \bar{g}(T)\, T^{1/2} \int_{\substack{\text{volume} \\ \text{of source}}} n_e^2\, dV, \qquad \text{(luminosity; W)} \qquad (5.41)$$

where we take T to be a constant throughout the volume. The luminosity increases with temperature as does j. It is also proportional to the integral of n_e^2 summed over the volume.

Specific intensity (resolved sources)

The *specific intensity* $I(\nu, T)$ (W m^{-2} Hz^{-1} sr^{-1}) is the quantity used by an observer to describe the emission from an extended object in the sky. By extended, we mean a source larger in angular size than the angular resolution of the telescope–detector system. It follows from the units that I is the energy flux detected per unit frequency interval per unit solid angle.

The product, $I(\theta, \phi, \nu, T)\,d\nu\, d\Omega$, represents the measured energy flux (W/m^2) detected at frequency ν in the interval $d\nu$ arriving from the celestial direction described by polar and azimuthal angles θ, ϕ in the increment of solid angle $d\Omega = \sin\theta\, d\theta\, d\phi$. We often suppress the variables θ, ϕ in the argument of I, but one should not forget that I is a function of the direction in space described by two angles.

The specific intensity measured for a certain angular position on a given source is identical in magnitude at any frequency to the quantity known as the *surface brightness*, $B(\nu, T)$ (W m^{-2} Hz^{-1} sr^{-1}). The latter quantity describes the emission radiating into unit solid angle from unit area (normal to the radiation direction) of that same portion of the observed surface. That is, $B(\nu, T) = I(\nu, T)$. This equivalence is discussed in terms of Liouville's theorem in Section 3.3.

In general, the specific intensity follows from the volume emissivity if the emission is assumed to be isotropic:

$$I(\nu, T) = \int_0^\Lambda \frac{j_\nu(r, \nu, T)}{4\pi}\, dr = \frac{j_{\nu,\mathrm{av}}(\nu, T)}{4\pi}\Lambda.$$ (Specific intensity; W m^{-2} Hz^{-1} sr^{-1})   (5.42)

The volume emissivity $j_\nu(r, \nu, T)$ (W m^{-3} Hz^{-1}) is taken to be a function of the radial position r along the line of sight as well as of frequency and temperature. The reader can confirm that this integral relation is valid (see AM, Chapter 8). The quantity $j_{\nu,\mathrm{av}}$ is the average value of j_ν along the line of sight through a cloud of thickness Λ (Fig. 1).

Emission measure

The expression for j_ν (37) may be substituted into the middle term of (42). If the plasma cloud is *isothermal* (i.e., if the temperature is constant along the line of sight), and if it consists solely of hydrogen so that $Z = 1$ and $n_i n_e = n_e^2$, we have

➡ $$I(\nu, T) = \frac{C_1}{4\pi}\, g(\nu, T)\, \frac{e^{-h\nu/kT}}{T^{1/2}} \int_0^\Lambda n_e^2\, dr \qquad \text{(W m}^{-2}\text{ Hz}^{-1}\text{ sr}^{-1}\text{)}$$
$$C_1 = 6.8 \times 10^{-51} \text{J m}^3 \text{ K}^{1/2}.$$ (5.43)

Rewrite (43) in terms of the average of n_e^2 for a plasma of thickness Λ along the line of sight:

$$I(\nu, T) = \frac{C_1}{4\pi}\, g(\nu, T)\frac{e^{-h\nu/kT}}{T^{1/2}} \langle n_e^2 \rangle_{\mathrm{av}} \Lambda.$$ (5.44)

This is the result anticipated in (2).

The integral in (43) is known as the *emission measure*, EM,

➡ $$\int_0^\Lambda n_e^2\, dr = \langle n_e^2 \rangle_{\mathrm{av}} \Lambda \equiv \text{Emission Measure (EM)}.$$ (m^{-5})   (5.45)

This is another example of a column line-of-sight integral; see (42). We see from (43) that the emission measure may be obtained from a measurement of $I(\nu, T)$ at some frequency ν if the temperature T is known.

Determination of T and EM

The function (43) may be considered to have two unknown parameters, the temperature T and the factor $\int n_e^2\, dr = $ EM. Measurement of $I(\nu)$ at two frequencies (e.g., at C and D in Fig. 6c) can yield these two parameters if the radiation is known to be thermal bremsstrahlung. For the assumption of $g = 1$, a simple fit to these two points would yield the entire exponential spectrum for T_2. The frequency ν at which the function has dropped to $e^{-h\nu/kT} = e^{-1}$ of its low-frequency intercept value gives T because, at this frequency, $h\nu = kT$ and therefore $T = h\nu/k$. With this value of T, any single measurement of I applied to (43) yields $\int n_e^2\, dr$, the EM.

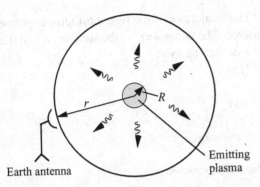

Figure 5.9. Geometry for obtaining the spectral flux density S (W m^{-2} Hz^{-1}) for an optically thin spherical and isotropically radiating source of radius R and distance r. If the telescope's angular resolution exceeds the angular size of the source, the source is detected as a "point" source.

If the frequency variation of the Gaunt factor is known and properly included, the spectrum has a unique shape for each temperature. In this case also, the temperature and the EM may be obtained from measurements at two frequencies.

Of course, this determination of T and EM is only possible if the source is resolved by the antenna or if the solid angle subtended by the source is independently known. Otherwise, the specific intensity (flux per steradian) on which this logic is based is not known. The situation is further complicated if there are significant magnetic fields in the plasma.

Spectral flux density S (point sources)

The specific intensity $I(\nu)$ can not be measured directly for a source with angular size smaller than the telescope resolution (i.e., a point source). However, one can use the spectral energy flux density $S(\nu)$ (W m^{-2} Hz^{-1}) to describe the radiation from such a source. This is the energy received per square meter at the telescope at some frequency ν in unit bandwidth $\Delta\nu = 1$ Hz. Formally, it is the specific intensity integrated over the solid angle encompassed by the source:

$$S(\nu, T) = \iint I(\nu, T)\, d\Omega. \qquad \text{(Spectral flux density, W m}^{-2}\text{ Hz}^{-1}\text{)} \qquad (5.46)$$

This will exhibit the same frequency dependence as I (for uniform conditions in the source), albeit with different proportionality constants.

Measured flux from volume emissivity

The spectral flux density S can be obtained directly from the volume emissivity j_ν. Consider a spherical emitting source of radius R at a (possibly unknown) distance r from the observer with average – over the volume of the source – volume emissivity $j_{\nu,\text{av}}(\nu, T)$ (Fig. 9). The spectral flux density is, from its elementary definition (energy per unit area),

$$S(\nu, T) = \frac{L_\nu}{4\pi r^2} = \frac{j_{\nu,\text{av}}(\nu, T)\, 4\pi R^3/3}{4\pi r^2}, \qquad \text{(Spherical source)} \qquad (5.47)$$

where L_ν is the luminosity per hertz. The numerator of the rightmost term expresses L_ν in terms of $j_{\nu,\text{av}}$, and the volume of the source. The factor $4\pi r^2$ is the surface area of the sphere surrounding the source at the distance r of the observer.

More generally, one would write (47) as

$$\rightarrow \quad S(\nu, T) = \frac{1}{4\pi r^2} \iiint j_\nu(\nu, T)\, dV, \tag{5.48}$$

where the integral is over the volume of the source.

Specific intensity and flux density compared

What information can one gain about the source itself from a measurement of S or I? Substitute j_ν (37) into (47) to obtain, after rearranging the terms with R,

$$S(\nu, T) = C_1\, g(\nu, T)\frac{e^{-h\nu/kT}}{T^{1/2}}\, n_e^2 \frac{R}{3\pi}\left(\frac{\pi R^2}{r^2}\right), \quad \begin{array}{l}\text{(Spectral flux density; spherical}\\ \text{source; W m}^{-2}\,\text{Hz}^{-1})\end{array} \tag{5.49}$$

where we again take $Z = 1$ and $n_i = n_e$, for a hydrogen plasma. Compare this with the expression (43) for specific intensity $I(\nu, T)$, which we rewrite for a measurement through the center of the sphere (i.e., for $\Lambda = 2R$):

$$I(\nu, T) = C_1\, g(\nu, T)\frac{e^{-h\nu/kT}}{T^{1/2}}\, n_e^2 \frac{2R}{4\pi}. \quad \begin{array}{l}\text{(Specific intensity through center of}\\ \text{spherical source; W m}^{-2}\,\text{Hz}^{-1}\,\text{sr}^{-1})\end{array} \tag{5.50}$$

With these two equations, (49) and (50), the relative merits of measuring S and I become apparent. The frequency dependence is the same in the two cases. In either instance the temperature can be extracted from two measurements as described above, after (45), and this specifies $g(\nu, T)\exp(-h\nu/kT)$ if the appropriate Gaunt function is known.

The same two measurements also yield the value of a second "unknown" – namely, the product of the other unknown terms in the expression. In the case of the I measurement (50), this product is $n_e^2\, 2R$, the emission measure. In the case of the S measurement (49), it is $n_e^2\, R\Omega$, where $\Omega = \pi R^2/r^2$ is the solid angle of the source. One can not find the emission measure because, by our terms, Ω is not known. If it were, we would measure I and use (50).

One learns more from the I measurement, but such a measurement is only possible if the telescope's resolution is sufficient to determine the angular size of the source and hence its solid angle Ω. The source must be of sufficient angular size to fill the "beam" of at least one pixel in the image plane of the telescope.

Problems

5.2 Hot plasma

Problem 5.21. (a) Formally write the requirement on temperature in degrees kelvin implied by the stipulation that the electrons in a thermal plasma not be relativistic. Require that the average kinetic energy of the particles (that obey the Maxwell–Boltzmann distribution) be much less than the rest energy mc^2 of the electron. Use SI units. (b) A plasma emits most of its energy in x rays in the energy range 1–20 keV. If the average particle energy is comparable to the photon energies, will the classical approximation be appropriate for this plasma? [Ans. $\sim 10^9$ K; –]

5.3 Single electron-ion collision

Problem 5.31. (a) In a thermal nonrelativistic hydrogen plasma, by what factor will the root-mean-square (rms) velocity of the electron exceed the rms velocity of the protons? (b) In a given electron-proton collision, by what factor will the acceleration of the electron exceed that of the proton? (c) Do you think electron-electron collisions are important sources of radiation? Why? Hint: think about electric fields. [Ans. ~ 40; ~ 2000; no]

Problem 5.32. Derive Larmor's formula (10) beginning with the expression for the radiated transverse electric vector (4). Follow the suggestions in the text and fill in the missing steps and calculations, including demonstrating that the expression for Poynting vector (5) follows from the energy densities given for electric and magnetic fields.

5.4 Thermal electrons and single ion

Problem 5.41. Convert the Maxwell–Boltzmann momentum probability $\mathcal{P}(p)$ (3.11) to the velocity probability $\mathcal{P}(v)$ (29) and (30). Follow the suggestions in the text. Argue from your result that the probability of finding vector velocity $\boldsymbol{v}$ is that given in (30).

Problem 5.42. Substitute the values of $\vartheta_v(\nu, v)$, $\mathcal{P}(v)$ and v_{min} given in (27), (30), and (33) into (35). Carry out the integration to obtain the spectral distribution $j_v(\nu)$ (36). By what factor does your answer differ from (36)? (Assume the Gaunt factor is precisely unity, $g = 1$.) This is an indication of the effect our approximations had on the final result. [Ans. ~ 0.4]

5.5 Spectrum of emitted photons

Problem 5.51. A hydrogen plasma from a companion star accretes (flows) onto a white dwarf star of radius 8000 km and mass 0.5 $M_\odot$. The rate of plasma flow onto the white dwarf is 10^{-9} $M_\odot$ per year. The plasma is guided to one pole of the white dwarf by the magnetic field in such a way that it impinges on only 1% of the star's surface. The kinetic energy of the plasma gained in the fall from "infinity" is suddenly reduced to near zero as the matter is abruptly slowed in a shock just above the surface; the matter then settles slowly down to the surface. The thin (1-m deep) region just below the shock effectively absorbs all the infall energy. This thin region thus contains a very hot plasma; it is optically thin with mass density $\rho = 10^{-2}$ kg/m^3. (a) What is the number density of ions n_i in the shock region? Assume a plasma of pure hydrogen. (b) Calculate the potential energy lost per second (J/s) by the accreting material as it falls from "infinity." (c) This energy is converted to thermal energy in the thin postshock region. What is the power (J/s) deposited in 1 m^3 of this region? (d) The radiated power from this region equals the input accretion power (J/s) in the steady-state condition. Use $j(T)$ (39) to find the equilibrium temperature of the plasma. Let the Gaunt factor be unity. What is the band of radiation (radio? gamma ray?, etc.) that corresponds to this temperature? [Ans. $\sim 10^{25}$ m^{-3}; $\sim 10^{27}$ W; $\sim 10^{14}$ W/m^3; $\sim 10^8$ K]

Problem 5.52. (a) Consider the volume emissivity of an optically thin plasma emitting thermal bremsstrahlung radiation at temperature T. Integrate the expression for $j_v(\nu)$ (37) from frequency ν_1 to $f\nu_1$ to obtain the power radiated in a fixed logarithmic frequency interval; if

$f = 10$, your result would give the power in one decade as a function of frequency. Demonstrate that the power drops rapidly at both $h\nu \ll kT$ and $h\nu \gg kT$. Let $g = Z = 1$. (b) Find the frequency at which the power in a fixed logarithmic interval is at a maximum as a function of f. To obtain a final solution, let the interval factor f approach unity, $f = 1 + \varepsilon$ for $\varepsilon \ll 1$. How does $h\nu$ compare with kT at the maximum as $\varepsilon \to 0$? [Ans. $-$; $h\nu \approx kT$]

Problem 5.53. (a) Verify the result of the integration of $j_\nu(\nu, T)$ (37) to obtain $j(T)$ for $g = 1$; see (39). (b) The Orion nebula, an H II region, is radiating by thermal bremsstrahlung. Consider it to be spherical (radius $R = 8$ LY), optically thin, and at temperature $T = 8000$ K throughout. Let $Z = 1$, $g = 1$, and $n_e = n_i = 6 \times 10^8$ m^{-3}. Find the luminosity (W) of the entire nebula in terms of solar luminosities. (c) In what wavelength band or bands will the power from the Orion nebula be radiated? [Ans. $-$; $\sim 10^4 \, L_\odot$; IR]

5.6 Measurable quantities

Problem 5.61. Consider a cylindrical nebula at distance r with the circular end (radius $R \ll r$) facing the observer and with length $3R$ along the line of sight. It is optically thin and has uniform volume emissivity $j_{\nu,0}$ (W m^{-3} Hz^{-1}) throughout. (a) Use (42) and (46) to obtain an expression for the spectral flux density $S(\nu)$ as a function of $j_{\nu,0}$, R, and r. Hint: how does the solid angle depend on R and r? (b) Now, find again the spectral flux density $S(\nu)$ by first calculating the specific luminosity L_ν (W/Hz) of the nebula. Hint: note (47). You should obtain the same answer as in (a). [Ans. $\sim j_{\nu,0} R^3/r^2$; $-$]

Problem 5.62. (a) Find the specific intensity function $I(\nu)$ (W m^{-2} Hz^{-1} sr^{-1}) you would expect to measure from the Orion nebula from thermal bremsstrahlung radiation. Evaluate it for $\nu \to 0$ (i.e., for the flat part of its spectrum on a log-log plot where the plasma is optically thin). Use the temperature, composition, electron density, size, and Gaunt factor as given in the statement of Prob. 53 above. Begin by finding the volume emissivity. Assume a cloud thickness along the line of sight equal to its diameter 16 LY. (b) Find the spectral flux density function $S(\nu)$ (W m^{-2} Hz^{-1}) for the nebula as a whole. Again, evaluate for $\nu \to 0$. The angular diameter of the nebula is $\sim 35'$. Assume your answer to (a) is valid over the entire angular extent of the nebula. Compare your answer with the measured value of 4400 Jy at 15 GHz. (1 Jy (jansky) $= 1.0 \times 10^{-26}$ W m^{-2} Hz^{-1} sr^{-1}). (c) Find the spectral flux density S of this plasma in the optical V band (effective frequency 5.5×10^{14} Hz). What is the expected V magnitude of the nebula? ($V = 0$ corresponds to 3600 Jy; see AM, Chapter 8). Compare with the actual value, $V \approx 4$. [Ans. $\sim 10^{-19}$ W m^{-2} Hz^{-1} sr^{-1}; ~ 3000 Jy; ~ 4 mag]

6

Blackbody radiation

What we learn in this chapter

A **photon gas** in perfect thermal equilibrium with its surroundings at some temperature T will exhibit an energy spectrum of a specific amplitude and shape known as the **blackbody spectrum**, which was first proposed by **Max Planck** in 1901. In its form as a **specific intensity** $I(\nu)$ (W m^{-2} Hz^{-1} sr^{-1}), the blackbody spectrum peaks at a frequency proportional to its temperature. At low frequencies (the **Rayleigh–Jeans approximation**), it increases linearly with temperature and quadratically with frequency. At high frequencies (the **Wien approximation**), it decreases quasi-exponentially. The **energy density**, $\propto T^4$, and **photon number density**, $\propto T^3$, follow directly from $I(\nu)$. The former is closely related to the **pressure** of a photon gas, whereas the latter is closely related to the **distribution function**, the number density in six-dimensional phase space. Calculation of the **average photon energy** yields $2.70\,kT$.

The **total energy flux (W)** passing in one direction through a unit surface is proportional to T^4. A normal gaseous (spherical) star emits a spectrum that approximates (roughly) that of a blackbody, which allows the **luminosity** to be expressed in terms of the **stellar radius** and an **effective temperature**. Momentum transfer by the photons to a hypothetical surface yields a **pressure** that is one-third the energy density. The blackbody flux is the **maximum intensity** that can be obtained from a thermal body.

The **universe** is permeated by photons with a blackbody spectrum of temperature 2.73 K, the **cosmic microwave background** (CMB) radiation. In the expanding universe, this radiation cools **adiabatically** while **maintaining the spectral shape and intensity of a blackbody**. Its **temperature** scales inversely as the **scale factor** of the expansion.

6.1   Introduction

Blackbody radiation is pervasive in astrophysics. The surfaces of "normal" stars emit a spectrum that approximates blackbody radiation, and the 3-K cosmic microwave background radiation (CMB) exhibits a nearly perfect blackbody spectrum (Fig. 1). Also, radio spectra from emission nebulae manifest the rising power-law character of blackbody radiation at low frequencies.

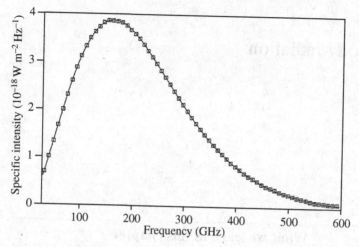

Figure 6.1. Plot of the CMB spectrum measured with the COBE satellite. The squares indicate measurements, and the solid line is the best-fit blackbody spectrum. The error bars are smaller than the squares. The fit is incredibly good over a large range of frequencies. The temperature is found with high precision to be 2.725 ± 0.002 K. Note that this is a linear-linear plot. [NASA/COBE Science Team; J. Mather *et al.*, *ApJ* **354**, 37 (1990)]

Blackbody radiation arises when matter is optically thick, and photons thereby scatter many times before encountering an observer. Under such conditions, the particles and photons continually share their kinetic energies. In perfect thermal equilibrium, the average particle kinetic energy will equal the average photon energy, and a unique temperature T may be defined. For nonrelativistic monatomic particles, the definition is

$$\frac{3}{2}kT = \left(\frac{1}{2}mv^2\right)_{av} = h\nu_{av}. \qquad \text{(Defines } T; \text{ J)} \qquad (6.1)$$

Perfect thermodynamic equilibrium is obtained (in theory) within a container when the radiation and the container walls are all at the same temperature T. In astrophysics, the 3-K microwave background spectrum originated under such conditions; all space was at the same temperature aside from tiny fluctuations that later led to galaxy formation. This radiation reflects the optically thick character of the universe just before protons and electrons combined to form hydrogen.

In contrast, in the atmospheres of stars, the temperature varies with height in the atmospheres. In this case, one can consider the gas and photons to be in thermodynamic equilibrium only in local regions; this is known as *local thermodynamic equilibrium* (LTE). The radiation spectrum in a small region of temperature T will approximate the blackbody form. The spectrum in another nearby region will approximate the blackbody form at the somewhat different temperature of that region.

The average and peak photon energies $h\nu_{av}$ and $h\nu_{peak}$ of the blackbody spectrum are approximately equal to kT within a factor of a few, $h\nu_{av} \approx h\nu_{peak} \approx kT$. Most of the emitted power in thermal bodies resides in this general region of the spectrum.

It is useful to know some conversions based on the equality $h\nu = kT$ and the numerical values of the Boltzmann and Planck constants as follows:

$$\nu(\text{Hz}) = 2.084 \times 10^{10}\, T(\text{K}). \qquad \text{(for } h\nu = kT) \qquad (6.2)$$

For example, the sun's surface at $T \approx 6000$ K gives photons of frequency $\sim 10^{14}$ Hz, which is in the optical band. Particle and photon energies are often given in electronvolts (1.0 eV = 1.602×10^{-19} J). Thus, one can write $T(\text{K}) = (kT(\text{eV}) e(\text{J/eV}))/k(\text{J/K})$, which gives

$$T(\text{K}) = 11\,605 \times kT(\text{eV}), \tag{6.3}$$

and, if we again make use of $h\nu = kT$,

$$T(\text{K}) = 11\,605 \times h\nu(\text{eV}). \qquad \text{(for } h\nu = kT) \tag{6.4}$$

This tells us that the solar surface ($T \approx 6000$ K) emits photons of energy $\sim 1/2$ eV. We emphasize that the relations (2) and (4) are based on the assumed equality $h\nu = kT$ and hence only approximate physical relations – for example, those relating ν_{av} and ν_{peak} to kT.

Classical ideas were not successful in explaining the spectral intensity function as observed from optically thick bodies. Max Planck introduced a quantum hypothesis of discrete states and obtained an expression in 1901 that successfully modeled the observed spectrum.

The Planck function can be derived with Bose–Einstein statistics that apply to integer spin particles (bosons). Unlike fermions, there is no a priori limit to the number of particles allowed in any given state, but there is a limit on the total energy available for the photons to share.

The Bose–Einstein statistics are used to find the most probable distribution of photons as a function of their energy $h\nu$. This can be expressed as the number density in six-dimensional (6-D) phase space, that is, the *distribution function f*. The distribution function is discussed in Chapter 3 in the context of Maxwell–Boltzmann statistics (3.14) and Fermi–Dirac statistics (3.53). Recall (3.14) that it is related to the momentum probability distribution $\mathcal{P}(p)$ as $f = n\mathcal{P}(p)$, where n is the particle number density.

For a blackbody cavity, the distribution function (not derived here) turns out to be

$$f = \frac{2}{h^3(e^{h\nu/kT} - 1)}. \qquad \begin{array}{l} \text{(Distribution function for massless bosons;} \\ \text{Bose–Einstein statistics; (J s)}^{-3}) \end{array} \tag{6.5}$$

This is the average number of photons in one 6-D phase-space cell divided by the volume h^3 of the cell as a function of photon energy $h\nu$ (Fig. 3.4 illustrates a 2-D phase-space cell). In other words, f is the density in 6-D phase space averaged over all the cells that are at the frequency $\nu = pc/h$ in phase space. (Recall from Section 3.3 that phase space is a 6-D momentum – physical (x, y, z) space and that the momentum of a photon is $p = h\nu/c$.) The average number in each cell, $2/(e^{h\nu/kT} - 1)$, varies from infinity at $\nu = 0$ to zero at $\nu \to \infty$. This is quite different from the Fermi–Dirac distribution (3.53), where the maximum number per cell is 2. The derivation of (5) entails rather subtle statistical arguments that we do not present.

In practice, one is usually interested in the number density of photons in physical (x, y, z) space as a function of energy. This depends, as will be demonstrated, on the number of available phase-space states at each energy.

The material in this chapter is fundamentally based on this distribution function. We have learned (3.26) that the specific intensity of propagating photons is directly related to the distribution function as $I(\nu) = (h^4\nu^3/c^2)f$. The two quantities, I and f, are equivalent; the characteristics of the radiation can be derived from either one of them.

We first present the Planck function in its form as the specific intensity $I(\nu, T)$. From this, we derive several characteristics of the radiation, including the energy and number

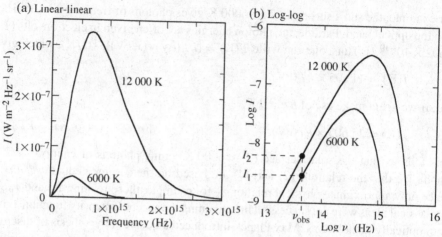

Figure 6.2. Sketches of blackbody spectra for two temperatures on both linear and log-log plots from (6). The temperatures differ by a factor of two. Note the straight-line power-law behavior at low frequencies in (b), the rapid decrease at high frequencies caused by the exponential term, the frequency of the maximum intensity increasing with temperature, and the rapid growth as a function of temperature. The areas under the two curves in (a) differ by a factor of 16 in accord with the integral $\int I \, d\nu \propto T^4$. At a specified (low) frequency, the temperature T may be used as a shorthand for the specific intensity I because $I \propto T$; see dashed lines in (b) at frequency ν_{obs}.

densities. We then derive the spectrum of an adiabatically expanding photon gas, which is of cosmological relevance.

6.2   Characteristics of the radiation

Several characteristics of blackbody radiation are derived from the blackbody specific-intensity, I (W m^{-2} Hz^{-1} sr^{-1}).

Specific intensity

The functional form of the specific intensity $I(\nu, T)$ for blackbody radiation follows directly from the distribution function (5) according to $I(\nu) = (h^4 \nu^3/c^2)f$, (3.26):

$$I(\nu, T) = \frac{2h\nu^3}{c^2} \frac{1}{e^{h\nu/kT} - 1}. \quad \text{(Planck radiation law; W m}^{-2}\text{ Hz}^{-1}\text{ sr}^{-1}\text{)} \quad (6.6)$$

This function is known as the *Planck radiation law* or the *Planck function*. It is plotted in Fig. 2 for two temperatures on linear-linear and log-log scales. It is our convention to use frequency units (intensity per hertz) rather than wavelength units (intensity per unit wavelength). We suppress the subscript ν that is sometimes used to indicate frequency units: $I_\nu(\nu, T) \to I(\nu, T)$.

The expression (6) specifies the absolute magnitude of the intensity at each frequency and hence the overall shape of the spectrum. This expression is the intensity that would be measured if one were immersed in the radiation. It is also the intensity of a distant diffuse source emitting blackbody radiation that would be measured with an antenna beam smaller in angular size than the source. Furthermore, the measured blackbody intensity would be equal to the *surface brightness* B (W m^{-2} Hz^{-1} sr^{-1}) of the emitting surface of the source (3.27).

Rayleigh–Jeans and Wien approximations

At frequencies well below the peak ($h\nu \ll kT$), the Planck function takes on a simple power-law form. This low-frequency limit is obtained from the Taylor expansion

$$e^{h\nu/kT} = 1 + \frac{h\nu}{kT} + \cdots \tag{6.7}$$

substituted into (6):

➡ $$I(\nu, T) \approx \frac{2\nu^2 kT}{c^2} \propto \nu^2 T. \quad (h\nu \ll kT; \text{Rayleigh–Jeans approximation}) \tag{6.8}$$

This yields a quadratic dependence on frequency and a linear dependence on temperature. Radio astronomers often use this relation to report a specific intensity at a given frequency as an *antenna temperature*. If the radiation is blackbody, this temperature will be that of the radiation (Fig. 2b). If it is not blackbody, the antenna temperature is simply a parameter proportional to the specific intensity.

At frequencies well above the peak, $h\nu \gg kT$, the exponential in (6) is much larger than unity,

$$e^{h\nu/kT} \gg 1, \quad\quad\quad (h\nu \gg kT), \tag{6.9}$$

and thus (6) becomes

➡ $$I(\nu, T) \approx \frac{2h\nu^3}{c^2} e^{-h\nu/kT}. \quad (h\nu \gg kT; \text{Wien approximation}) \tag{6.10}$$

This is known as *Wien's law*, or the *Wien approximation*. In this regime, as frequency ν increases, the ν^3 term drives the function up and the $\exp(-h\nu/kT)$ term drives it down. The latter is a much stronger variation when $h\nu/kT \gg 1$, and so it dominates and the function decreases rapidly.

The function $I(\nu, T)$ is plotted quantitatively in log-log format in Fig. 3 for six different temperatures. Each curve shows the variation with frequency ν for a given temperature. The straight-line behavior with slope 2 at low frequencies is due to the power-law character ($I \propto \nu^2$) of I (8). As temperature increases, the intensity increases at all frequencies, and so I is a monotonic function of T at each frequency ν. Also, the peak flux moves to higher frequency and the total intensity integrated over frequency (W m^{-2} sr^{-1}) increases rapidly; compare the areas of the two curves in Fig. 2a.

Peak frequency

The frequency at which $I(\nu, T)$ reaches its peak value is designated ν_{peak}. It can be demonstrated with some calculus (Prob. 23a) that $h\nu_{\text{peak}}$ varies with temperature as

➡ $$h\nu_{\text{peak}} = 2.82\, kT. \quad\quad (\text{Wien displacement law}) \tag{6.11}$$

The peak of the curve moves to higher frequency for higher temperatures. This expression is obtained in the usual manner for finding the maximum of a function: take the derivative of (6) with respect to ν, set it to zero, and solve for ν_{peak}. The result is a transcendental equation

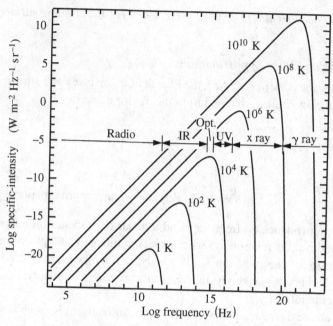

Figure 6.3. Log-log plot of blackbody spectra for a wide range of temperatures with horizontal scale expanded a factor of 2 relative to the vertical. At the low-frequency ($h\nu \ll kT$) end, the straight lines with slope 2 and the vertical separation of the curves indicate that $I \propto \nu^2 T$, which is the Rayleigh–Jeans approximation. At high frequencies ($h\nu \gg kT$), the function becomes $I \propto \nu^3 \exp(-h\nu/kT)$, the Wien approximation.

of order "3," which has the root 2.82 (see Table 2 in Section 4 below). Substitute the values of the constants h and k into (11) to obtain

$$\nu_{peak}(\text{Hz}) = 5.88 \times 10^{10} \times T(\text{K}).$$

$$(6.12)$$

You may compare this with (2) to see the effect of the factor of 2.8. For the CMB ($T = 2.73$ K), the peak occurs at $\nu_{peak} = 16.1 \times 10^{10}$ Hz $= 161$ GHz.

As stated in Section 1, most of the power of a blackbody spectrum is emitted in photons with frequencies in the vicinity of ν_{peak}. The intensity is high, and the bandwidth per decade of frequency is larger here than at lower frequencies; see the discussion of thermal bremsstrahlung radiation after (5.37). The rapid exponential decrease beyond the peak diminishes power at higher frequencies. From (12), we see that the frequencies that carry most of the power increase with temperature. This is in accord with common experience; a red-hot body is not as hot as a "white-hot" (bluish) object. This holds for a star or for a heated cannonball.

Wavelength units

The specific intensity may be converted to wavelength units, I_λ (W m^{-2} ($\Delta\lambda$)$^{-1}$ sr^{-1}), if one uses the equality of energy flux $I_\lambda d\lambda = I_\nu d\nu$, where $I_\nu \equiv I$ is the expression (6) and the wavelength and frequency are related as $\nu = c/\lambda$. The result (Prob. 23b) is

$$I_\lambda(\lambda, T) = \frac{2hc^2}{\lambda^5} \frac{1}{e^{hc/(\lambda kT)} - 1}.$$

(Planck function; wavelength units; W m^{-3} sr^{-1})

$$(6.13)$$

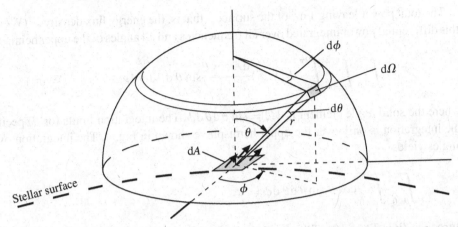

Figure 6.4. Physical surface element dA in the surface of a (quasi) blackbody radiator (e.g., of a star). Emission into the solid angle $d\Omega$ at θ, ϕ is shown. The total emission from this physical element into the upper hemisphere, $\mathscr{F} = \sigma T^4$ (W/m²) (18), is obtained from integration of the specific intensity over all angles of the hemisphere and over all frequencies while taking into account the changing projected surface area.

The peak of I_λ turns out to be at a wavelength λ_{peak} that is somewhat different than c/ν_{peak}. At temperature T, one finds that (Prob. 23b)

$$T\lambda_{peak} = 2.898 \times 10^{-3} \text{ K m.} \qquad (\lambda \text{ in m}; T \text{ in K}) \qquad (6.14)$$

For example, at $T = 2.73$ K, the peak is at $\lambda_{peak} = 1.06$ mm, which is equivalent to a frequency of 282 GHz, a factor of 1.8 greater than the 161-GHz value of ν_{peak} obtained from (12). This is a consequence of the different functional forms of I_λ and $I(= I_\nu)$.

Luminosity of a spherical "blackbody"

The specific intensity of radiation is equal to the surface brightness $B(\nu, T)$ of the object itself, $I = B$ (3.27). The surface brightness is defined as the energy flux density (per Hz-sr) emitted from a unit surface that lies normal to the view direction. The surface element is thus different for each view direction.

Energy flux density through a fixed surface

It is useful to know the total energy flux passing through unit area of a fixed surface. This surface could be a mathematical surface immersed within a blackbody cavity, and one would ask the rate of energy passing from one side to the other (e.g., from left to right). Within a cavity, an equal amount would flow from right to left to yield a zero net flux. Alternatively, the surface could be a small part of the surface of a star, and one could ask for the rate of energy passing from below the surface to above it given a perfect blackbody spectrum.

Consider the radiation (Fig. 4) at angle θ, ϕ that passes through the surface element dA in 1 s. The power at ν in $d\nu$ passing through dA into $d\Omega$ is the surface brightness $B(\nu, T)$ multiplied by the projected area $dA \cos \theta$, by the solid angle $d\Omega$, and by the frequency interval $d\nu$. Divide by the area dA to obtain the (differential) power in $d\Omega$ and $d\nu$ per unit area of the stellar surface.

The total power leaving 1 m^2 of the surface – that is, the energy flux density $\mathscr{F}$ (W/m^2) – is this differential power integrated over all frequencies and all angles of the upper hemisphere:

$$\mathscr{F} = \int_{\nu=0}^{\infty} \int_{\theta=0}^{\pi/2} \int_{\phi=0}^{2\pi} B(\nu, T) \frac{dA \cos \theta}{dA} \sin \theta \, d\theta \, d\phi \, d\nu, \qquad \text{(W/m}^2) \qquad (6.15)$$

where the solid angle element is $d\Omega = \sin \theta \, d\theta \, d\phi$. The integration limits for θ specify that the integration is only over the upper hemisphere shown in Fig. 4. The integration over the angles yields

$$\int_{\theta=0}^{\pi/2} \int_{\phi=0}^{2\pi} \cos \theta \sin \theta \, d\theta \, d\phi = \pi. \qquad (6.16)$$

Introduce $B(\nu, T) = I(\nu, T)$ from (6) into (15):

$$\mathscr{F} = \pi \int_0^{\infty} \frac{2h\nu^3}{c^2} \frac{d\nu}{e^{h\nu/kT} - 1}. \qquad (6.17)$$

This is an integral over frequency alone. Change the variable to $x = h\nu/kT$ to obtain an integral of the form $\int [x^3(e^x - 1)^{-1}] dx$. Solutions of this type of integral include the *Riemann zeta function*, which is defined and tabulated in Section 4; see (73), (74), and Table 1. The result of the integration is (Prob. 23c)

➡ $$\mathscr{F} = \sigma T^4, \qquad \text{(Energy flux of blackbody; W/m}^2) \qquad (6.18)$$

where σ is the Stefan–Boltzmann constant,

$$\sigma = \frac{2\pi^5 k^4}{15c^2 h^3} = 5.670 \times 10^{-8} \, \text{W m}^{-2} \, \text{K}^{-4}. \quad \text{(Stefan–Boltzmann constant)} \qquad (6.19)$$

We repeat that the calculated flux (18) is that which passes in one direction through a surface immersed in blackbody radiation. If a star were to emit a perfect blackbody spectrum, this would be the flux leaving 1 m^2 of its surface.

Effective temperature

The result (18) suggests that one can obtain the luminosity of a star of radius R by multiplying the flux (18) by the entire surface area $4\pi R^2$ of the spherical star. Thus, one might write $L = 4\pi R^2 \sigma T^4$, but this would be in error because a stellar atmosphere is not in perfect thermodynamic equilibrium; the temperature in the photosphere decreases with height. Also, the decreasing temperature with height leads to absorption lines in the spectrum.

Thus, if the true temperature ($\sim$6500 K) of the solar photosphere (at optical depth $\tau = 1.00$ and wavelength $\lambda = 500$ nm) is used in $4\pi R^2 \sigma T^4$, one finds a luminosity greater than the measured value. Nevertheless, spectra of stellar surfaces do roughly approximate blackbody spectra, and, as noted after (1), local thermodynamic equilibrium (LTE) may be assumed because the outward flux through a horizontal area element only slightly exceeds the inward flux.

It is customary, therefore, to invert the logic. Under the approximation that the star emits a blackbody spectrum, the measured luminosity and radius of a star are used to define its effective temperature T_{eff}:

$$L \equiv 4\pi R^2 \sigma T_{eff}^4. \qquad \text{(Definition of effective temperature; W)} \qquad (6.20)$$

The effective temperature for the sun is 5777 K, which is 11% less than the photospheric temperature.

Radiation densities

The specific intensity readily leads to the physical content of blackbody radiation – namely, the energy and photon number densities.

Energy density

The energy density of blackbody radiation is the sum of the energies $h\nu$ of photons contained at one instant in 1 m^3. The energy density may be defined as either $u_\nu(\nu, T)$, the energy density per hertz, or as $u(T)$, the total energy density summed over all frequencies:

$$u_\nu(\nu, T): \quad \text{Spectral energy density} \qquad \text{(J m}^{-3}\text{ Hz}^{-1}\text{)} \qquad (6.21)$$

$$u(T): \qquad \text{Energy density} \qquad \text{(J/m}^3\text{)}$$

Because photons of all frequencies travel at speed c, the frequency dependence of the spectral energy density is the same as that of the specific intensity I. It does not matter whether the radiation sampled is that in 1 m^3 at a fixed time or that impinging on a 1-m^2 surface in 1 s.

The relation between the two quantities, I and u_ν, follows from their definitions. Consider first the spectral energy density u_ν. It includes photons moving at speed c isotropically in all directions into all 4π sr. Divide by 4π to obtain the energy per unit volume (shaded in Fig. 5) flowing into 1 sr, $u_\nu/4\pi$. Multiply this by the speed c of the photons to obtain the energy flux per steradian passing through 1 m^2 in 1 s, which is the specific intensity I:

$$I(\nu, T) = u_\nu(\nu, T)\frac{c}{4\pi}. \qquad \text{(Conversion } u_\nu \text{ to } I\text{)} \qquad (6.22)$$

The expression for $u_\nu(\nu, T)$ is therefore, from (6) and (22),

$$u_\nu(\nu, T) = \frac{8\pi h\nu^3}{c^3}\frac{1}{e^{h\nu/kT} - 1}. \qquad \text{(Spectral energy density; J m}^{-3}\text{ Hz}^{-1}\text{)} \qquad (6.23)$$

There is another term in the energy density that could be included in (23). It is called the *zero-point energy*, which is a theoretical and normally unobservable energy density present in a vacuum even in the absence of photons. We do not consider it further.

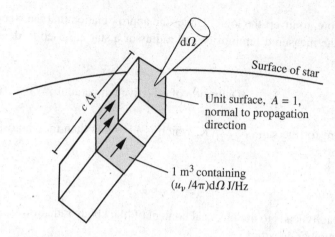

Figure 6.5. Geometry for the relation (22) between the energy density $u_\nu(\nu, T)$ (J m^{-3} Hz^{-1}) and the specific intensity $I(\nu, T)$, (W m^{-2} Hz^{-1} sr^{-1}) of the radiation. Unit volume (shaded region) of radiation flowing into solid angle $d\Omega$ contains $(u_\nu/4\pi)\, d\Omega$ joules per hertz. The column of length $c\Delta t$ crosses the unit surface A in time Δt.

The *total* energy $u(T)$ in a unit volume is simply $u_\nu(\nu, T)$ integrated over all frequencies. From (22),

$$u(T) = \int_0^\infty u_\nu(\nu, T)\, d\nu = \frac{4\pi}{c} \int_0^\infty I(\nu, T)\, d\nu. \qquad \text{(J/m}^3) \qquad (6.24)$$

The right-hand integral of (24) is identical to the integral (17), which was found (18) to equal $\sigma T^4/\pi$. Thus,

$$\Rightarrow \qquad u(T) = \frac{4}{c}\, \sigma T^4 = aT^4, \quad \text{(J/m}^3\text{; energy density of blackbody radiation)} \qquad (6.25)$$

where, from (19),

$$a = \frac{4\sigma}{c} = \frac{8\pi^5 k^4}{15 c^3 h^3} = 7.566 \times 10^{-16}\, \text{J m}^{-3}\, \text{K}^{-4}. \qquad (6.26)$$

The coefficient a is a derived physical constant. Energy densities for the CMB and the photosphere of the sun are worked out in Probs. 22 and 31.

The total energy density $u(T)$ is thus strictly a function of temperature. Given a cavity containing blackbody radiation of a given temperature T, the energy content in each cubic meter is determined uniquely by (25). It is a very strong function of temperature ($\propto T^4$), which increases by a factor of 16 for each doubling of T.

Spectral number density

The *spectral number density* of photons at a given frequency, $n_\nu(\nu, T)$ (photons m^{-3} Hz^{-1}), is simply the spectral energy density divided by the energy $h\nu$ of a single photon. From (23),

$$\Rightarrow \qquad n_\nu(\nu, T) = \frac{u_\nu(\nu, T)\, d\nu}{h\nu} = \frac{8\pi\nu^2}{c^3} \frac{1}{e^{h\nu/kT} - 1}. \qquad (\text{m}^{-3}\, \text{Hz}^{-1}) \qquad (6.27)$$

The number density of photons n_ν is driven toward zero by the ν^2 term at low frequency and the exponential term at high frequency.

Cells in phase space

We can gain some insight into this expression if we regroup the terms and multiply both sides by the bandwidth $d\nu$ as follows:

$$n_\nu(\nu, T)\, d\nu = \left(\frac{4\pi\nu^2}{c^3}\, d\nu\right)\left(\frac{2}{e^{h\nu/kT} - 1}\right). \qquad \text{(Spectral number density; m}^{-3}) \qquad (6.28)$$

In the context of phase space, the three terms of (28) can be described as

$$\begin{pmatrix} \text{Number } N \text{ of} \\ \text{photons in unit} \\ \text{physical volume} \\ \text{at } \nu \text{ in } d\nu \end{pmatrix} = \begin{pmatrix} \text{Number } Z_c \text{ of} \\ \text{phase-space cells,} \\ \text{each of volume } h^3 \\ \text{at frequency } \nu \text{ in} \\ d\nu \text{ in unit physical} \\ \text{volume} \end{pmatrix} \begin{pmatrix} \text{Average number} \\ <n_c> \text{ of photons} \\ \text{per cell at} \\ \text{frequency } \nu \text{ (two} \\ \text{polarizations)} \end{pmatrix}. \qquad (6.29)$$

Here "physical volume" means volume in ordinary x, y, z space.

The central term of (29) is the number Z_c of cells in a six-dimensional (6-D) x, p phase space with frequency ν in $d\nu$. It increases as ν^2 owing to the increasing volume of spherical shells in momentum space. This term is the product of the shell volume in momentum space, $4\pi p^2 dp$, where $p = h\nu/c$, and the specified unit physical volume divided by the volume h^3 of a single cell in 6-D phase space.

The final term shows that the average number $<n_c>$ of photons per cell at frequency ν decreases with energy $h\nu$ owing to the limitation on total energy. The factor of 2 takes into account photons of two polarizations (e.g., right and left circular). This term is simply the phase-space number density f (the distribution function) given in (5) multiplied by the volume h^3 of one cell in 6-D phase space – again for unit physical volume.

The distribution of photons (28) can be visualized as in Fig. 6, in which three photon energy levels, $h\nu_1$, $h\nu_2$, and $h\nu_3$, are shown. The three levels have energies in the ratios of 1: 2: 4. Each level contains Z_c phase-space cells, and the number of cells in fixed bandwidth $d\nu$ increase as ν^2. Each such cell contains the indicated numbers of particles, which average to $(e^{h\nu/kT} - 1)^{-1}$ for all the cells at that energy in accord with the last term of (28). We drop the factor of 2 because we assume only one polarization for the figure.

The rows in Fig. 6 contain the cells that would occupy different radial shells in a three-dimensional momentum space (Fig. 3.1c). The illustrated energy levels are well separated in energy; but in reality there is a continuum of energy states. Each cell can, in principle, contain any number of photons (bosons) from zero to infinity, but the fixed total energy limits the numbers.

The average number per cell at a given frequency in Fig. 6 decreases as frequency increases (28). Nevertheless, at low frequencies, the total photon numbers increase with frequency owing to the rising number of phase-space cells. Extrapolation to higher frequencies or energies would reveal a turnover to decreasing numbers of photons and decreasing total energy per level caused by the strong effect of the exponential term (Prob. 25).

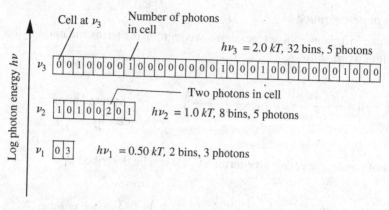

Figure 6.6. Sketch of phase-space cells in energy space for three different energy states $h\nu$ for a photon gas of temperature T, where $\nu_2 = 2\nu_1$ and $\nu_3 = 2\nu_2$. The number of photons in each cell is indicated for unit spatial volume. Photons at frequency ν_1 have energy $h\nu_1$ and are in the bottom row, those at frequency ν_2 have energy $h\nu_2$ and are in the middle row, and so on. In accord with (28), the number of cells at each frequency increases as ν^2, for fixed bandwidth, referenced to two cells at ν_1. We adopt one polarization rather than two, and so the average number of photons per cell is $1/(e^{h\nu/kT} - 1)$.

Total number density

The total number density of photons in blackbody radiation is the integral of (27) over frequency:

$$n(T) = \int_0^\infty n_\nu(\nu, T)\, d\nu = \frac{8\pi}{c^3} \int_0^\infty \frac{\nu^2}{e^{h\nu/kT} - 1}\, d\nu. \quad \text{(Number density; all frequencies; m}^{-3}\text{)} \quad (6.30)$$

Again, change variables and use the zeta function (Table 1) to evaluate the integral:

$$\Rightarrow \quad n(T) = 16\pi \left(\frac{kT}{hc}\right)^3 \times 1.202 = 2.029 \times 10^7\, T^3\, \text{m}^{-3}. \quad \text{(Photon number density; } T \text{ in K)} \quad (6.31)$$

At $T = 1$ K, there are 20 million photons per cubic meter. At $T = 3$ K, there are 27 times more, which works out to 0.55 photons per cubic millimeter. As noted (14), the peak of the wavelength distribution at 3 K is at $\lambda \approx 1.0$ mm. Thus, at 3 K, each cubic millimeter contains ~ 1 photon of wavelength ~ 1 mm.

At other temperatures, the same result is found; a cube of a size about equal to the peak wavelength will contain about one photon. This readily follows from $h\nu = kT$, giving $\lambda_{\text{peak}} \propto T^{-1}$. The volume of a cube of size λ_{peak} thus scales as T^{-3}. Multiplication by the density $n(T) \propto T^3$ (31) yields a photon number independent of temperature. Because the number is unity at $T = 3$ K, it is unity at all temperatures.

Average photon energy

Finally, the average photon energy $h\nu_{\text{av}}$ is the energy density u divided by the number density n. From $u = aT^4$ (25) and (31) with reference to (26), we find that

$$\Rightarrow \quad h\nu_{\text{av}} = \frac{u(T)}{n(T)} = \frac{\pi^4}{30 \times 1.202} kT = 2.70\, kT. \quad \text{(Average photon energy; J)} \quad (6.32)$$

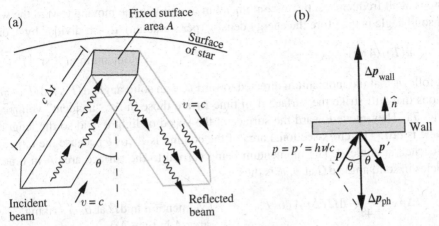

Figure 6.7. Geometry for derivation of pressure due to photons impinging on a fixed surface of area A. (a) Photons in the column of length $c\Delta t$ and cross section $A\cos\theta$ impinge on the surface A in time Δt. (b) Vector diagram showing the initial momentum $\boldsymbol{p}$ of an incident photon, its reflected momentum $\boldsymbol{p}'$, the change in photon momentum $\Delta\boldsymbol{p}_{\mathrm{ph}}$, and the equal and opposite momentum transferred to the wall $\Delta\boldsymbol{p}_{\mathrm{wall}}$. The pressure is force per unit area (N/m^2) or, equivalently, the momentum transferred per unit area and per unit time.

The average photon energy is 2.70 times the energy kT. This value is close to, but not exactly equal to, the energy corresponding to the peak of the frequency distribution, which is $2.82\,kT$ (11).

Radiation pressure

Photons carry momentum as well as energy. If they strike a surface they transfer momentum to it. The momentum transfer per unit time represents a force on the surface, and the force per unit area on the surface is a pressure. Thus, a gas of photons can exert a pressure on the surface of its container. The concept of pressure is also valid within the gas even if there is no physical surface; a mathematical surface is equivalent. As noted after (3.34), pressure is most generally a tensor, but it behaves as a scalar (with no direction) within an isotropic gas.

Beam of photons

The magnitude of the pressure due to a photon gas may be calculated from its momentum transfer in a way similar to our derivation in Section 3.4 of the pressure exerted by an ideal gas (see Fig. 3.3). Here, however, we properly calculate the effect of particles at all incidence angles.

Consider a beam of photons moving toward a surface element of area A at angle θ (Fig. 7a). The energy per unit volume flowing into unit solid angle in this direction is

$$u(T)/4\pi. \qquad \text{(Energy density per steradian)} \qquad (6.33)$$

The energy carried by a given photon of frequency ν is $E = h\nu$, and its momentum is (from special relativity) $p = E/c = h\nu/c$, where c is the speed of light. Because this is true for photons at all frequencies, the total momentum in unit volume moving toward the wall into unit solid angle is therefore the energy density (per steradian), $u/4\pi$, divided by c, or

$$u(T)/(4\pi c). \qquad \text{(Momentum m}^{-3}\text{ sr}^{-1}) \qquad (6.34)$$

It follows that the momentum directed toward θ, ϕ in solid angle $d\Omega$ is $u\, d\Omega/(4\pi c)$. The photons that will strike the surface A in time Δt are those in the "incident" volume of gas in Fig. 7a. They move toward the surface at the speed of light c, and so the length of the volume is $c\Delta t$. The cross-sectional area of the column is $A\cos\theta$ because the target area is approached at angle θ. The momentum being carried to the surface area A in time Δt by particles in solid angle $d\Omega$ at θ, ϕ is thus

$$\Delta p_A = \frac{u}{4\pi c}\, d\Omega\, c\Delta t\, A\cos\theta. \qquad \begin{array}{l}\text{(Momentum in } d\Omega \text{ at } \theta, \phi \text{ striking} \\ \text{area } A \text{ in time } \Delta t)\end{array} \qquad (6.35)$$

Momentum transfer

The momentum transferred by the photons to the surface depends on whether the photons are absorbed or reflected. The surface is assumed to be in thermal equilibrium with the photon gas. Thus, averaged over time, it emits photons with the same magnitudes of momenta (or energies) as any it might absorb. The net effect is the same as if each incoming photon underwent a specular reflection from the surface (as from a mirror) with the same magnitude of momentum with which it arrived. In this simplification, the change of momentum of our group of photons Δp_{ph} in terms of its initial momentum p and final momentum p' is

$$\Delta p_{ph} \equiv p' - p, \qquad (6.36)$$

where the magnitudes of p and p' are equal.

The component of momentum that is normal to the surface, $\Delta p_A\cos\theta$, contributes to the pressure. There is no momentum change parallel to the surface in a specular reflection or averaged over many reflections from a rough surface. The net momentum change of the photons Δp_{ph} (Fig. 7b) will thus be normal to the surface (i.e., downward). For the group of photons containing momentum Δp_A (35), it is

$$\Delta p_{ph} = -2\Delta p_A \cos\theta\, \hat{n}, \qquad \text{(Momentum transfer to particle; downward)} \qquad (6.37)$$

where $\hat{n}$ is the unit vector normal (outward) to the wall. The magnitude is just twice the normal component of the incident momentum. The momentum transferred to the wall is in the opposite direction to Δp_{ph}:

$$\Delta p_{wall} = +2\Delta p_A\cos\theta\, \hat{n}. \qquad \text{(Momentum transfer to wall by particle)} \qquad (6.38)$$

Substitute Δp_A from (35) into (38) to obtain

$$\Delta p_{ph} = \frac{2u}{4\pi c}d\Omega\, c\Delta t\, A\cos^2\theta\, \hat{n}. \qquad \begin{array}{l}\text{(Photon momentum in } d\Omega \\ \text{transferred to surface } A \text{ in } \Delta t)\end{array} \qquad (6.39)$$

This is the momentum transferred to area A in time Δt by photons in $d\Omega$ at θ, ϕ.

Photon pressure

The pressure on the surface is defined as the momentum transfer *per unit area and per unit time* (momentum $m^{-2}\ s^{-1} = N\ m^{-2}$) due to the reflecting photons. Thus, the magnitude of the differential pressure dP due to our photons is the magnitude of (39) divided by $A\,\Delta t$:

$$dP = \frac{2u}{4\pi c}\frac{d\Omega\ c\Delta t\ A\cos^2\theta}{A\Delta t} = \frac{u}{2\pi}\cos^2\theta\ d\Omega. \tag{6.40}$$

The total pressure is obtained by summing (integrating) over all arrival directions in one hemisphere as follows:

$$P = \iint_{\text{hemisphere}} \frac{u}{2\pi}\cos^2\theta\ d\Omega. \tag{6.41}$$

Substitute into (41) the energy density $u = aT^4$ (25) and the solid angle in terms of the coordinates of Fig. 4, $d\Omega = \sin\theta\,d\theta\,d\phi$, and finally evaluate the integral over the hemisphere to obtain the pressure P:

$$P = \int_{\theta=0}^{\pi/2}\int_{\phi=0}^{2\pi}\frac{aT^4}{2\pi}\cos^2\theta\sin\theta\ d\theta\ d\phi, \tag{6.42}$$

$$\Rightarrow \qquad P = \frac{aT^4}{3}, \qquad \text{(Equation of state for photons; J/m}^3\text{ or N/m}^2\text{)} \tag{6.43}$$

where, from (26), $a = 4\sigma/c = 7.566\times 10^{-16}\ J\ m^{-3}\ K^{-4}$.

The pressure (43) is simply one-third the energy density aT^4 (25). It happens that energy density has the same units as pressure, and so this result is not unreasonable from a dimensional point of view. The dimensional agreement can be seen as follows:

$$\text{Energy density (J/m}^3\text{)} \rightarrow \text{Energy/vol} \rightarrow \text{Work/vol} \rightarrow \tag{6.44}$$

$$\text{(Force}\times\text{distance)/vol} \rightarrow \text{Force/area (N/m}^2\text{)} = \text{Pressure.}$$

Equation (43) is the *equation of state* for a photon gas, as noted previously (3.44). It shows that, for a photon gas with a blackbody spectrum, the pressure depends only on temperature; no other parameter is necessary. The temperature uniquely determines the numbers and momenta of the photons in the gas, and these uniquely determine the magnitude of the pressure applied by the gas to an adjacent surface.

Summary of characteristics

We summarize the characteristics of blackbody radiation we have derived from the specific intensity (6), which in itself followed from the distribution function (5), in the list below.

(i) Rayleigh–Jeans and Wien approximations, which are, respectively, $I \propto \nu^2 T$ for $h\nu \ll kT$ (8) and $I \propto \nu^3 \exp(-h\nu/kT)$ (10) for $h\nu \gg kT$

(ii) Frequency at the peak of the specific intensity function in energy units, $h\nu_{\text{peak}} = 2.82\,kT$ (11)

(iii) Blackbody spectrum in wavelength units, $I_\lambda(\lambda, T)$ (W m^{-3} sr^{-1}) (13)

(iv)   Total flux (all frequencies and all angles) emitted from 1 m^2 of the surface of a "black-body," $\mathscr{F} = \sigma T^4$ (W/m^2) (18), where σ is the Stefan–Boltzmann constant.

(v)   Spectral photon energy density $u_\nu(\nu, T) = 4\pi I/c$ (J m^{-3} Hz^{-1}) (23)

(vi)   Photon energy density (u_ν integrated over all frequencies), $u(T) = aT^4$ (J/m^3), where a is a constant (25)

(vii) Spectral number density, $n_\nu(\nu, T) = u_\nu/h\nu$ (m^{-3} Hz^{-1}) (27)

(viii) Number density of photons (n_ν integrated over all frequencies), $n(T) \propto T^3$ (m^{-3}) (31)

(ix)   Average photon energy $h\nu_{av} = 2.70 kT$ (32)

(x)   Pressure of the radiation, $P = aT^4/3$ (N/m^2) (43)

Limits of intensity

The amplitude of the specific intensity of a blackbody emitter at a given frequency is a strict function of the temperature. The specific intensity $I(\nu, T)$ at a given ν and T can not be changed by adding or removing radiating particles as is the case for optically thin plasmas.

An optically thin plasma of a given temperature will emit less than a blackbody. In fact, the blackbody specific-intensity (or flux σT^4) is the maximum that can be emitted by a thermal source. The blackbody specific-intensity is therefore sometimes called the *blackbody limit* of specific intensity. This is the upper limit to the intensity obtained at a given frequency from a thermal source at a specified temperature.

In this section, we also discuss limits to the temperature of a thermal body.

Particles added

The effect on the overall spectrum (log-log plot) of adding particles to an optically thin gas while holding the temperature and volume constant is illustrated qualitatively in Fig. 8. The approximately exponential expression for the optically thin emission from unit volume is given in (5.2). It has the form $I \propto g(\nu, T)n_e n_i \Lambda T^{-1/2} \exp(-h\nu/kT)$, where Λ is the cloud thickness (m), g is the slowly varying Gaunt factor, and $n_e n_i$ are the electron and ion number densities, respectively.

As the particles are added at the temperature T, only the factor $n_e n_i$ of the thermal bremsstrahlung spectrum increases, and so all points of the spectrum displace upward vertically. As this occurs, the curve first encounters the blackbody limit (upper dark curve) at low frequencies; it becomes optically thick at these frequencies. The break point is called the *low-frequency cutoff*. As more and more numbers are added, the intensity at higher and higher frequencies reaches this limit. Finally, the cloud becomes optically thick at all frequencies of interest, and the spectrum becomes identical to the blackbody function.

Surface of last scatter

One can gain some physical insight into this limit. Consider an observer at the edge of an optically thin plasma cloud detecting photons from it (Fig. 5.1). As the densities of ions and electrons increase, the rate of electron-ion collisions (and hence emitted photons) increases as $n_i n_e$. If the cloud is optically thin, the intensity measured by the observer similarly increases.

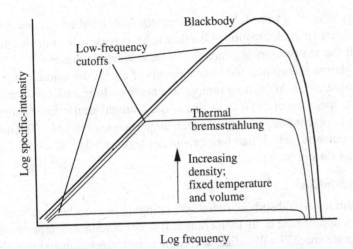

Figure 6.8. Qualitative sketch of optically thin (thermal bremsstrahlung) spectra as ions and electrons are added at fixed temperature and volume. The source increases in intensity except at low frequencies, where it has reached the blackbody intensity. The point of saturation increases in frequency as particles are added until the blackbody shape is attained. This low-frequency cutoff is typical of "optically thin" spectra.

As the particle density continues to increase, eventually photon-electron collisions prevent photons from leaving the plasma unimpeded. They may be scattered one or more times before escaping the plasma. The plasma is becoming optically thick, and our observer "sees" a distance of only one optical depth into the cloud to the *surface of the last scatter* (on average).

The distance to this surface is determined by the column density (particles/square meter) required to give unit optical depth. As the particle density increases, the physical distance (meters) of unit optical depth becomes less and less. Hence, a larger and larger fraction of the ion-electron collisions that give rise to the photons become invisible because they lie more than about one optical depth from the observer. Inasmuch as the column density (electrons/square meter) to the surface of last scatter does not vary with density, the observed intensity holds constant as density is changed.

The flux density $\mathscr{F} = \sigma T^4$ (18) and the luminosity $L = 4\pi R^2 \sigma T_{\mathrm{eff}}^4$ (20) for a spherical blackbody of radius R were both derived from the specific intensity I; see (15). Thus, they too are independent of particle density in the emitting body. If a star must dump energy into deep space with a blackbody spectrum at a faster rate than σT^4 per square meter, the expression (20) tells us that the star must increase either its surface temperature or its surface area (by expanding).

Temperature limit

Temperatures of celestial thermal sources are not expected to exceed $\sim 10^{12}$ K. At this temperature, typical particle energies are $\sim 10^8$ eV, or 100 MeV (4). At such energies, the nuclear particles (protons and neutrons) will have sufficient kinetic energies to create new nuclear-active particles in nuclear collisions. In particular, at these energies the unstable meson, the pion with a rest energy of $mc^2 \approx 140$ MeV, can be created. Additional energy input into a plasma thus goes largely into particle creation rather than to an increase in the kinetic energies of the particles. Recall that temperature is a measure of these kinetic energies.

In principle, if the plasma is confined and huge amounts of energy are injected, the temperature could rise indefinitely, next creating protons and neutrons ($mc^2 \approx 1$ GeV) and then disassociating them into their constituent quarks, and so on. Astrophysical systems, however, are neither expected to be so confined nor to have such large energies available. The notable exception is the early universe, which had temperature as high as $\sim 10^{27}$ K at the time of inflation ($\sim 10^{-34}$ s after the big bang).

Black and gray bodies

A perfect blackbody surface will absorb all radiation impinging on it and will radiate with the blackbody intensity appropriate to its temperature. If it has a nonzero temperature and faces into zero temperature space, it will radiate according to its temperature and will absorb no radiation.

In general, if the body is in an environment of temperature different from its own temperature, the rates of energies being absorbed and emitted will differ. If the body has a finite thermal capacity, it will change temperature until the rates become equal at an equilibrium temperature. A spacecraft with a black surface facing the sun will rapidly heat up owing to the impinging sunlight but will not heat up to solar temperatures because it will also be viewing the cold space surrounding the sun.

A *gray body* has an efficiency of emission and absorption at any given frequency that is less than the 100% of that for a blackbody surface. The temperature of a spacecraft can be regulated by the appropriate choice of materials or coatings for its outer surfaces, Again, the net rate of radiant heat transfer to or from the spacecraft depends on its relative temperatures and the regions it views. The latter is dependent on spacecraft orientation, which may expose it to the sun, the earth, or both as well as to cold space.

6.3   Cosmological expansion

The expanding universe contains a thermal distribution of photons of temperature 2.73 K known as the cosmic microwave background radiation (CMB). The radiation originated $\sim 3 \times 10^5$ yr after the big bang when electrons and protons combined to form hydrogen. This occurred when the expanding and cooling universe reached about 3000 K.

Before the formation of hydrogen, the photons and free electrons were in thermal equilibrium and interacted frequently. Afterward, the neutral hydrogen presented a very small cross section to the photons, and thus the radiation was *decoupled* from the particles. The photons released in this way had a blackbody spectrum of temperature ~ 3000 K and were then able to propagate freely. How does the 3000 K photon blackbody spectrum evolve in the expanding universe? We will find that the expansion is, in effect, adiabatic and that the spectrum remains perfectly blackbody in shape and amplitude as it cools to its current value of 2.73 K.

Adiabatic expansion

Consider the expansion of a photon gas during which no heat is added or subtracted ($\delta Q = 0$); this is known as an adiabatic expansion. We will find the pressure-volume relation for such a gas and compare it with a particulate gas.

Photons

Let the photon gas be confined to a spherical insulated volume of radius R so no heat can enter or leave the sample. If the radius increases, the volume increases as R^3. Hence, the energy density would appear to decrease as R^{-3}, but the gas also does work on the expanding walls and thus loses internal energy. Photons colliding with a receding wall recoil with reduced energy or frequency. These collisions are shown below (59) to cause the energy density of the photon gas to decrease by an additional factor of R^{-1}. Together, the two factors lead to an overall R^{-4} dependence of the energy density on the radius R,

$$u_{\text{rad}} \propto R^{-4}. \tag{6.45}$$

The relation between volume V and pressure P is thus

$$V \propto R^3 \propto \left(u^{-1/4}\right)^3 \propto P^{-3/4}, \tag{6.46}$$

where we invoked (45) and also $P \propto u_{\text{rad}}$ from (43) and (25). This then allows us to write

$$\Rightarrow \qquad P V^{4/3} = \text{constant.} \qquad \text{(Adiabatic expansion; photon gas)} \tag{6.47}$$

Comparison with particles

The expression (47) is highly reminiscent of the pressure-volume relation for an adiabatically expanding particulate gas – namely $PV^\gamma = \text{constant}$ (4.10) – where the constant γ is the ratio of specific heats for constant pressure expansion and constant volume heating, $\gamma \equiv C_P/C_V$.

For a monatomic gas with its three degrees of freedom per particle, the kinetic energy per particle is $3kT/2$. The internal energy per mole is thus $U = 3N_0kT/2 = 3RT/2$, where N_0 is Avogadro's number and $R \equiv N_0k$ the universal gas constant. A change in temperature yields $\Delta U = 3R\Delta T/2$. Specific heat is defined as the heat input δQ per unit temperature change. The first law of thermodynamics (2.84) tells us that, at constant volume ($\delta W = 0$), the heat absorbed equals the increase in internal energy, $\delta Q = dU$; thus, $C_V \equiv (\delta Q/dT)_V = dU/dT = 3R/2$.

In a constant pressure expansion, extra heat is required to make up for the work done by the gas on the expanding walls. It turns out that this requires another factor of R, and so we have $C_P \equiv (\delta Q/dT)_P = 5R/2$. Thus, the ratio of specific heats for a monatomic gas is $\gamma = C_P/C_V = 5/3$. A diatomic gas has five degrees of freedom per particle, three translational and two rotational. The specific heats in this case are $C_V = 5R/2$ and $C_P = 7R/2$; hence, $\gamma = 7/5$.

The specific heats just described are independent of temperature. In contrast, the energy required to raise a photon gas one degree at constant volume is temperature dependent: $dU/dT = d(aT^4)/dT \propto T^3$; see (25). Specific heat is thus not a particularly useful concept for a photon gas.

Nevertheless, we see from (47) that the pressure-volume relation for adiabatic expansion is similar to that of a particulate gas. The constant $4/3$ appears in the role of the ratio of specific heats, where

$$\gamma = 4/3 = 1.33. \qquad \text{(Photon gas)} \tag{6.48}$$

This value is low compared with the values $5/3 = 1.67$ and $7/5 = 1.40$ for monatomic and diatomic particulate gases, respectively. In fact it is clear by extrapolation that $\gamma = 8/6 = 4/3$ is the value for a hypothetical particulate gas with six degrees of freedom.

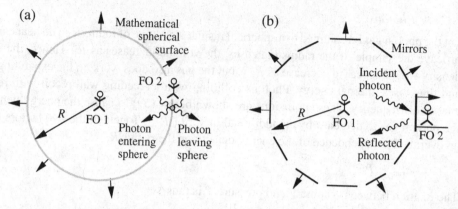

Figure 6.9. (a) Mathematical spherical surface expanding with the universe. The surface is at rest relative to fundamental observers (e.g., FO 2), each of whom experiences an isotropic universe. (b) Photons reflecting off imaginary mirrors that are located at the mathematical sphere. A reflection at the mirror in (b) is equivalent to a photon's leaving the sphere in (a) being matched by an incoming photon of the same frequency. No net energy enters the sphere in either case; the expansion is therefore adiabatic.

Sphere of receding mirrors

The reddening of radiation in an expanding universe may be obtained quantitatively from classical considerations.

Hubble expansion and fundamental observers

An expanding universe with galaxies is similar to a large expanding loaf of raisin bread in an oven. All raisins move away from one another such that an observer on any given galaxy (raisin) sees all other galaxies (raisins) receding with speeds that increase linearly with distance. The view is the same at some "cosmic time" for all observers who are at rest relative to a local raisin. Such observers are called *fundamental observers* (FO). Each FO sees an isotropic expanding universe; it looks on average to be the same in all directions. The expansion is described with the *Hubble constant H_0*, which gives the speed v of recession of a galaxy at distance r, namely $v = H_0 r$ (AM, Chapter 9).

Radiation that permeates such a universe will also be isotropic for each FO. As the universe expands and the FOs move apart from each other, it will continue to be isotropic for them. However, as we shall demonstrate, the frequencies (energies) of the photons decrease. This can be understood as the result of photon reflections from hypothetical receding mirrors.

Reflections from mirrors

Consider at some time t an imaginary sphere centered at some arbitrary point in an expanding universe and locate FO 1 at its center. (Fig. 9a). At an arbitrary radius (*scale factor*) R, the sphere's surface will intersect the positions of many FOs, including FO 2 in the figure. As the expansion continues, the surface of the sphere remains at rest relative to these FOs. The sphere thus expands with the Hubble expansion, and its radius $R(t)$ may be taken as a scale factor that describes the expansion.

Observer FO 2 at the edge of the expanding sphere (Fig. 9a) is, by definition, at rest with respect to that part of the universe. To this observer everything is isotropic, including the radiation. FO 2 thus observes that there are as many photons entering the sphere as there are leaving it. There is no net energy (i.e., heat) flowing through the surface of the mathematical sphere. It is as if the photons within the sphere were completely insulated from its surroundings. The expansion is adiabatic.

The physics would be equivalent if the inside surface of the expanding sphere were to consist of numerous flat mirrors with the reflecting surfaces facing inward (Fig. 9b). Our fundamental observer FO 2 is at rest with respect to the right-hand mirror in the figure. In the frame of reference of the mirror (and of FO 2), a photon arriving at the mirror from the inside will be reflected with the same energy it had before the reflection. The reflected photon takes the place of a photon that would have entered from outside the sphere if the mirrors had not been there. The photon content and energies inside the sphere are thus the same in both scenarios, with and without mirrors.

In the frame of reference of FO 1 at the center of the mirrored room, a photon leaving the center would eventually encounter a receding mirror, be reflected, and arrive back with reduced energy (frequency); think of balls bouncing off a receding wall. In the no-mirror scenario, FO 1 would compare the frequency of an outgoing photon with that of an incoming photon a short time later. The equivalence of the mirror scenario tells us that the latter photon must have a reduced frequency.

After the first reflection, a photon will continue to reflect back and from opposing mirrors, losing energy with each reflection. Thus the photons in the expanding universe become progressively redshifted through many "reflections."

We use the mirror scenario to estimate the magnitude of the reddening in terms of the mirror separation. If the room of mirrors is quite small relative to the size of the universe, space will appear Euclidean and mirror speeds will be low ($v_\mathrm{m} \ll c$). Hence, the reflections off the mirrors may be described with classical Doppler shifts.

Wavelength and scale factor

The calculation of the frequency shift for photons involves two Doppler-shift calculations for each reflection. The first is the redshift of the frequency of the outgoing photon in the frame of FO 2 relative to the frequency in the FO 1 frame. The second is the redshift of the reflected photon in the FO 1 frame relative to frequency in the FO 2 frame. The two redshifts yield the total redshift (frequency shift) for a simple reflection according to FO 1.

Let us do this calculation with reference to Fig. 10 in which the mirrors move outward at speed v_m relative to FO 1, the separation between the mirrors is $L(t) = 2R(t)$ (see Fig. 9), and a photon originating at FO 1 reflects back and forth between the left and right mirrors. The classical Doppler shift may be written as

$$\frac{\Delta\nu}{\nu} \equiv \frac{\nu' - \nu_0}{\nu_0} = -\frac{v_\mathrm{m}}{c} \equiv -\beta_\mathrm{m}, \tag{6.49}$$

where the minus sign appears because, for recession, the velocity is taken to be positive and $\nu' < \nu_0$.

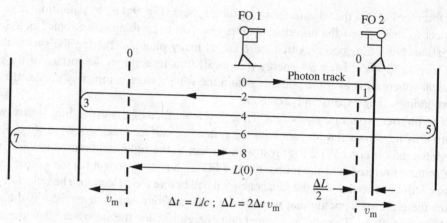

Figure 6.10. Two walls of the mirrored room showing a photon traveling back and forth between them. The mirror locations at the numbered times of initiation (dashed lines) and reflections (solid lines) are shown. The mirror separation $L(t)$ is twice the radius $R(t)$ of the sphere of Fig. 9. The mirror motion is shown greatly exaggerated.

Consider the first reflection ("1") from the right-hand mirror. Observer FO 2 is receding (with the mirror) away from FO 1 at speed v_m and thus observes the incoming (incident) photon to have the reduced frequency v', from (49),

$$v' = v_0(1 - \beta_m), \qquad \text{(Incident photon)} \qquad (6.50)$$

where v_0 is the frequency observed by FO 1 when the photon left the center. According to FO 2, this photon is reflected with the same (lowered) frequency v' it had in his frame before the reflection. Observer FO 1 then sees the reflected photon redshifted to even lower frequency v'' because the v' frequency was measured in the receding frame of the mirror. Thus,

$$v'' = v'(1 - \beta_m) = v_0(1 - \beta_m)^2 \approx v_0(1 - 2\beta_m), \quad \text{(Reflected photon)} \qquad (6.51)$$

where the final approximation is invoked because $\beta_m \ll 1$. Reorganizing the terms again, we obtain

$$\frac{\Delta v}{v} = -2\beta_m, \qquad (\beta_m \ll 1) \qquad (6.52)$$

where $\Delta v \equiv v'' - v_0$ is the total frequency shift according to FO 1 from a single reflection. The fractional frequency shift is twice the mirror recessional speed factor, $\beta_m = v_m/c$.

Now write β_m in terms of the change of mirror separation ΔL between bounces. Consider the fractional changes in separation and frequency to be small (i.e., $\Delta L/L \ll 1$ and $\Delta v/v \ll 1$). Let $\Delta t = L/c$ be the crossing time of a photon from one mirror to the other. In this same time, the right mirror moves outward a distance $v_m \Delta t$, and the mirror separation changes by twice this amount because both mirrors are moving at speed v_m; thus,

$$\Delta L = 2v_m \Delta t = 2v_m \frac{L}{c} = 2\beta_m L \qquad (6.53)$$

and $\beta_m(= v_m/c)$ becomes

$$\beta_m = \frac{\Delta L}{2L}. \tag{6.54}$$

The mirror speed parameter β_m equals half the fractional length change in the photon crossing time $\Delta t = L/c$.

Finally, eliminate β_m from (52) and (54) to find the desired relation between length and frequency:

$$\Rightarrow \quad \frac{\Delta \nu}{\nu} = -\frac{\Delta L}{L}. \qquad \text{(One bounce)} \tag{6.55}$$

This shows that the fractional frequency shift from one reflection is determined solely by the fractional change of length; it is independent of the speed of recession and the absolute value of L.

Many bounces across a small room or a few bounces across a large room lead to the same result: the fractional frequency shift is a direct measure of the change of the fractional length L if the fractions are much less than unity.

Integration of (55) provides the desired relation between L and ν for large changes in L as follows:

$$\int_{\nu_0}^{\nu(t)} \frac{d\nu}{\nu} = -\int_{L_0}^{L(t)} \frac{dL}{L} \tag{6.56}$$

$$\ln \frac{\nu(t)}{\nu_0} = -\ln \frac{L(t)}{L_0} \tag{6.57}$$

$$\frac{\nu(t)}{\nu_0} = \left(\frac{L(t)}{L_0}\right)^{-1} \tag{6.58}$$

$$\Rightarrow \quad \nu(t) \propto L(t)^{-1} \propto R(t)^{-1}. \qquad \text{(Frequency decrease)} \tag{6.59}$$

The frequency of the radiation in the cavity decreases inversely with the size $L(t)$ and also with the scale factor $R(t)$ because $L(t) = 2R(t)$ in our example (Figs. 9 and 10). Because $\nu = c/\lambda$ in a vacuum,

$$\Rightarrow \quad \lambda(t) \propto L(t) \propto R(t). \qquad \text{(Wavelength expansion)} \tag{6.60}$$

The wavelength scales as the size of the cavity or as the scale factor $R(t)$.

Thus, as the spherical volume of Fig. 9a expands, the energy $E = h\nu$ of each photon therein is gradually reduced as R^{-1}. This is the extra factor of R that leads to the energy density's decreasing as $u \propto R^{-4}$, as stated in (45).

In general relativity, the proper way to view the continuing redshift is that the waves are stretched to longer wavelengths simply because they are propagating in an expanding universe. The reddening follows directly from the metric that describes the intervals between space-time events. Our expanding-sphere-of-mirrors argument shows that this is what one would expect classically given the expansion of the universe.

Spectral evolution

In an expanding universe, the photon frequencies decrease. If, furthermore, the photon number is fixed, the photon number density will decrease. This suggests that an observer at a later time would detect "cooler" radiation. It is not at all evident, though, that the spectral shape would remain blackbody with precisely the correct spectral shape and corresponding amplitude. Here we demonstrate that a photon gas (e.g., the cosmic background radiation CMB) indeed remains thermal as it cools in an expanding universe.

Equal photon numbers

The specific intensity I (W m^{-2} Hz^{-1} sr^{-1}) of blackbody radiation at the time of decoupling (i.e., at the time of its emission) may be written from (6) as

$$I(\nu_{em}, T_{em}) = \frac{2h\nu_{em}^3}{c^2} \frac{1}{\exp(h\nu_{em}/kT_{em}) - 1}. \qquad \text{(W m}^{-2}\text{ Hz}^{-1}\text{ sr}^{-1}) \qquad (6.61)$$

The energy density per unit frequency interval u_ν (J m^{-3} Hz^{-1}) is proportional to the specific intensity according to (22):

$$u_\nu(\nu_{em}, T_{em}) = \frac{4\pi}{c} I(\nu_{em}, T_{em}). \qquad \text{(J m}^{-3}\text{ Hz}^{-1}) \qquad (6.62)$$

Finally, as we saw in (27), the photon number density per unit frequency interval n_ν is simply the energy density divided by the energy of a single photon,

$$n_\nu(\nu_{em}, T_{em}) = \frac{u_\nu(\nu_{em}, T_{em})}{h\nu_{em}}. \qquad \text{(m}^{-3}\text{ Hz}^{-1}) \qquad (6.63)$$

Substitute (62) into (63) and multiply by a frequency interval $d\nu_{em}$ to obtain

$$n_\nu(\nu_{em}, T_{em}) \, d\nu_{em} = \frac{4\pi}{c} I(\nu_{em}, T_{em}) \frac{d\nu_{em}}{h\nu_{em}}. \qquad \text{(m}^{-3}\text{ in } d\nu \text{ at } \nu) \qquad (6.64)$$

This is the number of photons per unit volume in frequency interval $d\nu_{em}$ at frequency ν_{em}. A similar expression could be written for the photon number density $n_\nu(\nu_{ob}, T_{ob}) \, d\nu_{ob}$ observed at a later time such as the present epoch.

Now let the universe expand with scale factor $R(t)$ and ask what happens to the radiation. Recall that, in the expanding sphere of Fig. 9a (a *comoving volume*), the overall number of photons remains constant. If one considers only photons at frequency ν_{em} in a band $d\nu_{em}$, the number is still conserved as the volume expands *if* the observed frequency ν_{ob} and band $d\nu_{ob}$ are allowed to vary so that they follow and encompass the photons in the sphere as they decrease in frequency. The number in the band is the product of the number density (at ν in $d\nu$) and the volume, $\propto R^3$. The constancy of the number is given by

$$\rightarrow \qquad n_\nu(\nu_{ob}, T_{ob}) \, d\nu_{ob} \, R(t_{ob})^3 = n_\nu(\nu_{em}, T_{em}) \, d\nu_{em} R(t_{em})^3.$$

$$\text{(Equal photon numbers)} \qquad (6.65)$$

This conservation of numbers is the essence of this development.

Temperature and intensity

Write both sides of (65) in terms of I. Substitute from (64) into the right side of (65) and the counterpart of (64) for $n_\nu(\nu_{ob}, T_{ob})$ into the left side:

$$\frac{4\pi}{c} I(\nu_{ob}, T_{ob}) \frac{d\nu_{ob}}{h\nu_{ob}} R_{ob}^3 = \frac{4\pi}{c} I(\nu_{em}, T_{em}) \frac{d\nu_{em}}{h\nu_{em}} R_{em}^3. \tag{6.66}$$

Our goal is to determine the observed spectrum $I(\nu_{ob}, T_{ob})$ in terms of the observed variables ν_{ob} and T_{ob}.

The fractions $d\nu_{ob}/h\nu_{ob}$ and $d\nu_{em}/h\nu_{em}$ are equal as we now demonstrate. The photon frequency ν undergoes the cosmological redshift (59) – that is, it varies inversely as R^{-1}:

$$h\nu_{ob} = \frac{R_{em}}{R_{ob}} h\nu_{em}. \qquad\qquad \text{(Photon energy decrease)} \tag{6.67}$$

Take the differential to find that, at fixed R_{em}/R_{ob}, the element $d\nu_{ob}$ also varies as R^{-1}:

$$d\nu_{ob} = \frac{R_{em}}{R_{ob}} d\nu_{em}. \qquad\qquad \text{(Bandwidth decrease)} \tag{6.68}$$

As R_{ob} increases with the cosmological expansion, the bandwidth, $d\nu_{ob}$, decreases. This decrease is due to the compression of the frequency band as the photons are redshifted. Consider photons emitted over a band from 160 to 200 GHz for a bandwidth $\Delta\nu_{em} = 40$ GHz. If the universe expands a factor of two, the 160-GHz photons are redshifted down to 80 GHz and the 200-GHz photons to 100 GHz. The bandwidth at observation is thus $\Delta\nu_{ob} = 20$ GHz, which represents a factor of two reduction, as expected from (68).

The ratio of (68) over (67) demonstrates the equality of the fractions in (66), which therefore cancel one another, leaving us with

$$I(\nu_{ob}, T_{ob}) = \frac{R_{em}^3}{R_{ob}^3} I(\nu_{em}, T_{em}). \tag{6.69}$$

Thus, the specific intensity scales as R_{ob}^{-3}.

The manipulations leading to (69) took into account (*i*) that the reduction of photon energy $h\nu$ during expansion would lower the observed specific intensity, and (*ii*) that, at the same time, the reduced bandwidth $d\nu$ would increase the specific intensity, which is a measure of flux *per unit bandwidth*. These two effects canceled one another. What remains, though, is the increase in volume occupied by the fixed number of photons. This leads to a net reduction of specific intensity (69).

The specific intensity function (6) may now be introduced into the right side of (69) while eliminating the variable ν_{em} with (67):

$$I(\nu_{ob}, T_{ob}) = \frac{R_{em}^3}{R_{ob}^3} \frac{2h}{c^2} \left(\nu_{ob}\frac{R_{ob}}{R_{em}}\right)^3 \left[\exp\left(\frac{h\nu_{ob}\frac{R_{ob}}{R_{em}}}{kT_{em}}\right) - 1\right]^{-1}. \tag{6.70}$$

The cubed ratios of radii cancel, and the observed frequency ν_{ob} appears as expected in the Planck function (6); that is, it is cubed and is also an argument of the exponential. In fact, the distribution (70) is the Planck function for the temperature

$$\Rightarrow \quad T_{ob} = T_{em}\frac{R_{em}}{R_{ob}}. \qquad \text{(Observed temperature)} \qquad (6.71)$$

The correspondence to the Planck function thus becomes exact:

$$\Rightarrow \quad I(\nu_{ob}, T_{ob}) = \frac{2h\nu_{ob}^3}{c^2}\frac{1}{\exp(h\nu_{ob}/kT_{ob}) - 1}. \quad \text{(Observed specific intensity)} \qquad (6.72)$$

We find in (72) that the observed specific intensity obeys the Planck blackbody function for the new lower temperature T_{ob} both in spectral shape and in magnitude, which is not intuitively apparent. Also, the temperature T is found in (71) to decrease inversely with R.

Indeed the measured CMB spectrum closely adheres with great precision to the Planck function as measured by the COBE satellite (Fig. 1), This is a compelling argument that the radiation did originate 10^5 years ago in the decoupling of radiation and matter when they were last in thermal equilibrium. Since then, the radiation spectrum has evolved greatly. The peak frequency and the amplitude at low frequencies have both decreased by a factor of ~ 1000, but the spectral shape remains precisely in agreement with that proposed by Professor Planck.

6.4   Mathematical notes

Riemann zeta function

The Riemann zeta function $\zeta(z)$ is defined as

$$\zeta(z) \equiv \sum_{n=1}^{\infty}\frac{1}{n^z}. \qquad \text{(Zeta function)} \qquad (6.73)$$

One can find $\zeta(z)$ tabulated in various references. Several values are given in Table 1.

It can be shown that, for $\mathrm{Re}(z) > 1$,

$$\zeta(z) = \frac{1}{\Gamma(z)}\int_0^{\infty}\frac{x^{z-1}}{e^x - 1}\,dx, \qquad (6.74)$$

where the gamma function is related to the factorial function as $\Gamma(z) \equiv (z-1)!$ This enables one to evaluate integrals of the type (74).

Roots of a transcendental equation

The following transcendental equation is useful for determining the maximum of the blackbody distribution function:

$$\frac{x_n}{1 - e^{-x_n}} = n. \qquad (n = 2, 3, 4, \ldots) \qquad (6.75)$$

It can be shown to have roots x_n, as given in Table 2.

Table 6.1. *Riemann zeta function*

z	$\zeta(z)$
1	diverges
2	$1.644\,934\ldots = \pi^2/6$
3	$1.202\,057\ldots$
4	$1.082\,323\ldots = \pi^4/90$

Table 6.2. *Roots of transcendental equation (75)*

n	x_n
1	no solution
2	1.593 624
3	2.821 439
4	3.920 690
5	4.965 114

Problems

6.2 Characteristics of the radiation

Problem 6.21. Why does a heated body never become bluer than "white-hot" as it rises in temperature (i.e., why is there no "green-hot" or "blue-hot")? Assume blackbody radiation.

Problem 6.22. (a) What is the effective temperature of the sun given its photospheric radius of 6.955×10^8 m and luminosity of 3.845×10^{26} W? (b) What is the energy density of blackbody radiation of temperature $T = 3$ K in J/m³ and eV/cm³? Repeat for the photospheric temperature of the sun, 6500 K. Note the quantities that are near unity to help visualize the energy content. (c) Use (31) to find an expression for the number of photons that reside in a cubic volume with sides equal to the peak wavelength given in (14). Is it a function of temperature? Comment. [Ans. $\sim$5800 K; $\sim$1/2 eV/cm³, $\sim$1 J/m³; $\sim$1/2]

Problem 6.23. (a) Demonstrate that $h\nu_{\text{peak}} = 2.82\,kT$ (11) for the frequency blackbody spectrum (6). (b) Convert the blackbody distribution in frequency space (6) to the distribution in wavelength space (13). Use the latter to find $\lambda_{\text{peak}} = 2.9 \times 10^{-3}\,T^{-1}$ (14). (c) Carry out the integrations (16) and (17) to obtain the flux in the form $\mathcal{F} = \sigma T^4$ (18). (d) Carry out the integration (30) to obtain the photon number density given in (31). You will need the Riemann zeta function (Table 1). In each case follow the suggestions in the text.

Problem 6.24. Find the equilibrium temperature(s) of a spherical planet of radius r at distance $D = 1$ AU from the sun for the cases listed below. Take into account the solar energy absorbed by the planet and its reradiation into cold (assume 0 K) space. Ignore internal energy being released by the planet. Assume that the surfaces of both objects are perfect absorbers and emitters and that the atmosphere is completely transparent. Hint: equate the absorbed and reradiated fluxes. Useful variables and their numerical values are $T_{\text{eff},\odot} = 5800$ K, $R_\odot = 7 \times 10^8$ m, $R_E = 6.4 \times 10^6$ m, and $D = 1.5 \times 10^{11}$ m. In each case, find the numerical value or range of values of temperature. The several conditions of spin and conductivity follow:

(a) The surface thermal conductivity of the planet is infinite. The incoming energy is rapidly distributed over the entire surface of the planet, and so the entire planetary surface has a uniform temperature.

(b) The thermal conductivity is zero, and the planet is not spinning. Your answer will be a function of location on the planet defined as the polar angle ϕ measured from the subsolar point. Make a sketch of an isothermal region on the planet's surface.

(c) The thermal conductivity is again zero, but the planet is spinning rapidly with spin axis normal to the planet-sun line. Your answer will again be a function of ϕ. Again sketch an isothermal temperature region.

(d) Discuss the applicability of these cases to the earth and moon.

[Ans. ∼300 K; ∼400 cos$^{1/4}$ ϕ; ∼300 cos$^{1/4}$ ϕ; −]

Problem 6.25. Consider the distribution of photons shown in Fig. 6 for temperature $T_1 = T$ and one polarization. (a) Check the values illustrated in Fig. 6 with the parenthetical expressions in (28). Specifically, make a table that gives, for the three energy states shown and for three additional states, $h\nu_4 = 4kT_1$, $h\nu_5 = 8kT_1$, $h\nu_6 = 16kT_1$, the following quantities: (i) the number of cells Z_c, given a fixed bandwidth and two cells in frequency state 1, (ii) the average number $<n_c>$ of photons per cell, (iii) the total number of photons $N = <n_c> Z_c$, and (iv) the total energy $E = Nh\nu$ in terms of kT_1. Is Fig. 6 consistent with your results? (b) Make a new table for the situation in which the temperature is doubled, $T_2 = 2T_1$. Consider the same six states (in terms of T_1), $h\nu_1 = 0.5kT_1$, $h\nu_2 = 1.0kT_1$, and so on. Again calculate and tabulate Z_c, $<n_c>$, N, and E, the latter again in terms of kT_1. Before tabulating, consider first which of the tabulated quantities should change values when the temperature increases. Comment on the differences in the two tables, including the energy $h\nu$ (in terms of kT_1) at which the total energy peaks. [Ans. $E(\nu_4, T_1) = 9.6 kT_1$; $E(\nu_6, T_2) = 11.0 kT_1$]

Problem 6.26. A small celestial radio source projects onto the sky as a uniformly filled circle of angular diameter $\theta_d = 5 \times 10^{-4}{}''$. What is the maximum spectral flux density $S(\nu)$ (W m^{-2} Hz^{-1}) that one could plausibly expect from it at the radio frequency 10 GHz under the assumption that it is a thermal source and that the maximum possible temperature is 10^{12} K? [Ans. ∼10^{-25} W m^{-2} Hz^{-1}]

6.3   Cosmological expansion

Problem 6.31. The cosmic microwave background (CMB) is observed in the current epoch at a temperature of $T = 2.73$ K. (a) What is its energy density in J/m^3 and in MeV/m^3? Compare your answer with the energy densities of cosmic rays and magnetic fields in the Galaxy, ∼1.0 and 0.2 MeV/m^3, respectively (Table 10.3). (b) Find, from the relation between scale factor and frequency (67) and from the definition of redshift z,

$$z \equiv \frac{\lambda_{ob} - \lambda_{em}}{\lambda_{em}}, \quad \text{or} \quad \frac{\lambda_{ob}}{\lambda_{em}} = 1 + z,$$

the often quoted relation between the scale factors R_{ob} and R_{em} and redshift z. (c) What was the energy density of the CMB when light now arriving at the earth left a distant quasar with redshift $z = 4.0$? (d) What was the CMB energy density at the time of decoupling at $z = 1088$ (according to a WMAP result)? [Ans. comparable; $1 + z = R_{ob}/R_{em}$; ∼200 MeV/m^3; ∼10^{12} MeV/m^3.]

7

Special theory of relativity in astronomy

What we learn in this chapter

Albert Einstein postulated that the speed of light has the same value in any inertial frame of reference or, equivalently, that there is no preferred frame of reference. The consequence of this postulate is the **special theory of relativity**, which yields nonintuitive relations between measurements in different inertial frames of reference. We demonstrate the **Lorentz transformations** for space and time (x, t) and the compact and invariant **four-vector** formulation. From this, the four-vectors for **momentum-energy** (p, U) and **wave propagation-frequency** (k, ω) are formed, and these in turn yield the associated Lorentz transformations. The transformations for **electric and magnetic field** vectors are also presented. Examples of each type of transformation are given. The **relativistic Doppler shift** of wavelength or frequency is derived from **time dilation** and also directly from the k, ω transformations. The latter yield the transformation of radiation direction (**aberration**) from one inertial frame to another. Stellar aberration explains the displaced celestial positions of stars due to the earth's motion about the sun.

Astrophysical jets often emerge from objects that are accreting matter such as **protostars**, **stellar black holes** in binary systems, and **active galactic nuclei** (**AGN**) of galaxies. With our special-relativity tools, we study three aspects of the **jet phenomenon**: the **beaming of radiation** from objects traveling near the speed of light, the associated **Doppler boosting** of intensity, and **superluminal motion**. The latter refers to measured transverse velocities of **radio-emitting ejections** that seem to exceed the speed of light. An example is the galactic **microquasar** GRS 1915+105.

With **relativistic dynamics** we calculate the **relativistic cyclotron frequency**, which enters the derivation of **synchrotron radiation** by high-energy electrons in Chapter 8. **Particle collisions**, relativistically calculated, yield the threshold energy, ~ 1 PeV (10^{15} eV), for $e^{+}e^{-}$ production in interactions of gamma rays with photons of the **cosmic background radiation** (CMB). This limits astronomy above these energies to galactic distances. At lower energies, **TeV astronomy** (10^{12} eV) probes extragalactic distances with numerous detections of **supernovae** and **AGN**. Collisions of **cosmic-ray protons** with interstellar protons in gaseous clouds of the galactic plane yield **diffuse emission** of ~ 100 MeV gamma rays through π^{0} production. Extremely high-energy protons, $\gtrsim 10^{20}$ eV, collide with CMB photons to create pions (the **GZK effect**), degrading the proton energy and limiting the energies to which **cosmic-ray astronomy** can reach.

7.1   Introduction

Einstein's special theory of relativity profoundly affects the way we think about speeds, momenta, and energies for particles that have speeds approaching that of light, c. According to the theory, only massless particles like photons can travel at the speed of light. Particles with a finite mass are always destined to travel at slower speeds, though it could be at $v = 0.99997\,c$. The theory yields nonintuitive predictions. For example, observers in two different inertial frames moving with respect to each other can not agree on the time interval between two events or on the length of a rod.

Here we briefly present the essentials of the theory as a needed tool for topics in this and other chapters. Gaining familiarity with the concepts of special relativity requires a fair amount of exposure. Readers may find reference to less compact developments helpful. Those familiar with the theory may choose to skip Section 3. Special relativity does not take into account the effects of gravity and its relation to inertial forces in accelerated frames. That is the province of general relativity (GR), some aspects of which appear in this text; see Sections 4.4 and 12.3.

7.2   Postulates of special relativity

The theory is based on a single postulate, the *principle of relativity*: all physics is the same in any inertial frame of reference. In other words, Maxwell's equations are valid in any inertial frame of reference. From this follows a corollary: the speed of light measured in any inertial frame of reference is the same as in any other such frame. This corollary is the basis of the formal development of special relativity; it is often called a second postulate of the theory. In addition, the theory assumes that all points in space and time are equivalent with respect to transformations between frames; that is, space is homogenous and isotropic.

The statement that light travels at the same speed in any frame of reference is indeed strange relative to our day-by-day experience. Consider the following thought experiment. A student climbs into a rocket ship and travels at two-thirds the speed of light away from the professor. The professor points a laser in the direction of the departing spaceship and triggers a single flash. As a check, a lab assistant standing with the professor measures the speed of the light flash as it leaves and finds $v = 3.0 \times 10^8$ m/s (i.e., $v = c$) as expected.

The student on the spaceship, knowing that the professor would do this, sets up measuring equipment and measures the speed of the wave front as it overtakes and passes the spaceship. All parties expect the student to measure $v = c/3$ in the frame of the rocket ship because it is moving away from the professor at $v = 2c/3$. Their experience with sound waves seems to demand it. The student, however, measures $v = c = 3.0 \times 10^8$ m/s! They are astonished and puzzled. Einstein simply accepts this as fact and proceeds from there.

Speed measurements require rulers and clocks, and so it is no wonder that the consequences of his theory are that observers in different frames of reference can not agree on such things as the length of a stick or an interval of time. Experiment has verified these strange effects, and special relativity is now accepted as a fundamental part of physics. When bodies are moving at high speed, one must modify Newtonian physics with special relativity, or with GR if gravity is important.

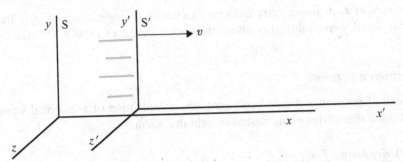

Figure 7.1. Inertial frames of reference used for Lorentz transformations in this chapter. Frame S′ moves down the positive axis of frame S at speed v with x- and $x′$-axes superimposed but shown here slightly separated. The origins coincide when the times are zero (i.e., $x = x′ = 0$ at $t = t′ = 0$).

7.3 Lorentz transformations

Here we present the transformation equations that allow conversion of the position x and time t of an event from the coordinates of one inertial frame to the coordinates of another inertial frame. These are called Lorentz transformations. The *four-vector* nature of position and time coordinates is demonstrated. From this four-vector, other four-vectors can be constructed with different sets of physical variables that may also be transformed with Lorentz equations. The sets we present here are momentum–energy and propagation-vector–frequency. We also present without proof the Lorentz transformations for electric and magnetic fields.

Two inertial frames of reference

Consider two frames of reference, S and S′, moving relative to each other with a constant velocity v (Fig. 1). The S′ frame moves down the positive x-axis with the x- and $x′$-axes superimposed and times set equal and to zero, $t = t′ = 0$, when the origins coincide, $x = x′ = 0$. With constant velocities, S and S′ are *inertial frames of reference*. It is only their relative motion that is pertinent, but it may be helpful to think of one of them, S, as being at rest, the so-called *laboratory frame of reference*, and the other S′ as the *moving frame*.

Let the S frame of reference have a system of observers spread throughout its space. They set up a Cartesian coordinate (x, y, z) system in S to permit position measurements with a unit based on an atomic wavelength they can measure. They then set up a time system with units based on a measurable atomic frequency and synchronize their watches or clocks by sending signals back and forth to each other.

An event that takes place at some place and time in S is known as a *space-time event*. Its occurrence can be noted and recorded with its coordinates x and t by the observer who happens to be at that position. (We usually assume $y = z = 0$.) This result can be communicated to other S observers, who will find no reason to contest it.

Another set of observers in S′ will similarly set up a Cartesian $(x′, y′, z′)$ system with the same length units as in S and establish a time standard of synchronized clocks with the same time units. The S′ observers can record the coordinates $x′$, $t′$ of space-time events in S′. The coordinates of a space-time event are generally not equal in the two frames.

The observers in each frame carry out experiments on a single beam of light traveling along the x-axis and, remarkably, they obtain the same value, $c = 2.998 \times 10^8$ m/s.

Position and time

The discussion of position and time begins with the examination of a spherical wave front emanating from a source that emits flashes of light in a vacuum.

Spherical wave front

An isotropic flash of light at the origin of S ($x = y = z = 0$) at $t = 0$ will propagate outward as a spherical wave at speed c in all directions. At a given time t, the coordinates x, y, z of any point on the sphere must satisfy

$$x^2 + y^2 + z^2 = c^2 t^2, \qquad \text{(Spherical wave in S)} \qquad (7.1)$$

where ct is the radius of the sphere.

The origin ($x' = y' = z' = 0$) of S' (Fig. 1) is coincident with the S origin at $t = t' = 0$. Because the speed of light in S' has the same value c in any frame and space is isotropic, the wave front in S' at some time t' must again be spherical; hence, we also have

$$x'^2 + y'^2 + z'^2 = c^2 t'^2. \qquad \text{(Spherical wave in S')} \qquad (7.2)$$

We use the same speed c in both (1) and (2) in accord with the postulate of special relativity presented above in Section 2.

Transformations

To obtain the transformations, specify that the frames S and S' are inertial frames of reference (neither is accelerating) and that S' moves down the positive x-axis of S with speed v such that the origins coincide at $t = t' = 0$. Also require that neither frame be preferred over the other in any way, the principle of relativity, and that space be homogenous. The latter dictates that the transformations be linear.

Under these conditions, one can find linear relations $x'(x, t)$ and $t'(x, t)$ that will satisfy both (1) and (2). Try $x' = a_1(x - vt)$, $y' = y$, $z' = z$, and $t' = a_2 t + a_3 x$, where a_1, a_2, and a_3 are unknown coefficients. Substitute into (2) and organize the terms in the form of (1). Require that the coefficients of each term of the result be equal to those of (1). The result will be three simultaneous equations that can be solved for the constants a_1, a_2, and a_3 (Prob. 33).

The transformations so obtained are presented as (x1.1) and (x1.4) in Table 1. They may be solved for the inverse expressions (x1.5) and (x1.8). The transverse directions y and z are not affected by the motion in x.

For compactness, the transformations make use of two parameters, γ and β, which are functions of the speed v and are defined as

$$\gamma \equiv \left(1 - \frac{v^2}{c^2}\right)^{-1/2} = (1 - \beta^2)^{-1/2}; \quad \beta \equiv v/c, \quad \begin{array}{l} \text{(Lorentz factor; } 1 \leq \gamma < \infty; \\ 0 \leq \beta < 1) \end{array} \qquad (7.3)$$

Table 7.1. *Lorentz transformations*[a,b]: x, t

Frame S to frame S′	Frame S′ to frame S
(x1.1) $x' = \gamma(x - \beta ct)$	(x1.5) $x = \gamma(x' + \beta ct')$
(x1.2) $y' = y$	(x1.6) $y = y'$
(x1.3) $z' = z$	(x1.7) $z = z'$
(x1.4) $t' = \gamma(t - \beta x/c)$	(x1.8) $t = \gamma(t' + \beta x'/c)$

[a] $\beta \equiv v/c;\ \gamma \equiv (1 - \beta^2)^{-1/2}$.
[b] The x'-axis of S′ is aligned with the x-axis of S, and S′ moves in the $+x$ direction at speed v in the frame of S. The origins coincide at $t = t' = 0$.

where the limits quoted imply that v ranges from 0 to c. The factor γ is called the *Lorentz factor*; it is unity at $v = 0$ and increases without limit for greater velocities, becoming infinite at $v = c$.

The transformations were constructed to ensure that the speed of light is c in both S and S′. However, their utility is more general – namely, the transformation of any single *space-time event* from one frame to another. Thus, with the expressions (x1.1)–(x1.4), an event in the S frame at x, y, z, t or (x, t) can be expressed in terms of the coordinates x', t' that an S′ observer would measure. At low velocities ($v \ll c$, $\beta \ll 1$, and $\gamma \approx 1$), second-order terms in v/c may be dropped. The transformations then become Galilean. For (x1.1)–(1.4), they become $x' = x - vt$, $y' = y$, $z' = z$, and $t' = t$.

The similarity of the two sets of equations in Table 1 illustrates the fundamental similarity of the two frames. The situation for the transformations from S′ to S differs from the inverse only in that, relative to S′, the S frame is moving in the *negative x'* direction. Hence, the transformations (x1.5)–(x1.8) are identical to (x1.1)–(x1.4) except that $\beta \to -\beta$. The speed factor β itself is a positive value in all the transformations.

The transformations of Table 1 are valid for our basic arrangement that S′ moves down the positive x-axis of S and that the origins coincide at $t = t' = 0$. We adhere to this arrangement in this chapter.

Time dilation

Consider the two frames of Fig. 1 with a strobe lamp fixed at position x'_1 in S′ (Fig. 2). Let the light flash twice at times t'_1 and t'_2 so that the time interval between them in S′ is $\Delta t' \equiv t'_2 - t'_1$. Let us now use the Lorentz transformations (Table 1) to find the time interval between the events in S, $\Delta t \equiv t_2 - t_1$.

The flashes, F1 and F2, are two space-time events that are recorded in both S′ and S. Each event must be transformed separately. Both events are at the same position $x'_1 = x'_2$ in S′. To transform t'_1 and t'_2, apply (x1.8) and set $x'_2 = x'_1$:

$$t_1 = \gamma \left(t'_1 + \frac{\beta x'_1}{c} \right)$$

$$t_2 = \gamma \left(t'_2 + \frac{\beta x'_1}{c} \right). \tag{7.4}$$

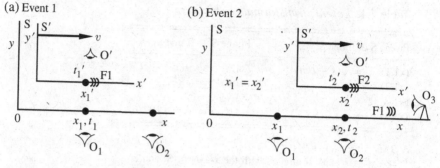

Figure 7.2. Time dilation and Doppler observers in frames shown in Fig. 1 with x'-axis shown displaced upward for clarity. A stationary source at position $x_1' = x_2'$ in moving frame S' emits two flashes, F1 and F2, at times t_1' and t_2'. The flashes are recorded in S by observers O_1 and O_2, whose clocks are synchronized and who are located, respectively, at the positions of the two space-time events. The time interval between the flashes, according to these two observers in S, is found to be greater than that measured by observer O' in S'. If a single observer (O_3) in frame S detects both flashes, the measured interval is further modified by Doppler compression.

Subtract the relation for t_1 from that for t_2 to find that the x' terms cancel. The result is the *time dilation* relation

$$t_2 - t_1 = \gamma(t_2' - t_1')$$
$$\Delta t = \gamma \Delta t'. \qquad \text{(Relativistic time dilation)} \qquad (7.5)$$

Why did we choose (x1.8) rather than x1.4? Both contain the two quantities needed, t and t'; however, x1.8 also contains x', the quantity common to both space-time events, which could thereby be removed by subtraction.

This result shows that the interval between the pulses in the frame S is greater than the interval between pulses measured by an observer in the S' frame; recall that $1 \le \gamma < \infty$ where γ is a measure of speed of the frame (3). Hence, the greater the speed of the S' frame, the greater the dilation of time in S. The frame S' is a very special frame for these two events because they occur at the same position; in this frame, the time interval is less than in all other possible frames.

Time dilation is detected in the increased half-lives of unstable particles moving rapidly in an accelerator or in the cosmic-ray flux. (For astronaut aging, see Prob. 32.)

Length contraction

Another nonintuitive result follows from the measurement of length in two frames of reference. Consider a stick that lies along the x'-axis of S', where again Fig. 1 defines the relative motion of S and S'. The ends of the stick are at positions x_1' and x_2'; hence, its length is $\Delta x' = x_2' - x_1'$ according to observers in S'.

How is the stick measured in S? A host of S observers at many different positions along the x axis have synchronized watches. At an agreed upon time t, the S observers who happen to be at the positions of the ends of the fast-moving stick mark the end positions on the x-axis of their frame (S). The markings at the two ends at time t are the two space-time events; event 1 has coordinates x_1, t_1, and event 2 has coordinates x_2, t_2, where $t_1 = t_2 = t$. (S' observers would agree that these two events properly marked the two ends of the rod but would claim they had occurred at two different times.)

The Lorentz transformation (x1.1) gives a relation between the two coordinates x and x' as a function of the common time t. We choose this transformation for both events rather than (x1.5) because, as before, it allows us to eliminate the common parameter – in this case t. Write (x1.1) for both events, subtract, and rearrange to yield

➡️ $$x_2 - x_1 = \frac{x'_2 - x'_1}{\gamma}. \qquad \text{(Relativistic length contraction)} \qquad (7.6)$$

Thus, we find that the length $x_2 - x_1$ measured in S is reduced by the factor γ relative to the length $x'_2 - x'_1$ measured in S'.

This effect is known as *relativistic length contraction*. The rod is found to be shortened in all inertial frames moving relative to S' at some speed $|v| > 0$. Conversely, the length in the rest frame of the stick S' is greater than in all other inertial frames.

Space-time invariant

The arrival at some x, y, z, t of a pulse of light that left the origin at $t = 0$ is a space-time event that obeys both (1) and (2). The former may be written as $x^2 + y^2 + z^2 - c^2 t^2 = 0$. Similarly, $x'^2 + y'^2 + z'^2 - c^2 t'^2 = 0$. Apply transformations x1.1–x1.4 to $x^2 + y^2 + z^2 - c^2 t^2$ to find that, even for non-zero values of the summations,

$$x^2 + y^2 + z^2 - c^2 t^2 = x'^2 + y'^2 + z'^2 - c^2 t'^2 \ [= -s^2]. \qquad (x, t \text{ invariant}) \qquad (7.7)$$

The left equality demonstrates that the quantity $x^2 + y^2 + z^2 - c^2 t^2$ is an invariant of the transformation. This means that it has the same value for a given space-time event in the coordinates of S and S' or of any other such frame. We assign the parameter $-s^2$ as the value of the invariant; the minus sign is conventional. The parameter is equal to zero for the wavefronts of (1) and (2).

An event at arbitrary position and time x, t is likely to have a nonzero value of $-s^2$. For example, an event in S on the x-axis at $x = +1$, $y = z = 0$ at time $t = 0$ gives $-s^2 = +1$. The S' coordinates of this event are obtained from the transformations (x1.1)–(1.4), namely, $x' = \gamma$, $y' = z' = 0$, and $t' = -\gamma\beta/c$. Substitute these into $x'^2 + y'^2 + z'^2 - c^2 t'^2$ and recall the definition $\gamma = (1 - \beta^2)^{-1/2}$ to demonstrate that this summation yields the same value of the invariant, namely, $-s^2 = +1$.

Are the transformations of Table 1 that were used to find (7) valid for space-time events that do not yield $-s^2 = 0$? Postulate that the quantity $x^2 + y^2 + z^2 - c^2 t^2$ for any such event is invariant from frame to frame; that is, the equality (7) is valid with a nonzero value of $-s^2$. If one were to search again for linear transformations from S to S' but for arbitrary $-s^2$, the result would be just the transformations given in Table 1; the nonzero invariant has no effect on the result. (Review your solution to Prob. 33 to see this.) Thus, in summary, any space-time event x, t will transform with the Lorentz transformations and will have the same value $-s^2 \ (= x^2 + y^2 + z^2 - c^2 t^2 \text{ in S})$ in any frame.

Space-time intervals: proper time and distance

Consider two events that are measured in the two frames S and S'. Visualize the transformation equations (x1.1)–(1.4) with subscript "1" applied to all variables for event 1. Similarly, visualize an identical set of equations with subscripts "2" for the coordinates of event 2. Take the differences of the matching equations in each set. Because the Lorentz transformations are linear, one finds that the difference quantities, $\Delta x = x_2 - x_1$, $\Delta x' = x_2' - x_1'$, $\Delta t = t_2 - t_1$, and $\Delta t' = t_2' - t_1'$ also satisfy the Lorentz transformations. (We drop the y and z coordinates from this discussion for simplicity.)

Thus, analogous to (7), the quantity

$$(\Delta x)^2 - c^2 (\Delta t)^2 = -(\Delta s)^2$$

(7.8)

is also an invariant of the transformation (i.e., its value is the same in any inertial frame). This permits us to relate space-time intervals from frame to frame. There are three cases as follows:

(i) $(\Delta s)^2 = 0$: The two events are separated by the distance $\Delta x = c\Delta t$ and hence could represent a light signal traveling at speed c. The two events mark the positions and times at two points on its path. The "interval" between the events is thus called a *lightlike interval*.

(ii) $(\Delta s)^2 < 0$: The two events are separated by $\Delta x > c\Delta t$, and so a light signal would not have time to traverse the distance between the two events in the time between the events. Thus, one event could not cause the other to occur. Such intervals are called *spacelike* because the spatial part dominates the time part.

(iii) $(\Delta s)^2 > 0$: The two events are separated by $\Delta x < c\Delta t$. In this case, one event could cause the other to happen. A light signal could be transmitted from event 1 to location 2, where it would start a timer that would trigger event 2. These are *timelike intervals*.

In case (iii), where $c\Delta t > \Delta x$, the invariant interval $\Delta s/c$ is defined as the *proper time interval* $\Delta \tau$,

$$(\Delta \tau)^2 \equiv (\Delta t)^2 - \left(\frac{\Delta x}{c} \right)^2 .$$

(Proper time interval)   (7.9)

The invariance of $\Delta \tau$ implies that the proper time between two events, defined as in (9) with the coordinates of the frame in question, will have the same value in any other inertial frame. If the events are at the same position x' in one particular frame S', then $\Delta x' = 0$ and the proper time is simply $\Delta t'$:

$$(\Delta \tau) = (\Delta t)' = \frac{\Delta t}{\gamma} .$$

$(\Delta x' = 0)$ (7.10)

The proper time interval is thus the one measured in that special frame S', where the two events have the same position. In (10), we invoked (5), the time-dilation expression, to obtain the proper time in terms of the interval Δt in any other frame S moving with Lorentz factor γ relative to S'. Note that Δt, for any of these frames, is larger than the proper time interval. This is the time dilation effect in a different context.

In case (ii), one defines a *proper distance interval* $\Delta \sigma$ by setting $(\Delta \sigma)^2 = -(\Delta s)^2$ in (8):

$$(\Delta \sigma)^2 \equiv (\Delta x)^2 - (c\Delta t)^2 .$$

(Proper distance interval)   (7.11)

Because Δs is an invariant under transformation, so is the proper distance $\Delta \sigma$. The proper distance is the distance between two events in the special frame S' where they occur at the same time, $\Delta t' = 0$, and so $\Delta \sigma = \Delta x'$. Invoking (6), we find $\Delta x = \Delta \sigma / \gamma$; the spatial distance between the events in all other frames is less than the proper distance. This is length contraction in a different context.

In our discussion of length contraction, we could have marked the ends of the stick simultaneously in its rest frame S' (rather than in S) with two space-time events. In this case, the distance between the events in S' would be the proper distance, or, more familiarly, the rest length of the stick. The measured stick length in all other frames would be less. In our derivation of length contraction, we used events that occurred simultaneously in a frame S moving relative to the stick. Their spatial separation does not represent the length of the stick at rest but rather the length of some other (shorter) object at rest in S.

The invariance of $(\Delta s)^2$ in (8) tells us that a spacelike interval in one inertial frame will be a spacelike interval in any other inertial frame. The same is true for timelike and lightlike intervals.

Four-vector

The position and time of a space-time event may be described with a *four-vector* $[x, ct]$ that has components x, y, z, ct. The invariant quantity associated with this four-vector is $x^2 + y^2 + z^2 - c^2t^2$, which can be viewed as a length squared of the four-vector. It is similar to the dot product of the four-vector with itself but with a minus sign inserted for the fourth component:

$$[x, y, z, ct] \cdot [x, y, z, ct] \rightarrow x^2 + y^2 + z^2 - c^2t^2. \quad \text{(Invariant "length squared"}$$
$$\text{of a four-vector)} \quad (7.12)$$

The four-vector $[x, ct]$ is thus a compact reference to the invariant quantity $x^2 + y^2 + z^2 - c^2t^2$.

Four-vectors can be formed from other sets of physically measurable variables. The definition of such a vector is that its components transform according to Lorentz transformations. The invariant quantity, given such transformations, is readily obtained with the rule (12). Other examples will be developed in the following sections.

Momentum and energy

Here we find a four-vector in momentum and energy simply by finding the differential four-vector $[\Delta x, c\Delta t]$ and then multiplying it by a scalar.

Four-vector

Consider again the difference of two four-vectors. Its components transform according to the Lorentz transformations because, as for (8), the transformations of the individual components are linear in position and time. The difference is thus also a four-vector. More specifically, the difference of the two four-vectors $[x_2, ct_2]$ and $[x_1, ct_1]$ is $[x_2 - x_1, c(t_2 - t_1)]$ or simply $[\Delta x, c\Delta t]$. This four-vector will transform according to the Lorentz transformations of Table 1 but with $\Delta x, \Delta y, \Delta z,$ and Δt replacing x, y, z, t, and similarly for the primed components.

One can also obtain a new four-vector by multiplying each component of a known four-vector by a scalar. The resultant four-vector will transform according to the Lorentz transformations – again because the transformations are linear. By definition, a scalar is invariant under transformations.

Choose such a scalar, namely the ratio $m/\Delta \tau$, where m is a scalar and the proper time interval $\Delta \tau$ (9) has been shown to be an invariant under Lorentz transformations. Invoking the expressions (10) for $\Delta\tau$, we have $m/\Delta\tau = \gamma m/\Delta t$, or $\gamma m/dt$ in differential form. Finally, multiply the (differential) four-vector $[dx, c\,dt]$ by the scalar $\gamma m/dt$ to obtain a new four-vector as follows:

$$\Rightarrow \quad [dx, c\,dt] \times \left(\frac{\gamma m}{dt}\right) = \left[\gamma m \frac{dx}{dt}, \frac{\gamma mc^2}{c}\right] \to \left[p, \frac{U}{c}\right]. \quad (p, U \text{ four-vector}) \quad (7.13)$$

The components of this new four-vector will transform from frame to frame with the associated Lorentz transformations.

The physical significance of (13) is indicated in its rightmost term as we now explain. The parameter γ in (13) is the function of velocity given in (3), dx is the spacing between two events, dt is the time interval between them. The scalar m can be set to be the *mass* of a particle measured by an observer in the inertial frame where the particle is at rest. We further choose to interpret the two events as two locations of the particle at two different times; thus, $dx/dt = v$ is the instantaneous velocity of the particle.

The term $\gamma m(dx/dt)$ in the second set of brackets of (13) thus equals γmv, which reduces to the classical momentum mv for low velocities because $\gamma \to 1$ as $v \to 0$. This justifies our interpretation of m as the mass. We call the three-vector γmv the *relativistic momentum*,

$$\Rightarrow \quad p = \gamma mv = \gamma m\beta c, \qquad\qquad \text{(Relativistic momentum)} \quad (7.14)$$

where $\beta = v/c$ is the vector form of β (3).

The fourth component of the four-vector (13) contains the factor γmc^2,

$$\gamma mc^2 = \left(1 - \frac{v^2}{c^2}\right)^{-1/2} mc^2 \xrightarrow[v \ll c]{} \left(1 + \frac{1}{2}\frac{v^2}{c^2}\right) mc^2 = mc^2 + \frac{1}{2}mv^2. \quad (7.15)$$

At low velocities it becomes the classical kinetic energy added to the quantity mc^2. The latter quantity is known as the *rest energy*. It is the energy associated with the mass of a particle of mass m when it is at rest ($v = 0$). The sum of the two terms is the *total energy* of the particle in the classical limit of low speeds. In classical problems, the rest energy is usually suppressed as an underlying constant.

The term γmc^2 is thus called, analogously, the *relativistic total energy U*, or simply the "total energy,"

$$\Rightarrow \quad U \equiv \gamma mc^2 = E_{\text{rest}} + E_{\text{kinetic}}. \qquad\qquad \text{(Total energy)} \quad (7.16)$$

These definitions of p (14) and U (16) allow us to write the four-vector (13) as a *momentum-energy four-vector* $[p, U/c]$.

The energy equation (16) yields another "definition" of the Lorentz factor γ that is quite physical and is often used:

$$\gamma = \frac{U}{mc^2} = \frac{E_{\text{rest}} + E_{\text{kinetic}}}{mc^2}. \qquad\qquad \text{(Lorentz factor)} \quad (7.17)$$

The γ factor is simply the ratio of the total energy to the rest energy mc^2. It approaches infinity as $U \to \infty$ and, from (3), as $v \to c$. Because infinite energy is unattainable, the speeds of material objects are limited to $v < c$.

Invariant

Recall that the invariant quantity under transformation is the dot product of the four-vector with itself. In the case of the $[p, U/c]$ four-vector, we find the invariant to be the difference $p^2 - (U/c)^2$:

$$\left[p, \frac{U}{c} \right] \cdot \left[p, \frac{U}{c} \right] \rightarrow p_x^2 + p_y^2 + p_z^2 - \frac{U^2}{c^2} = p^2 - \frac{U^2}{c^2}. \qquad (p, U \text{ invariant}) \qquad (7.18)$$

For a single particle or a system of particles, this quantity will have the same value in different inertial frames of reference. In one frame it might have less momentum and less energy than in another, but the difference (18) will be the same from frame to frame. In our S and S' frames, for example, we would have $p^2 - (U/c)^2 = p'^2 - (U'/c)^2$, where the momenta p and p' are measured in the S and S' frames, respectively, and likewise for the energies U and U'.

This invariant quantity turns out to be related to the particle mass for the case of a system consisting of a single particle at rest in the zero-momentum frame. In this frame, the particle is at rest, giving $p^2 = 0$ and $\gamma = 1$. Thus, from (16), $U = mc^2$. The invariant (18) is therefore

$$p^2 - \frac{U^2}{c^2} = 0 - \frac{(mc^2)^2}{c^2} = -m^2 c^2. \qquad \text{(Frame of particle)} \qquad (7.19)$$

For any other frame (i.e., one in which the particle has nonzero momentum), the invariant will, by definition, be the same, $-m^2 c^2$; hence, in general,

$$\Rightarrow \qquad U^2 - (pc)^2 = (mc^2)^2, \qquad \begin{array}{l} \text{(Energy-momentum invariant;} \\ \text{single particle of mass } m) \end{array} \qquad (7.20)$$

where we multiplied (19) through by c^2.

In this form, the invariant quantity, $U^2 - (pc)^2$, is equal to the rest energy squared $(mc^2)^2$ of the particle. In the frame where the momentum is zero, the U^2 term equals the rest energy squared; there is no kinetic energy. This expression (20) is a fundamental and powerful relation between the momentum, total energy, and mass of a single particle.

Photons

Photons have no mass and can not be at rest in any frame. Their energy-momentum relation follows from (20). Set $m = 0$ to obtain

$$U = pc, \qquad \text{(Photon)} \qquad (7.21)$$

or, because the energy of a photon is $U = h\nu$,

$$\Rightarrow \qquad p = \frac{h\nu}{c}. \qquad \text{(Photon momentum)} \qquad (7.22)$$

Invariance for system of particles

For a system of particles, the invariant is, from (18), $U^2 - (pc)^2$, where U and p are the total system energy and momentum, respectively. The value of the invariant takes on a somewhat different meaning in this case. It is not simply the sum of the rest energies but rather, from (20), the total energy squared in the zero-momentum frame. The momentum-energy invariant is thus equal to all the system energy, rest plus kinetic, in the zero-momentum frame.

Table 7.2. *Lorentz transformations*[a]: p, U

Frame S to frame S′	Frame S′ to frame S
(x2.1) $p'_x = \gamma(p_x - \beta U/c)$	(x2.5) $p_x = \gamma(p'_x + \beta U'/c)$
(x2.2) $p'_y = p_y$	(x2.6) $p_y = p'_y$
(x2.3) $p'_z = p_z$	(x2.7) $p_z = p'_z$
(x2.4) $U' = \gamma(U - \beta c p_x)$	(x2.8) $U = \gamma(U' + \beta c p'_x)$

[a] $\beta \equiv v/c$; $\gamma \equiv (1 - \beta^2)^{-1/2}$; S′ moves in the $+x$ direction of S at speed v.

In relativistic interactions of particles and photons, total energy and total momentum are typically conserved. Thus, the quantity $U^2 - (pc)^2$ has the same value before and after the interaction and is as well invariant from inertial frame to inertial frame. This is true even when particles are created or destroyed in an interaction (collision). These two invariants are highly useful in solving for the results of such interactions.

Transformations

The momentum-energy four-vector components must transform from one inertial frame to another according to the Lorentz transformations; the four-vector was constructed (13) so that they would do so. A set of transformations for $[p, U/c]$ may be obtained directly from the $[x, ct]$ transformations (Table 1) simply by replacing x, y, z, ct with $p_x, p_y, p_z, U/c$. The results are given in Table 2.

One can transform energies and momenta from inertial frame to inertial frame according to these relations. The momentum and energy are linked in the same manner as space and time. They trade off against each other as one transforms between frames, just as x and t do.

Consider a particle of mass m at rest in frame S′; its momentum is zero, and its total energy consists only of its rest energy,

$$p'_x = 0; U' = mc^2. \qquad \text{(Particle at rest in } S') \qquad (7.23)$$

Let S′ move down the $+x$-axis of frame S with some speed $v = \beta c$ (Fig. 1). An observer in S notes the particle moving to the right along the x-axis at this speed. It has energy and momentum given by (x2.5) and (x2.8) in Table 2 as follows:

$$p_x = \gamma \left(0 + \frac{\beta}{c} mc^2 \right) = \gamma m \beta c \qquad (7.24)$$

$$U = \gamma(mc^2 + 0) = \gamma mc^2. \qquad (7.25)$$

These reproduce the expressions (14) and (16) we had inferred from comparisons with the low-velocity limits of the four-vector components (13). Similarly, the invariant relation $U^2 - (pc)^2 = m^2 c^4$ (20) follows immediately from (24) and (25) and the relation $\gamma = (1 - \beta^2)^{-1/2}$.

The transformations in Tables 1 and 2 change only the components of position or momentum along the direction of motion of one frame relative to the other; the transverse components are unchanged by the transformation. A momentum vector exists in momentum space (not x, y, z space) with components p_x, p_y, p_z. The transformation of the four-vector is thus in

Table 7.3. *Lorentz transformations^a:* k, ω

Frame S to frame S′	Frame S′ to frame S
(x3.1) $k'_x = \gamma(k_x - \beta\omega/c)$	(x3.5) $k_x = \gamma(k'_x + \beta\omega'/c)$
(x3.2) $k'_y = k_y$	(x3.6) $k_y = k'_y$
(x3.3) $k'_z = k_z$	(x3.7) $k_z = k'_z$
(x3.4) $\omega' = \gamma(\omega - \beta c k_x)$	(x3.8) $\omega = \gamma(\omega' + \beta c k'_x)$

^{a} $\beta \equiv v/c$; $\gamma \equiv (1 - \beta^2)^{-1/2}$; S′ moves in the $+x$ direction of S at speed v.

a four-space with momentum-energy coordinates. The Lorentz transformations in this p, U space are identical to those in x, t space.

Wave-propagation vector and frequency

Yet another four-vector may be created – this time for wave propagation. It is derived from the momentum-energy four-vector $[p, U/c]$. Multiply the momentum magnitude p by the factor $2\pi/h$, where h is the Planck constant, and recall from (22) that the momentum of a photon is $p = h\nu/c$. Thus, for a vacuum,

$$p \times \frac{2\pi}{h} \xrightarrow[\text{photon}]{} \frac{h\nu}{c}\frac{2\pi}{h} = \frac{2\pi\nu}{c} = \frac{2\pi}{\lambda} \equiv k. \tag{7.26}$$

The quantity $k \equiv 2\pi/\lambda$ is the magnitude of the *wave-propagation vector* $k = (2\pi/\lambda)\hat{k}$, where $\hat{k}$ is the unit vector in the propagation direction. Multiplication of each component of p by $2\pi/h$ thus yields the components of the vector k, which has the direction of the photons or of the propagating wave.

Now multiply the fourth term of $[p, U/c]$ by the same factor and use the energy of a photon, $U = h\nu$,

$$\frac{U}{c} \times \frac{2\pi}{h} \rightarrow \frac{h\nu}{c}\frac{2\pi}{h} = \frac{\omega}{c}, \tag{7.27}$$

where ω is the angular frequency ($\omega = 2\pi\nu$) of the radiation.

Transformations

Because all four components of $[p, U/c]$ were multiplied by the same factor, the result is another four-vector $[k, \omega/c]$ that (by definition) has components that transform according to the Lorentz transformations. Again, we simply change the variables of Table 2, in this case, from $[p, U/c]$ to $[k, \omega/c]$. The result is in Table 3.

With these transformations, it is possible to find how a propagation vector and the frequency ω of a given wave appear when observed in a different inertial frame. The change in frequency (Doppler shift) and the change in direction of propagation (aberration) follow from these relations.

Four-vectors

This four-vector $[k, \omega/c]$ when dotted with itself yields the invariant quantity

$$\rightarrow \quad [k, \omega/c] \cdot [k, \omega/c] \rightarrow k_x^2 + k_y^2 + k_z^2 - \frac{\omega^2}{c^2} = 0, \qquad (k, \omega \text{ invariant}) \tag{7.28}$$

Table 7.4. *Lorentz transformations*[a]: $\boldsymbol{B}, \boldsymbol{E}$

$(\boldsymbol{E}, \boldsymbol{B})$ to $\boldsymbol{E}'$	$(\boldsymbol{E}, \boldsymbol{B})$ to $\boldsymbol{B}'$
(x4.1) $E_x' = E_x$	(x4.4) $B_x' = B_x$
(x4.2) $E_y' = \gamma(E_y - \beta c B_z)$	(x4.5) $B_y' = \gamma(B_y + \beta E_z/c)$
(x4.3) $E_z' = \gamma(E_z + \beta c B_y)$	(x4.6) $B_z' = \gamma(B_z - \beta E_y/c)$

[a] $\beta \equiv v/c$; $\gamma \equiv (1 - \beta^2)^{-1/2}$; S' moves in the $+x$ direction of S at speed v.

where we have set the invariant to zero. Consider the simple case of a wave propagating in the x direction so that $k_x = 2\pi/\lambda$, $k_y = 0$, $k_z = 0$, and $\omega = 2\pi\nu$. Substitute into (28) to find that the invariant is zero for a wave traveling at speed c (i.e., for the condition $\lambda\nu = c$).

The dot product of two different four-vectors also yields an invariant. The product of $[\boldsymbol{x}, ct]$ and $[\boldsymbol{k}, \omega/c]$ is of interest to astronomers:

$$[\boldsymbol{x}, ct] \cdot [\boldsymbol{k}, \omega/c] \rightarrow \boldsymbol{k} \cdot \boldsymbol{x} - \omega t. \qquad \text{(Invariant under transformation)} \qquad (7.29)$$

This appears in the argument of expressions such as

$$E(x, t) = E_0 \sin(k_x x - \omega t), \qquad (7.30)$$

which describes a sinusoidal traveling wave as we now explain. If the argument in parentheses is held constant, one obtains a fixed value of E even as x and t vary. As time progresses, the position that yields a given (fixed) value of E moves to greater x. The differential of the fixed argument is $k_x \Delta x - \omega \Delta t = 0$; hence, $\Delta x/\Delta t = \omega/k_x$. All points on the wave thus move to the right a distance ω/k_x in 1 s (i.e., at a speed $v = \omega/k_x$). Thus (30) is indeed a traveling wave.

The argument $(k_x x - \omega t)$ gives the phase in radians of a particular position (e.g., a maximum) on this moving wave. Its invariance means that S and S' observers will agree on the value of the phase and hence that it is, for example, a maximum of the sine function. They will also agree on how many cycles are in a wave train.

Electric and magnetic fields

Electric $\boldsymbol{E}$ and magnetic $\boldsymbol{B}$ fields are interrelated and play off one another from frame to frame much as x and t do. We present their transformations but do not derive them.

Transformations

Electric and magnetic fields $\boldsymbol{E}$ and $\boldsymbol{B}$ each have three components for a total of six. In the case of Tables 1, 2 and 3, there were four parameters to convert, the four components of the four-vector – for example, $[\boldsymbol{x}, ct]$ or $[\boldsymbol{k}, \omega/c]$. It turns out that the electromagnetic field can be described by an antisymmetric four-tensor, which is fully defined by six elements, namely, the six components of $\boldsymbol{E}$ and $\boldsymbol{B}$. The components of $\boldsymbol{B}$ and $\boldsymbol{E}$ transform in SI units with the expressions given in Table 4.

Table 4 contains only the transformations from the S to the S' system for each of the six components. The inverse transformations may be derived algebraically from these or simply inferred from the equivalence of the two frames. The direction of the motion is reversed, and so $\beta \rightarrow -\beta$. Note particularly that it is the components of the fields normal to the direction of motion that are changed in the transformation; the components along the direction of motion remain unchanged. Here again, we see that the magnitude and direction of one quantity (e.g., E) can be traded off against those of the other (B) in the transformation.

Magnetic field transformed

As an example, take a case presented in the derivation of synchrotron radiation in Chapter 8. An observer in the (stationary) S frame sets up a B field in the z direction (out of the paper in Fig. 1), and there is no E field. The x, y, z components of each field are thus

$$B: 0, 0, B_z; \quad E: 0, 0, 0. \qquad \text{(Fields in S frame)} \qquad (7.31)$$

Another observer at rest in the frame S′ passes by at a high speed v down the $+x$-axis and encounters the transverse (to the frame motion) magnetic field B_z. The components in S′ are, from the six transformations of Table 4,

$$E': 0, -\gamma \beta c B_z, 0; \quad B': 0, 0, \gamma B_z. \qquad \text{(Fields in S′ frame)} \qquad (7.32)$$

The observer in S′ thus experiences a (larger) B field in the z direction by a factor of γ and *also* an E field in the $-y$ direction! The B field in S transforms into a combination of E' and B' in S′.

From where did the electric field in (32) come? According to a stationary observer (in S), a test charge at rest in S′ must experience a magnetic force $F = qv \times B$ because it is passing through the magnetic field. In our example, the force would be in the $-y$ direction for positive charge q. In S′, the charge is at rest ($v = 0$), and so it can experience no magnetic force. Nevertheless, the S′ observer notes the acceleration in the $-y$ direction and concludes that an electric field (force per unit charge) must be acting on it in that direction. A Newtonian observer would expect its magnitude in the y' direction to be $E_y' = F_y'/q = -vB_z = -\beta c B_z$.

This is in accord with (32) except for the factor γ, which enhances the electric field E' at high speeds. This latter is an effect of special relativity alone, and a Newtonian observer would not have anticipated it.

Field lines

The electric field of a charge moving uniformly is isotropic for speeds $v \ll c$, but at higher speeds $v \approx c$, the lines become highly bunched in the transverse direction (Fig. 3a,b). This follows immediately from the Lorentz transformations for B and E.

Consider the geometry of Fig. 1 in which frame S′ moves down the $+x$-axis of S but with a charge q at rest at the origin of S′. The field lines of the charge in S′ are isotropic (Fig. 3a) because it is at rest in S′. In the $x'-y'$ plane, the components of E' are E_x', E_y', 0 (Fig. 3c) because $E_z' = 0$ in the $x'-y'$ plane. Assume zero magnetic field in S′.

Transform the vector E' shown in Fig. 3c from the S′ to the S frame. We require the inverse of the transformations in Table 4, and so we must change the sign of β everywhere it occurs. From (x4.1), (x4.2), and (x4.6), the fields in the $x-y$ plane of S are

$$
\begin{aligned}
E_x &= E_x' \\
E_y &= \gamma E_y' \\
B_z &= \gamma \beta E_y'/c.
\end{aligned}
\qquad \text{(Transformations to S, } x-y \text{ plane)} \qquad (7.33)
$$

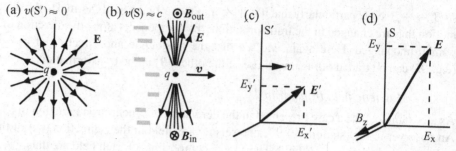

Figure 7.3. (a) Isotropic electric field lines for a positive electric charge q at rest at the origin in frame S'. (b) Field lines for this charge if it were moving at speed v approaching c relative to the observer in frame S. Note the circulating B field. (c, d) Transformation from frame S' to frame S (Fig. 1) of electric field vector E' vector lying in the first quadrant of the E'_x–E'_y plane. The result in S is a rotated and increased E vector with a magnetic field in the B_z direction (see Table 4). This leads to the bunched E field lines with circulating B field shown in (b). Be careful to distinguish the field lines in (a, b) and the electric vectors of (c, d). The former represent the magnitude (lines/m²) and directions of the vectors at each point in space.

The other components are zero. The results are valid in the x–y plane, which is coincident with the x'–y' plane.

Consider first the electric field components (33). In the frame S, the longitudinal component E_x is the same as in S', but the transverse component E'_y is enhanced by the factor γ. This increases the overall magnitude of E and also rotates the field vector toward the transverse direction (Fig. 3d). All E vectors in the x, y plane will thus be rotated toward the transverse direction. Coulomb's law remains valid ($F \propto r^{-2}\hat{r}$), and so the electric vectors and the field lines remain radial relative to the position of the charge.

The result is that the field lines emerging from the moving charge are bunched in the transverse direction in frame S (Fig. 3b). The bunching is symmetric about the direction of motion. Because the density of field lines (lines/m²) represents the field strength, this bunching is a visualization of the increased magnitude of E in the transverse direction as just discussed. The fields weaken with distance from the charge according to Coulomb's law ($\propto r^{-2}$).

The results (33) also give a value of B_z in the x–y plane, which is positive (out of the paper) when E'_y is positive and negative when E'_y is negative (Fig. 3b). This and the components of B at positions off the x–y plane in S (where $E'_z \neq 0$) indicate that a magnetic field circulates about the x-axis position of q in S. From the point of view of an observer in S, the charge moving down the $+x$-axis is a small instantaneous current. Ampere's law tells us the current produces a circulating magnetic field in the direction derived; recall the right-hand rule.

We thus see from the transformations that the electric field lines of a rapidly moving charge are bunched in the transverse direction and accompanied by a circulatory magnetic field in the stationary S frame. The bunching and the magnitudes of the field components are not intuitively obvious. Both fields, E and B, are proportional to γ; they can be huge if the speed of the passing charge is nearly c.

This discussion pertains only to charges moving at constant velocities. Acceleration of charges yields electromagnetic radiation, which is a quite different phenomenon; see Fig. 5.2a.

7.4   Doppler shift

The Lorentz transformations yield a Doppler shift that differs from the classical expression if the relative velocity of the source and observer is significant compared with the speed of light.

Derivation

We first present the classical Doppler frequency shift for a head-on approach of an emitter and then derive the relativistic shift, taking into account different angles of approach. We obtain the relativistic shift in two ways: from time dilation – and then more directly but less intuitively – from the k, ω transformations.

Classical Doppler shift

Radiation emitted from a moving source will be shifted in frequency by the well-known Doppler effect. Consider a source of sound waves approaching a stationary observer at a velocity v that is less than the speed c_s of sound in the medium that carries the waves. The Doppler effect arises because a later wave crest is emitted closer to the observer than an earlier wave crest. This compresses the wave and leads to a shorter time interval between the crest arriving at the observer. The rate of crest arrivals (i.e., the frequency) is thus increased. Here we consider the wave speed to be independent of frequency.

The classical Doppler shift for a moving source (speed v) and stationary observer – motions relative to the propagation medium – is readily demonstrated to be

$$\nu = \frac{\nu_0}{1 - \beta_s \cos \theta}, \qquad \text{(Classical Doppler shift; moving source)} \qquad (7.34)$$

where ν is the observed frequency, ν_0 the emitted frequency, θ the angle of the observer from the source referenced to the velocity direction (Fig. 4), and $\beta_s \equiv v/c_s$. For $\beta_s \ll 1$, and for $\theta = 0$ or π, the expression (34) may be approximated as

➡ $$\frac{\nu}{\nu_0} \approx 1 \pm \beta_s, \text{ or } \frac{\Delta \nu}{\nu} \approx \pm \frac{v}{c_s}. \qquad \text{(Classical Doppler shift; } v \ll c_s; \text{ (+) approaching; (−) receding)} \qquad (7.35)$$

An approaching source (+) leads to an increase in frequency, and a receding source (−) leads to a decrease in frequency. (As before, β_s is a positive definite quantity, and $\Delta \nu \equiv \nu - \nu_0$.) Finally, note that if the source were moving through the medium toward a stationary observer, the relation (34) would become $\nu = \nu_0(1 + \beta_s \cos \theta)$, which differs from (34) only in the second order of β.

The Doppler shift can manifest itself in the interval between two pulses (or flashes) of energy emitted by the source. For an approaching source ($|\theta| < \pi/2$), the time interval between the pulses detected by a stationary observer will be decreased, and the "frequency" thereby increased, as given in (34).

Relativistic Doppler shift

In our time-dilation discussion, the times of the two flashes in S′ were measured by observers O_1 and O_2 at the flash locations in frame S (Fig. 2). In contrast, the Doppler shift pertains to

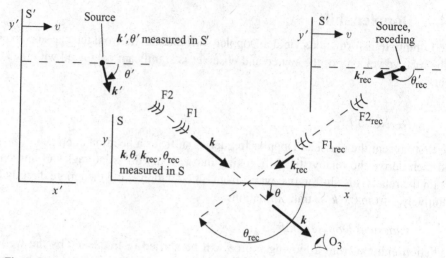

Figure 7.4. Aberration and Doppler shift for approaching source (upper left) and receding source (upper right). In each case, the source emitted the two flashes shown in quick succession, and so the source motion was negligible during the interval between them. The radiation in S′ has propagation vector k' (or k'_{rec}) at angle θ' (or θ'_{rec}). In frame S, the propagation vector k (or k_{rec}) is rotated forward to angle θ (or θ_{rec}). The frequency in S is greater than in S′ when the radiation is approaching from the far left. It becomes the same and then less so at some time before the radiation arrives from overhead owing to the second-order Doppler effect. When the radiation is arriving from the right, at θ_{rec}, the frequency in S is lower than in S′.

measurement by a third observer O_3 in frame S (Fig. 2b) with a frame-S synchronized clock. This observer records the arrival time of *both* flashes. Their interval Δt_3 is then compared with the S′ interval $\Delta t'$. (All S′ observers agree on $\Delta t'$ because they are at rest relative to the source.)

Two effects come into play: the relativistic time dilation $\Delta t = \gamma \Delta t'$ (5), and the different light travel times from the two flash positions to O_3 (Fig. 2). The latter is nothing more than the classical Doppler effect. We will take into account both effects to find the ratio of time intervals, $\Delta t_3/\Delta t'$; the desired frequency ratio, ν_3/ν', is the inverse of this.

The geometry for an arbitrary angle of approach is shown in Fig. 4. The propagation vector k measured by S-frame observers is at angle θ from the S′ velocity direction. The angle is measured in S clockwise from the velocity direction of the source, or equivalently, from the positive x-axis of S. This angle increases from 0 to π radians as the source moves from left to right. (The usual right-handed definition of θ is counterclockwise, but we choose clockwise to keep angles <180°. The results are not affected because the trigonometric functions of θ and $\theta - 360°$ are the same.) The angles measured by observers in the S′ frame differ from those in S owing to aberration (Section 5); note the k' vectors in the figure.

Two source positions are shown in Fig. 4. Consider the approaching case (upper left) from the viewpoint of an S observer. The source (upper left) emits two sequential flashes, F1 and F2, in quick succession. Let the interval between them Δt (in S) be sufficiently small that the source position, and hence angle θ, changes negligibly during the interval. The two pulses of light thus move along nearly the same track, at angle θ, as shown. Their nominal separation $c\Delta t$ is reduced by the component of the source's displacement along the track during the interval – namely $(v \cos \theta) \Delta t$.

The spacing between the two pulses along the track in S is thus $(c - v \cos \theta)\Delta t$. Divide by c to obtain the time interval Δt_3 measured by O_3 and also express Δt in terms of the interval in the emitter frame S':

$$\Delta t_3 = (1 - \beta \cos \theta)\Delta t$$
$$= (1 - \beta \cos \theta)\gamma \Delta t'. \tag{7.36}$$

The latter step introduces the time dilation effect (5).

The time intervals in (36) could be the time between two adjacent peaks of a sinusoidal wave – namely its period P. The frequency of the wave is then $\nu = P^{-1} = (\Delta t)^{-1}$. We thus take the reciprocal of (36) to obtain the general result for the Doppler shift in terms of the emitted and detected frequencies,

$$\Rightarrow \quad \nu = \frac{1}{\gamma(1 - \beta \cos \theta)}\nu_0 = \frac{(1 - \beta^2)^{1/2}}{1 - \beta \cos \theta}\nu_0, \qquad \text{(Relativistic Doppler shift in terms of } \theta) \tag{7.37}$$

where $\nu_0 = \nu'$ is the emitted frequency measured by an S' observer, and $\nu = \nu_3$ is that detected by a single observer O_3 in S (Fig. 2b).

The angle θ is measured in S; it describes the direction of the radiation or, equivalently, the position of the source at the time it emitted the two pulses. The distinction between the approaching and receding cases is controlled by the cosine function; the $\pm$ sign of (35) is not required for this case.

A source approaching head-on, $\theta = 0$, gives, if one notes that $1 - \beta^2 = (1 - \beta)(1 + \beta)$,

$$\Rightarrow \quad \nu = \left(\frac{1 + \beta}{1 - \beta}\right)^{1/2}\nu_0, \qquad \text{(Relativistic Doppler shift; source approaching head-on)} \tag{7.38}$$

where ν and ν_0 are again the observed and emitted frequencies. As expected, the shift is toward higher frequency for radiation emitted from an approaching source. If the source were receding directly from the observer, we would have $\cos \theta = -1$, and the $+/-$ signs in (38) would be interchanged. In this case the frequency would be downshifted. The symmetry of the two frames is apparent in (38). It is only the relative velocity of the two frames of reference that enters; neither frame is preferred.

The relativistic Doppler (37) shift reduces to the nonrelativistic form (34) for $\beta \ll 1$,

$$\frac{\nu}{\nu_0} \approx \left(1 - \frac{\beta^2}{2} + \cdots\right)(1 + \beta \cos \theta + \cdots) \approx 1 + \beta \cos \theta \ldots, \quad (\beta \ll 1) \tag{7.39}$$

where we have retained the angular dependence. This is in close agreement with (34); see also discussion below (35).

Earth-orbiting satellite

Consider the Doppler shift of a constant-frequency signal emanating from a low-altitude (~ 500 km) earth-orbiting satellite approaching from the left, passing overhead, and then receding to the right (Fig. 4). The orbital speed is large (~ 7.6 km/s), but because $\beta = 2.5 \times 10^{-5}$, we can ignore the β^2 term in (37), or, equivalently, use the classical expression (39). For simplicity, assume straight-line motion.

While the satellite approaches, the function $\cos \theta$ is positive, and, according to (39), the signal in S is blueshifted to the higher frequency, $\nu/\nu_0 > 1$. As the component of velocity along the line of sight begins to decrease, $\cos \theta$ decreases and so does the frequency ν. The ratio ν/ν_0 passes through unity (no shift) when the satellite is overhead at $\theta \approx 90°$. The function $\cos \theta$ then becomes negative, and so the frequency becomes redshifted, $\nu/\nu_0 < 1$. The frequency approaches an asymptotic redshift, $\nu/\nu_0 = 1 - \beta$, as $\cos \theta$ approaches -1.

Second-order Doppler shift

Now consider that a space warship moves relativistically, $v \approx c$, from left to right, as in Fig. 4. In this case, the β^2 term in (37) plays a significant role. When the satellite signal is arriving in S from directly overhead ($\theta = 90°$), substitution of $\cos \theta = 0$ into (37) yields

$$\frac{\nu}{\nu_0} = \frac{1}{\gamma} = (1 - \beta^2)^{1/2}. \qquad (\theta = \pi/2; \text{ second-order Doppler shift}) \qquad (7.40)$$

We find, surprisingly, a redshift to lower frequency ($\nu/\nu_0 < 1$) in contrast to the classical case, which yields no shift. This is known as the *transverse or second-order Doppler effect*. This effect becomes apparent only if β^2 is significant compared with unity.

Examination of (37) shows that the numerator can drive the ratio below unity even when $\cos \theta > 0$. The redshift can thus extend to positions well before overhead passage. The closer β^2 is to unity, the earlier the redshift begins. See more on this below in our discussion of Fig. 10 and in Prob. 41.

The second-order Doppler effect is solely a result of special relativity; it does not enter into a classical description. In fact, it is nothing more than the time dilation effect.

Doppler from k, ω transformations

The expression (37) can be found directly from the transformation of the $[k, \omega/c]$ four-vector (Table 3) because $\omega = 2\pi\nu$ is the quantity we wish to compare from frame to frame. Recall the propagation vector $k = (2\pi/\lambda)\hat{k}$, where $\hat{k}$ is the unit vector in the propagation direction. Maxwell's equations for a vacuum give the *dispersion relation*, the relation between k and ω,

$$\frac{\omega}{k} = c, \qquad \text{(Dispersion relation for vacuum)} \qquad (7.41)$$

where $k = (k_x^2 + k_y^2)^{1/2}$ in the two-dimensional geometry of Fig. 4. From the definitions of ω and k, this is simply the familiar relation $\lambda\nu = c$. In a medium described by an index of refraction η, one replaces c with c/η. For a vacuum, $\eta = 1$.

Again, we adopt the geometry of Fig. 4, assume transmission of light waves in a vacuum, and solve for the frequency in S. The desired comparison will be between the frequencies in the two frames, ω and ω'. We choose the transformation for ω' (x3.4) of Table 3 because k_x in that expression is an S-frame quantity, and the angle θ measured in the S frame is given by

$$\cos \theta = \frac{k_x}{k}. \qquad \text{(Propagation angle in S)} \qquad (7.42)$$

Divide the transformation (x3.4) by ω:

$$\frac{\omega'}{\omega} = \gamma\left(1 - \frac{\beta c k_x}{\omega}\right). \qquad \text{(Lorentz transformation)} \qquad (7.43)$$

Eliminate ω in the right-hand term with (41) and then invoke (42):

$$\frac{\omega'}{\omega} = \gamma(1 - \beta \cos\theta). \qquad (7.44)$$

Solve for the S-frame frequency to find the desired Doppler shift:

$$\omega = \frac{\omega'}{\gamma(1 - \beta \cos\theta)}. \qquad \text{(Doppler shift in terms of } \theta) \qquad (7.45)$$

This expression is identical to our earlier result (37) because $\omega = 2\pi\nu$, and $\omega' = 2\pi\nu_0$.

We will find it useful below to express the ω, ω' relation in terms of θ', the propagation angle in S'. The Table 3 transformation (x3.8) yields, after similar manipulations (Prob. 42), the frequency ω' observed in S' at angle θ' as a function of the frequency ω,

$$\omega' = \frac{\omega}{\gamma(1 + \beta \cos\theta')}. \qquad \text{(Doppler shift in terms of } \theta') \qquad (7.46)$$

It is identical in form to (45) except for the sign preceding β. This arises because, in S', the frame S moves down the negative x'-axis, whereas, in S, frame S' moves down the positive x-axis. The factor β is, we again state, taken to be a positive quantity.

Doppler shifts in astronomy

The frequencies of spectral lines from celestial sources are often shifted owing to the motions of the emitting object: gaseous clouds, stars, or galaxies. The Doppler shifts may be due, for example, to the orbital motions of a star in a binary system, to the motions of stars in the Galaxy relative to the sun, to the motions of stars in other galaxies, and to galaxy recession in an expanding universe, though the latter is more properly viewed as a cosmological redshift.

In most cases, the velocities are sufficiently low that the classical Doppler shift is adequate. However, relativistic particles have been discovered in a host of objects such as protostars, binary systems, supernova remnants, and extragalactic jets. Also, extragalactic objects in the expanding universe, such as quasars, can be sufficiently distant to have relativistic recession speeds.

Astronomical sign convention

In the classical Doppler shift, the observed shift of frequency reflects only the component of the velocity along the line of sight, the radial component v_r. The astronomical convention is that v_r be positive if it is directed outward and negative if it is directed inward. Let us further define $\beta_r \equiv v_r/c$. Here, both v_r and β_r carry sign, unlike β (see our previous discussion). The classical Doppler shift (34) thus takes the form

$$\frac{\nu - \nu_0}{\nu_0} = -\frac{v_r}{c}, \quad \text{or} \quad \frac{\nu}{\nu_0} = 1 - \beta_r, \quad (|\beta_r| \ll 1; \beta_r > 0 \text{ for recession}) \qquad (7.47)$$

where ν and ν_0 are again the observed and emitted frequencies, respectively, and where v_r is the radial component of the velocity.

Optical astronomers often work in wavelength units rather than frequency units. For a vacuum, the relation $\lambda = c/\nu$ yields $\lambda/\lambda_0 = \nu_0/\nu$, giving, for $|\beta_r| \ll 1$,

$$\frac{\lambda}{\lambda_0} = 1 + \beta_r. \qquad (\beta_r \ll 1) \qquad (7.48)$$

The relativistic version of this, for strictly radial motion, is modified from (38) for the astronomical sign convention ($\beta_r > 0$ for recession),

$$\boxed{} \qquad \frac{\lambda}{\lambda_0} = \left(\frac{1 + \beta_r}{1 - \beta_r}\right)^{1/2}. \qquad \text{(Wavelength ratio)} \qquad (7.49)$$

The observed wavelength λ in (49) is greater than the emitted wavelength λ_0 for a receding emitter. As $\beta_r \to 1$, the ratio λ/λ_0 can grow indefinitely (i.e., as recessional speed v_r approaches c). In this case, the radiation is greatly reddened and the frequency decreases toward zero, as do the photon energies $h\nu$.

Redshift parameter

The optical spectra of distant luminous objects in the universe called quasars have spectral lines shifted by large amounts to lower frequencies. If these redshifts are interpreted as Doppler shifts, they indicate recession velocities approaching the speed of light. These velocities are due to the expansion of the universe; the expansion is such that the more distant the object, the faster it recedes (AM, Chapter 9).

Astronomers define the "redshift" parameter z as

$$\Rightarrow \qquad z \equiv \frac{\lambda - \lambda_0}{\lambda_0} = \frac{\lambda}{\lambda_0} - 1, \qquad \text{(Definition of redshift } z\text{)} \qquad (7.50)$$

where again λ_0 is the emitted wavelength. Substitute the ratio of wavelengths (49) into this,

$$\Rightarrow \qquad z + 1 = \left(\frac{1 + \beta_r}{1 - \beta_r}\right)^{1/2}, \qquad \text{(Redshift–speed relation)} \qquad (7.51)$$

to obtain a relation between z and recessional speed for our Doppler-shift interpretation of the redshift. As the recession speed approaches c, $\beta_r \to 1$ and z increases indefinitely.

The most distant quasars known are at redshifts $z \approx 6$. At this redshift, $\lambda/\lambda_0 = 7$; the observed wavelength is seven times the rest wavelength in the quasar frame! An ultraviolet emission line at $\lambda_0 = 121.5$ nm (Lyman α) would be shifted almost into the near infrared at 850.5 nm. In this case, the relation (51) yields a speed factor $\beta_r = 0.960$. The quasar is receding at 96% the speed of light.

We remind the reader that special relativity is not really appropriate to our universe with its changing rate of expansion. In GR, redshifts are not viewed as Doppler shifts. Instead they are intrinsic to the expansion of the universe. The proper view is that light waves (and hence their wavelengths) are stretched as they travel through the expanding intergalactic space en route to earth from the quasar. The relation (51) would apply only to a freely expanding universe with no matter content and hence no deceleration.

On the other hand, the redshift parameter z (50) is a widely used observational parameter that is independent of any theory of the expansion. Cosmologists construct theories to match the measured redshifts of distant galaxies.

In the following discussions, we will continue to use the self-consistent Lorentz relations wherein the speed factor β is a positive definite quantity insofar as frame S' moves down the positive x-axis of frame S.

7.5 Aberration

The apparent direction of radiation will differ according to observers in two frames of reference that are moving with respect to each other. This phenomenon is known as aberration. This is not the transverse bunching of electric field lines (Fig. 3a,b). Rather, aberration refers to the propagation directions of electromagnetic waves.

Aberration results in displaced positions of stars on the celestial sphere and in the beaming of radiation from astronomical jets. One can explain stellar aberration for small frame speeds $v \ll c$ with a simple "falling-rain" argument (AM, Chapter 4). Now we have the tools to calculate the aberration properly in the context of special relativity.

Transformation of k direction

The task is simply to compare the angle of the propagation vectors measured in S' and S; that is, θ' and θ, respectively (Fig. 4). The comparison of angles in the two frames is obtained through the Lorentz transformations for $\mathbf{k}$, ω (Table 3).

The dispersion relation for radiation in a vacuum (41) is valid in any inertial frame, and so we can write the equality

$$\frac{\omega}{k} = \frac{\omega'}{k'} = c, \qquad \text{(Dispersion relation)} \qquad (7.52)$$

where $k = (k_x^2 + k_y^2)^{1/2}$, $k' = (k_x'^2 + k_y'^2)^{1/2}$, and $k_z = k_z' = 0$. The quantities of interest are

$$\cos \theta = \frac{k_x}{k}; \quad \cos \theta' = \frac{k_x'}{k'}. \qquad (7.53)$$

Invoke the transformation for k_x (x3.5) of Table 3 and divide by k':

$$\frac{k_x}{k'} = \gamma \left(\frac{k_x'}{k'} + \frac{\beta}{c} \frac{\omega'}{k'} \right). \qquad (7.54)$$

Apply (52) to the first and last terms of (54):

$$\frac{k_x}{k} \frac{\omega}{\omega'} = \gamma \left(\frac{k_x'}{k'} + \beta \right). \qquad (7.55)$$

Eliminate ω/ω' with the Doppler relation (46) and express the ratios k_x/k and k_x'/k' as the cosine functions (53):

$$\Rightarrow \qquad \cos \theta = \frac{\cos \theta' + \beta}{1 + \beta \cos \theta'}. \qquad \text{(Transformation of k vector directions)} \qquad (7.56)$$

This is the desired expression that relates the directions of k and k'. This relation does not require that the radiation be emitted from frame S' (as in our examples). It simply compares the directions of any bit of propagating radiation according to observers in two frames, S and S'. The angle in S can be calculated from the angle in S' and the frame speed parameter β. This result could also have been obtained directly from the two Doppler relations (45) and (46) (Prob. 51).

The general effect of (56) is to rotate a propagation vector of radiation emanating from a source toward the forward direction of the source motion, and so $\theta < \theta'$ or $\cos \theta > \cos \theta'$ (Prob. 53). The following limiting cases of aberration are easily extracted from (56):

$$\beta = 0 \qquad \cos \theta = \cos \theta' \qquad (7.57a)$$
$$\theta' = 90°, 270° \quad \cos \theta = \beta \qquad (7.57b)$$
$$\beta \to 1 \qquad \cos \theta \to 1. \qquad (7.57c)$$

In the first case, at $\beta = 0$, the two observers are effectively in the same inertial frame, and thus the angles θ and θ' do not differ. The second case will be applied to stellar aberration as an illustrative example. The third results in beaming, or the "headlight" effect, which is an extreme case of aberration. We discuss these latter two cases, respectively, in the next section immediately below and in Section 6 (under "Beaming ('headlight effect')").

Stellar aberration

The earth's motion in its orbit about the sun leads to an annual variation of the propagation direction of starlight according to an earth observer. This causes star positions to appear slightly displaced ($\leq 20''$) from their cataloged positions in the sky. The effect is small because the earth's orbital speed is only 29.8 km/s, which is much less than the speed of light.

Earth as stationary frame

To be in accord with our previous examples, we first place the emitting star in the "moving" frame S' and the earth observer in the "stationary" frame S (Fig. 5a). Consider the limiting case (57b) in which the radiation is emitted at exactly 90° or 270° in S' and the angle of k (in S) is specified by $\cos \theta = \beta$, a small positive value. Thus, in S, the k vector is rotated forward of 90°, toward positive x, as shown.

In our case, $v = 29.79$ km/s, and $\beta = 0.9937 \times 10^{-4}$. Because the angle $\theta - 90°$ is small, we write $\cos \theta = -\sin(\theta - 90°) \approx -(\theta - 90°)$. Taking care to use radians, we have from (57b) that $-(\theta - (\pi/2)) \approx \beta$, or

➡
$$\theta = \frac{\pi}{2} - \beta = \frac{\pi}{2} - (0.9937 \times 10^{-4}) \quad \text{rad}$$
$$= 90° - 20.50''. \qquad \text{(Stellar aberration at } \theta' = 90°) \quad (7.58)$$

The propagation vector in S is rotated 20.5″ toward the direction of the source motion.

To receive such radiation, a telescope on S (Fig. 5a) would have to be pointed opposite to the propagation vector and tilted to the left, as shown. This is the stellar aberration effect for a star that lies ~90° from the star velocity direction. The effect at other angles could be obtained here, but let us first reframe the discussion.

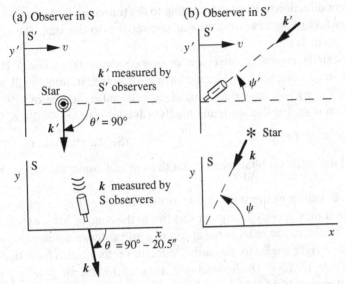

Figure 7.5. Stellar aberration. (a) Observer at rest in "stationary" S frame and emitting star in "moving" S' frame. The k' vector that lies normal to the direction of motion of S' is rotated in S by a small amount toward the direction of the S' motion. The rotation is 20.5" if S' moves at the speed of the earth in its orbit about the sun; $\beta_{earth} = 0.9937 \times 10^{-4}$. The telescope in S is tilted toward the left to intercept the ray (i.e., in the direction S moves relative to S'). (b) Star at rest in "stationary" S frame at elevation ψ. The earth and astronomer are at rest in a (moving) S' frame with the star at elevation ψ'. Again, the telescope must be tilted in the direction of the (earth) motion relative to the star frame (S) – in this case to the right – to the smaller angle $\psi' < \psi$.

Stars as stationary frame

The stellar aberration problem is usually described with the astronomer on a moving earth that is rushing through starlight from a stationary source. Also, conventionally, the angles are defined to be the directions *to* the star, or more precisely, the negative of the propagation directions k and k'.

Let us then use our standard frames S and S', placing the star in the "stationary" S and the astronomer in the "moving" S' and further defining the angles ψ and ψ' to the star (Fig. 5b). Because the S' frame moves down the positive x-axis of S, the transformation of the k directions in (56) remains valid. Again, we desire to find the astronomer's angle in terms of the angle in the emitter frame. Thus, we need to modify (56) to give $\cos \theta'$ in terms of $\cos \theta$. One can solve (56) for $\cos \theta'$, or equivalently, interchange primes and nonprimes while reversing the signs preceding each β.

The directions θ and ψ in Fig. 5 are directly opposed, and so are θ' and ψ'. Thus, $\psi = 180 - \theta$ and $\psi' = 180 - \theta'$, giving $\cos \theta = -\cos \psi$ and $\cos \theta' = -\cos \psi'$. These variable changes applied to the expression for $\cos \theta'$ just described yield (Prob. 52)

$$\cos \psi' = \frac{\cos \psi + \beta}{1 + \beta \cos \psi}, \qquad \text{(Transformation of direction to a star)} \qquad (7.59)$$

where ψ' is the apparent direction to the star according to the (moving) earthbound observer and ψ is the direction according to an observer at rest relative to the star. The resultant equation has the same form as (56).

The magnitude of the stellar aberration offset $\alpha \equiv \psi' - \psi$ depends on the angle ψ. Because the earth's velocity is small compared with c, we have $\beta \ll 1$ and, in turn, small α. Noting that $\cos \psi' = \cos(\psi + \alpha)$, we separately expand the left and right side of (59), make approximations based on $\alpha \ll 1$ and $\beta \ll 1$, and finally solve for α to obtain (Prob. 52)

➡ $$\alpha = -\beta \sin \psi = -20.5'' \sin \psi, \qquad \text{(Stellar aberration)} \qquad (7.60)$$

where, again, we used the earth's orbital speed about the sun and converted radians to arcseconds.

The angle ψ is the direction to the star relative to the direction of the earth's velocity vector. The offset angle α thus applies to any star that lies in the cone of half-angle ψ about the x-axis in Fig. 5b. The aberration reduces the apparent angle ψ by an amount that ranges from 20.5″ for a source at right angles to the earth's velocity vector to zero for a star in the direction of the velocity vector ($\psi = 0$). Accordingly, a star at the ecliptic pole ($\psi = \pi/2$) will trace out a circle of radius 20.5″ in a year, whereas a star on the ecliptic ($0 < \psi < 2\pi$) will move back and forth along a small 41″ segment of the ecliptic. At intermediate ecliptic latitudes, the star position will trace out an ellipse.

These displacements are relative to an inertial coordinate system. An image of a small sky region does not reveal the effect because all stars in a given region are displaced by nearly the same amounts. Precision pointing of a telescope to a given star thus requires absolute coordinates that are corrected for aberration. The corrections are a function of the time of year.

The 20.5″ magnitude of stellar aberration for $\psi = 90°$ can be understood, as noted above, in a deceptively simple classical calculation – namely, the forward tilt of an umbrella required to keep one dry while running at speed v through a vertical rainfall descending at speed c.

7.6 Astrophysical jets

Photographs of galaxies occasionally show a small jet emerging from the nucleus. Examples are 3C273 and M87. The latter is shown in Fig. 6 in three wavebands with contemporary instruments. The jet is ~5000 LY long and exhibits bright knots of emission. Polarization indicates the emission is synchrotron emission arising from electrons spiraling around magnetic field lines. The electrons emitting the radiation have lifetimes at their current energies significantly less than 5000 yr, and so they must be continuously reenergized by matter ejected from the central object.

The visible knots may be blobs of plasma moving outward along the jet – possibly at relativistic velocities. Alternatively, they may be quasi-stationary regions of high density (i.e., *shocks*) in a continuously flowing jet of plasma, where the shocks are caused by irregularities in the magnetic field. The acceleration and collimation of the jet material may be due to electromagnetic fields arising from the rotational and gravitational energy of the material accreting onto or into the central object or from the rotation of the compact object itself.

A simplistic schematic of such a system is shown in Fig. 7. An accretion disk of material spiraling inward toward the compact object, a jet, and the compact object are the primary features. In the case of the *active galactic nucleus* (AGN) of a galaxy, the disk may be fed by

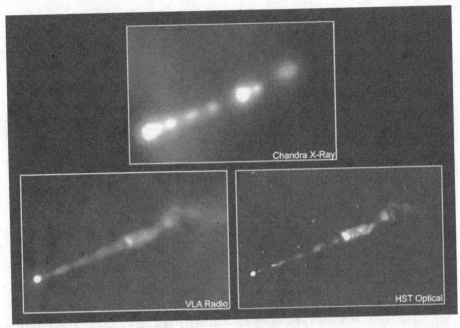

Figure 7.6. Jet of the elliptical galaxy M87 in three energy bands. The elliptical galaxy M87 at distance 50 MLY with angular diameter 4′ dominates the Virgo cluster of galaxies. The jet is embedded within the galaxy and extends about 5000 light years, or ~20″, from the nucleus, which is at the lower left of each panel. Material ejected from the nucleus energizes electrons in the bright knots of radiation to relativistic energies. These electrons radiate as they spiral around magnetic field lines, providing the observed synchrotron radiation. [Chandra X-ray Center (CXC Web site). X-ray: H. Marshall (MIT) *et al.*, CXC, NASA; Radio: F. Zhou, F. Owen (NRAO), J. Biretta (STScI); Optical: E. Perlman (UMBC) *et al.*, STScI, NASA]

neighboring gases from disrupted stars, and the central object is most likely a massive black hole of 10^6 to 10^8 $M_\odot$. Some of the infalling material is ejected along the angular momentum axis to form the relativistic jet of ejected plasma. The ejection can be continuous or in the form of discrete outbursts. The appearance to an observer varies greatly as determined by the observer's location relative to the jet direction.

Radio astronomers observe these outflows as resolved images of steady, large-scale jets or as variable, unresolved sources whose fluxes and spectra vary as expected for an expanding plasma. With high-resolution interferometry, discrete sources of radio emission can be observed moving away from the nucleus.

Stellar objects also exhibit accretion disks and jets, notably protostars and x-ray–emitting black-hole binary systems. Protostars are embryonic stars in their formative stages before settling on the main sequence. The black-hole binaries are sometimes called *microquasars* because they exhibit relativistic jets similar to those of extragalactic quasars. In the binary accreting systems, the accreted matter is provided by a normal gaseous companion star. The masses of the black holes in microquasars are thought to be of order 10 $M_\odot$. Relativistic ejections from a microquasar are shown in Fig. 8.

Here we evaluate three characteristics of the radiation from a blob of emitting plasma moving outward from a compact object at relativistic speeds: the *beaming of the radiation*,

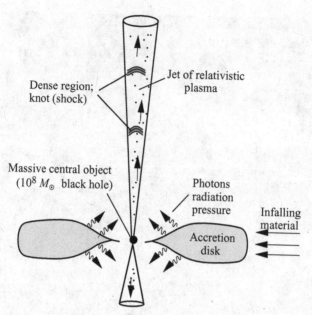

Figure 7.7. Sketch of a jet of emitting plasma emerging along the rotation axis of an active galactic nucleus (AGN). Energy to power the jet could be extracted from electromagnetic fields arising from the rapid rotation of the inner accretion disk or through acceleration of particles by the rotating axisymmetric magnetic field (if one exists) of the black hole. The jet may include shocks that appear as visible bright knots of radiation such as those seen in Fig. 6.

the *Doppler boosting* of the specific intensity, and *superluminal motion*. The latter refers to knot motions that seem to imply speeds greater than that of light. As a necessary step in the determination of the specific intensity of beamed radiation, we also demonstrate the Lorentz invariance of the distribution function (phase-space density) of propagating photons.

Beaming ("headlight effect")

The radiation from a blob of plasma moving relativistically is highly beamed in the direction of the velocity vector of the blob according to a stationary observer. This follows from the third limiting case of aberration (57c) in which S' is moving relativistically ($\beta \approx 1$). In this limit, the numerator and denominator of (56) tend to the same value, and so $\cos \theta \to 1$, or $\theta \to 0$. According to S observers, a given propagation vector k' in S' would thus be rotated to almost the forward direction regardless of the emission direction θ'. An example is shown in Fig. 9a.

Consider an isotropically and continuously emitting optical source (Fig. 9b) to be at rest in S', which is moving to the right at relativistic speed. In S', the light radiates into all directions uniformly, but in S, the radiation from S' would be concentrated in the forward direction. It would appear as if S' had a narrow-beam headlight. Radiation in the rearward hemisphere would be greatly weakened.

Define the beam angle θ_b to be the half-angle of the cone that includes one-half of the rays (Fig. 9b). In S', the polar angle of the cone that encloses half the rays is $\theta' = 90$. Thus, θ_b is the propagation angle in S that corresponds to a ray in S' at $\theta' = 90°$.

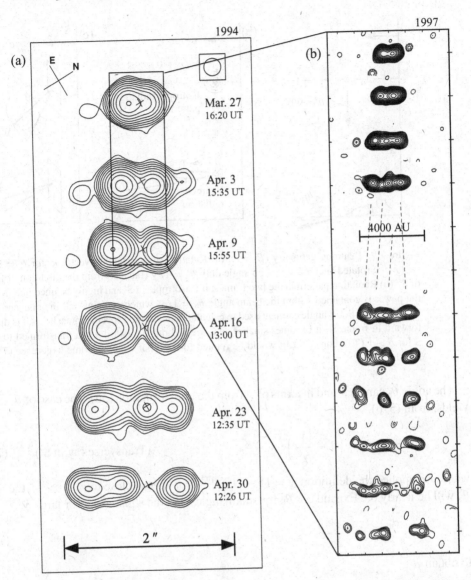

Figure 7.8. Radio images of the galactic microquasar GRS 1915+105 showing expansion of radio knots away from the compact object. (a) VLA images at 8.6 GHz made over a period of 34 days during a 1994 outburst. The compact object is located at the cross. (b) Higher-resolution MERLIN images at 5 GHz during a 1997 outburst over a period of 12 days. The source is most likely an accreting black-hole binary star system in the Galaxy; it is distant ~41 000 LY with black-hole mass ~15 $M_\odot$. The smaller rectangle superimposed on the left panel indicates the vertical and horizontal extent of the large rectangle of the right panel. One of the moving knots in each case is moving at apparent speed greater than the speed of light. [R. Fender and T. Belloni, *ARAA* **42**, 317 (2004) with permission; (a) F. Mirabel and V. Rodriguez, *Nature* **371**, 46 (1994); (b) R. Fender et al., *MNRAS* **304**, 865 (1999)].

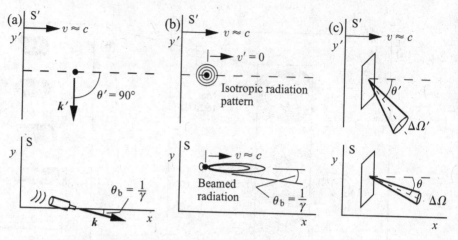

Figure 7.9. Beaming geometry ($\beta \approx 1$). (a) Rotation of propagation vector k' for $\theta' = 90°$. If $v \approx c$, the rotated vector k lies at an angle only $\theta_b = 1/\gamma$ (radians) from the direction of motion of S'. (b) Radiation pattern in the two frames; it is isotropic in S' and highly beamed in S. Most of the power is contained within the beam angle $\theta_b = 1/\gamma$, which contains half the rays emitted in S'; see (a). (c) Solid-angle element used to calculate transformed specific intensity. If it is directed forward in frame S', it becomes smaller and directed more forward when transformed to frame S for $\delta > 1$ (70). For $\delta < 1$, it would be rotated forward but would become larger; see (75) and Fig. 10.

The angle θ_b can be found in terms of γ from the transformation (56). The case of $\theta' = 90°$ yields, from (57b),

$$\cos \theta_b = \beta = \left(1 - \frac{1}{\gamma^2}\right)^{1/2}, \qquad \text{(Transverse ray in S')} \qquad (7.61)$$

where we applied the definition $\gamma = (1 - \beta^2)^{-1/2}$. For speeds close to c ($\gamma \gg 1$), the angle θ_b will be nearly 0°. Expand $\cos \theta_b$ for small angles and the square root for large γ,

$$1 - \frac{\theta_b^2}{2} = 1 - \frac{1}{2\gamma^2}, \qquad (7.62)$$

to obtain

$$\theta_b \approx \frac{1}{\gamma}. \qquad \text{(Beam polar angle; } \gamma \gg 1; \text{ rad)} \qquad (7.63)$$

The opening half-angle of the headlight beam, according to the S observer, is simply the reciprocal of the Lorentz factor for a highly relativistic source. Remember that the beam, so defined, contains the radiation emitted into the forward hemisphere in S'.

Most of the radiated energy in S is emitted within this cone of angles because the forward-emitted rays in S' are Doppler shifted to very high frequencies (high energies) in S compared with the rearward emitted rays, and these frequency shifts can be huge for large γ.

If the source of the radiation is a fast-moving particle of mass m, one can express γ in terms of the mass and total energy of the particle, $\gamma = U/mc^2$, from (17). Thus, from (63), the opening half-angle in radians is the ratio mc^2/U. An electron spiraling around magnetic field lines in a young supernova remnant might have energy 10^{11} eV in the observer's frame (S).

In the frame of reference in which it is at rest, it will radiate with the dipole radiation pattern, which is roughly isotropic (actually, it is a toroid; Fig. 5.2b). In the laboratory frame S, the radiation will be strongly peaked in the forward direction into a half-angle of only 1″,

$$\theta_b = \frac{mc^2}{U} = \frac{5 \times 10^5 \text{eV}}{10^{11} \text{eV}} = 5 \times 10^{-6} \, \text{rad} = 1.0''. \qquad (7.64)$$

This forward beaming reappears in our study of *synchrotron radiation* (Chapter 8).

We reiterate that this beaming refers to the electromagnetic radiation that has been emitted by a source. It is not the bunching of the electric field lines of a fast, uniformly moving charge as described earlier (Fig. 3). If a moving charge undergoes acceleration, the field lines become distorted. These distortions travel outward at speed c in a vacuum and become the propagating electromagnetic waves whose frequencies are Doppler shifted and whose directions are beamed as described here.

Lorentz invariance of distribution function

The invariance of the distribution function $f(x, p)$ under Lorentz transformations is demonstrated here. Recall that the distribution function is the particle density in six-dimensional phase space; see Section 3.3. We will use this invariance to find the specific intensity of a moving source (e.g., in a jet) in the next section ("Doppler boosting").

First demonstrate that a phase-space volume element $d\mathcal{V}'$ moving with a group of particles through phase space is a Lorentz invariant. In the S′ frame,

$$d\mathcal{V}' = d^3x' \, d^3p' = dx'dy'dz'dp_x'dp_y'\,dp_z'. \qquad (7.65)$$

The particles at the center of this comoving element will be at rest simply because we chose a comoving frame. Off-center particles within the differential elements will have small nonzero momenta and hence slow motions in x, y, z space. Let the x- and x'-axes be aligned with the direction of S′ motion so that the transformations of Tables 1 and 2 apply. The transverse components dy', dz', dp_y', and dp_z' thus equal their unprimed counterparts in the S frame.

The dx interval is smaller than dx' as the result of length contraction. Thus, from (6),

$$dx = \frac{dx'}{\gamma}. \qquad (7.66)$$

For the dp_x' term, take the differential of transformation (x2.5) of Table 2:

$$dp_x = \gamma(dp_x' + \beta \, dU'/c). \qquad (7.67)$$

The dU' energy term is vanishingly small compared with dp_x' because the velocities are near zero and the kinetic energy is quadratic in velocity. Substitute these primed position and momentum elements into (65) to find that the phase-space volumes in the S and S′ frames are equal:

$$d\mathcal{V}' = dx \, dy \, dz \, dp_x \, dp_y \, dp_z \qquad \text{(Lorentz invariant)} \qquad (7.68)$$
$$d\mathcal{V}' = d\mathcal{V}.$$

Now the phase-space density f is defined as the number of particles dN per unit volume of phase space, that is, $f \equiv dN/d\mathcal{V}$. The quantity dN is a countable number that is itself an invariant; observers in two frames would agree on the number. Because $d\mathcal{V}$ is also an invariant, one can conclude that f is also a Lorentz invariant:

$$\Rightarrow \quad f = \frac{dN}{d\mathcal{V}} = \text{Lorentz invariant.} \tag{7.69}$$

The density of particles in phase space is thus the same from the viewpoint of observers in any inertial frame of reference no matter its speed or direction of travel relative to the rest frame of the particles.

This result is not immediately applicable to photons because they travel at the speed of light in any frame of reference; there is no frame of reference in which they are at rest or moving slowly. The argument, however, does apply to particles moving at any speed (e.g., $0.99c$ or even $0.999\,99c$).

As particles become increasingly relativistic, they behave more and more like photons in their speed, momentum, and energy characteristics. If f is invariant for particles traveling arbitrarily close to the speed of light, one can argue validly that it must also be conserved for photons. It is possible to prove so directly by transforming the spatial volume containing a sample of photons. This is worked out in the addendum to this chapter, Section 8.

This result will allow us to obtain directly the specific intensity in Doppler-boosted (beamed) radiation.

Doppler boosting

The observer of radiation from a relativistically moving source may be within or outside the beam opening angle, which can be quite narrow. The intensity within the beam can be greatly enhanced ("boosted") owing to both the beaming and the Doppler shifting of the frequency. Outside the beam it can be either boosted or deboosted as determined by the angular location of the observer and the speed of the source. Here we calculate the change in the measured specific intensity ($\text{W m}^{-2} \text{ Hz}^{-1} \text{ sr}^{-1}$) and related quantities due to these effects.

Doppler factor δ

Consider again a stationary, continuously emitting source at rest in S'. Define for convenience the factor δ by which the frequency is Doppler shifted, from (37),

$$\Rightarrow \quad \delta \equiv \frac{\nu_{ob}}{\nu_{em}} = \frac{(1 - \beta^2)^{1/2}}{1 - \beta \cos\theta} \xrightarrow[\theta=0]{} \left(\frac{1 + \beta}{1 - \beta}\right)^{1/2}, \quad \text{(Relativistic Doppler shift)} \tag{7.70}$$

where $\beta \equiv v/c$ is the speed parameter of the frame S' as well as that of the radiating source, as measured in frame S. The propagation vector k in S is at angle θ relative to the source velocity (Fig. 4). The limit for an observer at $\theta = 0$ is also given in (70). For relativistic speeds, $\beta \approx 1$, or equivalently, $\gamma \gg 1$, the $\theta = 0$ result becomes

$$\delta = \left(\frac{1 + \beta}{1 - \beta}\right)^{1/2} \left(\frac{1 + \beta}{1 + \beta}\right)^{1/2} = \frac{1 + \beta}{(1 - \beta^2)^{1/2}} \xrightarrow[\beta \approx 1]{} 2\gamma, \quad (\theta \approx 0; \beta \approx 1) \tag{7.71}$$

where γ is the Lorentz factor (3). This relation, $\delta \approx 2\gamma$, is a useful approximation for highly relativistic jets viewed nearly head-on.

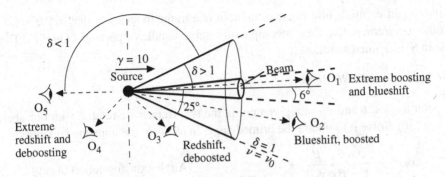

Figure 7.10. Five observers in a stationary frame viewing radiation from a fast-moving ($\gamma = 10$) source that emits isotropically in its rest frame. The "beam," defined in (63), includes half the rays that, in the observer frame, lie within the inner cone of half-angle, $\sim 1/\gamma = 5.7°$. The outer cone of half-angle $25.2°$ represents a null Doppler shift; $\delta = 1$, and so $\nu = \nu_0$ (70). A randomly located observer will most probably view redshifted and deboosted radiation, though it could be below detection threshold.

In general, the Doppler factor (70) tells us whether radiation is blueshifted to higher frequencies ($\delta > 1$) or redshifted to lower frequencies ($\delta < 1$) as a function of the source speed and the detected propagation angle. We will find that it also tells us that the specific intensity $I(\nu)$ and flux density $S(\nu)$ are boosted to higher values if $\delta > 1$ and deboosted to lower values if $\delta < 1$.

Boosting and deboosting angles

In Fig. 10, five observers view radiation from a fast-moving ($\gamma = 10$) source. A pulse is emitted isotropically in the source frame. The beam cone has angular radius $\theta_b \approx 1/10 \, \text{rad} = 6°$, from (63), and contains all photons emitted into the forward hemisphere in the source frame. The outer $25°$ cone contains all photons that are blueshifted. Its half-angle is obtained by setting $\nu/\nu_0 = 1$ in the Doppler formula (70) (Prob. 64).

Observers O_1 and O_2 see a blueshifted signal with enhanced (boosted) intensity as we shall demonstrate; see (78) below. Observer O_3 sees a redshifted and deboosted signal even though, at the time of emission, the source was approaching the observer. (The radiation arrives from the left.) This is less puzzling than might at first appear to be the case if one realizes that all detected photons outside the $6°$ beam are emitted in the rearward direction in the source rest frame. Observers O_4 and O_5 view emission from a receding source, which is highly reddened and strongly deboosted.

Solid angle

The detected specific intensity is greatly affected by the solid angle into which the radiation is beamed. We have seen that a relativistic isotropic radiator will beam half its photons into a cone of angle $\theta_b \approx 1/\gamma \, \text{rad}$ (63) in the frame of a stationary observer. For large γ, this dramatically reduces the solid angle for these rays and hence greatly increases the emitted power per steradian – that is, the surface brightness (W m^{-2} Hz^{-1} sr^{-1}). The detected specific intensity is similarly increased because it is equivalent to the surface brightness (3.27).

Find the differential solid angle element in one frame in terms of that in the other frame with the requirement that they encompass the same bundle of photons (Fig. 9c). The solid angle in S′ is defined as follows:

$$d\Omega' = \sin\theta' d\theta' \, d\phi' = -d(\cos\theta') \, d\phi'. \tag{7.72}$$

The polar angles, θ and θ', of a given ray in the two frames are related with our aberration formula (56). Solve it to obtain the primed angle in terms of the unprimed quantities:

$$\cos\theta' = \frac{\cos\theta - \beta}{1 - \beta\cos\theta}. \qquad \text{(Angle transformation of ray)} \tag{7.73}$$

We discussed this above in the logic leading to (59).

Substitute (73) into (72) and take the differential to obtain, after some manipulation,

$$d\Omega' = \frac{1 - \beta^2}{(1 - \beta\cos\theta)^2} \sin\theta \, d\theta \, d\phi = \delta^2 d\Omega, \tag{7.74}$$

where δ is the Doppler shift parameter defined in (70) and where we took note that the azimuthal angle is invariant under the transformation, that is, $d\phi = d\phi'$. (This is the angle measured about, and normal to, the translation direction.) Thus, the transformation of solid angles tells us that a differential solid angle element in the observer frame (S) is less (for $\delta > 1$) than that in the emission frame (S′) precisely by the factor δ^2:

$$\Rightarrow \qquad d\Omega = \frac{d\Omega'}{\delta^2}. \qquad \text{(Transformation of solid angle)} \tag{7.75}$$

The two solid angles, $d\Omega$ and $d\Omega'$, contain the very same rays (or photons) because this result (75) is based on (73), the transformation for a single ray. The two solid angles lie at the appropriately different angles in the two frames. The element $d\Omega'$ in S′ will thus be folded forward and compressed in S because of the beaming effect if $\delta > 1$.

The δ factor is the Doppler shift factor by which frequencies in the two frames differ (70). As noted after (71), it can be greater than, or less than, unity. If less than unity, the solid angle in S is folded forward (as for $\delta > 1$), but it is larger in S than in S′. This occurs for solid-angle elements that are in the redshifted zone in S (Fig. 10). Note again that these elements are all directed into the rearward hemisphere in S′.

Specific intensity

Now, let us compare the specific intensities, $I'(\nu', \theta')$ and $I(\nu, \theta)$ (W m^{-2} Hz^{-1} sr^{-1}) in the two frames of reference. We have found (3.26) that the relation between specific intensity and the distribution function is $I = h^4 \nu^3 f / c^2 \propto \nu^3 f$. Because f is a Lorentz invariant (69), we also have

$$\frac{I(\nu)}{\nu^3} = \text{Lorentz invariant}, \tag{7.76}$$

or, in the two frames,

$$\frac{I(\nu)}{\nu^3} = \frac{I'(\nu')}{(\nu')^3}. \tag{7.77}$$

Because the ratio of the frequencies in the two frames is, by definition, $\delta = \nu/\nu'$ (70), we have from (77)

➧ $$I(\nu, \theta, \phi, t) = \delta^3 I'(\nu', \theta', \phi', t'), \qquad \text{(Specific-intensity transformation;} \atop \text{W m}^{-2} \text{ Hz}^1 \text{ sr}^{-1}) \qquad (7.78)$$

where δ can lead either to a boost or a deboost of the intensity as determined by the angle and frame speed. This applies to a single, continuously radiating source traveling at relativistic speed, where the primed and unprimed variables are related by the transformations (37), (56), and x1.4 of Table 1.

Some insight into the factor δ^3 may be gained from the following. For convenience, we consider radiation in a forward direction, where $\delta > 1$, so that frequency and intensity are boosted to higher values.

(i) The transformed solid angle is rotated forward and is reduced by δ^{-2}, that is, $\Delta\Omega = \delta^{-2}\Delta\Omega'$ (75). Thus, the power per unit solid angle increases by a factor δ^2.

(ii) The frequencies of the individual photons are Doppler boosted by the factor δ (37). Hence, the detected energies, $h\nu$, and therefore the intensity, are increased by a factor δ (i.e., $\nu/\nu' = \delta$).

(iii) The time interval during which a group of photons arrives is compressed by the factor δ because of relativistic Doppler compression, $\Delta t_{\text{obs}} = \delta^{-1}\Delta t'$ (36), thus increasing the rate of detected photons, and hence the intensity, by a factor δ.

(iv) The bandwidth $\Delta\nu'$ in the source frame is increased proportionally with the photon frequency, and hence by a factor δ (i.e., $\Delta\nu/\Delta\nu' = \delta$). This changes (reduces) the intensity (power per unit frequency interval) by a factor δ^{-1}.

These four items produce the net δ^3 change given in (78).

Photon conservation

The astute reader may realize that we might have used these factors in a different approach to (78) entailing the conservation of numbers of a specific set of photons, as we did for blackbody radiation (6.65). Describe the number N of a specific set of photons in terms of the specific intensities in the primed and unprimed frames of reference and equate them, setting $N = N'$:

$$\frac{I(\nu, \theta, \phi, t)}{h\nu} \Delta t \, \Delta\nu \, \Delta\Omega \, \Delta A_n = \frac{I'(\nu', \theta', \phi', t')}{h\nu'} \Delta t' \Delta\nu' \Delta\Omega' \Delta A_n'. \qquad (7.79)$$

The division by $h\nu$ converts from energy intensity to number intensity. The differential quantities follow directly from the definition of specific intensity (W m^{-2} Hz^{-1} sr^{-1}). The area factors, ΔA_n and $\Delta A_n'$, are surfaces normal to the respective lines of sight (Fig. 6.5).

Solve (79) for $I(\nu)$ and apply the four items listed above to the various ratios (e.g., $\Delta t'/\Delta t = \delta$). The desired relation (78) follows if one further assumes the ratio of source areas normal to the line of sight is unity $\Delta A_n'/\Delta A_n = 1$. This ratio is shown to equal unity through the Lorentz transformation of sample photon trajectories in the addendum to this chapter (Section 8). In fact, the derivation here, through the Lorentz invariance of the distribution function, effectively demonstrates that this ratio of areas must be unity.

Meaning of "boosting factor"

It is important to realize that the transformation (78) refers to the same photons in each frame. In other words, I' is the intensity in the source frame of photons emitted into θ', ϕ' at ν', whereas I is the intensity at the forward-tilted angular position θ, ϕ and shifted frequency ν of these same photons.

Suppose observer O_3 (Fig. 4) records intensity $I(\theta)$ from the approaching source, which is shown at its position when the indicated radiation was emitted. One might be tempted to think that I' (78) is the intensity O_3 would measure if the source were, hypothetically, at rest in the observer frame and out there at the same indicated location. In fact, the radiation emitted in direction θ' from this stationary source would pass far to the left of O_3; note the k' vector in the figure. Only if the source were farther along its track would the radiation intercept O_3.

In the special case of isotropic emission in S', the emission I' in (78) can be interpreted as the specific intensity that would be detected if the source were at rest at its emission position. However, the prevalence of anisotropic jets in many systems makes this an unwarranted assumption.

The point to remember is that the expression (78) relates the observed intensity I to a hypothetical intensity I' – either that measured by an observer racing along with the source at the appropriate angle or that measured by a stationary observer with the source artificially at rest at the appropriate position along its path. In either case, the hypothetical S' detector must be set to the appropriate unshifted frequency.

Spectral flux-density

Often the angular size of a source can not be measured, and so the specific intensity $I(\nu)$ can not be determined, whereas the spectral flux-density $S(\nu)$ (W m^{-2} Hz^{-1}) can be. Do not confuse spectral flux-density, $S(\nu)$ and $S'(\nu')$, (italic type) with the frame of reference designations, S and S', (roman type).

Consider the specific intensity at the telescope from the viewpoint of observer O_3 in the stationary frame S (Fig. 4). The specific intensity I is flux per unit solid angle. Thus, multiply the (unknown) I by the (unknown) solid angle of the source $\Delta\Omega_s$ to get an expression for the spectral flux-density,

$$S(\nu) = I(\nu, \theta)\Delta\Omega_s = I(\nu, \theta)\frac{\Delta A_n}{D_A^2}. \qquad \text{(Spectral flux-density; W m}^{-2}\text{ Hz}^{-1}\text{)} \qquad (7.80)$$

The solid angle is written here in terms of the surface area of the radiating source, ΔA_n, normal to the line of sight and the angular-diameter distance D_A to the source in the frame S of O_3. (See more on D_A after (90) and after (92)).

We know that $I = \delta^3 I'$ (78), where I' is the intensity in the rest frame (S') of the source. This can be invoked if the source is viewed from the emission direction θ' in frame S'. We also can write $\Delta A_n = \Delta A_n'$ because the area element normal to the line of sight is a Lorentz invariant; see Section 8.

If, also, the distance D_A' from telescope to source in frame S' is taken to be the same numerical value as the distance in frame S (i.e., D_A' equals D_A, *not* the transformation of D_A), then we would have equal solid angle elements, $d\Omega_s = d\Omega_s'$. Apply these expressions for I and $\Delta\Omega_s$ to (80) and note that $I'\Delta\Omega_s' = S'$:

$$S(\nu) = \delta^3 I'(\nu', \theta')\Delta\Omega_s' \qquad \text{(Spectral flux-density transformation;}$$
$$\text{➡} \quad S(\nu) = \delta^3 S'(\nu'). \qquad \text{W m}^{-2}\text{ Hz}^{-1}\text{)} \qquad (7.81)$$

The flux density thus transforms, with these restrictions, in the same manner as the specific intensity (78) – that is, as δ^3.

This is the result we desire, but note again, that, for a given observed flux $S(\nu, \theta)$, the flux $S'(\nu', \theta')$ in (81) is rather hypothetical because the observer in frame S' must measure the flux at the correct angle θ' and frequency ν', and from the correct distance $D'_A = D_A$ (e.g., at 100 MLY if it is measured at 100 MLY in S.)

Flux density

Now, consider the transformation of the energy flux-density $\mathscr{F}$ (W/m^2). This quantity is the product of the spectral flux-density $S(\nu, \theta)$ and the chosen frequency band $\Delta\nu$,

$$\mathscr{F}(t) = S(\nu)\Delta\nu = \delta^3 S'(\nu')\,\delta\Delta\nu', \qquad (7.82)$$

where we invoked the transformations for $S(\nu)$ (81) and $\Delta\nu$, the latter from the differential of (70). Over a broad band of frequency, one would integrate, and so $\mathscr{F} = \int S\,d\nu$.

Recognize $\mathscr{F}' = S'\Delta\nu'$ in the right-hand term of (82) to obtain the desired flux transformation

➡ $$\mathscr{F}(t) = \delta^4 \mathscr{F}'(t'). \qquad \text{(Flux-density transformation; W/m}^2\text{)} \qquad (7.83)$$

The additional factor of δ in (83) may be viewed as being due to the inapplicability of the δ^{-1} bandwidth factor described in item (*iv*) in the list following (78). In measuring $\mathscr{F}$, one does not divide the received power by the frequency bandwidth.

The flux density $\mathscr{F}'$ includes all photons in some chosen band in frame S', whereas $\mathscr{F}$ includes all the same photons over the associated broader (or narrower) frequency band in frame S. Consider a radiating knot of plasma traveling in a jet with $\delta \approx 10$. The specific intensity would be increased a factor of 1000 and the flux density a factor of 10 000 relative to that detected if the source were at rest and the observer properly placed with detector properly tuned to detect the same photons.

K correction

The transformations (78) and (81) yield the intensity or flux at the transformed observer frequency in terms of the intensity or flux at the untransformed source frequency. It is sometimes useful or convenient to compare the fluxes at the *same* frequency in both frames. For example, one may wish to use observations in frame S at some frequency (e.g., $\nu = 5$ GHz) to estimate the flux in the source frame S' at the same frequency ($\nu' = 5$ GHz).

In this case, different photons are being observed in the two frames because different portions of the emitted spectrum are sampled. Transformation formulas must therefore contain information about the spectral shape. This adjustment for spectral shape is known as the *K correction*.

What happens to the shape of a spectrum, $S'(\nu')$, upon transformation to the observer's frame? Consider the case of Doppler upshift of frequency ($\delta > 0$). According to (81), the shifted magnitude of the spectrum is a factor of δ^3 greater than the unshifted magnitude if it is measured at a frequency greater by a factor of δ.

On a logarithmic plot, a change of a value by a given factor is a fixed distance at any value. In our case, δ and δ^3 serve as fixed multiplicative factors. Thus, on a log-log spectral plot of $\log S$ versus $\log \nu$, each point of the curve will shift to the right by a fixed amount, $\log \delta$, and upward by the fixed amount, $\log \delta^3$. This preserves the spectral shape on a log-log plot, as illustrated in Fig. 11a.

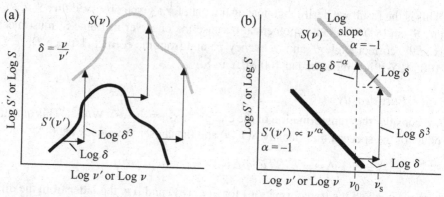

Figure 7.11. Spectral transformations plotted against coincident log ν and log ν' abscissas. (a) Arbitrary spectral shape. On a log-log plot, the emitted spectrum (dark line) is shifted by fixed factors in both horizontal and vertical directions by the relativistic Doppler transformation (81); see text. The shape of the transformed spectrum (shaded line) is thus preserved on such a plot. (b) Power-law spectrum similarly transformed. In the latter case, comparison of spectral flux-densities $S'(\nu')$ and $S(\nu)$ at the same frequency $\nu' = \nu = \nu_0$ introduces an additional factor of $\delta^{-\alpha}$ known as the K correction. These diagrams are for Doppler factors that upshift the frequency ($\delta > 0$).

A spectral shape often found in practice is a power law, $S' \propto \nu'^{\alpha}$, where α is a constant; this spectrum is a straight line with slope α on a log-log plot. Synchrotron sources often have power-law spectra (Chapter 8) with negative values of α wherein the flux density decreases with energy. (The *spectral index*, which is the negative of the logarithmic slope, is often used; pay attention to the definition of α.)

For example, consider the decreasing ($\alpha < 0$) power-law spectrum $S' \propto \nu'^{\alpha}$ of Fig. 11b, where $\alpha = -1$ and again $\delta > 0$. Carry out the transformation, as in Fig. 11a, to obtain the upper spectrum, which is also a power law with logarithmic slope -1.

Now, let us compare fluxes of the shifted and unshifted spectra at some emission frequency ν_0, our initial objective. Rather than quoting the value of S at the shifted frequency $\nu_s = \delta \nu_0$, one quotes the higher value at the lower frequency ν_0. Decreasing the frequency by the factor δ invokes a flux adjustment by a factor $\delta^{-\alpha}$.

Apply this additional correction to the transformation of the spectral flux density (81) and generalize to any frequency ν:

➡ $$S(\nu) = \delta^{3-\alpha} S'(\nu).$$ (Transformation; fixed frequency
for power-law spectrum)   (7.84)

Carefully note the unprimed frequency arguments in (84); in this case, the spectral flux-density is measured at the same frequency in both the source (primed) and observer (unprimed) frames.

Superluminal motion

Ejections of material in jets can be steady or episodic. Sources with radio lobes often show evidence of multiple outbursts from the central object. We found in Fig. 8 that, with the high angular resolution of radio interferometry, astronomers are sometimes able to follow a bright

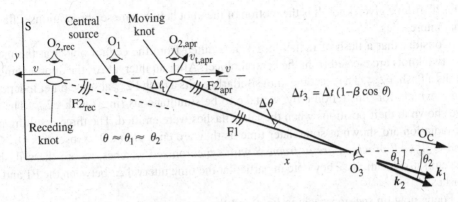

Figure 7.12. Geometry for superluminal transverse motion with five observers, four in one frame S with synchronized watches and one, O_C, at a cosmological distance. The observers O_1, $O_{2,apr}$, and O_3 are those seen in Fig. 2 except that, in this case, O_3 is offset from the source track.

knot as it moves away from a central object. This occurs on time scales of days for galactic objects to a few years for extragalactic objects.

The example in Fig. 8 shows hot spots or "knots" of radio emission moving away from the *microquasar* GRS 1915+105 over days. The two principal knots in the left panel may have been ejected from the black hole in opposing directions. If this is the case, and if the central object is not moving rapidly, the distance to one is lessening whereas that to the other is increasing.

In some cases, such as in Fig. 8, the angular rate of motion of the emitting knot, together with the independently determined distance of the source, suggests motions transverse to the line of sight that are faster than the speed of light. This *superluminal motion* is an apparent violation of special relativity.

We now understand that this phenomenon is due simply to the effect of beaming if the emitting plasma is moving relativistically almost directly toward the observer. The approaching source leads to a compression of time intervals (the classical Doppler effect), and this gives an exaggerated measure of angular velocity. This explanation takes place entirely in the observer's frame S; it does not invoke the Lorentz transformations.

Apparent transverse velocity

Let us find the apparent transverse velocity as a function of the physical knot speed $\beta = v/c$ and the angle θ by which the radiation propagation direction is offset from the knot velocity vector (Fig. 12). The figure shows a stationary (in the frame S) central object (dark circle) and two simultaneously ejected knots of material (open ovals), one moving to the left and one to the right, each with velocity v, again in frame S. The former is receding from the observer O_3, whereas the latter is approaching.

Ejected knots will typically radiate continuously as they move away from the central object. For our purposes, though, we consider that they emit flashes of radiation that will be detected, after transit delays, by O_3. The radiation received during any single observation, such as that at 16:20 UT on March 27 (Fig. 8), can be considered to be the detection of a "flash" from a given knot. It is the motion of the knot between these observations ("flashes") that interests us.

Consider that a flash of radiation, F1, is emitted from the central object at the moment the two knots are ejected from the central object. After an interval Δt, the approaching knot emits a flash, $F2_{apr}$. The receding knot similarly emits a flash $F2_{rec}$ at some other independent time, which, for illustration only, we take to be simultaneous (in S) with $F2_{apr}$. The knots are shown at their positions when the latter flashes were emitted. The three resulting pulses of radiation are shown at some later time as they are en route to O_3. Observers O_1, $O_{2,rec}$, and $O_{2,apr}$, *all of whom are in frame S* with synchronized watches, record these flashes at the emission locations. They note in particular the time interval Δt between the F1 and $F2_{apr}$ flashes.

Focus now on the approaching knot and the two pulses from it, F1 and $F2_{apr}$. (For our purposes, the earlier F1 flash could equally well have been emitted by the knot after ejection.) The two pulses follow the tracks shown in Fig. 12 and are eventually detected by the single observer O_3, who is also in frame S with a watch synchronized to the others. Observer O_3 measures a (Doppler-compressed) time interval Δt_3 and slightly displaced propagation directions, $\Delta\theta = \theta_2 - \theta_1$.

The two propagation directions are at almost the same angle $\theta \approx \theta_1 \approx \theta_2$ because we assume that the knot moved very little in the sky between the flashes. Nevertheless, they are sufficiently displaced from one another so that their angular separation is detectable with radio interferometry. (In Fig. 4, the two tracks are drawn as a single dashed line.)

The linear velocity component transverse to the O_3 line of sight for the approaching knot is the transverse displacement, $\Delta\ell_t = v \sin\theta \, \Delta t$, divided by the time interval between flashes, Δt, or

$$v_{t,apr} = \frac{\Delta\ell_t}{\Delta t} = \frac{v \sin\theta \, \Delta t}{\Delta t} = v \sin\theta, \qquad \text{(Actual transverse velocity)} \qquad (7.85)$$

where the subscript "apr" refers to the approaching knot.

Observer O_3 at a single location well removed from the two flash events measures the Doppler-compressed interval, Δt_3 (36), between the arriving pulses,

$$\Delta t_3 = (1 - \beta \cos\theta) \Delta t. \qquad \text{(Doppler-compressed time interval)} \qquad (7.86)$$

Observer O_3 also measures the angular displacement $\Delta\theta$ of the two arrival directions and thus with a distance to the (central) source is able to arrive at an *apparent tranverse velocity*, $v_{t,a,apr}$. This is worked out below (91).

It is $v_{t,a,apr}$ that can exceed the speed of light c. One can see this by writing $v_{t,a,apr}$ in terms of the velocity parameters of the approaching knot, from (85) and (86), as follows:

$$v_{t,a,apr} \equiv \frac{\Delta\ell_t}{\Delta t_3} = \frac{v \sin\theta \, \Delta t}{(1 - \beta \cos\theta) \Delta t} = \frac{v \sin\theta}{1 - \beta \cos\theta}. \qquad \begin{array}{l}\text{(Apparent transverse} \\ \text{velocity for } O_3)\end{array} \qquad (7.87)$$

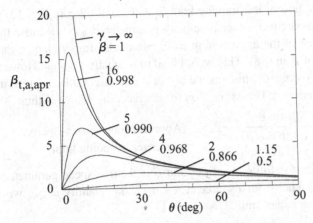

Figure 7.13. Plot of the apparent transverse velocity parameter $\beta_{t,a,apr}$ (88) versus angle θ between the direction of the emitter velocity and the propagation direction (to observer) for various values of (approaching) cloud velocities v, where $\beta = v/c$ and $\gamma = (1 - \beta^2)^{-1/2}$. For moderately small angles θ and large values of β, the apparent motion is superluminal (i.e., $\beta_{t,a,apr} > 1$).

Note how this differs from the actual transverse component (85). Divide (87) by the speed of light to obtain $\beta_{t,a,apr}$, the apparent transverse velocity parameter of the approaching knot,

$$\beta_{t,a,apr} \equiv \frac{v_{t,a,apr}}{c} = \frac{\beta \sin \theta}{1 - \beta \cos \theta}, \qquad \text{(Apparent transverse velocity parameter; approaching knot)} \qquad (7.88)$$

where $\beta = v/c$ is the true speed parameter of the knot and θ specifies the radiation propagation direction relative to the direction of the knot velocity (Fig. 12, where $\theta \approx \theta_1 \approx \theta_2$). For certain values of β and θ (Fig. 13), $\beta_{t,a,apr}$ can take on values greater than unity, giving the appearance of superluminal motion, $v_{t,a,apr} > c$.

Figure 13 is a plot of $\beta_{t,a,apr}$ for several values of β, or $\gamma = (1 - \beta^2)^{-1/2}$. The expression (88) is driven to zero at $\theta = 0$ by the $\sin \theta$ term as expected because there is no transverse motion at $\theta = 0$. At angles, $\theta \approx \pi/2$, it takes on the value β, again as expected, inasmuch as the motion in this case is purely transverse; there is no Doppler compression. These limiting cases are "subluminal," meaning $\beta_{t,a,apr} < 1$.

In contrast, at moderate values of θ (e.g., 5°–10°), and for values of β approaching unity, the denominator can drive $\beta_{t,a,apr}$ above unity, and the motion can appear superluminal. For example, if $\beta = 0.990$, the expression (88) yields a maximum value of $\beta_{t,a,apr} = 7.0$ at $\theta = 8.1°$. The apparent transverse velocity is seven times the speed of light! Interestingly, if $\beta < 1/\sqrt{2}$, it can never appear superluminal (Prob. 62).

In essence, the superluminal effect arises because of the quickly decreasing path length between a rapidly approaching knot and the observer. The detected pulses are observed at greatly reduced intervals by O_3 because the second pulse has a shorter distance to travel. The artificially shorter time interval makes it seem as if the cloud is moving faster in the transverse direction than it really is. We emphasize again that our calculations were exclusively in the observer's frame of reference S; relativistic time dilation did not enter into consideration.

Two flashes from the receding or left-moving knot (e.g., F1 and F2$_{rec}$ in Fig. 12) yield a similar expression for the apparent transverse velocity parameter $\beta_{t,a,rec}$. Because the knot velocity is reversed from that of the approaching case, one could simply change the signs preceding the appearances of β in (88). This would lead to a negative $\beta_{t,a,rec}$. However, we choose to define $\beta_{t,a,rec}$ as positive definite, as we do for β. To effect this, we refrain from changing the sign in the numerator. The expression for the receding source is thus

➡ $$\beta_{t,a,rec} = \frac{v_{t,a,rec}}{c} = \frac{\beta \sin\theta}{1 + \beta \cos\theta},$$ (Apparent transverse velocity parameter; receding knot)   (7.89)

where $\beta_{t,a,rec}$ and β are positive definite quantities and β is the true speed parameter of the receding knot. For a receding knot from a source at $\theta < \pi/2$, the quantity $\beta_{t,a,rec}$ will never exceed unity; it can not appear superluminal.

Knot speed and direction

The quantities $\beta_{t,a,apr}$ and $\beta_{t,a,rec}$ are useful because they can be obtained directly from measurements. In turn, they provide, through (88) and (89), constraints on the possible values of β and θ. For example, if $\beta_{t,a,apr} = 16$, Fig. 13 tells us that γ must be greater than ~16 and θ less than ~7°. If two ejections are assumed to be back-to-back with identical speeds (as in Fig. 12), measurement of both $\beta_{t,a,apr}$ and $\beta_{t,a,rec}$ and application of (88) and (89) lead to unambiguous values of β and θ (Prob. 63).

Measured quantities

The final step in our logic is to ascertain what measured quantities determine $\beta_{t,a,apr}$ or $\beta_{t,a,rec}$. Consider first our observer O$_3$ who is in the general vicinity, cosmologically speaking, of the central source. Assume a flat Euclidean universe.

The observer O$_3$ can measure an angular displacement $\Delta\theta = \theta_2 - \theta_1$ (Fig. 12) of the moving approaching knot in the interval Δt_3 between the two pulse detections. If the (angular-diameter) distance D_A to the source from the observer is independently known, the transverse displacement inferred by O$_3$ is

$$\Delta \ell_t = D_A \Delta\theta.$$
(7.90)

The *angular-diameter distance* D_A is the ratio of the physical and angular diameters of an astronomical object. It is one of several distance measures in GR.

The apparent transverse speed parameter (88) can thus be written in terms of measurable parameters as

$$\beta_{t,a,apr} = \frac{\Delta \ell_t}{c \Delta t_3} = \frac{D_A \Delta\theta}{c \Delta t_3}.$$ (Apparent transverse velocity parameter for "local" observer)   (7.91)

This is the measured quantity with which one enters (88) to deduce information about β and θ. For extragalactic objects, the distance D_A is obtained from the redshift. For galactic objects, it must be obtained indirectly with other methods (see AM, Chapter 9).

Cosmological correction

A cosmological correction is needed for an extragalactic object. Consider a cosmologically distant, earth-located observer, O$_C$, off to the right in Fig. 12, and take into account the expansion of the universe according to GR: Two effects come into play: time intervals and redshift-distance relations.

Time intervals are stretched as photons travel in the expanding universe. For a source at a redshift distance $z \equiv (\lambda - \lambda_0)/\lambda_0$, wavelengths are increased by the factor $1 + z$ (50) as are time intervals (e.g., the time between peaks of a wave.) The cosmological observer O_C thus detects longer intervals than does the "local" observer O_3. Conversely, the local observer measures a shorter interval than does O_C, $\Delta t_3 = \Delta t_C/(1 + z)$.

The transverse displacement $\Delta \ell_t$ can be written in terms of parameters measured by O_C, $\Delta \ell_t = D_{A,C} \Delta \theta_C$. Thus we can rewrite $\beta_{t,a,apr}$ (91) as

$$\beta_{t,a,apr} = \frac{D_{A,C} \Delta \theta_C}{c \Delta t_C}(1 + z), \tag{7.92}$$

The distance $D_{A,C}$ is a function of z that depends on the expansion model of the universe. For a universe that is critical (flat geometry) and with zero cosmological constant, an angular-size distance is related to redshift z as

$$D_A = \frac{c}{H_0} \frac{2}{(1 + z)} \left[1 - \frac{1}{(1 + z)^{1/2}} \right], \qquad \begin{array}{l}\text{(Angular-diameter distance} \\ \text{for } \Omega_0 = 1, \Lambda = 0)\end{array} \tag{7.93}$$

where H_0 is the Hubble constant at the present epoch. (We encounter this relation again in (12.51)).

Substitute (93) into (92) to obtain the apparent transverse velocity $\beta_{t,a,apr}$ *for a local observer (O_3)* in terms of the quantities $\Delta \theta_C, \Delta t_C$, and z measured by a cosmologically distant earth-based observer; that is,

$$\beta_{t,a,apr} = \frac{\Delta \theta_C}{\Delta t_C} \frac{2}{H_0} \left[1 - \frac{1}{(1 + z)^{1/2}} \right], \qquad (\Omega_0 = 1, \Lambda = 0) \tag{7.94}$$

where H_0 and Δt_C have units (s^{-1}) and (s) respectively, whereas $\Delta \theta_C$ (rad) and z are dimensionless.

This, again, is the parameter $\beta_{t,a,apr}$ that may be used in (88) for an approaching knot at redshift z in a critical universe with Hubble constant H_0 and zero cosmological constant. This too can be used to place limits on β and θ.

Comment on other jet models

The jet model discussed in this chapter is quite basic, yet fundamental. It consists of one or two discrete radiating sources moving with relativistic speeds. There are numerous variants that would yield different results. Nevertheless, the principles illustrated here must be incorporated into such models.

7.7   Magnetic force and collisions

Forces and collisions play important roles in astrophysics. Examples are the frequency with which an energetic particle circulates in a magnetic field and the energy thresholds for absorption of extremely high-energy photons or cosmic-ray protons by the low-energy photons of the cosmic microwave background (CMB). Special relativity is required to calculate these quantities properly, and the results profoundly affect the kinds of astronomy that can be carried out.

Relativistic cyclotron frequency

Here we find the angular frequency of a particle circulating in a uniform magnetic field at a relativistic speed. At speeds $v \ll c$, a particle will circulate or spiral about the field lines with a *cyclotron frequency* $\omega_B = qB/m$, which is independent of the particle energy as can easily be demonstrated. (Equate the magnetic force qvB to the centripetal force mv^2/r and note that $v = \omega r$, where r is the radius of the circular path.) At relativistic speeds, the cyclotron frequency is no longer independent of particle energy. This is pertinent to the observed frequency dependence of synchrotron radiation (Chapter 8).

Equation of motion

A particle of charge q moving with velocity v in a magnetic field B will experience the force

$$\boldsymbol{F} = q\boldsymbol{v} \times \boldsymbol{B}. \tag{7.95}$$

The cross product indicates that there is no force either in the direction of the magnetic field or in the direction of the velocity. In a uniform magnetic field, therefore, the electron will drift along the field with constant speed, but the force will modify the components of motion normal to the magnetic field direction. The particle will spiral around the field line. If it has no velocity component in the direction of the field, its track will be a simple circle.

Relativistic dynamics must be employed if the particle has a speed that is a significant fraction of the speed of light. In that case, we write the equation of motion, $\boldsymbol{F} = \mathrm{d}\boldsymbol{p}/\mathrm{d}t$, as

$$q(\boldsymbol{v} \times \boldsymbol{B}) = \mathrm{d}\boldsymbol{p}/\mathrm{d}t, \qquad \text{(Newton's second law)} \tag{7.96}$$

where $\boldsymbol{p}$ is the relativistic vector momentum (14),

$$\boldsymbol{p} = \gamma m \boldsymbol{v} = \gamma \boldsymbol{\beta} mc. \tag{7.97}$$

Here, m is the particle mass, γ is the relativistic Lorentz factor (3), and $\boldsymbol{\beta} = \boldsymbol{v}/c$.

For the case of circular (not spiral) motion in a uniform magnetic field, the velocity v is perpendicular to B. The cross product is in the radial direction toward the center of the circle. From the geometry in Fig. 14, the magnitude of the momentum change may be written as

$$\Delta p = p\Delta\theta, \tag{7.98}$$

and its rate of change as

$$\frac{\mathrm{d}p}{\mathrm{d}t} = p\frac{\mathrm{d}\theta}{\mathrm{d}t} = p\omega, \tag{7.99}$$

where ω is the angular velocity. The equation of motion (96) thus becomes, for circular motion,

$$\qquad qvB = p\omega, \qquad \text{(Equation of motion; circular track)} \tag{7.100}$$

where we equated the magnitudes of the vector quantities.

Angular velocity

The angular velocity of rotation of the relativistic particle follows directly from (100) and the relativistic momentum (97),

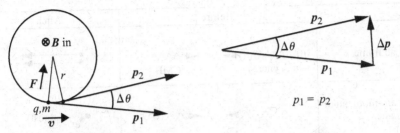

Figure 7.14. Change of momentum vector Δp for a positively charged ($q > 0$) mass m in circular motion at constant speed v in a uniform magnetic field B directed into the paper. The geometry yields $dp/dt = p\omega$.

$$\omega = \frac{qvB}{p} = \frac{qvB}{\gamma mv} = \frac{qB}{\gamma m}, \tag{7.101}$$

or, for the case of an ion of charge $q = Ze$, where Z is the atomic number and e the electron charge,

$$\omega = \frac{ZeB}{\gamma m}. \qquad \text{(Relativistic cyclotron angular frequency)} \tag{7.102}$$

This expression is also valid for a spiraling electron (i.e., one with the component of momentum along the field line). This follows if one considers the speed v and momentum p in this derivation to be the components normal to the field line, $v_\perp$ and $p_\perp = \gamma m v_\perp$, where $\gamma = U/mc^2$ represents the full energy of the particle. The terms $v_\perp$ cancel as in (101). The frequency of spiraling is thus independent of pitch angle.

We find from (102) that the angular frequency is the cyclotron frequency (qB/m) divided by the relativistic Lorentz factor γ. The angular frequency decreases with increasing energy as $\gamma^{-1} = (U/mc^2)^{-1}$, whereas it is independent of energy at nonrelativistic speeds.

The original cyclotrons (particle accelerators) were designed for nonrelativistic speeds. The oscillating electric fields that accelerated the circulating protons could remain at constant frequency as a group of protons were accelerated to their maximum energy. The next generation of cyclotrons, striving to reach higher energies, featured a variable frequency that decreased at the rate dictated by special relativity so the protons could be accelerated to relativistic speeds.

CMB opacity to high-energy photons and protons

Relativistic particle collisions are often important in astrophysics. Here we consider a very high-energy photon, a gamma ray, colliding with a very low-energy photon of the cosmic microwave background (CMB) to produce a pair of electrons. We also consider, through Prob. 73, the collision of a high-energy cosmic-ray proton with a CMB photon to produce a pi meson. In each case, we calculate the threshold energy for the reaction but not the cross sections. We discuss the implications for astronomy in each case.

Photon absorption through pair production

Photons must be able to traverse the Galaxy and beyond if one wishes to do extragalactic astronomy with them. The universe is quite transparent for much of the electromagnetic spectrum with the notable exceptions of the scattering of optical photons in dusty regions

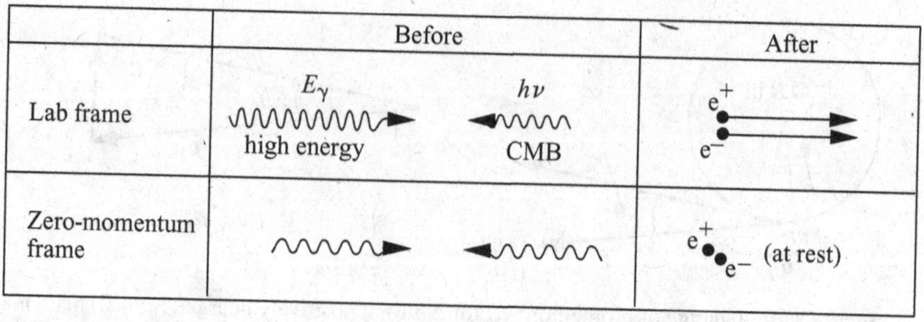

Figure 7.15. Interaction of a gamma ray with energy E_γ and a CMB photon with energy $h\nu$ before and after a head-on collision in the laboratory frame of reference and in the zero-momentum frame of reference. An electron-positron pair (e$^+$, e$^-$) is produced. At threshold, they are at rest in the zero-momentum frame.

and the absorption of ultraviolet photons and low-energy x rays by photoelectric absorption in interstellar gases.

We will find here that it is energetically possible for CMB photons to absorb high-energy photons, those above about 10^{15} eV (1 PeV), through the creation of an electron-positron pair (*pair production*); that is,

$$\gamma_{\mathrm{he}} + \gamma_{\mathrm{cmb}} \rightarrow \mathrm{e}^+ + \mathrm{e}^-. \qquad \text{(Absorption in CMB)} \qquad (7.103)$$

The cross section for this reaction (not derived here) is sufficient to limit extragalactic photon astronomy to energies below the ~1 PeV threshold we calculate here. Do not confuse the usual symbol for a photon, gamma (with subscript label and not italicized here), with the Lorentz factor (italicized without subscript).

The CMB radiation permeates the universe. It has a blackbody spectrum of 2.73 K. The universe became transparent to these photons at the time of hydrogen recombination about 400 000 yr after the big bang when the universe had a temperature of ~3000 K. The expansion of the universe since then has cooled the radiation to the presently observed 2.73 K. Its spectrum follows quite precisely that of a blackbody of this temperature (Fig. 6.1).

Energy threshold

Our special-relativity skills allow us to calculate the threshold gamma-ray energy E_γ required for the production of an electron-positron pair in a head-on collision with a CMB photon of energy $h\nu \approx 10^{-3}$ eV. We make use of the energy-momentum invariant $U^2 - p^2c^2$ (20). In our case, the total energy of the interacting system is U, and p is the total momentum of the system. We will examine the case in which there is just sufficient energy to create the electron-positron pair (Fig. 15) in the zero-momentum inertial frame of reference.

Recall that this quantity $U^2 - p^2c^2$ is not only invariant from one inertial frame to another but that, in any given inertial frame, it is conserved during an interaction because energy and momentum are separately conserved. We thus can equate this quantity's value before the interaction in the laboratory frame to its value after the interaction in the zero-momentum inertial frame:

$$[U^2 - (pc)^2]_{\mathrm{lab,before}} = [U'^2 - (p'c)^2]_{\mathrm{zero\,mom,after}}. \qquad (7.104)$$

We chose the left term because it contains the known incoming energies and momenta and the right because it can be expressed in terms of the known masses of the created electrons.

Before the collision, in the lab frame, the total energy is the sum of the two photon energies, $U = E_\gamma + h\nu$ (Fig. 15). The total momentum, for the head-on collision of the two photons, is $p = (E_\gamma/c) - (h\nu/c)$ because the momenta are oppositely directed. After the interaction at threshold, there must just be sufficient energy in the zero-momentum frame to create the rest-mass energies of two electrons with no excess kinetic energy. Thus, $U_{\text{zero mom}} = 2m_e c^2$, where m_e is the electron (or positron) mass. By definition, the momentum in the zero-momentum frame is zero. The equality (104) thus becomes

$$(E_\gamma + h\nu)^2 - (E_\gamma - h\nu)^2 = (2m_e c^2)^2. \tag{7.105}$$

Solve for E_γ, the threshold energy of the incident gamma ray:

$$4\, h\nu\, E_\gamma = 4\, m_e^2 c^4 \tag{7.106}$$

$$E_\gamma = \frac{m_e^2 c^4}{h\nu}, \qquad \text{(Threshold } \gamma \text{ energy for } e^+ e^- \text{ pair production)} \tag{7.107}$$

where m_e is the electron mass and $h\nu$ is the energy of the target CMB photon. This is the desired threshold energy of the incident gamma ray in a head-on interaction.

Substitute into (107) the rest-mass energy of an electron, $m_e c^2 = 0.51\,\text{MeV}$, and the average photon energy (6.32) for a 2.73-K blackbody, $\langle h\nu \rangle_{\text{av}} = 2.70 \times kT = 6.4 \times 10^{-4}\,\text{eV}$:

$$E_\gamma = \frac{(0.51 \times 10^6)^2}{6.4 \times 10^{-4}} = 4 \times 10^{14}\,\text{eV} = 0.4\,\text{PeV}.$$

$$\text{(Threshold for } e^+, e^- \text{ pair production in } \gamma_{\text{ph}} + \gamma_{\text{cmb}} \text{ interaction)} \tag{7.108}$$

At energies substantially below this, the interaction can not occur, and astronomy would not seem to be adversely affected. At higher energies, it should be seriously impaired.

The actual threshold calculation must take into account all photon energies and directions in the CMB radiation. The absorption increases rapidly as the gamma-ray energy increases beyond 0.4 PeV because more and more of the blackbody photons are subject to this interaction.

Our calculation does not yield the cross section. It turns out that the mean free path is about 26 000 LY at 1 PeV. This is comparable to the 25 000 LY distance to the center of the Galaxy. Thus, astronomy can be carried out at these energies within the closer parts of the Galaxy, but extragalactic astronomy should not be possible.

Gamma rays originating at great distances in our expanding universe will exhibit absorption at lower energies than indicated in (108). Gamma rays and CMB photons observed locally were previously more energetic because their energies decrease (redden) in the expanding universe. Thus, gamma rays we might have observed below threshold at the earth would have been above the absorption threshold energy in distant regions. Their journey to us would likely have been interrupted by a photon-photon collision.

MeV to TeV astronomy

This all may be quite hypothetical because discrete sources of PeV photons are unlikely to be discovered in this era. The decreasing spectra of known sources and the difficulty of detection in known backgrounds argue against it.

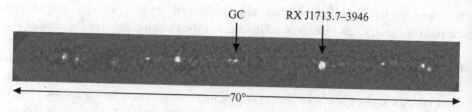

GC   RX J1713.7–3946

←————————————————70°————————————————→

Figure 7.16. Map of the central portion of the galactic plane ($-35° < \ell < +35°$, $-4° < b < +4°$) at gamma-ray energies above about 0.1 TeV by the High-Energy Stereoscopic System (HESS). This instrument is one of a new class of TeV observatories now coming on line that use multiple telescopes to detect the Cerenkov radiation in extensive air showers initiated by energetic gamma rays. Discrete sources cluster on the galactic plane, some with sizes exceeding the telescope resolution of ∼0.1°. The indicated RX source is a supernova remnant, and GC is at the galactic center region of Sgr A*. Some of the others are likely remnants of supernova explosions; the nature of some are not yet known. [Courtesy HESS group; see F. Aharonian *et al.*, *Science* **307**, 1938 (2005); astroph/0504380]

At the lower, but still very high, TeV energies ($\sim 10^{12}$ eV), there are ∼50 known gamma-ray sources at this writing. The gamma-ray photons are emitted mostly from high-energy electrons in supernova remnants and in the jets of active galactic nuclei (AGN). A TeV gamma ray will initiate an extensive air shower (EAS) of photons and electrons in the atmosphere, and the resultant highly directional Cerenkov radiation is detectable with large reflective telescopes. This is known as *TeV astronomy*. Experiments with great sensitivity are now coming on line: the European HESS, the Australian CANGAROO III, and the American VERITAS. The galactic plane as viewed with HESS in 2005 is shown in Fig. 16.

A more local source of gamma-ray photons is relativistic collisions of cosmic ray protons with protons of the interstellar medium. These collisions produce π^0 mesons (rest energy 135 MeV) that immediately decay into two gamma rays, each of 67.5 MeV in the π^0 rest frame. This should produce a glow of ∼100 MeV gamma rays from interstellar clouds lying along the galactic plane. Such gamma rays have been mapped by the EGRET experiment on the Compton orbiting gamma-ray observatory (CGRO; 1991–2000). A followup mission, GLAST, should extend its work. The threshold energy for the production of these π^0 mesons can be calculated (Prob. 72) in a manner similar to that for electron pair production (108).

Cosmic-ray protons and the CMB

Relativistic collisions of cosmic-ray protons of $\gtrsim 100$ EeV (10^{20} eV) with CMB photons produce π mesons, the GZK (Greisen, Zatsepin, Kuzmin) effect (Prob. 73). This limits the travel distance of such protons to ∼150 MLY, which is ∼1% of the Hubble radius. Protons of energies up to ∼300 EeV have been detected through the extensive air showers they create in the atmosphere (AM, Chapter 12). The limited mean free path at these extreme energies indicates that they likely originate in our local region of the universe.

At these extreme energies in these limited distances, the proton paths should not be greatly deviated by galactic and intergalactic magnetic fields. Thus, measurements of the arrival directions of these protons should reveal the source locations if sufficient numbers of the proton EAS events can be recorded (e.g., with the Pierre Auger Observatory, the huge EAS experiment in Argentina with 3000 km² effective area). Large detector areas are needed to acquire these extremely rare events. There is only ~1 EAS per 100 km² per year at energies greater than 100 EeV. Early results from Auger indicate the sources could be nearby active galactic nuclei (AGN).

7.8 Addendum: Lorentz invariance of distribution function (photons)

Here we demonstrate directly that a volume element in x, p phase space containing a given set of photons is a Lorentz invariant. This, because of the fixed photon number, tells us that the photon density in phase space – that is, the distribution function (69) – is also a Lorentz invariant. This is more direct than the approach in Section 6 in which the invariance was demonstrated for particles and then inferred for photons in the limit of $v \rightarrow c$.

The development of this result also demonstrates that the radiating area normal to the propagation direction is also a Lorentz invariant. This was used in the photon-conservation method (79) for transforming the specific intensity.

Invariance of phase-space volume element

The approach here is first to find a general formula for transforming the world line of a photon. The world line $x(t)$ is the photon's track in x, t space. This will allow us to define a volume in x, y, z space and to observe it in two frames of reference. The transformation of a volume element in momentum space is then combined with that of the spatial element to obtain the transformation of a phase-space volume element that follows the photons (i.e., that contains a specified set of photons).

General formula for transforming a photon world line

Consider a set of photons of frequency ν' moving in frame S′ (defined in Fig. 1) at fixed z' at angle θ' from the x'-axis (Fig. 17a). The world line of such a photon is described in three dimensions by

$$x' = x_0' + (\cos\theta')ct' \qquad \text{(World line; frame } S') \qquad (7.109)$$
$$y' = y_0' + (\sin\theta')ct'$$
$$z' = z_0',$$

where θ' is the angle of the photon track in S′ on an x', t' plot. The subscripted coordinates are the position coordinates at $t' = 0$. Note that the distance traveled in time t' is ct', and hence the speed along the track is c, as it should be for a photon.

Rewrite the variables x', t' in terms of the unprimed coordinates x, t using the Lorentz transformations (x1.1) and (x1.4) of Table 1; retain the constants x_0' and $\cos\theta'$, for we will find the new constants x_0 and $\cos\theta$ in terms of them. Solve for $x(t)$. Collect the constant terms and, separately, the coefficients of ct to find (Prob. 81a)

$$x(t) = \left[\frac{x_0'}{\gamma(1 + \beta\cos\theta')} \right] + \left[\frac{\cos\theta' + \beta}{1 + \beta\cos\theta'} \right] ct. \qquad \text{(World line } x \text{ component in S)} \qquad (7.110)$$

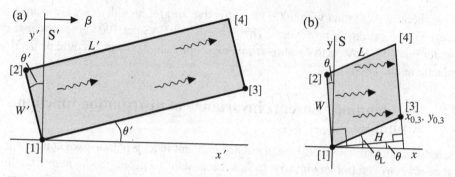

Figure 7.17. (a) Rectangular side of volume element in frame S' ($\beta = 0.8$) with sides parallel and normal to the photon propagation direction and with $L'/W' = 3$ and $\theta' = 14.9°$. The z dimension of the element is a fixed value, $\Delta z' = \Delta z$. (b) Transformed rectangle in S frame. It is a parallelogram with $L = 0.358 L'$, $W = W'$, $\theta = 5.0°$, and $\theta_L = 23.9°$. The photon trajectories are at the angles θ' and θ in the S' and S frames, respectively, and, in each case, are normal to the W (or W') side. (Courtesy A. Levine)

This completely specifies the x-component of the world line in S of the photon described with (109) in S'. In (110), it is expressed in terms of the constants describing the world line in S'.

Now, the components of a world line in frame S must have the forms

$$x = x_0 + (\cos \theta)\, ct \qquad\qquad \text{(World line in S)} \qquad (7.111)$$
$$y = y_0 + (\sin \theta)\, ct$$
$$z = z_0,$$

where θ is the angle of the photon track in S on an x, t plot, and the subscripted coordinates are the position coordinates at $t = 0$. Note that the speed along the track is again c, as it must be for electromagnetic waves in a vacuum.

Comparison of the x-component in (110) with that in (111) gives us the values of x_0 and $\cos \theta$ in terms of the primed parameters

$$x_0 = \frac{x_0'}{\gamma(1 + \beta \cos \theta')}; \quad \cos \theta = \frac{\cos \theta' + \beta}{1 + \beta \cos \theta'}. \qquad (7.112)$$

Note that we have rederived the angle transformation (56) that was applied to stellar aberration.

Similarly, transform the y'-component (109) with the transformations (x1.2) and (x1.4). Eliminate x with (110), solve for $y(t)$, and collect constants and also coefficients of ct to find

$$\Rightarrow \quad y(t) = \left[y_0' - \frac{\gamma\beta \sin \theta' x_0'}{\gamma(1 + \beta \cos \theta')} \right] + \left[\frac{\sin \theta'}{\gamma(1 + \beta \cos \theta')} \right] ct.$$

$$\text{(World line } y \text{ component in S)} \qquad (7.113)$$

This specifies the y-component in frame S. Comparison with (111) indicates that the left bracket equals y_0 and the right bracket equals $\sin \theta$:

$$y_0 = y_0' - \frac{\gamma\beta \sin \theta' x_0'}{\gamma(1 + \beta \cos \theta')}; \quad \sin \theta = \frac{\sin \theta'}{\gamma(1 + \beta \cos \theta')}. \qquad (7.114)$$

The expression for $\sin\theta$ could have been derived directly from the transformation of $\cos\theta'$ (112); recall the identity $\sin^2\theta + \cos^2\theta = 1$. The expression for $\sin\theta$ allows us to rewrite the y-component (113) more compactly as

$$y(t) = y_0' - \gamma\beta x_0' \sin\theta + (\sin\theta)\, ct. \qquad (7.115)$$

Finally, the z-component becomes, for a photon traveling at fixed z' at $z' = z_0'$,

➡ $z(t) = z' = z_0'.$ (World line z component in S)   (7.116)

The expressions for $x(t)$, $y(t)$, and $z(t)$ completely specify the world line in frame S of any photon traveling with no z motion in terms of its parameters in frame S'.

Transformation of a rectangular volume element

Take a rectangular spatial volume element at rest in (moving) frame S' containing a fixed number of photons at $t' = 0$. One side, shown in Fig. 17a, lies in the x'–y' plane at $z' = 0$. The dimension not shown extends toward the reader a fixed distance $z' = \Delta z$. The volume is chosen so that the photons interior to it are all traveling parallel to its long direction at angle θ from the axis. Our task is to find the boundaries of the volume containing the same photons in the (stationary) frame S at some fixed time (e.g., at $t = 0$).

The result of the transformation to S, as developed in this section, is shown in Fig. 17b for $\beta = 0.8$. The photon propagation angle is less in S ($\theta = 5°$) than in S' ($\theta' = 15°$). This is our familiar aberration or beaming phenomenon (Sections 5 and 6). The angles here are related according to (112) for $\beta = 0.8$. The rectangular area element of S' turns out to be a parallelogram in S with the ends again normal to the (altered) photon propagation direction but with its long axis not parallel to it.

To obtain this result, we specify the locations of three corners, [1], [2], and [3], of the rectangle of Fig. 17a as the positions of three of our photons at $t' = 0$. We then transform the three world lines from S' to S and find the locations at $t = 0$. These latter positions will be the corners of the transformed rectangle. The fourth corner [4] is not calculated because its position in S relative to [3] should be the same as for [2] relative to [1]. Imagine moving the origin of the coordinates to [3].

The initial ($t' = 0$) coordinates in S' for the three positions are, from Fig. 17a,

	x_0'	y_0'	(Coordinates; three positions in S') (7.117)
[1]	0	0	
[2]	$-W'\sin\theta'$	$W'\cos\theta'$	
[3]	$L'\cos\theta'$	$L'\sin\theta'$.	

Because we are interested only in the positions in S at time $t = 0$, it is sufficient to find the values of x_0 and y_0, as given in (112) and (114), in terms of W' and L' and the unprimed coordinate θ. The latter permits us to plot the points in S. The needed transformations of angles are those given in (112) and (114) and their inverses. Recall that an inverse transformation is obtained by changing the sign of β and interchanging θ' and θ. Also useful will be the relation

$$\frac{1}{\gamma(1 - \beta\cos\theta)} = \gamma(1 + \beta\cos\theta'), \qquad (7.118)$$

which follows from the expression for $\cos\theta$ (112), the definition of the Lorentz factor, $\gamma \equiv (1 - \beta^2)^{-1/2}$, and the identity $\cos^2\theta + \sin^2\theta = 1$.

Now proceed as outlined. Substitute the initial conditions (117) into the expressions for x_0 (112) and y_0 (114) for each position. In each case, use the angular transformations as needed. The resultant coordinates in frame S at $t = 0$ are

	x_0	y_0	$(x_0^2 + y_0^2)^{1/2}$
[1]	0	0	0
[2]	$-W'\sin\theta$	$W'\cos\theta$	W'
[3]	$L'\gamma(\cos\theta - \beta)$	$L'\gamma\sin\theta$	$L'\gamma(1 - 2\beta\cos\theta + \beta^2)^{1/2}$.

(7.119)

Parallelogram in frame S

The results (119) tell us first that the origin [1] is at the origin in S at $t = 0$. The coordinates of [2] have the same form as in S' (117) except that they are in terms of θ. Thus the [1]–[2] segment is tilted relative to the y-axis by the angle θ and is therefore normal to the photon propagation direction in S as well as in S'. We also give the lengths of the segments [1]–[2] and [1]–[3] in (119), and so we have the parallelogram dimensions in S as

$$W = W'$$
$$L = L'\gamma(1 - 2\beta\cos\theta + \beta^2)^{1/2}. \qquad \text{(Parallelogram dimensions)} \qquad (7.120)$$

Finally, the tangent of the angle θ_L of the long side of the parallelogram (Fig. 17b) is, from (119),

$$\tan\theta_L = \frac{y_{0,3}}{x_{0,3}} = \frac{\sin\theta}{\cos\theta - \beta}. \qquad \text{(Angle of length)} \qquad (7.121)$$

This angle is a function only of quantities fixed in the problem; it is independent of W' and L'. Thus, if we had chosen a portion of our sample of photons, with, say, one-half the length, the angle would have been the same; this assures us the segment [1]–[3] is a straight line. Similarly, the segment [1]–[2] is a straight line.

This, together with the equal lengths and angles of the two segments [1]–[2] and [3]–[4] (as argued just before (117)), tells us that we have a parallelogram. The drawing (Fig. 17b) is for the particular values $\theta = 5°$ and $\beta = 0.8$. Note that the length L is drastically shortened by the transformation but that the width remains the same. Keep in mind that the spatial volume element we discuss has a (fixed) z-dimension; it runs from $z = 0$ to $z = \Delta z$ over the entire rectangle of Fig. 17a and from $z' = 0$ to $z' = \Delta z' = \Delta z$ over the parallelogram of Fig. 17b.

Area in two frames

The area of a parallelogram is the base times the height, or

$$A' = W'L'$$
$$A = WH, \qquad (7.122)$$

where the height H (Fig. 17b) in the S frame may be written as

$$H = x_{0,3}\cos\theta + y_{0,3}\sin\theta, \qquad (7.123)$$

which becomes apparent if one drops a vertical line from corner [3] to the x-axis. Substitute in the coefficients $x_{0,3}$ and $y_{0,3}$ from (119) to find $H = \gamma L'(1 - \beta \cos \theta)$. The area A follows from (122) and $W = W'$ (119):

$$\Rightarrow \quad A = A'\gamma(1 - \beta \cos \theta). \qquad \text{(Area conversion)} \qquad (7.124)$$

This describes the reduced area apparent in Fig. 17b.

Phase-space volume invariant

A phase-space volume consists of two parts, the volume in x, y, z space and a momentum part, p_x, p_y, p_z. The former may be written as

$$\Delta V_x = \Delta x \Delta y \Delta z = A'\gamma(1 - \beta \cos \theta) \Delta z', \qquad \text{(Spatial volume)} \qquad (7.125)$$

where we invoked (124) and the invariance of the Δz interval (x1.3). The volume element in momentum space may be written as (Fig. 3.2b), for frame S,

$$\Delta V_p = p^2 \, dp \, d\Omega_p = \left(\frac{h}{c}\right)^3 \nu^2 \, d\nu \, d\Omega, \qquad (7.126)$$

where we applied the relation for a photon, $p = h\nu/c$, and took the axes of the spatial and momentum coordinates to be coaligned so that $d\Omega = d\Omega_p$.

Recall (70) that we defined the Doppler shift parameter, $\delta \equiv [\gamma(1 - \beta \cos \theta)]^{-1}$, and that $\nu = \delta \nu'$ (70) and $d\Omega = d\Omega'/\delta^2$ (75) so that $\nu^2 d\Omega = \nu'^2 d\Omega'$. Because $d\nu = \delta d\nu'$, we can rewrite the momentum volume element in terms of primed components and the Doppler factor, which we write in terms of θ as

$$\Delta V_p = \left(\frac{h}{c}\right)^3 \nu'^2 \, d\Omega' \, d\nu' \frac{1}{\gamma(1 - \beta \cos \theta)}. \qquad \text{(Momentum-space volume)} \qquad (7.127)$$

Phase-space volume is the product of the spatial and momentum volumes. Thus, from (125) and (127),

$$\Delta V_{ph} = \Delta V_x \Delta V_p = A'\gamma(1 - \beta \cos \theta) \Delta z' \left(\frac{h}{c}\right)^3 \nu'^2 \, d\Omega' \, d\nu' \frac{1}{\gamma(1 - \beta \cos \theta)}$$

$$= A'\Delta z' \left(\frac{h}{c}\right)^3 \nu'^2 \, d\Omega' \, d\nu' = \Delta V_x' \Delta V_p' = \Delta V_{ph}', \qquad (7.128)$$

where we invoked $\Delta V_x' = A'\Delta z'$ and also $\Delta V_p'$, the primed analog of (126). Thus,

$$\Rightarrow \quad \Delta V_{ph} = \Delta V_{ph}'. \qquad \text{(Phase-space volumes equal in frames S and S')} \qquad (7.129)$$

The phase-space volume is the same in both frames of reference. We conclude that phase-space volume containing a fixed set of photons is indeed a Lorentz invariant, and hence, so is the distribution function.

Invariance of radiating area

The transformation of positions [1] and [2] given in (117) and (119) and the invariance $\Delta z = \Delta z'$ provide the invariance of the radiating area, which is taken to be that of the surface normal to the propagation direction (Fig. 17). We found (120) that the width W is an invariant, $W = W'$, and that the width segment [1]–[2] is normal to the photon propagation direction in both frames. The two rays originating at [1] and [2] will therefore be separated by the same physical distance in both frames even though the direction of the photons has changed because of aberration.

If the two rays are from the outer edges of a celestial source, observers in S and S′ would agree on the width of the source. Other rays from the edges out of the x, y plane could define the entire periphery, and their separations would also be Lorentz invariants. The observers in the two frames would thus agree on the apparent physical area of the source projected normal to their lines of sight. This justifies the statement made after (79) regarding the transformation of specific intensity that the emitting area element normal to the line of sight is an invariant, that is,

➡ $$\Delta A_n = \Delta A_n' \qquad \text{(Normal area is Lorentz invariant)} \qquad (7.130)$$

We also used this in our discussion of the transformation of spectral flux density (80).

Problems

7.3 Lorentz transformations

Problem 7.31. Two events occur at the same location in frame S at times t_1 and t_2. They are observed in another inertial frame S′ that moves down the positive x-axis of S at speed v. Find $\Delta t'$, the time interval in S′, as a function of Δt, the time interval in S. Compare with (5) and explain how both results can be valid.

Problem 7.32. (a) How much less does an astronaut age by traveling eastward for 1 year in a satellite, which is in a low earth equatorial orbit (altitude 400 km), compared with a ground-based controller located on the equator? The earth's equatorial radius is 6400 km, and its mass is $M = 6 \times 10^{24}$ kg. Consider the astronaut and controller to be in inertial reference frames so that Lorentz transformations are applicable. You will need to find the satellite and earth-equatorial velocities. (b) The general theory of relativity introduces another effect, the *gravitational redshift*. In this, a photon originating with frequency ν_r at distance r from a point mass M will lose energy en route to an observer at a larger distance. An observer at infinity will detect the photon with frequency ν_∞, according to (4.35),

$$\frac{\nu_\infty}{\nu_r} = \left(1 - \frac{2GM}{c^2 r}\right)^{1/2}.$$

What is the aging difference of the astronaut and controller (the former in orbit and the latter on the earth's surface) if this effect is added to your time dilation result in (a)? [Ans. ~9 ms; ~8 ms]

Problem 7.33. (a) Construct the two Lorentz transformations $x'(x, t)$ (x1.1) and $t'(x, t)$ (x1.4) of Table 1 from the spherical-wave equation in S given in (1). The transformations must satisfy (2). Proceed as suggested in the text and as follows. Obtain the following simultaneous equations in the three unknown coefficients:

(i) $a_1^2 - a_3^2 c^2 = 1$

(ii) $a_1^2 v + a_2 a_3 c^2 = 0$

(iii) $a_1^2 v^2 - a_2^2 c^2 = -c^2$.

Solve the three equations for the unknown coefficients in terms of the speeds v and c and substitute into the original trial expressions; compare with (x1.1) and (x1.4). Suggestion:

eliminate a_1 in (*i*) and (*iii*) with (*ii*). Then use the second of the resulting equations to eliminate a_3 in the first to obtain a_2. (b) Demonstrate, in general, that the equality (7) between frames for an arbitrary x, t event is maintained by the x, t Lorentz transformations (Table 1) even when the four-vector $[x, ct]$ is nonzero.

Problem 7.34. (a) What is the kinetic energy E_k of a particle in terms of γ, m, and c? (b) What is the speed parameter β in terms of γ? (c) Consider a particle of mass m. For each of the Lorentz factors ($\gamma = 1.01, 1.1, 1.5, 2, 3, 10$, and 100), find (*i*) the values of E_k in units of mc^2, (*ii*) the total energy U in units of mc^2, and (*iii*) the numerical value of the speed parameter $\beta = v/c$. Present your results in tabular form. [Ans. $(\gamma - 1)mc^2$; –; $\beta = 0.94$ at $\gamma = 3$]

Problem 7.35. Show that the invariance of $U^2 - p^2c^2$ is in accord with the Lorentz transformations for energy and momentum (Table 2). Specifically, show that $U_1^2 - p_1^2c^2 = U_2^2 - p_2^2c^2$ is satisfied by the transformations, where the quantities are measured in frames S_1 and S_2, the latter of which moves down the positive x-axis of the former with speed v.

Problem 7.36. Find a relativistically correct expression for vector velocity v in terms of vector momentum p, total energy U, and physical constants. Hint: see (14) and (17). [Ans. $c^2 p/U$]

7.4 Doppler shift

Problem 7.41. A broadcasting spaceship moving with speed v passes directly over a stationary observer at altitude 500 km. Consider the motion to be along a straight line from left to right as shown in Fig. 4. (a). Find an expression for the angle $\theta = \theta_0$ (propagation direction measured in the stationary S frame as in Fig. 4) at which the Doppler shift is zero as a function of $\beta = v/c$. What are the values of θ_0 at $\beta \ll 1$, $\beta \to 0$, and $\beta \to 1$? (b) Tabulate the values of θ_0 and the arrival angle of the radiation from the vertical ($\phi_0 = \theta_0 - 90°$) for $\beta = 0.01, 0.1, 0.5, 0.9, 0.99, 0.999, 0.9999$. (c) Consider a low–earth-orbiting satellite that passes overhead at 7.7 km/s. How far from the vertical is it when its Doppler shift is zero? Again, assume straight-line motion. (d) Find the actual angular position of the satellite (not the signal direction) relative to the vertical when the null signal is detected by the observer. Will the satellite have passed the zenith at this time? [Ans. –; $\phi_0 \approx -3°$ for $\beta = 0.1$; $\sim -3''$; $\sim 3''$]

Problem 7.42. (a) Use the k, ω transformations to derive (46), the Doppler shift in terms of the angle θ' of the propagation vector k' in S'. (b) Refer to Fig. 4. Confirm that the expression (46) yields the expected relation between ω' and ω at $\theta' = 0$; see (38). (c) A signal is emitted in S' at frequency ω' vertically downward at $\theta' = 90°$. What is the frequency ω measured in S? Why is your result not the "second-order" Doppler *red*shift given in (40)?

7.5 Aberration

Problem 7.51. Obtain the aberration formula (56) directly from the Doppler expressions (45) and (46).

Problem 7.52. (a) Convert the expression (56) for the transformation of k-vector direction θ for radiation *from* a star to the transformation (59) of direction ψ *to* a star. Follow the text, except begin by solving (56) for $\cos \theta'$. (b) Use (59) to obtain the expression (60) for the angular displacement $\alpha \equiv \psi' - \psi$, given $\beta \ll 1$, and hence for $\alpha \ll 1$.

Problem 7.53. (a) Consider the angle transformation (56) for the aberration of light emitted from a source in S' (Fig. 5a). Make a table of values of θ (in units of $\pi/8$ radians) for the following values of θ': $0, \pi/8, 2\pi/8, 3\pi/8. \ldots 7\pi/8$, and π, each for the following values of β: 0.01, 0.1, 0.5, 0.9, 0.99, and 0.999. (Program your hand calculator.) What general statement can you make about the relative values of θ and θ'? (b) Indicate the table cells where $\theta < \theta'/10$ (i.e., when beaming is most pronounced). Indicate the cells where the radiation in S in directed rearward, $\theta > \pi/2$. Comment on each. Make a sketch patterned after Fig. 5a showing qualitatively, for each of the two categories, the vectors k' (in S') and k (in S). (c) "Beaming" is an extreme case of stellar aberration. Describe or otherwise illustrate the changing appearance and positions of the stars in the sky for an observer traveling in a straight line at increasing speeds up to and beyond about $v = 0.9\,c$. Include the appearance of the rearward hemisphere.

7.6 Astrophysical jets

Problem 7.61. Consider a source at rest in S' but moving relative to S. It emits an isotropic flash of N photons of energy $h\nu'$. The source speed is described with the factor $\beta = v/c$, which can approach unity. (a) Calculate the total radiated energy in S by integrating over all rays in S' but with the appropriate Doppler-shifted frequencies (or energies) they have in S. How does your result compare with the total energy $Nh\nu'$ in S'? Hint: use (46) and integrate over angles in the S' frame. (b) How much of this total energy in S is within the "beam" defined in S' as $\theta' < \pi/2$? What is this beam energy for $\beta \to 0$ and for β approaching unity? Are your answers to (a) and (b) reasonable? Why? [Ans. $\gamma Nh\nu'$; limits are $0.5\,Nh\nu'$, $0.75\gamma Nh\nu'$]

Problem 7.62. (a) Find the angle θ_{max} as a function of β for which superluminal motion $\beta_a \equiv \beta_{t,a,apr}$ (88) is a maximum; see Fig. 13. Also, express your answer in terms of γ. (b) What is the value of $\beta_{a,max} \equiv \beta_{t,a,apr,max}$ at this angle in terms of β, γ, or both? (c) Is there a limit on β below which the source will never appear superluminal? If so, what is that limit? Use your answer to (b). (d) What are the values of θ_{max} and $\beta_{a,max}$ for the case of $\gamma = 10$? [Ans. $\cos\theta_{max} = \beta$; $\beta\gamma$; ~ 0.7; $\sim 6°$, ~ 10]

Problem 7.63. Consider the 1994 outburst of microquasar GRS 1915+105 (Fig. 8a). Assume the microquasar (at cross in figure) simultaneously ejected the two major knots of equal power in opposite directions at the same time with identical but unknown speed parameters $\beta = v/c$ and at unknown angle θ relative to the line of sight. One knot is thus receding from the observer and the other is approaching, as illustrated in Fig. 12. (a) From a superficial examination of the figure, which of the two lobes is likely the one approaching the observer? Explain. (b) Find the apparent transverse velocity parameters $\beta_1 \equiv \beta_{t,a,apr}$ and $\beta_2 \equiv \beta_{t,a,rec}$ of the approaching and receding knots from measurements of the figure and data given in the caption. Is either one superluminal? (c) Find the common knot velocity parameter β and angle of ejection θ. Hints: apply (88) and (89) and find an expression for $b \equiv \beta \cos\theta$ in terms of β_1 and β_2 and then find $\tan\theta$ in terms of b and β_1. [Ans. –; ~ 1.2, ~ 0.6; ~ 0.9, $\sim 70°$]

Problem 7.64. Consider Fig. 10 in which a relativistic source transmits isotropically. (a) Find, from (37) in terms of the Lorentz factor $\gamma = (1 - \beta^2)^{-1/2}$ alone, an expression for $\cos\theta_N$, where θ_N is the angle corresponding to a null Doppler shift in the observer's frame ($\nu/\nu_0 = 1$). That is, find the half-angle of the outer cone in the figure. (b) Tabulate the values of θ_N and the beam half-angle θ_b (61) for the following values of γ: 1.000, 1.001, 1.1, 2, 5, 10, 20, 50100, 1000. (c) Find the ratio of θ_N/θ_b for the limiting case $\gamma \gg 1$. Check it against a tabulated value. Is there any γ for which the radiation within the beam ($\theta < \theta_b$) is redshifted? [Ans. $(\gamma - 1)(\gamma^2 - 1)^{-1/2}$; see Fig. 10 for $\gamma = 10$; $(2\gamma)^{1/2}$]

7.7 Magnetic force and collisions

Problem 7.71. Consider a magnetic field line at about 1 earth radius from the surface of the earth at the magnetic equator with a strength of 4 μT containing electrons of (relativistic) kinetic energy $E_k = 2 m_e c^2$. (a) What is the cyclotron frequency ν of rotation (Hz) of an electron about the field lines? Does it apply to all pitch angles? (See text.) (b) Consider a particle with velocity directed close to the field line (pitch angle $\phi \approx 0°$). How long does it take the electron to pass from one magnetic pole to the other? (Electrons will reflect between the two poles owing to the convergence of the field lines at the poles.) Let the distance between reflection points be 4 earth radii ($R_E = 6400$ km). Approximately how many cycles of cyclotron rotation about the field line occur in that time? (Assume, unrealistically, a constant magnetic field.) (c) What is the radius R of the spiral path of the electron about the field line if $\phi = 5°$? [Ans. ~40 kHz; ~3 000; ~100 m]

Problem 7.72. (a) Consider the reaction $p + p \rightarrow p + p + \pi^0$, where a high-energy proton interacts with a stationary proton to create a neutral pion. (Baryon conservation requires that there be two protons in the final state.) Find the threshold kinetic energy of the incident proton in units of the proton rest energy $m_p c^2$ and in MeV ($m_p c^2 = 938$ MeV; $m_{\pi^0} c^2 = 135$ MeV). Hints: follow the procedure leading to (107) while making use of (24) and (25) for particle momentum and energy. (b) What is the threshold kinetic energy required to produce an antiproton, a proton of negative charge with the same mass as a proton? To conserve baryon number, a proton must be produced with the antiproton. Thus, the final state consists of three protons and one antiproton. [Ans. ~300 MeV; ~6 GeV]

Problem 7.73. (a) Find the energy threshold $U_{p,thr}$ in electronvolts for a high-energy proton interacting with a CMB photon of average energy in a head-on collision to create a neutral pion, $p + h\nu \rightarrow p + \pi^0$; $m_p c^2 = 938$ MeV; $m_{\pi^0} c^2 = 135$ MeV; $\langle h\nu \rangle_{av,CMB} = 6.4 \times 10^{-4}$ eV. Find the Lorentz factor and the total energy in electronvolts of the incident proton. What is its kinetic energy? (b) What fraction of this energy is retained by the proton after the collision? (c) What is the threshold if the CMB photon has twice the average photon energy? [Ans. ~10^{20} eV; ~90%; –]

7.8 Addendum: Lorentz invariance of distribution function (photons)

Problem 7.81. Write out the derivations of this section, filling in the missing steps.

Problem 7.82. Carefully draw the area elements of Figs. 17a and 17b for the later times, $t' = L'/2c$ for Fig. 17a and $t = L/2c$ for Fig. 17b. No calculations are required.

Problem 7.83. Plot (as in Fig. 17) the x, y area in S and S' for several different orientations θ' of the rectangle in S' and frame velocities β. For example, use the several combinations of $\beta = 0.1, 0.5, 0.9, 0.99$ and $\theta' = 5°, 45°, 90°, 135°$. Program your calculator to ease the task. Be careful to handle negative signs property. Use the same ratio of $L'/W' = 3$ in each case. In each instance, make sketches for the appearance in S and in S'. Summarize the trends with β and θ' as you see them.

8

Synchrotron radiation

What we learn in this chapter

Intense light emanating from electrons circulating in an accelerator (**synchrotron**) was the first evidence that radiation from **relativistic electrons** can be much more intense than expected classically. A few years later, in 1954, light from the **Crab nebula** was found to be **polarized**, thus demonstrating that the nebula contains highly relativistic electrons **spiraling around magnetic field lines**. This radiation is now called **synchrotron radiation** or **magnetic bremsstrahlung**.

A relativistic electron circulating around magnetic field lines radiates primarily into a narrow beam in the forward direction owing to aberration, the so-called **headlight effect** (Chapter 7). The **characteristic frequency** of the detected radiation is simply the inverse of the time the beam takes to sweep over an observer. This frequency turns out to be about equal to the Lorentz factor squared times the cyclotron frequency; it thus increases as the electron energy squared.

The **power radiated by a circulating electron** is found from the classical Larmor radiation formula applied in an inertial frame of reference in which the electron is momentarily at rest. The power transformed back to the observer frame grows as the square of both the electron energy and the magnetic field. The electron energy divided by this power yields the **characteristic lifetime of electrons** at this energy. For example, in the Crab nebula, spiraling electrons emitting optical photons have a lifetime of only ~100 yr, and those emitting x rays live only a few years. Such electrons could not have been accelerated in the 1054 CE supernova collapse that spawned the Crab nebula. Their **energy source** was a puzzle until the discovery of the **Crab pulsar** in 1968.

An **ensemble of radiating electrons** with a decreasing power-law energy spectrum is common in astrophysical systems (e.g., the cosmic ray flux). Such an ensemble in the presence of a magnetic field will emit synchrotron radiation with a **volume emissivity** ($J m^{-3} Hz^{-1}$) that is also a descending power law. Such spectra are called **nonthermal** to contrast them with the rising power-law "thermal" spectrum of blackbody radiation at low frequencies. Integration of the volume emissivity along the line of sight through the emitting region then yields the **specific intensity** ($W m^{-2} Hz^{-1} sr^{-1}$).

Electrons streaming outward along curved field lines from the magnetic pole of a **spinning neutron star** will be accelerated by this curvature and emit a version of synchrotron radiation called **curvature radiation**. A **group of N electrons** traveling together will radiate in phase as a single charge. The radiated power for this **coherent radiation** scales as N^2. Curvature radiation by bunches of electrons is a likely source of the radiation detected from **radio pulsars**.

8.1 Introduction

The radiation from certain classes of celestial objects or celestial regions turns out to be *linearly polarized*. That is, the electric vector of the electromagnetic wave detected with an antenna lies preferentially in one plane.

The phenomenon that most often gives rise to polarization is called *synchrotron radiation* or *magnetic bremsstrahlung* (braking). This radiation is due to relativistic electrons ($v \approx c$) spiraling around magnetic fields. The magnetic force $q(v \times B)$ causes the electrons to be accelerated.* Accelerated charges radiate photons and lose some of their kinetic energy (Fig. 1a). This is analogous to the braking of electrons in collisions with the ions in the thermal bremsstrahlung process (Chapter 5). In that case, electric fields provide the accelerating forces, whereas in synchrotron radiation the forces are provided by magnetic fields.

The specific intensity, $I(\nu)$ (W m^{-2} Hz^{-1} sr^{-1}), of such radiation depends on the shape of the electrons' energy spectrum. The photon spectrum usually decreases with increasing frequency, which is behavior opposite to the increase expected from blackbody radiation at radio frequencies. It has thus come to be called "nonthermal" radiation. This name is appropriate because the originating electrons are not in thermal equilibrium with their surroundings; they are much too energetic.

In this chapter, expressions that describe the radiation from a single accelerating electron are developed. These include the frequency of the radiation, the power radiated by the electron, and the characteristic time for the electron to lose a substantial portion of its energy. We then explore the characteristics of radiation from ensembles of many particles and also a variant of synchrotron radiation known as curvature radiation.

8.2 Discovery of celestial synchrotron radiation

Puzzling radiation from the Crab nebula

The possibility that relativistic nonthermal electrons might be a significant source of radiation from celestial objects was overlooked for many years. Now the phenomenon is known to be quite prevalent and is quite well understood.

Bluish diffuse light

A large part of the supernova remnant called the Crab nebula appears as an amorphous bluish haze on the sky, and the origin of this light was, for a long time, a big mystery. Its spectrum extends all the way from the radio to the gamma-ray band (Fig. 2), and its overall luminosity

---

* In this chapter, the vector B (or B_z, its line-of-sight component) refers, for the most part, to a large-scale, relatively static magnetic field – not, for example, the oscillating magnetic field of an electromagnetic wave.

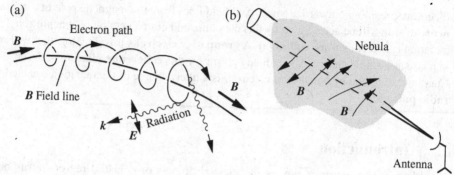

Figure 8.1. (a) Electron spiraling around a magnetic field line emitting polarized radiation in the electron's forward direction. (b) Line of sight through a nebula with partially ordered magnetic fields.

of $\sim 10^5\, L_\odot$ is huge. The optical radiation from the nebula also includes reddish filaments that exhibit hydrogen Balmer emission spectral lines, forbidden nebular lines, and others, indicating temperatures of $\sim 10^4$ K.

The bluish luminosity in the optical band might have been considered to signify the presence of a hot, optically thin thermal plasma (Chapter 5) with a temperature of $\sim 50\,000$ K. This temperature was inferred from the decreasing intensity $I(\nu)$ of the spectrum in the optical band and that expected for thermal bremsstrahlung emission (Fig. 5.5). Such a plasma, though, would radiate pronounced optical emission lines from excited atoms in the plasma similar to the x-ray lines from a much hotter plasma in Fig. 5.8. In fact, no spectral lines are observed; the spectrum is a smooth continuum. Another difficulty was that the inferred mass of the plasma was found to be uncomfortably large ($\sim 15\, M_\odot$) for the residue of a stellar outburst or supernova. The nature of the bluish emission was indeed a puzzle.

Spectral energy distribution (SED)

Spectra are sometimes plotted as $\nu S(\nu)$ on a log-log plot, as in Fig. 2, where $S(\nu) = \int I \, d\Omega$ is the spectral energy flux-density (W m^{-2} Hz^{-1}). The units of $\nu S(\nu)$ are thus W/m^2. This type of plot is known as a *spectral energy distribution* (SED). Remember that, in this text, we drop the subscript ν, thus, $S \equiv S_\nu$. We sometimes call $S(\nu)$, simply, "flux density" or "flux."

In the SED presentation, the logarithmic slope of a power-law spectrum is increased by unity. A photon spectrum $S \propto \nu^{-1}$ with logarithmic slope $\alpha = -1$ would have a flat SED, $\nu S_\nu \propto \nu^0$. For such a flat spectrum, direct integration shows that equal logarithmic intervals contain equal energy fluxes. That is, one decade of frequency in a low-energy x-ray band would have the same energy flux, $\mathscr{F} = \int S \, d\nu$ (W/m^2), as that in a decade of frequency at some higher x-ray or gamma-ray energy. The SED effectively presents the distribution of energy across the wave bands.

The spectrum of the Crab nebula is plotted as an SED in Fig. 2. Note that it is quite flat from ~ 10 eV in the ultraviolet to ~ 100 keV in the hard x-ray region. It would appear even flatter if the compressed horizontal scale were uncompressed. The spectrum $S(\nu)$ in this region is thus just slightly steeper than $\alpha = -1$ that is, $S \propto \nu^{-1.1}$. In the radio region, it is about $S \propto \nu^{-0.25}$ (Prob. 23).

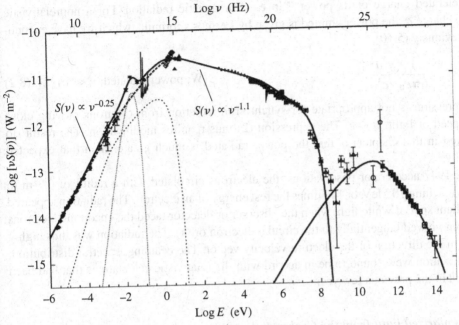

Figure 8.2. Spectral energy distribution (SED), $\nu S(\nu)$, for photons from the Crab nebula over 19 decades of frequency reaching from the radio to TeV gamma rays with compressed horizontal scale. The straight-line segments and polarization measurements indicate that most emission up to about 100 MeV is synchrotron emission. The solid lines are a fit to the data for a synchrotron self-Compton (SSC) model (Section 9.4) for a nebular magnetic field of 16 nT with contributions from a millimeter synchrotron (radio) region and from nebular dust (narrow and broader light dashed peaks, respectively). The heavy dashed line is the SSC contribution without these two components. The peak at TeV energies is attributed to synchrotron photons that have been upscattered via inverse Compton scattering (Chapter 9). [Compiled by HEGRA team from HEGRA TeV data and the literature; F. Aharonian *et al.*, *ApJ* **614**, 897 (2004)]

Electron accelerators (synchrotrons)

The first detection of synchrotron radiation arose from visual observations of the electrons circulating in an electron accelerator (*synchrotron*) in 1947. In this device, the electrons spiral around a circular track and are gradually accelerated to higher and higher energies by electric fields. The term synchrotron signifies that the accelerating electric fields are timed to be synchronized with the decreasing circulation frequency of the electrons as the energy increases (7.102).

The circular motion of the electrons in the synchrotron requires an inward centripetal force, $F = dp/dt$. (Here we neglect the tangential accelerating force because it is small by comparison.) Magnets along the orbit provide the required force,

$$F = q(v \times B). \qquad \text{(N)} \qquad (8.1)$$

An accelerated charge emits power $\mathscr{P}$ in electromagnetic radiation. For a nonrelativistic ($v \ll c$) charge q, the power emitted is given by Larmor's formula, which we have encountered previously (5.10),

$$\mathscr{P} = \frac{1}{6\pi\varepsilon_0} \frac{q^2 a^2}{c^3}. \qquad \text{(W; power radiated; } v \ll c\text{)} \qquad (8.2)$$

This expression is not appropriate for synchrotron radiation from electrons moving close to the speed of light, $v \approx c$. The expression (2) thus requires modification. The result (to be derived in this chapter) is that the power radiated is much greater than that expected classically.

In the 1947 accelerator observations, the electrons circulated with a radius of 0.3 m to energies reaching 70 MeV, or 140 times the rest energy of an electron. The radiation appeared as a brilliant spot of white light when the observer or detector faced the oncoming electrons (i.e., when viewed tangentially to the circular electron orbit). The radiation was thus highly beamed in the direction of the electron velocity vector. The beaming, spectral distribution, and polarization were found to be in accord with the relativistic calculations that had been developed in the previous few years.

Polarized light from the Crab nebula

The puzzle of the source of radiation from the Crab nebula was solved dramatically when a Russian scientist (I. S. Shklovsky) postulated in 1953 that the bluish radiation from the Crab nebula might be the same as that previously discovered in man-made electron accelerators. He therefore suggested that optical astronomers look to see if the radiation from the Crab is polarized.

They did (in 1954) and found it to be so. The nebula was observed at the telescope through a Polaroid™ filter transmitting only one polarization of incident light. It was found that regions of the nebula appeared brighter (relative to other regions) at some orientations of the polaroid filter than they did at other orientations (Fig. 3).

In this way the nature of the amorphous bluish radiation came to be understood. The model naturally explains the absence of atomic spectral lines. Also, the radiation by the electrons is so efficient that relatively few electrons are required to provide the observed luminosity; a large mass of plasma is not required. Nevertheless, this explanation was quite a surprise; it required that the radiating electrons have energies in excess of 10^{11} eV. This of course raised the question of how they attained such energies. Much later (1969), it was found that the energy source is a spinning neutron star, the *Crab pulsar*, as we discuss in Section 4 under "Crab nebula."

The net polarization of the radiation coming from a single position (resolution element) on the nebula is a sum of the emissions by many electrons at many positions along the line of sight (Fig. 1b). A net polarization is most likely if (*i*) the magnetic fields in the nebula are organized so that, along a given line of sight through the (optically thin) nebula, there is a dominant direction, and (*ii*) the electrons are relativistic so that the radiation observed is dominated by approaching electrons (the headlight effect; Section 7.6). Nonrelativistic electrons approaching and receding on opposite sides of their orbit would radiate equally toward the observer with opposite polarizations, thus yielding a negligible net polarization.

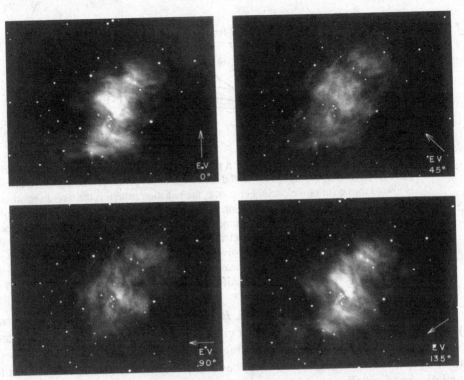

Figure 8.3. Photographs of the Crab nebula in polarized light with the polarizer at different orientations. The arrows show the directions or planes of the transmitted transverse electric vector. Note the changing brightness pattern from photo to photo. The nebula has angular size $4' \times 6'$ and is ~6 000 LY distant from the solar system. North is up and east to the left. The pulsar is the southwest (lower right) partner of the doublet at the center of the nebula. [Palomar Observatory/CalTech]

The degree of magnetic field organization is reflected in the degree of the polarization. At the radio frequency of 10 GHz, the degree of linear polarization from the Crab nebula as a whole is ~5%. Radiation from local regions of the Crab would be more highly polarized.

Synchrotron radiation is not limited to young supernova remnants such as the Crab nebula. The general galactic background of radio emission is polarized owing to electrons spiraling in galactic magnetic fields. The lobes of radio-emitting plasma previously ejected from the active nuclei of galaxies (AGN) are also synchrotron emitters as are jets of relativistic particles from AGN (Section 7.6).

Polarization of radiation can arise for other reasons. For example, the interstellar grains (*dust*) in the Galaxy tend to align themselves with the local magnetic field and thereby to slightly polarize the visible starlight passing through them (AM, Chapter 10).

8.3 Frequency of the emitted radiation

The $q(v \times B)$ force on a charge will cause it to spiral around the magnetic field lines in a corkscrew pattern if there are velocity components both along and normal to the local B field lines. If there is no component along the field lines, and if the field is uniform in that region of space, the motion will be circular.

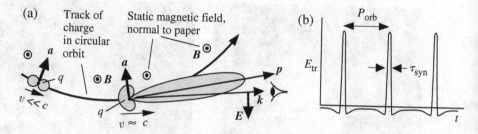

Figure 8.4. (a) Radiation patterns for a positive charge moving with speed v in a circle in the presence of a uniform magnetic field B directed out of the paper. The instantaneous acceleration is toward the center of the circle. For $v \ll c$, the radiation pattern is the toroid shown to the left and in Fig. 5.2b. For $v \approx c$, the radiation is peaked strongly in the forward direction. (b) Plot of the transverse component of E versus time detected by the observer in (a) for a single charge with $v \approx c$. The dominant frequencies detected are approximately equal to the inverse of the duration τ_{syn} of each pulse. Note the finite negative field between pulses and the sudden field reversals at each edge of a pulse. The widths of the relativistic forward lobe in (a) and that of the individual pulses in (b) are greatly exaggerated. [After V. Ginzburg and S. Syrovatskii, *ARAA* **3**, 297 (1965)]

Our discussion here will address the $v \perp B$ (pitch angle $\phi = 90°$) case in which the motion is circular because this is the motion that gives rise to acceleration and radiation. Later we will introduce the effect of lesser pitch angles. The moving charge q will be treated as if it were positive. The charge $q = -e$ can be substituted into the derived expressions for the typical case of radiating electrons.

We now present a semiquantitative derivation of the dominant frequency of the emitted radiation by a single particle.

Instantaneous radiation patterns

The instantaneous angular distribution of radiated power emitted from the orbiting electron depends greatly on the speed of the electron. The distributions are quite different for nonrelativistic and relativistic particles.

Classical radiation pattern ($v \ll c$)

Consider the effect of circular motion of a charge q at nonrelativistic speed, ($v \ll c$), along the circular track of Fig. 4a. The acceleration vector a due to the circular motion is shown (left end of the track) together with the associated toroidal (doughnut-shaped) dipole radiation pattern of Fig. 5.2b. The pattern represents the magnitude of the radiated electric field (at large distances); it is at maximum in the equatorial plane of the toroid (i.e., normal to a).

The direction of the radiated E vector along some line of sight is transverse to the propagation direction and, for a positive charge, is in the direction opposite to the acceleration vector a projected onto a plane normal to the line of sight. See Fig. 5.2a for a positive charge and Fig. 5.3a,b for an electron.

This is all included in the mathematical description of the radiated transverse electric vector E (5.4). It tells us that the magnitude at large distance is $E_{tr}(t) \propto [qa(t')/r] \sin \theta$, where r is the distance from the accelerating charge q to the observation point, θ is the angle between the acceleration vector and the line of sight, and a is the magnitude of the acceleration a. The acceleration that gives rise to E_{tr} at time t occurs at the earlier time $t' = t - (r/c)$.

As the charge in Fig. 4a executes its circular orbit with angular frequency ω (rad/s), a distant observer, stationary in the orbital plane, would detect an E vector lying in the plane of the orbit and oscillating sinusoidally in magnitude with the orbital frequency ω (rad/s). The radiation is *linearly polarized* because E is consistently in the orbital plane. In contrast, an observer at the pole of the orbit would perceive a rotating E vector or *circular polarization*.

In either case, the characteristic frequency of the electromagnetic wave would be that of the orbit, ω. In our nonrelativistic situation, the electron is freely orbiting in a static magnetic field B, and so the frequency is simply the *cyclotron frequency*,

$$\omega_B = \frac{qB}{m}, \qquad \text{(Cyclotron angular frequency; rad/s)} \qquad (8.3)$$

which is readily deduced from the magnetic force (1) and Newton's second law with acceleration $-\omega^2 r$ for circular motion (Prob. 33).

Relativistic radiation pattern ($v \approx c$)

A charged particle moving relativistically ($v \approx c$) in a straight line has a pattern of E vectors that bunch in the transverse direction (Fig. 7.3b). This was formally demonstrated (7.33) from Lorentz transformations of E, B fields. The more relativistic the particle is, the more the field lines are concentrated in the transverse direction.

As in the nonrelativistic case (Section 5.3), acceleration of relativistic particles leads to distortions of the field lines that propagate at speed c in vacuum and become transverse electromagnetic waves at large distances. For a relativistic charge, these waves are highly beamed in the forward direction (Fig. 4a), and so the observer sees a sharp pulse of E field once each rotation of the charge (Fig. 4b).

This beaming is due to extreme aberration (the headlight effect) derived previously from the relativistic transformation of the propagation vector k (7.63). For $\beta \equiv v/c \approx 1$, the radiation propagates as if it were from a narrow, laserlike headlight on the radiating electron (7.63) with little power propagating in the backward and sideward directions.

Field lines for relativistic circular motion

The instantaneous E field lines of a relativistic, positive charge moving in a circle at constant speed, $v = 0.85c$, are shown for a fixed moment in time in Fig. 5. The curved and kinked field lines are the acceleration-distorted E *field* lines. Line features close to the charge were recently emitted, and those farther out were emitted earlier. The double kinks in each line arise from the two sudden reversals of the field at each pulse of Fig. 4b. Remember that a field line direction gives the direction of E at that position and that the density of lines gives the field magnitude. Thus, the closely spaced lines at the kinks (small dark circle) represent a region of intense electric field.

As time progresses, the electron *and* the radiation pattern rotate together about the center of rotation (at the cross "$+$"). The charge position and velocity are indicated by the vector v. The Poynting vector $\mathscr{F}_P$ at the charge is indicated. You can recreate the circular motion by making

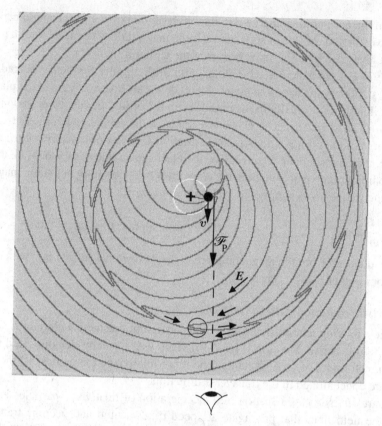

Figure 8.5. Instantaneous E field lines of a relativistic, positively charged particle (filled circle) at the time it is at the position shown. The particle has been moving for some time along a circular path (white) centered at "+" with speed $v = 0.85\,c$ in a magnetic field emerging from the paper. The field patterns shown arise from accelerations at earlier times. They travel outward at the speed of light. At the moment shown, the charge is moving downward with velocity v and is radiating Poynting flux $\mathscr{F}_P$ (W/m^2) also downward like a headlight. This radiation will be detected later by the observer. The field lines chosen for this figure, for illustrative purposes, are isotropic at the charge in the laboratory (observer) frame contrary to reality. The closely spaced transverse reversed field lines at the kinks (open circle) represent a strong, right-directed electric field of the headlight beam from the previous cycle of the charge. Additional field lines would show a similar kink slightly to the right heading directly toward the observer. [Courtesy S. Olbert]

a copy of this figure (without the observer), pinning it at its center (the cross), and rotating it clockwise. A stationary observer will experience a sharp pulse of transverse radiation once each revolution of the charge. The detected pulses were radiated by the electron at the earlier times when it was moving directly toward the observer. This is synchrotron radiation.

Electric field waveform, $E(t)$

The electric waveform, $E(t)$, detected by the observer determines the dominant frequencies that will be observed. If, for example, the amplitude of the electric vector as a function of time is a perfect sinusoid with a period P, the frequency of the detected electromagnetic radiation would be $1/P$. A more complicated waveform is detected in the relativistic case. This gives rise to a wide range of frequencies but with most of the radiated power in a restricted band. It is the typical frequency of this restricted band that we seek.

Brief pulses of radiation

As the relativistic charge moves around the circle (Fig. 4a), the observer in the plane of the orbit detects the headlight beam sweeping by once each cycle. The brief pulse of radiation is detected as a transverse electric vector that quickly increases to a large value and then rapidly decreases to zero, reverses direction, and takes on small negative values for the rest of the orbit (Fig. 4b). See also the instantaneous field line directions in Fig. 5.

The negative values in Fig. 4b represent the effect of the small backward-facing lobe in Fig. 4a. The separation of the peaks is simply the time it takes the electron to complete the circle. The electric vector E undergoes one cycle of the repetitive pattern as the electron completes a single orbit.

The duration of the *observed* electric pulse according to the stationary observer in Fig. 4a is the inverse of the dominant detected frequency. The sharp spikes of radiation can be synthesized with a large contribution from a sine wave having a period comparable to the very short duration of one of the sharp pulses of Fig. 4b. This corresponds to a very short wavelength or a very high frequency – much higher than the electron orbital frequency.

Several factors enter into the determination of the observed peak's duration. First, recall that the half-angle width of the headlight beam was shown (7.63) to be approximately

$$\theta_b \approx mc^2/U = 1/\gamma, \qquad \text{(Angular half-width of beam; rad)} \qquad (8.4)$$

where m is the electron mass, U is the electron's *total* (rest + kinetic) energy, and $\gamma \equiv (1 - \beta^2)^{-1/2} = U/mc^2$ is the Lorentz factor (7.3). Electrons of energy 10^{11} eV that emit optical radiation from the Crab supernova remnant exhibit a beamwidth of only $\theta_b = 1.0''$. The beam can be very, *very* narrow!

The angular frequency at which a nonrelativistic particle of charge q moves in a circle in the presence of a magnetic field of magnitude B is the cyclotron frequency ω_B (3). If the particle is relativistic, the frequency is reduced by the factor γ (7.102):

$$\omega = \frac{1}{\gamma} \omega_B = \frac{1}{\gamma} \frac{qB}{m}. \qquad \text{(Relativistic cyclotron frequency; rad/s)} \qquad (8.5)$$

The factor γ increases linearly with energy, and so, as energy increases, the orbital frequency decreases and the cycle time increases.

The time τ_0 for the full width of the beam to pass over the observer would seem to be the ratio $2\theta_b/\omega$, or, from (4) and (5),

$$\tau_0 \approx \frac{2\theta_b}{\omega} = \frac{2m}{qB} = \frac{2}{\omega_B}. \qquad (s) \qquad (8.6)$$

Note, interestingly, that the factor γ dropped out – at least for now. The duration of the pulse is approximately the classical value because, as the particle energy increases, the lower angular velocity of the orbital motion just compensates for the narrower beamwidth. It seems that the radiation detected is not at a higher frequency after all. But wait, there is another important effect.

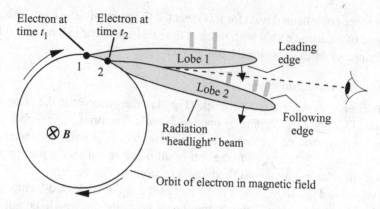

Figure 8.6. Sketch of the "headlight" beam of an electron at the two times its front and back edges traverse the observer. The photons emitted at t_2 have less distance to travel to the observer than those emitted at t_1. The observed duration of the pulse is therefore less by a factor of γ^2 than would otherwise be expected.

Charges chasing photons

Equation (6) gives the time interval τ_0 during which the observer would see the strong E pulse if it traveled at *infinite* speed from the charge to the observer. Consider the headlight beam of a car turning a corner at night sweeping over a distant burglar lurking in someone's yard. In calculating the time the burglar would be illuminated, one would naturally make this infinite-speed assumption because the headlight photons travel so much faster than the car.

In our case, the photons have the finite speed c, and the emitting charge is traveling very close to this speed (e.g., $v = 0.99c$). The net effect of this is that the observer sees a pulse much briefer than τ_0 owing to the Doppler compression of the pulse of radiation.

Consider the positions of the charge when the front and back edges of the headlight beam traverse the observer (Fig. 6). The first photons detected were emitted at time t_1 into lobe 1. The final photons detected were emitted at later time t_2 into lobe 2 when the charge was closer to the observer. The last photons thus had a smaller distance to travel, and the observer detects the back edge of the beam earlier (relative to the front edge) than would be the case for infinite photon speed. The electromagnetic pulse thus has shorter duration than τ_0 given in (6). This is a very big effect for an electron with $v \approx c$ because it is racing along right on the heels of the photons it has just emitted.

This compression of the interval between two events – as measured in a single frame – due to the motion of the source was encountered in the derivation of the relativistic Doppler effect. The time interval for an approaching source ($\theta = 0$) is reduced by the factor $1 - \beta$ (7.36). The time-dilation effect characterized by the factor γ in (7.36) does not enter into consideration because our discussion is totally from the perspective of an observer in the laboratory frame of reference; there is no transformation of events between inertial frames.

Now, for extremely relativistic particles ($v \approx c$), it follows from the definition $\gamma \equiv (1 - \beta^2)^{-1/2}$, that

$$1 - \beta = (1 - \beta)\frac{1 + \beta}{1 + \beta} = \frac{1 - \beta^2}{1 + \beta} \approx \frac{1}{2\gamma^2}. \qquad \text{(Doppler factor; } \beta \approx 1) \qquad (8.7)$$

The radiated pulse of radiation is thus shortened by the factor $2\gamma^2$, and so the observed "synchrotron" pulse duration τ_{syn} is $\tau_0/(2\gamma^2)$, which becomes, from (6),

$$\Rightarrow \qquad \tau_{syn} \approx \frac{\tau_0}{2\gamma^2} = \frac{1}{\omega_B}\frac{1}{\gamma^2}. \qquad\qquad \text{(Observed width; s)} \qquad (8.8)$$

The observed pulse duration is simply the inverse of ω_B reduced by the factor γ^2.

Observed frequency

The angular frequency ω_{syn} (rad/s) of the observed (synchrotron) radiation depends on the duration τ_{syn}; see "Brief pulses of radiation" above. The sine waves that best match the waveform (Fig. 4b) will represent the dominant frequencies. (Formally, one would calculate the Fourier transform of the waveform.)

Synchrotron frequency

As an approximation of ω_{syn}, we adopt the sine wave that changes by 1 rad in the time τ_{syn}. The angular frequency ω_{syn} of this sine wave is just the inverse of τ_{syn}. From (8), we obtain

$$\omega_{syn} \approx \frac{1}{\tau_{syn}} = \gamma^2 \omega_B. \qquad\qquad \text{(rad/s)} \qquad (8.9)$$

Introduce the expressions for ω_B and γ, introduce the pitch-angle factor $\sin\phi$, and divide by 2π to obtain the quantity we define as ν_{syn} (Hz):

$$\Rightarrow \qquad \nu_{syn} \equiv \frac{1}{2\pi}\left(\frac{U}{mc^2}\right)^2 \frac{qB}{m}\sin\phi. \qquad \begin{array}{l}\text{(Characteristic angular frequency}\\ \text{of synchrotron radiation; s}^{-1})\end{array} \qquad (8.10)$$

This approximates the dominant frequency of the electromagnetic radiation detected by the observer.

Pitch angle

The factor $\sin\phi$ in (10) accounts for the component of the electron velocity along the magnetic field line (Fig. 1a). The *pitch angle* ϕ is the angle between the velocity vector v and the magnetic field vector B. Heretofore, we assumed the particle was orbiting in a circle about the field line with v perpendicular to B ($\sin\phi = 1$).

For the more general spiraling motion, the frequency of rotation about the field line is independent of the pitch angle; it remains $qB/\gamma m$ (5), where $\gamma = U/mc^2$ represents the full energy (not "a component") of the electron; see discussion after (7.102). Furthermore, the compression term that leads to the γ^2 factor, (7) and (10), remains valid because the observer of the radiation pulse must be in the forward direction of the instantaneous electron motion; recall that the beam is forward directed and very narrow.

The observed frequency is nonetheless affected by the spiraling motion. In this case, the center of the (narrow) beam sweeps around the celestial sphere along a circle of angular radius $\phi < 90°$, which is to say along a *small* circle rather than a *great* circle ($\phi = 90°$). The circumference of the small circle ($\propto \sin\phi$) is less than that of a great circle. Thus, at any given time, the beam covers a greater fraction of the small-circle track. Because the cycle time and beam size are independent of the pitch angle, a beam at $\phi < 90°$ will therefore take longer to cross over an observer at a fixed location on the celestial sphere than would a beam at $\phi = 90°$. (In each case the observer would be located on the beam track.) The characteristic frequency ν_{syn} for a spiraling electron of a certain energy U is thus reduced by the factor $\sin \phi$ as given in (10).

We tacitly assume here that the beamwidth is less than the pitch angle, $\theta_b < \phi$. If not, the beam would continuously illuminate an observer in line with the magnetic field direction. This is illustrated in our approximate formula (10), which yields zero frequency when the motion is directed along the field line, $\phi = 0$.

Note the strong dependence of ν_{syn} (10) on the energy U of the charge, or equivalently on $\gamma = U/mc^2$. At low energies, the total energy U is about equal to the rest energy mc^2, and $\gamma \approx 1$. Hence, the (angular) frequency is just the cyclotron frequency $\omega_B = qB/m$ for $\phi = 90°$ and is quite independent of energy. As the electron begins to enter the relativistic domain, the observed frequency begins to increase. At $v \approx 0.3\,c$ or, equivalently, $\gamma \approx 1.05$, the frequency is $\sim 10\%$ greater.

If the particle's total energy U is much greater than the rest energy ($\gamma \gg 1$), the γ^2 dependence results in extremely high frequencies. For example, electrons of $U = 10^{11}$ eV and rest energy $m_e c^2 = 5 \times 10^5$ eV have $\gamma = U/(mc^2) = 2 \times 10^5$. The angular frequency of the emitted radiation is a factor of 4×10^{10} greater than the cyclotron frequency ω_B!

Electron energies in Crab nebula

Turn the problem around and use the existence of the polarized optical radiation to deduce the energy U of the emitting electrons in, for example, the Crab nebula. The magnetic field in the nebula is found to be about $B \approx 5 \times 10^{-8}$ T from equipartition of energy; see below (Section 4 under "Crab nebula"). Thus, the cyclotron frequency for an electron is

$$\nu_B = \frac{1}{2\pi} \frac{eB}{m} = 1.4 \times 10^3 \text{ Hz}, \quad \text{(Cyclotron frequency; Crab nebula)} \quad (8.11)$$

where m is the electron mass.

The observed radiation is in the optical band, and so we take its frequency to be typical of that band, $\nu_{syn} \approx 5 \times 10^{14}$ Hz. From (9), the ratio $\omega_{syn}/\omega_B = \nu_{syn}/\nu_B$ is about equal to γ^2. Thus, we can solve for the Lorentz factor γ of the individual electrons that give rise to the radiation:

$$\gamma \approx \left(\frac{\nu_{syn}}{\nu_B} \right)^{1/2} = \left(\frac{5 \times 10^{14}}{1.4 \times 10^3} \right)^{1/2} = 6.0 \times 10^5. \quad (8.12)$$

The electrons exceed their rest energy, $mc^2 = 0.5$ MeV, by a factor of $\sim 600\,000$. The total energy $U = \gamma mc^2$ of a typical electron giving rise to optical photons is thus $\sim 3 \times 10^{11}$ eV. The electrons in the Crab nebula are *extremely* relativistic.

Power spectrum shape

The power emitted by a single relativistic electron circulating in a magnetic field contains many frequencies other than ν_{syn} (10). The shape of this waveform in Fig. 4b is not a perfect sine wave of frequency ν_{syn}; it contains frequencies ranging down to the orbital period and up to roughly ν_{syn}. The actual spectral distribution of the radiation follows from an exact calculation of this waveform (Fig. 4b) and a Fourier analysis of it.

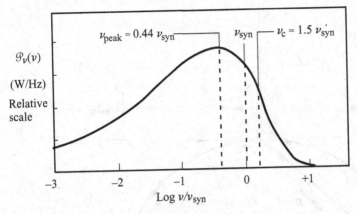

Figure 8.7. Distribution of radiated power $\mathscr{P}(\nu)$ (W/Hz) plotted with respect to $\log \nu/\nu_{\text{syn}}$ for a single orbiting electron. The vertical axis is a relative linear scale. This gives the relative power in the component frequencies of the waveform (Fig. 4b) of the radiated electric field. The peak and critical frequencies are indicated with their ratios relative to ν_{syn}. Note that ν_{syn} is a function of the pitch angle ϕ. [After B. Rossi and S. Olbert, *Introduction to the Physics of Space*, McGraw Hill, 1970, p. 42; data from V. Ginzburg and S. Syrovatskii, Table 1, *ARAA* **3** 297 (1965)]

The result is plotted in Fig. 7 with a logarithmic abscissa scale. The ordinate is a linear scale of emitted power per unit frequency (W/Hz) at frequency ν, namely, $\mathscr{P}_\nu(\nu)$. This is known as a *power spectrum*. Absolute values are not given; it is the shape of the function that interests us for now.

The quantity $\mathscr{P}_\nu(\nu)$ is our usual spectrum of power versus frequency of electromagnetic radiation. In this case, it is the power emitted from a *single* electron of well-defined energy U and pitch angle ϕ in a magnetic field $\boldsymbol{B}$. If the waveform were a perfect sinusoid, all the power would be at a single frequency, and the plot would exhibit a single spike (delta function). In our case (Fig. 7), the power extends over a broad range of frequencies with a peak at $\nu_{\text{peak}} = 0.44 \, \nu_{\text{syn}}$.

It is conventional to define a critical frequency $\nu_c = 1.5 \nu_{\text{syn}}$ that characterizes the frequency range of rapid descent of the function

$$\nu_c \equiv \frac{3}{2}\nu_{\text{syn}} = \frac{3}{2}\gamma^2 \frac{\omega_B}{2\pi}\sin\phi, \qquad \text{(Critical frequency; Hz)} \qquad (8.13)$$

where ν_{syn} is defined in (10). For the charge and mass of an electron, this becomes

$$\begin{array}{ll} \nu_c = 6.3 \times 10^{36} \quad U^2 \quad B \sin\phi & \text{(Critical synchrotron} \qquad (8.14) \\ \text{(Hz)} \qquad\qquad (\text{J}^2) \quad (\text{T}). & \text{frequency for electron)} \end{array}$$

Note that these frequencies, ν_{syn}, ν_{peak}, and ν_c, are all proportional to $\sin\phi$; they decrease with decreasing pitch angle.

The energy $h\nu_c$ of an emitted photon can never exceed the kinetic energy $E_k = U - mc^2 \approx U$ of the electron that emits it; otherwise, energy conservation would be violated. Because the critical photon energy $h\nu_c$ increases as the square of the electron energy U^2 (14), this violation would occur if the electron energy were sufficiently high.

For many practical cases of interest, this is not an issue. For example, in the field encountered at the earth's surface, $\sim$50 μT, this condition occurs at $U = 3 \times 10^{19}$ eV for $\phi = \pi/2$

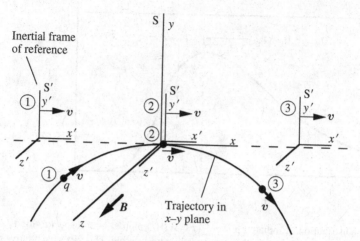

Figure 8.8. Circular motion of a positive charge q moving through a uniform magnetic field in the x–y plane of the stationary or laboratory frame S. The frame of reference S′ moves with the same speed v as the charge but along a straight line. The circled numbers 1, 2, 3 mark the positions of the electron and the S′ frame at three times (in S time). When the charge arrives at the origin of both S and S′, it is momentarily at rest in S′ and is accelerating downward. The classical Larmor radiation formula may therefore be applied in S′.

(Prob. 31). The most energetic cosmic rays detected are of comparable energies. In weaker magnetic fields, such as those of the interstellar medium (~0.3 nT), the energy limit is even higher. In the extremely high fields of some neutron stars, 10^8 T, the limit is much lower, $U \approx 15$ MeV, and it does become an issue, as discussed in Section 6.

8.4   Power radiated by the electron

In this section, we calculate the instantaneous total power radiated by a single charge moving at relativistic speed ($v \approx c$) in a magnetic field. Unfortunately, Larmor's formula for the power emitted applies only to the nonrelativistic case. We get around this by first observing the charge in its rest frame, where Larmor's formula is valid, and then by making the appropriate Lorentz transformations back to the stationary frame.

Two frames of reference

Stationary frame of reference

The circular trajectory of a positive charge q traveling with velocity v normal to the local static magnetic field B (i.e., $\phi = 90°$) is shown in Fig. 8 from the point of view of an observer in the stationary (laboratory) frame S. The B field is taken to be uniform everywhere in space and to be directed along the $+z$-axis.

Under the influence of the $q(v \times B)$ force, the charge follows a circular path at constant speed in the x–y plane at fixed z; the velocity v of the charge is always perpendicular to the magnetic field direction. The stationary coordinate system S is chosen so that the charge passes through its origin at the top of its orbit. The charge q is shown just as it passes the origin of S and also at earlier and later times indicated with circled numbers.

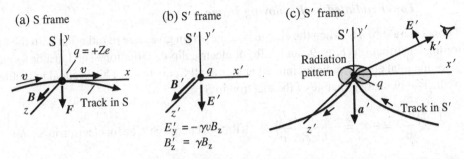

Figure 8.9. Charge $q = +Ze$ as viewed (a) by an observer in the S frame of reference at the instant it is at the origin of frames S and S' and (b,c) by an observer in the S' frame. At the origin in S', the stationary charge is momentarily at rest and hence encounters no magnetic force, but it does experience a downward electric field and hence a downward acceleration. The resultant radiation has the classical dipole (toroid) radiation pattern shown in (c). The track of the charge before and after its arrival at the origin is shown in (a) and (c).

The force on the charge q in the S frame (Fig. 9a) is, with (1) repeated,

$$F = q(v \times B). \tag{8.15}$$

At the origin of S, time circle 2 of Fig. 8, the force is in the negative y direction:

$$F_y = -qvB_z. \tag{8.16}$$

The force per unit charge is

$$F_y/q = -vB_z. \tag{8.17}$$

Unfortunately, we can not use this force to directly obtain the acceleration and power radiated because the charge is moving relativistically. We thus move to a frame of reference where the charge is momentarily at rest.

Moving frame of reference

Choose an inertial frame of reference S' that moves in a straight line at the same speed as the charge. Its axes, x', y', and z', are each parallel with those of S, and the x- and x'-axes are coaligned. Let the origin of S' arrive at the origin of S at exactly the time the charge is there. At that moment, the charge and S' are moving parallel to one another, and the charge is momentarily at rest in S'. The charge is shown at the moment it is at the origins of S and S' in Fig. 9a,b,c.

The track followed by the charge on its approach and departure from the origin is shown for the two frames in Fig. 9a,c. In S, it is the circular track of Fig. 8. In S', the charge approaches from the lower right and slowly rises toward the origin from below. It hesitates at the origin and then moves downward and off to the left (Prob. 44). Although the speed in S' is zero at the instant it is at the origin, the charge is undergoing downward acceleration, and hence it radiates electromagnetic radiation. The slow motion in S' allows us, as previously stated, to make use of the nonrelativistic Larmor radiation formula (2).

Power radiated in the moving frame

The acceleration in S′ when the charge is at the origin gives rise to radiation with the familiar toroidal distribution (Figs. 9c and 5.2b) of electric dipole radiation. The Larmor formula (2) gives the total energy radiated into all angles at all frequencies. This is equal to the negative of the rate of energy change of the electron in S′,

$$\frac{dU'}{dt'} = -\mathscr{P}' = -\frac{1}{6\pi\varepsilon_0}\frac{q^2 a'^2}{c^3}, \quad \text{(Rate of energy change for electron in S′; W)} \quad (8.18)$$

where the primed quantities remind us that they are measured in S′. The rate of energy change dU'/dt' of the electron is negative because it is losing (rather than gaining) energy. Our immediate objective is to evaluate (18). We do this by finding the acceleration a' in the S′ frame.

Electric field

The electric and magnetic fields in S′ for this case have been worked out in Chapter 7 from the relativistic transformations for such fields. In the S frame, we postulated a **B** field in the z direction (B_z), which we found (7.32) transforms to the S′ frame as an increased magnetic field, $B_z' = \gamma B_z$, and a "new" electric field E_y' in the negative y' direction:

$$E_y' = -\gamma\beta c B_z = -\gamma v B_z$$
$$B_z' = \gamma B_z. \quad\quad\quad\quad\quad \text{(Fields in S′)} \quad (8.19)$$

The **B′** and **E′** vectors are shown in Fig. 9b.

Consider the "new" downward electric vector. In S′, the charge is at rest at the origin, but it has a downward acceleration that arises (in S) from its interaction with the **B** field. This downward acceleration indicates that there must be a force $F' = dp'/dt'$ on the charge in S′, but it can not be a magnetic force ($F' = qv' \times B'$) because $v' = 0$; magnetic fields exert no force on a charge at rest. The S′ observer must therefore conclude that the force is due to an electric field because electric fields can act on a charge at rest. Indeed the Lorentz transformation tells us that such a field does exist in S′.

The S′ electric field, $E_y' = -\gamma v B_z$ (19), is, by definition, the force per unit charge. Compare this with the magnetic force per unit charge in frame S, which is $F_y/q = -v B_z$ (17). They are similar in form except that the Lorentz factor γ appears in S′; the downward force in S′ is larger by a factor of γ than the downward force in S. This is strictly an effect of special relativity and would not be expected classically.

Acceleration

The acceleration a' in the S′ frame can be expressed simply as the force-to-mass ratio, where the electric force is qE_y'. Use (19) and set $q = Ze$, where Z is the atomic number of the charge and e is the electron charge (a positive definite quantity) to obtain

$$a' = \frac{F_y'}{m} = \frac{qE_y'}{m} = -\frac{Ze\gamma v B}{m}, \quad\quad\quad\quad (8.20)$$

where we drop the subscripts for $B \equiv B_z$ and $a' \equiv a'_y$. For relativistic particles, one uses the relations $v \approx c$ and $U = \gamma m c^2$ to obtain

$$a' = -\frac{ZeBU}{m^2 c}. \qquad \text{(Acceleration in S'; } v \approx c \text{)} \qquad (8.21)$$

The quantities B and U are the given values that can be measured in S.

Energy-loss rate

Substitute (21) into (18) to obtain the rate of energy change of the electron in the S' frame in terms of the given quantities

$$\frac{dU'}{dt'} = -\frac{1}{6\pi \varepsilon_0} \frac{(Ze)^2}{c^3} \left(\frac{ZeBU}{m^2 c}\right)^2 \qquad (8.22)$$

$$\frac{dU'}{dt'} = -\frac{1}{6\pi \varepsilon_0} \frac{1}{c^5} \left(\frac{Ze}{m}\right)^4 U^2 B^2. \qquad \begin{array}{l} \text{(Rate of energy change} \\ \text{of electron in S'; W)} \end{array} \qquad (8.23)$$

This is the negative of the power radiated according to an S' observer if $\beta \approx 1$ and $\phi = \pi/2$.

Power radiated in the stationary frame

The final step is to determine dU/dt, the rate of electron energy change in the stationary (S) frame of reference.

Transformation to the stationary frame

Consider the packet of energy dU' emitted in S' in time dt'. The relativistic transformations for these differential quantities will be the same as those for the quantities U' and t' themselves because the transformations are linear. Because the energy dU' in S' is radiated in the $+x$ and $-x$ directions equally (see toroid in Fig. 9c), the total momentum p'_x of the radiated energy packet in S' is zero. The transformation (x2.8) in Table 7.2 thus yields

$$dU = \gamma \, dU'. \qquad (8.24)$$

Because all of the emission (over the period dt') takes place at the same position x' in S', one has $dx' = 0$. Thus, (x1.8) in Table 7.1 yields

$$dt = \gamma \, dt', \qquad (8.25)$$

and hence,

$$\frac{dU}{dt} = \frac{dU'}{dt'}. \qquad (8.26)$$

The rate of energy loss turns out to be the same in the S frame as it is in the S' frame. In the stationary frame S, the instantaneous rate of energy loss by the particle is, therefore, from (23),

$$\frac{dU}{dt} = -\frac{1}{6\pi \varepsilon_0} \frac{1}{c^5} \left(\frac{Ze}{m}\right)^4 U^2 B^2 \beta^2 \sin^2 \phi, \qquad \text{(Energy loss rate in S; W)} \qquad (8.27)$$

where U is the total particle energy (rest plus kinetic) of the radiating particle, m is its mass, Ze is its charge, $B = B_z$ is the magnitude of the static magnetic field, and ε_0 ($= 8.854 \times 10^{-12}$ SI units) is the vacuum permittivity. The pitch angle ϕ and the particle speed parameter $\beta = v/c$ have been inserted to yield the expression that a full derivation would have produced (Prob. 45). They allow for pitch angles other than 90° and for particle speeds less than c.

Electrons are by far the most efficient radiators of synchrotron radiation. This follows from the factor $(Z/m)^4$ in (27). The electron mass is 1836 times less than that of a proton, but they have the same charge ($Z = 1$). The $(Z/m)^4$ factor is thus ~10^{13} times greater for an electron than a proton. For bare atomic nuclei with charge $+Ze$, the ratio Z/m is comparable to that of a proton (within a factor of 2). In astrophysics, therefore, electrons are believed to be the only particles that give rise to measurable synchrotron radiation.

Numerical values for the fixed parameters in (27) may be introduced (Prob. 42) to yield convenient forms for the energy-loss equation for radiating *electrons* ($Z = 1$):

$$\frac{dU}{dt} = -2.37 \times 10^{12} \, U^2 B^2 \beta^2 \sin^2 \phi. \qquad (8.28)$$
$$\text{(W)} \qquad\qquad\qquad (\text{J}^2)(\text{T}^2)$$

We reiterate that this is the power loss from a *single* spiraling electron with pitch angle ϕ. A second form in terms of electron volts (eV) is

$$\frac{dU}{dt} = -3.79 \times 10^{-7} \, U^2 B^2 \beta^2 \sin^2 \phi. \qquad (8.29)$$
$$\text{(eV/s)} \qquad\qquad (\text{eV}^2)(\text{T}^2)$$

These formulas give the total rate of energy loss in the observer's frame. Remember that the energy is emitted anisotropically as the result of forward beaming (Fig. 4a).

Magnetic energy density as target

Equation (27) is the relation we set out to derive – namely the rate of energy loss by a synchrotron-emitting charge $q = Ze$. It is strongly dependent (as the square) on both the energy U of the charge and the magnetic field B. It is thereby proportional to the energy density of the magnetic field u_B, which in electromagnetic theory is

$$u_B = \frac{\boldsymbol{B} \cdot \boldsymbol{B}}{2\mu_0} = \frac{B^2}{2\mu_0}, \qquad \text{(Energy density of } \boldsymbol{B} \text{ field; J/m}^3) \qquad (8.30)$$

where μ_0 is the permeability constant ($4\pi \times 10^{-7}$ in SI units). We will use this just below in (34) to simplify the energy-loss formula. In effect, the electrons in the synchrotron process are colliding with, and being accelerated by, the local magnetic field in a way that is analogous to the collisions with electric fields in the thermal-bremsstrahlung process.

A random choice of possible directions v of the charge relative to the magnetic field line may be taken into account by averaging the $\sin^2 \phi$ term in (27) over all directions,

$$\langle \sin^2 \phi \rangle_{\text{av}} = \frac{1}{4\pi} \int_{\text{sphere}} \sin^2 \phi \; d\Omega = \frac{2}{3}, \qquad (8.31)$$

where the solid angle element is $d\Omega = 2\pi \sin \phi \, d\phi$. The Thomson cross section (2.49) may also be invoked,

$$\sigma_T = \frac{8}{3}\pi r_e^2 = 6.6525 \times 10^{-29} \text{ m}^2, \qquad \text{(Thomson cross section)} \qquad (8.32)$$

where the classical radius of the electron is

$$r_e = \frac{1}{4\pi\varepsilon_0}\frac{e^2}{mc^2} = 2.8179 \times 10^{-15} \text{ m} \qquad (8.33)$$

and where m and e are the mass and charge of the electron, respectively. With the aid of (30)–(33), $U = \gamma mc^2$, and finally $c^2 = (\mu_0\varepsilon_0)^{-1}$ from electromagnetic theory, the energy loss expression (27) may be written as

$$\Rightarrow \qquad \frac{dU}{dt} = -\frac{4}{3}\sigma_T c\beta^2\gamma^2 u_B, \qquad \begin{array}{l}\text{(Energy loss of electron in terms}\\ \text{of magnetic energy density)}\end{array} \qquad (8.34)$$

where γ is the Lorentz factor of the electron, $\beta \equiv v/c$, v is the electron speed, σ_T is the Thomson cross section, and u_B is the energy density (J/m^3) of the magnetic field. This is the energy loss rate by a relativistic single electron ($Z = 1$) averaged over randomly chosen pitch angles. Substitute in the constant factors:

$$\frac{dU}{dt} = -2.66 \times 10^{-20}\beta^2\gamma^2 u_B \qquad (8.35)$$
$$\text{(W)} \qquad\qquad \text{(J/m}^3\text{)}.$$

Electron energy lifetime

The characteristic lifetime τ of an electron of energy U that is losing energy to synchrotron radiation is approximately the energy it currently has divided by the rate at which it is losing energy. For $dU/dt \propto -U^2 B^2$ (27), we have the following proportionality:

$$\tau = -\frac{U}{dU/dt} \propto \frac{U}{U^2 B^2} \propto \frac{1}{U B^2}. \qquad (8.36)$$

Formally this is the time it would take the electron to lose all its energy if it continued to lose energy at the initial rate. However, when the electron has decayed to lower energies, its rate of energy loss ($\propto U^2$) is much lower. Thus, τ is a *characteristic* lifetime during which the electron loses an amount of energy comparable to its initial energy (but not 100%). The characteristic lifetime increases as the energy decreases; this is unlike exponential decay, in which the characteristic time remains constant.

A quantitative value for the lifetime can be obtained for the $\beta \approx 1$ and $\phi = \pi/2$ case if we use our expression for dU/dt (28) in (36) and introduce $mc^2 = 8.2 \times 10^{-14}$ J for the electron:

$$\Rightarrow \qquad \tau = \frac{5.16}{B^2}\frac{mc^2}{U} = \frac{5.16}{B^2}\frac{1}{\gamma}, \qquad \begin{array}{l}\text{(Electron decay time; s;}\\ \beta = 1; \phi = \pi/2; B \text{ in teslas)}\end{array} \qquad (8.37)$$

where $B = B_z$ is the magnitude of the magnetic field (assumed normal to the velocity vector), U is the total electron energy, m is the electron mass, and $\gamma = U/mc^2 = (1 - \beta^2)^{-1/2}$ is the Lorentz factor.

Table 8.1. *Synchrotron radiation (Crab nebula)*

	Photon frequency[a] ν_{syn} (Hz)	Electron energy[b] U (eV)	Electron lifetime[c] τ (yr)
Radio (0.5 GHz)	5×10^8	3.05×10^8	109,000
Optical (600 nm)	5×10^{14}	3.05×10^{11}	109
X ray (4.1 keV)	1.0×10^{18}	1.37×10^{13}	2.4
Gamma ray (41 MeV)	1.0×10^{22}	1.37×10^{15}	0.024 (9 d)

[a] Detected frequency, from left column, is taken to be ν_{syn} (10).

[b] Energy of an electron that emits radiation at the frequency ν_{syn} from (10) for pitch angle $\phi = \pi/2$ and magnetic field $B = 5 \times 10^{-8}$ T.

[c] Characteristic time it takes for the electron to lose a substantial portion of its energy (37).

Crab nebula

These several formulas can be applied to the Crab nebula to derive the approximate electron energy and the approximate lifetime of the emitting electrons for the radio, optical, and x-ray bands (Table 1). In each case, a typical frequency for the band is chosen, and a magnetic field strength of $B = 5 \times 10^{-8}$ T($= 50$ nT) is adopted. This field strength is deemed to be reasonable from an energy equipartition argument in which particle and magnetic field energy densities in the nebula are taken to be of comparable magnitudes. Another approach, based on modeling *synchrotron self-Compton* (SSC) emission (Section 9.4) yields a somewhat smaller value, 16 nT; see also the caption to Fig. 2.

The electron energies in Table 1 are impressive because they are so high. The 14-TeV electrons that give rise to the x-ray photons are twice the proton energy expected from the Large Hadron Collider (LHC) energy. The electrons giving rise to the gamma rays are ~100 times more energetic than the highest accelerator energies!

Short lifetimes

The lifetimes are equally surprising. For the optical, x, and gamma radiation, the electron lifetime is very much shorter than the 950-year "age" of the Crab nebula. There is no way that the supernova explosion in 1054 CE could have given the electrons the energy they *now* have; they would long ago have given up their energy to synchrotron photons. (Although the Crab is ~6000 LY distant, the characteristics we detect today on earth represent the nebula's state 950 years after the supernova event.)

This was one of the biggest puzzles in astrophysics in the years after the discovery that the continuum radiation is synchrotron radiation. It is clear that something, even today, is still feeding energy into the electrons that emit the synchrotron radiation. Because this radiation amounts to about 10^{31} W, or 10^5 times the solar luminosity, it is a very powerful energy generator.

Crab pulsar

This puzzle was resolved only with the discovery of radio pulsars, which were determined to be spinning neutron stars. One such object is in the Crab nebula. It is a neutron star spinning at 30 Hz (30 complete rotations per second). The power source for the synchrotron radiation from the Crab nebula is the rotational energy of this rapidly spinning Manhattan-sized neutron star. Energy from the initial implosion in 1054 CE was stored in the rapid rotation, which has gradually been slowing ever since. The neutron star's organized rotational energy (a big flywheel) is converted to very high energy electrons that radiate via synchrotron radiation. See the discussions of spinning neutron stars in Section 6 and of compact stars in Section 4.4.

8.5   Ensemble of radiating particles

The rate of electron energy loss and the frequency of the emitted radiation presented heretofore apply to the radiation from a *single* electron of energy U. Realistic situations involve the collective effects of many radiating electrons of differing energies. The summed radiation from all the electrons must be calculated to find the total synchrotron-radiation intensity.

The summation is particularly straightforward if the energy spectrum of the electrons is taken to be a power law and certain simplifying assumptions are made. In this case, the photon spectrum also turns out to be a power law. Power-law spectra are found frequently in nature, and so an examination of this spectral form is relevant to many astrophysical objects.

In this section, we find the spectral volume emissivity j_ν (W m^{-3} Hz^{-1}) and then the specific intensity I (W m^{-2} Hz^{-1} sr^{-1}) of synchrotron radiation from an electron population with a power-law spectrum that interacts with a uniform magnetic field of constant magnitude and direction. We then touch upon two examples: the diffuse radiation from the Crab nebula and the galactic radio background, both of which arise from the synchrotron process.

Power-law spectra

An ensemble of high-energy particles, in the interstellar medium or in the Crab nebula for example, is found typically to have a spectrum that is a power law over wide ranges of energy. The energy spectrum of the particles can be directly measured in the case of cosmic rays arriving at the earth or inferred from the radiated synchrotron spectrum in astrophysical sources. In general these spectra decrease with increasing energy.

The several ways one can present power-law *particle* spectra are summarized here. We encountered particle spectra previously in Section 3.3.

Number specific-intensity

A power-law energy spectrum, by definition, varies with energy U raised to a fixed power p. For a particle spectrum, we write

$$J(U)\,dU \propto U^p\,dU, \qquad \text{(Particles s}^{-1}\text{ m}^{-2}\text{ sr}^{-1} \text{ at } U \text{ in d}U)\qquad (8.38)$$

where $J(U)$ (particles s^{-1} m^{-2} J^{-1} sr^{-1}) is the *number specific-intensity* introduced in (3.17). The quantity U is the total (kinetic + rest) energy. It follows that the product $J(U)\,dU$ is the number of particles in the interval dU at U per unit (s m^2 sr). Note that particle spectra are typically expressed as functions of particle energy U, whereas photon spectra are generally expressed as functions of frequency, which is proportional to the photon energy $h\nu$.

Often the particle spectrum is very steep. For example, in the Crab nebula it is deduced to be $p \approx -3.0$ for large parts of the spectrum. For cosmic rays it is measured over large ranges of energy to be

$$p \approx -2.5 \qquad \text{(Power-law exponent; galactic cosmic rays)} \qquad (8.39)$$

with moderate changes in exponent at a "knee" and at an "ankle" in the spectral shape. On the other hand, we saw that the blackbody energy spectrum rises with photon energy (or frequency) as a power law $I \propto \nu^2$ (6.8). Its number spectrum, $I/h\nu$, would still rise with frequency, $\propto \nu^1$.

On a log-log plot, a power-law spectrum is a straight line with slope equal to the exponent p. The exponent is thus the *logarithmic slope*. The *spectral index* is often defined as the *negative* of the logarithmic slope, and so one must be careful to note the sign convention when working with such spectra.

Energy specific-intensity

A particle spectrum may also be described in terms of the energy flux rather than the number flux, that is, with an *energy spectrum* $I_p(U)$ (W m^{-2} J^{-1} sr^{-1}). This is obtained simply by multiplying the number of particles in each energy interval (38) by the energy, U, of each particle. Thus,

$$I_p(U)\,dU \propto U^{p+1}\,dU, \qquad \text{(W m}^{-2}\text{ sr}^{-1}\text{ at } U \text{ in } dU) \qquad (8.40)$$

where $I_p(U)$ (W m^{-2} J^{-1} sr^{-1}) is the *energy specific-intensity* of the particles. The product $I_p(U)\,dU$ is the energy contained in all the particles that individually have energy U in the interval dU per unit (s m^{2}sr). Note that the exponent is one power greater than that of the number spectrum. A steep decreasing number spectrum (negative p) becomes a less steep energy spectrum.

Spectra of energetic particles may be given either as a number spectrum or as an energy spectrum. If the particles are not highly relativistic, one must take care to distinguish kinetic and total energy. Here we assume the particles are highly relativistic, and thus the rest energy is negligible and the kinetic and total energies are essentially equal to one another.

Number density

The number of particles per cubic meter at U in dU, including all directions of travel, may be written as

$$n(U)\,dU \propto U^p\,dU, \qquad \text{(Particles/m}^3\text{ at } U \text{ in } dU) \qquad (8.41)$$

where we adopt the same functional dependence for $n(U)$ (m^{-3} J^{-1}) as for $J(U)$ (38). This is valid if the particles are all moving at the same speed v, as we now explain.

The particle specific-intensity $J(U)$ (s^{-1} m^{-2} J^{-1} sr^{-1}) may be written in terms of the particle density $n(U)$ (m^{-3} J^{-1}) and speed v, namely $J = nv/4\pi$, if the particle motions are isotropic; see discussion preceding (6.22). In our case, the speeds are assumed to be relativistic, $v \approx c$, and so the conversion from $n(U)$ to $J(U)$ is very nearly independent of speed, $J \approx nc/4\pi$. In this limit, $J(U)$ and $n(U)$ are linearly related. Their distributions with energy will thus have the same logarithmic slope p.

Define the proportionality constant of (41) to be the ratio of two constants, n_0 and U_0^p:

➡ $$n(U)\, dU = \frac{n_0}{U_0^p} U^p\, dU = n_0 \left(\frac{U}{U_0}\right)^p dU. \qquad \text{(Particles/m}^3 \text{ in } dU \text{ at } U) \qquad (8.42)$$

Set $U = U_0$ to see that n_0 is the density at U_0, that is, $n_0 = n(U_0)$. If U_0 is chosen to be an energy in the spectral region of interest, the coefficient n_0 will represent the actual particle number-density, per unit joule, at that energy.

Energy density

For completeness, we write the energy-density spectrum $u(U)$ (J m^{-3} J^{-1}). Multiply $n(U)\, dU$ (42) by the energy U of a single particle:

$$u(U)\, dU = U n(U)\, dU = U n_0 \left(\frac{U}{U_0}\right)^p dU = u_0 \left(\frac{U}{U_0}\right)^{p+1} dU.$$

$$\text{(Energy density in } dU \text{ at } U) \qquad (8.43)$$

In the last step, we multiplied by U_0/U_0 and defined $u_0 \equiv U_0 n_0$, the energy density per joule (J m^{-3} J^{-1}) at energy U_0. This yields the $p + 1$ exponent seen in energy flux expression (40), illustrating again the similar distributions of flux and density in the relativistic limit.

These several expressions describe only the particle-energy distributions. Now we turn to the radiative power they emit via synchrotron radiation and the spectrum of the latter.

Volume emissivity

Here, we sum the synchrotron power emitted by individual electrons. We restrict our consideration to highly relativistic electrons of charge e and mass m and immediately focus on the radiation emerging from unit volume.

Function of particle energy

The energy loss rate to synchrotron radiation for a *single* electron of energy U is, from (27), for $\beta \approx 1$ and $\phi = \pi/2$,

$$dU/dt \propto -U^2 B^2, \qquad \text{(Single particle; W)} \qquad (8.44)$$

where, for now, we do not carry along the constants to better illustrate the essence of the argument. From this we obtain the *photon* volume emissivity $j_u(U)$ (W m^{-3} J^{-1}) as a function of U. (The subscript "u" reminds us that the power is per unit energy interval.)

The total radiated power $j_u(U)\, dU$ of all the electrons of energy U in dU in unit volume is the negative of the rate of energy loss of a single electron (44) multiplied by the number of electrons at U in dU in a cubic meter (42):

$$j_u(U)\, dU = -(dU/dt) \times n(U)\, dU. \qquad \text{(W/m}^3 \text{ at } U \text{ in } dU) \qquad (8.45)$$

Invoke the abbreviated forms for dU/dt (44) and for $n(U)$ (41) to obtain

$$j_u(U)\, dU \propto U^2 B^2 U^p\, dU. \qquad (8.46)$$

and

$$j_u(U) \propto B^2 U^{p+2}. \qquad \text{(Volume emissivity of radiation; W m}^{-3}\text{ J}^{-1}) \qquad (8.47)$$

This is the power in radiation emerging from unit volume from electrons of energy U in unit joule bandwidth. For simplicity, as always, the magnetic field $\boldsymbol{B}$ has magnitude B and is taken to be a constant throughout the emission region.

Function of emitted frequency

Now find the volume emissivity in terms of the frequency of the emitted radiation, $j_\nu(\nu)$ (W m^{-3} Hz^{-1}). We encountered this quantity previously; see (5.1). Consider a simple model in which all the energy given up by an electron of energy U is emitted as photons of the critical frequency ν_c, from (14):

$$\nu_c \propto U^2 B. \qquad \text{(Characteristic frequency of synchrotron radiation)} \qquad (8.48)$$

This provides a one-to-one relation between the particle energy and the emitted frequency and avoids an integration over the emitted band of frequencies in Fig. 7. Our task then becomes a straightforward unit conversion.

The power radiated from electrons of energy U in dU in unit volume must equal the power radiated by the same electrons into photons of frequency ν in the frequency interval $d\nu$ that corresponds to the interval dU. Thus,

$$j_\nu(\nu)\, d\nu = j_u(U)\, dU. \qquad \text{(W/m}^3\text{ at }\nu\text{ in }d\nu\text{ or at }U\text{ in }dU) \qquad (8.49)$$

The desired photon (synchrotron) volume emissivity is therefore

$$j_\nu(\nu) = j_u(U)\frac{dU}{d\nu}. \qquad \text{(W m}^{-3}\text{ Hz}^{-1}) \qquad (8.50)$$

We will convert both right-hand factors to functions of ν with the aid of (48).
Set $\nu_c = \nu$ and rewrite (48) as

$$U \propto (\nu/B)^{1/2}. \qquad (8.51)$$

Substitute into (47) to obtain j_u as a function of frequency:

$$j_u(U) \propto B^2 \left[\left(\frac{\nu}{B}\right)^{1/2} \right]^{p+2} = B^{(2-p)/2} \nu^{(2+p)/2}. \qquad \begin{array}{l}\text{(Volume emissivity per}\\ \text{joule; W m}^{-3}\text{ J}^{-1})\end{array} \qquad (8.52)$$

The relation between dU and $d\nu$ is, from (51),

$$dU \propto B^{-1/2} \nu^{-1/2}\, d\nu, \qquad \text{(Relation between intervals)} \qquad (8.53)$$

where B is taken to be a fixed value. The relation between the intervals depends on the frequency ν (or equivalently on U) because U and ν are not linearly related.

Finally, substitute (52) and (53) into (50) to find $j_\nu(\nu)$:

$$j_\nu(\nu) \propto B^{(2-p)/2}\nu^{(2+p)/2}B^{-1/2}\nu^{-1/2}. \quad \text{(Volume emissivity; W m}^{-3}\text{ Hz}^{-1}) \qquad (8.54)$$

Collect terms and multiply by the bandwidth $d\nu$ to obtain, for $\beta \approx 1$ and $\phi = \pi/2$,

⇒ $$j_\nu(\nu)\,d\nu \propto B^{(1-p)/2}\nu^{(1+p)/2}d\nu. \qquad \text{(Synchrotron photon-energy}$$
$$\text{spectrum; W/m}^3 \text{ at } \nu \text{ in } d\nu) \qquad (8.55)$$

This gives the desired volume emissivity $j_\nu(\nu)$ of the emitted photons. Like the original electron-number spectrum (41), it is also a power law but is expressed in terms of electro-magnetic wave frequency ν rather than particle energy U.

The logarithmic slope of the photon-energy spectrum is the exponent of ν in (55), namely $(p+1)/2$, which we designate as α:

⇒ $$\alpha \equiv \frac{p+1}{2}. \qquad (8.56)$$

This relates the logarithmic slope p of the particle-*number* spectrum to the logarithmic slope α of the photon-*energy* spectrum. The photon-energy spectrum (55) may thus be written as

$$j_\nu(\nu)\,d\nu \propto B^{1-\alpha}\nu^\alpha d\nu, \qquad \text{(W/m}^3 \text{ at } \nu \text{ in } d\nu) \qquad (8.57)$$

where $j_\nu(\nu)$ is the volume emissivity with units W m^{-3} Hz^{-1}.

It is convenient to define the coefficient of (57) as a combination of terms,

$$j_\nu(\nu)\,d\nu = j_\nu(\nu_0)\left(\frac{B}{B_0}\right)^{1-\alpha}\left(\frac{\nu}{\nu_0}\right)^\alpha d\nu, \qquad (8.58)$$

where B_0 and ν_0 are constants with dimensions of magnetic field and frequency, respectively, and $j_\nu(\nu_0)$ is the volume emissivity for magnetic field B_0 at the frequency ν_0. It is usually convenient to choose values for B_0 and ν_0 to be round numbers that are close to the values of interest (e.g., $B_0 = 5.00 \times 10^{-8}$ T and $\nu_0 = 1.00 \times 10^{18}$ Hz for synchrotron x rays generated in the magnetic field of the Crab nebula). Then, $j_\nu(\nu_0)$ is a meaningful quantity close to the actual physical value of interest.

The particle-number slope p in most sources is more negative than -1, which ensures that the photon-energy slope $\alpha = (p+1)/2$ will also be negative. The particle-*energy* slope, as given in (40), is $p+1$, which is just twice the photon-*energy* slope α. This relation between the two energy spectra is easily recalled.

The relation between the several spectral log slopes discussed here is illustrated in the log-log representations of Fig. 10. The example shown has $p = -3$ for the electron number-density spectrum, $p+1 = -2$ for the electron energy-density spectrum, and $\alpha = (p+1)/2 = -1$ for the photon energy-density (volume-emissivity) spectrum. These spectra, sequentially, are less and less steep.

Our logic would have yielded the proportionality constants for $j(\nu)$ in (57) or (58) had we carried along the original coefficients for $n(U)$, dU/dt, and ν_c as given in (42), (27), and (13), respectively. If one carries through with this, adopting $Z = 1$ (for an electron),

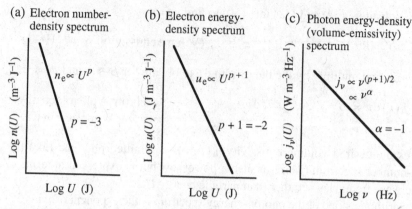

Figure 8.10. Power-law spectra for a synchrotron emitting system on log-log plots: (a) the number-density spectrum of the emitting electrons, (b) their energy-density spectrum, and (c) the photon energy-density spectrum. A logarithmic slope of -3 was adopted for (a); the other slopes follow from this. The logarithmic slopes of these three "density" spectra will be the same as the associated specific-intensity spectra if all the particles of (a) and (b) have $v \approx c$.

$\sin \phi = 1$, $\beta \approx 1$, and the definition of ν_B (11), the result (Prob. 52) is very close to the correct expression, which (averaged over all pitch angles ϕ) is

➡ $$ j_\nu(\nu) = 8\pi^2 \left(\frac{e^2}{4\pi\varepsilon_0} \frac{\nu_B}{c} \right) a(p)\,n_0 U_0 \left(\frac{mc^2}{U_0} \right)^{2\alpha} \left(\frac{2\nu}{3\nu_B} \right)^\alpha, \quad \begin{array}{l}\text{(Volume emissivity; } \beta \approx 1; \\ \text{W m}^{-3}\text{ Hz}^{-1})\end{array} \qquad (8.59) $$

where j_ν is the volume emissivity of radiation (W m^{-3} Hz^{-1} = J m^{-3}), ν is the frequency of the radiation, $\alpha = (p+1)/2$ is the logarithmic slope of the photon-energy spectrum, and n_0, U_0, and p are parameters of the emitting-particle spectrum (42).

The correct result (59) includes the dimensionless factor $a(p)$, which depends on the logarithmic particle spectral slope p and is of order 0.1. It arises from a proper calculation that incorporates the single-electron energy spectrum (Fig. 7) and takes into account all pitch angles in an isotropic distribution of velocities. Values of $a(p)$ are 0.283 at $p = -1$ ($\alpha = 0$), 0.103 at $p = -2$ ($\alpha = -1/2$), 0.0742 at $p = -3$ ($\alpha = -1$), and 0.0922 at $p = -5$ ($\alpha = -2$).

The coefficients incorporating n_0 and U_0 in (59) are in the form that appears in the particle spectrum (42) if we invoke the relation between α and p (56) to note that

$$ \frac{n_0}{U_0^{2\alpha-1}} = \frac{n_0}{U_0^p}. \qquad (8.60) $$

Knowledge of the log slope p and the coefficient n_0/U_0^p of the particle spectrum (42) allows one to write the expected synchrotron photon spectrum (59) directly given knowledge of the magnetic field $\boldsymbol{B}$ and hence ν_B.

The magnetic field dependence of the volume emissivity (59) becomes apparent if we expand $\nu_B = eB/2\pi m$ (11) as follows:

➡ $$ j_\nu(\nu) = \frac{e^3}{\varepsilon_0 mc} a(p)\,n_0 U_0 \left(\frac{mc^2}{U_0} \right)^{2\alpha} B^{1-\alpha} \left(\frac{4\pi m\nu}{3e} \right)^\alpha. \quad \begin{array}{l}\text{(Volume emissivity; } \beta \approx 1; \\ \text{W m}^{-3}\text{ Hz}^{-1})\end{array} \qquad (8.61) $$

Finally, substitute the numerical values into (61),

$$j_\nu(\nu) = 1.70 \times 10^{-24} a(p) n_0 U_0 \left(\frac{mc^2}{U_0}\right)^{2\alpha} B^{1-\alpha} \left(\frac{\nu}{4.2 \times 10^{10}}\right)^\alpha, \qquad (8.62)$$

where all quantities are in SI units.

Specific intensity and flux density

Recall that the volume emissivity j_ν (W m^{-3} Hz^{-1}) leads directly to the specific intensity I (W m^{-2} Hz^{-1} sr^{-1}) measured from an optically thin cloud according to the relation

$$I(\nu) = \int_0^\Lambda \frac{j_\nu(r, \nu)}{4\pi} dr = \frac{j_{\nu,av}(\nu)}{4\pi} \Lambda, \qquad \begin{array}{l} \text{(Specific intensity;} \\ \text{W m}^{-2}\text{ Hz}^{-1}\text{ sr}^{-1}) \end{array} \quad (8.63)$$

which was encountered previously (5.42) and developed in AM, Chapter 8. The integration is along the line of sight through the depth Λ of the cloud. Under the simple assumption that $j_\nu(\nu)$ is constant throughout a source and that the depth Λ is known, one need only substitute (59) into the right-hand term of (63) to obtain the synchrotron photon specific intensity I arising from a known particle spectrum. The shape of the spectrum $I(\nu)$ will be the same as that of $j_\nu(\nu)$ because the conversion involves only geometrical factors.

The expression (63) is applicable only if the radiation is emitted perfectly isotropically from each volume element and if that radiation reaches the observer without scattering or absorption. Even if all the electrons with different pitch angles and orbital phases were taken into account, it is expected that synchrotron radiation from a volume element would not be isotropic. Thus (63) is at best an approximation. If the magnetic fields in the source along the line of sight within the resolution element have many random directions (*tangled magnetic fields*), the approximation would be improved.

The spectral flux-density $S(\nu) = \int I\,d\Omega$ also has the same spectral form as $j_\nu(\nu)$ because it is, in turn, related to I by geometrical factors. Thus, one can take the flux-density spectral slopes for the Crab nebula (Fig. 2) in the radio and x-ray regions, $\alpha_r \approx -0.3$ and $\alpha_x \approx -1.1$, to find the particle-spectrum logarithmic slope $p = 2\alpha - 1$ (56), namely $p_r \approx -1.6$ and $p_x \approx -3.2$. It is just as if we were looking directly at the electrons in the Crab nebula, which in fact is what we are doing with our radio and x-ray vision. See Probs. 46 and 53.

Galactic radio synchrotron radiation

The diffuse radio emission observed over the whole sky (Fig. 11) is the result of the radiation by cosmic-ray electrons interacting with interstellar magnetic fields. It is concentrated toward the galactic plane, exhibits a power-law character, and has the expected polarization. The origin of this radiation, the cosmic-ray electrons, can be detected directly above the atmosphere with satellite-borne instruments. Thus, both the particle and photon components of the synchrotron process can be measured. Together, they provide information about the interstellar magnetic fields (Prob. 54).

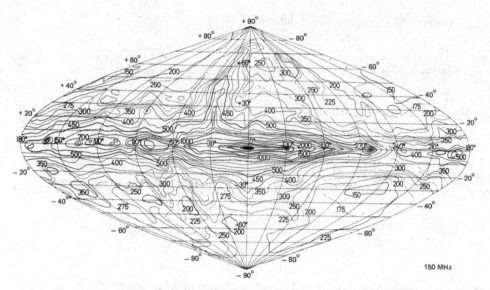

Figure 8.11. Continuum emission from the Galaxy at 150 MHz. The coordinates are galactic; the equator is the galactic equator, and the galactic center is at the center. The contours are brightness temperatures (K). Most of this radiation is synchrotron radiation from electrons spiraling around magnetic fields in interstellar space within the Galaxy. [T. Landecker and R. Wielebinski, *Australian J. Phys. Astrophys. Suppl.*, **16**, 1 (1970)]

8.6   Coherent curvature radiation

Relativistic electrons in a strong magnetic field radiate high frequencies, namely, $\nu_c = (3/2)\gamma^2\nu_B$ (13) if they cycle about the field lines with velocity approximately normal to the field line ($\phi \approx \pi/2$). As noted after (14), if the energy of the electrons or the magnetic field is sufficiently high, a synchrotron photon can have energy $h\nu_c$ approaching that of the emitting electron. This invalidates the classical (nonquantum) nature of our derivation. When the two energies are equal, a single emitted photon would remove all the emitting electron's energy!

On the other hand, if an electron's velocity is directed along the field line (i.e, with $\phi \approx 0$), only a very small fraction of the motion is projected normal to the field line, and the energy of the typical radiated photon is small. In this case electrons of extremely high energy can be sustained in a given magnetic field without catastrophic energy loss.

If the field is highly uniform, there would seem to be no limit to the electron energy if ϕ is sufficiently close to zero. If the field lines are not straight, however, such electrons will be guided along the curved track of the field lines; they will thus undergo acceleration normal to their velocity, as in circular motion, and they will radiate. This is called *curvature radiation*.

In the magnetic field of a neutron star, the radius of curvature of the field lines will be much larger than that of the electron track circulating about the magnetic field lines with $\phi = \pi/2$. Thus, at a given energy, the electron's acceleration is less, the radiated power is less, and the observed power might be undetectable. If, however, the electrons have extremely high energies and also move in bunches so that each bunch radiates like a single large charge, the power radiated can be huge. This is called *coherent curvature radiation*.

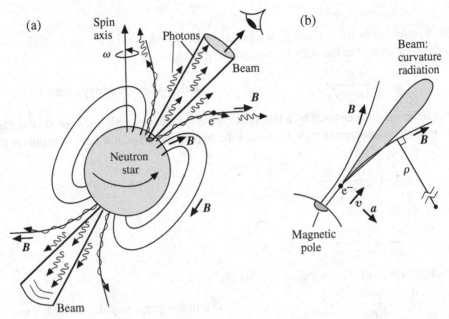

Figure 8.12. (a) Spinning isolated neutron star (radio pulsar) with offset magnetic pole. Electrons spiral outward along magnetic field lines and beam radiation outward in the magnetic polar direction. (The spiraling electrons illustrated would not radiate into the beam.) (b) Magnetic pole showing the radius of curvature ρ and curvature-radiation beam (shaded region) of a single emerging electron.

Our expressions for energy loss and frequency of synchrotron radiation can be adapted to provide the characteristics of this radiation. Such radiation is thought to be important in the radiation from rapidly spinning isolated neutron stars (i.e., radio pulsars), which can have magnetic fields reaching to and beyond 10^8 T.

Curved trajectory

A curved magnetic field line can be characterized by its radius of curvature ρ at any given location on the line. Imagine that the segment of the field line that contains the radiating electron is part of a large circle of radius ρ. For the magnetic field of a neutron star, the typical radius of curvature would be comparable to, or greater than, the radius of the neutron star itself, ~10 km (10^4 m); see Fig. 12.

As the electron moves along the field line, it would thus be on a circular path at least momentarily. The situation is therefore identical to that of Fig. 4a, and so the expressions we had for acceleration, frequency emitted, and power radiated in synchrotron radiation should be applicable here. The radiation depends only on the acceleration of the charge; it is immaterial what force causes the acceleration. In the present case, we will describe these characteristics as a function of ρ and γ, rather than B and γ, where γ represents the electron energy.

In the formalism earlier in this chapter, we characterized the synchrotron frequency emitted and power radiated in terms of the magnetic field controlling the radius of the orbit. This formalism can be used in the present case if we postulate a fictitious magnetic field $\boldsymbol{B}_f$ normal to the velocity ($\phi = \pi/2$), which would produce the given radius of curvature ρ for the specified electron energy $U = \gamma mc^2$. We again assume $v \approx c$.

Frequency emitted

The frequency of orbital rotation of an electron of charge e and mass m in a field of strength B_f directed normal to the velocity vector, is, from (5),

$$\nu_{orb} = \frac{eB_f}{2\pi\gamma m}.$$ (Relativistic cyclotron frequency)   (8.64)

This frequency is measured by a laboratory observer and is simply the inverse of the time for the particle to undergo one cycle around the hypothetical circular path of radius ρ at speed $v \approx c$. Thus,

$$\nu_{orb} = \frac{c}{2\pi\rho},$$ ("Orbital" frequency)   (8.65)

and therefore

$$\frac{c}{2\pi\rho} = \frac{eB_f}{2\pi\gamma m},$$ (8.66)

which allows us to solve for the fictitious B_f,

$$B_f = \frac{\gamma mc}{e\rho}.$$ (Fictitious perpendicular **B** field; T)   (8.67)

This field would provide the radius of curvature ρ for an electron energy characterized by the factor γ if the field were perpendicular to the orbit, as in Fig. 4a. The instantaneous acceleration of an electron circulating about the B_f field lines would be identical to that in our current curvature case. We can thus use the synchrotron expression (13) for the radiated critical frequency, for $\phi = \pi/2$,

$$\nu_c = \frac{3}{2}\gamma^2\nu_B,$$ (Critical frequency)   (8.68)

where ν_B is the cyclotron frequency that would be associated with B_f,

$$\nu_B = \frac{eB_f}{2\pi m} = \frac{\gamma c}{2\pi\rho}.$$ (8.69)

Here, we have eliminated B_f with (67). Substitution into (68) then yields the critical frequency of the curvature radiation in terms of ρ and γ,

→ $$\nu_c = \frac{3}{2}\gamma^3\frac{c}{2\pi\rho},$$ (Critical frequency for curvature radiation; Hz)   (8.70)

where $\gamma = U/mc^2$ is the Lorentz factor for the emitting electron and ρ is the instantaneous radius of curvature.

The factor $c/(2\pi\rho)$ in (70) is the frequency of the hypothetical orbit (cycles per second). The critical frequency for curvature radiation thus has a form similar to the synchrotron expression (68) except that it scales as γ^3 rather than γ^2.

The two expressions for ν_c differ in the parameters presented. One (68) is in terms of ν_B, or equivalently **B**, and varies as γ^2, whereas the other (70) is in terms of the radius of curvature ρ and varies as γ^3. This accounts for the different dependencies on the energy parameter γ. In the synchrotron case (68), as energy increases in the presence of a fixed field **B**, the orbital radius will increase and the orbital frequency will decrease as γ^{-1} (5). This offsets in part the

Doppler effect ($\propto \gamma^2$) and the beam narrowing ($\propto \gamma$) that led to the high emitted frequencies; see (8) and (4). This yields the γ^2 dependence. In the curvature case (70), ρ is held fixed, and the orbit by specification does not grow with energy. Because $v \approx c$, the orbital time does not change with energy, and thus the factor of γ^{-1} does not enter into consideration. The Doppler and beam-narrowing terms together yield the γ^3 dependence.

Finally, one can rewrite (70) to obtain the critical (or characteristic) photon *energy* for curvature radiation,

$$\Rightarrow \qquad E_c = h\nu_c = \frac{3}{2}\gamma^3 \frac{hc}{2\pi\rho}. \qquad \text{(Critical photon energy; J)} \qquad (8.71)$$

Equations (70) or (71) allow characterization of the radiation band as radio, optical, and so on. Equation (71) also allows us to compare the radiated photon energy with that of the emitting electron, $U = \gamma mc^2$.

It is sometimes helpful to remember the following factor needed to convert photon frequency ν to its energy $h\nu/e$ in electron volts and vice versa:

$$\frac{e}{h} = 2.418 \times 10^{14} \frac{\text{Hz}}{\text{eV}}. \qquad \text{(Conversion factor for photon)} \qquad (8.72)$$

Power emitted

The energy-loss rate of the electron, dU/dt, via curvature radiation is similarly obtained from our expression (27) for synchrotron radiation with the fictitious $B_f = (\gamma mc/e\rho)$ (67) used for the magnetic field B. With this substitution, and with $\phi = \pi/2$, $\beta = 1$, $Z = 1$ and $U = \gamma mc^2$, the energy-loss equation (27) becomes

$$\frac{dU}{dt} = -\frac{1}{6\pi\varepsilon_0}\frac{1}{c^5}\left(\frac{e}{m}\right)^4 (\gamma mc^2)^2 \left(\frac{\gamma mc}{e\rho}\right)^2 \qquad (8.73)$$

or

$$\Rightarrow \qquad \frac{dU}{dt} = -\frac{1}{6\pi\varepsilon_0}\frac{e^2}{c^3}\gamma^4 \left(\frac{c^2}{\rho}\right)^2. \qquad \text{(Energy-loss rate; curvature radiation; W)} \qquad (8.74)$$

The latter expression is cast into the form of the Larmor equation (2), where the "acceleration" c^2/ρ is reminiscent of the classical centripetal acceleration v^2/R for circular motion at speed v in a circle of radius R. The relativistic effects show up in the limiting speed $v = c$ and in the factor γ^4. The latter causes the energy-loss rate to increase rapidly with electron energy. On the other hand, the energy-loss rate can be greatly depressed if the radius of curvature ρ is large, as is the case for the magnetic field lines of a neutron star.

Note that the energy-loss rate dU/dt of synchrotron radiation (27) is proportional to $U^2 \propto \gamma^2$, whereas it is proportional to γ^4 for curvature radiation (74). The source of the difference between (27) and (74) is again the fixed radius of curvature parameter in the latter equation. As energy increases, the electrons are forced to follow the field lines with curvature ρ. They can not move out to larger orbits, which would tend to lessen the acceleration, as occurs in normal synchrotron radiation with fixed B. Thus, at a given energy, they experience greater acceleration and hence radiate more efficiently.

Coherent radiation from bunched electrons

The radiated power can be raised substantially if the electrons are bunched in a tight cluster smaller than the wavelength of the emitted radiation. All of the electrons in the bunch then radiate in phase, or *coherently*. The increase in emitted power can be dramatic.

Consider the classical (nonrelativistic) case. From Larmor's formula (2), the power radiated by an accelerating electron, summed over all angles and frequencies, is

$$\mathscr{P} \propto q^2 a^2. \tag{8.75}$$

This power is equal in magnitude to the electron energy loss rate, dU/dt. If there are N electrons radiating at random phases, the total power will be N times that radiated by a single electron:

$$\mathscr{P} \propto N e^2 a^2. \tag{8.76}$$

On the other hand, if a tight bunch of N electrons accelerate together and emit radiation of wavelength substantially longer than the bunch size, the radiation from each electron will be in phase with that of all the others. In other words, the electrons radiate as a single charge, $q = Ne$. Substitution into (75) yields

$$\Rightarrow \qquad \mathscr{P} \propto (Ne)^2 a^2 = N^2 e^2 a^2. \quad \text{(Coherent radiation for bunch of N electrons)} \tag{8.77}$$

The radiated power varies as N^2 rather than N. We find, in this classical argument, that 100 particles moving together will emit 10^4 times more power than a single electron.

This effect persists in our relativistic calculation of dU/dt (27) for ordinary synchrotron radiation. Represent a bunch of N electrons, each with energy U, mass m, and charge e, traveling together as a single particle of charge Ne, mass Nm, and energy NU. The increased charge is represented as $e \to Ne$. We find from (27), for B, β, and ϕ held constant and $Z = 1$, that

$$-\frac{dU}{dt} \propto \left(\frac{e}{m}\right)^4 U^2 \to \left(\frac{Ne}{Nm}\right)^4 (NU)^2 = N^2 \left(\frac{e}{m}\right)^4 U^2.$$

(Fixed B; coherent synchrotron radiation; one bunch of N electrons)   (8.78)

Thus, the power radiated is again N^2 times that for a single particle with $Z = 1$.

Make the same substitutions in the expression for energy loss in curvature radiation. From (74), for fixed radius of curvature, and noting that $\gamma = U/mc^2$, we obtain

$$-\frac{dU}{dt} \propto e^2 \gamma^4 = e^2 \left(\frac{U}{mc^2}\right)^4 \to N^2 e^2 \left(\frac{NU}{Nmc^2}\right)^4 = N^2 e^2 \gamma^4.$$

(Fixed ρ; coherent curvature radiation)   (8.79)

Again, we find that the rate of energy loss varies as N^2.

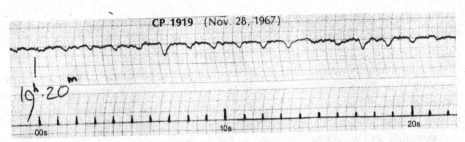

Figure 8.13. (top trace) Discovery data for the pulsar CP 1919 on 28 November 1967. The pen trace shows regular pulses at 1.3-s intervals on the upper trace. Increasing intensity is downward. The lower trace shows clock pulses at 1-s intervals. [Mullard Radio Astronomy Observatory, in A. Hewish, *Science* **188**, 1079 (1975); see also J. Bell, *Sky and Telescope*, March 1978, 220]

Spinning neutron stars

The discovery of radio pulsars in 1967 (Section 4.4, "Evidence for neutron stars") revealed the existence of the long-sought neutron star. A pulsar is a spinning neutron star that beams radiation toward the earth only once or twice each spin period much like a seaside lighthouse. The first detected pulses from the first discovered pulsar, CP 1919, are shown in Fig. 13.

As shown in Fig. 12a, the pulsing could arise from beamed radiation emitted from the magnetic pole of the neutron star. If the magnetic axis is offset from the spin axis, as it is for the earth, the beam would traverse an appropriately located observer once each spin period. The forward beaming of the radiation in this picture suggests curvature radiation from relativistic particles traveling along the magnetic field lines near the magnetic pole. Observations of spindown rates and also simple scaling arguments indicate that these "young" neutron stars have magnetic fields of order 10^8 T. Even higher magnetic fields are found in *magnetars*, neutron stars with magnetic fields reaching to $\sim 10^{11}$ T.

It can be shown (not here) that huge electric fields are generated along the open magnetic field lines of a spinning neutron star. Potentials in excess of $\sim 10^{15}$ volts between the star and the interstellar medium are expected for young Crab-like pulsars. These could give electrons streaming along the lines extremely high energies capable of producing x rays and gamma rays via curvature radiation.

The emitted radiation from pulsars exhibits a wide range of frequencies from the radio to gamma rays up to at least 100 MeV for the Crab pulsar. These energies are substantially greater than the 15-MeV upper limit for the energy of synchrotron photons emerging from a magnetic field of 10^8 T at $\phi = \pi/2$. At ~ 15 MeV, the emitted photon energy equals the electron energy, and the electrons would lose energy catastrophically; see discussion after (14).

In contrast, electrons of huge energies, up to beyond 1 TeV, can in principle be sustained in the magnetospheres of pulsars without catastrophic energy loss while they emit curvature radiation. For example, a 1.0-TeV electron following a track of $\rho = 10^4$ m will emit gamma rays of frequency $\nu_c = 5.4 \times 10^{22}$ Hz (~ 220 MeV); see (70). It would take the equivalent of ~ 4500 of these gamma rays to deplete the electron's 1.0-TeV energy.

The radio emission detected from such pulsars could arise from lower-energy electrons or from regions more removed from the star, where ρ is larger, or both. For example, 20 MeV electrons with $\rho = 10^4$ m would emit $\sim$430 MHz radio waves.

For completeness we note that electrons orbiting the field lines of such strong fields are constrained to *Landau* quantum energy levels. This quantization limits the energies, and hence orbital radii, to specific values. This leads to spectral lines that are observed in x-ray spectra of neutron stars in x-ray binaries.

A complete and widely accepted model for the emission processes from pulsars has yet to be developed by theorists, but, it is likely that the geometries and emission processes outlined here play significant roles in the creation of the observed radiation.

Problems

8.2 *Discovery of celestial synchrotron radiation*

Problem 8.21. Examine the photos (Fig. 3) of the Crab nebula taken with polarizing filters. (a) What can you say about the directions of motion and acceleration of the emitting electrons at various positions in the nebula? (b) How much can you say about the magnetic fields in the nebula? (c) Can you distinguish a B field directed to the north from one directed to the south? (d) Over what extent, roughly in light years, is the B field organized? See figure caption for useful information.

Problem 8.22. Derive a rough mass for a hypothetical optically thin hydrogen plasma of temperature $T = 50\,000$ K emitting continuum thermal-bremsstrahlung radiation from the Crab nebula. Proceed as follows: (*i*) Determine the specific intensity I (W m^{-2} Hz^{-1} sr^1) of the optical radiation at 5×10^{14} Hz from the flux density (Fig. 2) and the angular diameter of the nebula $\sim$5′. (*ii*) Estimate the line-of-sight path length Λ within the nebula if it is about as thick as it is wide. (*iii*) Use these results to obtain $j_\nu(\nu)\,d\nu$, the power emitted per unit volume of the plasma into $d\nu$ at ν (63). (*iv*) From this, determine the electron density in the postulated hydrogen plasma (5.37); adopt unity for the Gaunt factor, $g = 1$. (*v*) Use this and the volume of the nebula to obtain the total mass in kilograms and solar masses. Do not forget to include the protons. [Ans. $\sim$100 $M_\odot$; a 1942 paper reported $\sim$15 $M_\odot$]

Problem 8.23. Confirm that the two power-law exponents of the spectral flux densities $S(\nu)$ of the Crab nebula as indicated in Fig. 2 are correct. Make careful measurements on a blowup of the figure of the radio region (solid line, $-6 < \log E(\text{eV}) < -3$) and of the ultraviolet–x-ray region (solid line, $1 < \log E(\text{eV}) < 5$).

8.3 *Frequency of the emitted radiation*

Problem 8.31. (a) Consider an electron of energy 10^{13} eV moving with circular motion in a region with magnetic field of 10^{-7} T. (*i*) What is the expected frequency ν_{syn} (Hz) of the observed radiation? (*ii*) What is the associated photon energy in electronvolts and what kind of radiation is this? (b) Verify the numerical value shown in (14) for ν_c. (c) Find the electron energy U in electronvolts at which, according to (14) for $\sin \phi = 1$, the energy $h\nu_c$ of the emitted photons would equal the electron energy for (*i*) interstellar space ($B \approx 3 \times 10^{-10}$ T), (*ii*) the radio-flaring region of an active galactic nucleus ($B \approx 10^{-9}$ T), (*iii*) the Crab nebula ($B \approx 5 \times 10^{-8}$ T), (*iv*) the magnetic field of the earth ($B = 5 \times 10^{-5}$ T), and (*v*) a neutron star (10^8 T). Express your answers in electronvolts and in terms of the Lorentz factor $\gamma = U/(mc^2)$ and comment on them. [Ans. $\sim 10^{18}$ Hz; $-$; $\sim 10^{25}$ eV for (i)]

Problem 8.32. (a) The observer of Fig. 5 measures and plots the profile of E_{tr} from a single electron similar to that of Fig. 4b. Can this observer deduce the direction of the B field (in or out of the paper) from such data? Make two sketches similar to Fig. 5, but for a negative charge, one for each of the two magnetic field directions to illustrate your argument. Keep the observer at the bottom of the page and draw only one field line in each sketch, including its kinks. (b) From measurements of the pattern in Fig. 5 with a millimeter rule, deduce the speed of the electron relative to the speed of light, namely $\beta = v/c$. [Ans. $-$; see caption.]

Problem 8.33. Consider a charge q in a uniform magnetic field $B = B_z \hat{k}$ with a component of its velocity in the direction of the field; that is, the charge spirals around, and drifts along, the magnetic field lines with a pitch angle ϕ (angle between p and B). (a) For a nonrelativistic particle ($v \ll c$), find expressions for angular frequency ω and radius R of the circular motion projected into a plane normal to the magnetic field direction as functions of the momentum p, magnetic field B_z, and pitch angle ϕ. Start with the equation of motion $dp/dt = q(v \times B)$. What is the frequency ω_{obs} of the observed electromagnetic wave and does it depend on ϕ? (b) Find ω and R for relativistic motion of the charge with momentum $p = \gamma m\beta c$ and pitch angle ϕ. (c) Validate the expression for ω_{syn} given in (10) for relativistic motion at pitch angle ϕ. Modify the arguments in the text (for $\phi = 90°$) and make sketches as needed. [Ans. $q B_z/m$, $p \sin \theta/q B_z$; $q B_z/\gamma m$, $p \sin \theta/q B_z$; $-$]

8.4 Power radiated by the electron

Problem 8.41. (a) Find the rate of energy loss (eV/s) for a 10^{13} eV electron in a 10^{-8} T field from (29). See if you get the same answer from (34). Use the average of $\sin^2 \phi$. (b) Compare the rate of energy loss with synchrotron radiation of electrons, protons, and helium nuclei (alpha particles) by calculating the proton–electron and alpha–electron energy-loss ratios. Assume the same energy U and magnetic field B for each kind of particle. (c) What is the ratio of proton to electron total energies that would cause them to radiate similar power? [Ans. ~ 3 keV/s; $\sim 10^{-14}$; $\sim 10^6$]

Problem 8.42. (a) Demonstrate that (28) follows from the energy-loss equation (27), when the latter is applied to relativistic radiating electrons. (b) Then show that the mixed units expression (29) follows from (28). (c) Convert (27) to the form (34), which is in terms of the Thomson cross section and the magnetic field energy density. Follow the suggestions in the text.

Problem 8.43. (a) Integrate the rate-of-energy-loss formula for synchrotron radiation, $dU/dt = -CU^2$, where C is a constant, to obtain an expression for $U(t)$ if $U = U_0$ at $t = 0$. What is the value of U at the characteristic time defined as $\tau \equiv -U/(dU/dt)$ evaluated at $t = 0$? Express your answer in terms of the initial energy U_0. (b) Repeat (a) for a different energy loss formula, $dU = -CU dt$. (c) Comment on the difference in the results. [Ans. $U_0/2$; U_0/e; $-$]

Problem 8.44. (a) Show analytically that the trajectory of the charge q in the frame S$'$ in Fig. 9c has a cusp at the point of contact as shown. Procedure: Let angle θ indicate the position of the charge on the circular track in S relative to the contact point as subtended from the center of the circle. Find the horizontal x' and vertical y' displacements of the charge relative to the observer's position for small angles $\theta (\theta \ll 1)$ and from this find the function $y'(x')$. Demonstrate that this function yields a cusp. Ignore relativistic effects in your geometry. Would the effects of special relativity change your conclusion? Explain.

Problem 8.45. Consider an electron spiraling in a uniform magnetic field with speed factor $\beta = v/c$ and pitch angle ϕ, the angle between the velocity and magnetic vectors. Reconstruct the arguments that led to dU/dt, taking into account the pitch angle and speed factor in order to obtain the $\beta^2 \sin^2 \phi$ dependence in (27). Make appropriate sketches. Hints: consider an instant of time, define coordinate systems aligned with the particle velocity, and apply the Lorentz transformations as needed.

Problem 8.46. (a) Find, from the energy loss formula (28), expressions for the approximate number density n_e (m^{-3}) and the kinetic energy density u_k (J/m^3) of synchrotron-emitting electrons in the Crab nebula implied by the luminosity L_b (W) in a defined frequency band $\Delta \nu$ at some frequency ν. Let the physical volume of the nebula be $V = 3 \times 10^{50}$ m^3 with a magnetic field $B = 5 \times 10^{-8}$ T throughout the region. Assume all electrons of energy U are highly relativistic, have pitch angle $\phi = \pi/2$, and radiate solely at the associated critical frequency ν_c. Find expressions that are functions of ν_c, L_b, B, V, and the numerical coefficients of (14) and (28); call the latter a and b, respectively. (b) Evaluate your expressions for three frequency bands (radio, optical, x-ray) centered on the frequencies given in Table 1. Use the SED values, $\nu S(\nu)$, for these frequencies from Fig. 2 and adopt Crab distance $D = 6000$ LY. To obtain L_b, adopt one decade of bandwidth $\Delta \nu$ centered logarithmically on ν in each case and assume a constant S over the band. Compare the number density you find for the optical band with the $\sim 10^9$ m^{-3} obtained for a plasma model in Prob. 22. (c) From your results, estimate roughly a total u_k for the entire radio to x-ray range and compare with the magnetic energy density, $B^2/2\mu_0$ (J/m^3). [Ans. $n_e = aL_b(Vb\nu_c B)^{-1}$, $u_k = a^{1/2}L_b(bV\nu_c^{1/2}B^{3/2})^{-1}$; x ray $\sim 3 \times 10^{-7}$ m^{-3}; u_k and u_B agree within factor ~ 10]

8.5 *Ensemble of radiating particles*

Problem 8.51. (a) For the fixed value $\alpha = -1$ in (57), how does $j_\nu(\nu)$ depend on ν and B? What is the consequence of doubling the magnetic field? of doubling the frequency? (b) Integrate $j_\nu(\nu)$ from ν_1 to ν_2 for $\alpha = -1$ and with proportionality coefficient A to obtain the integrated volume emissivity j for the band. Evaluate j for the following decades of frequency: ν_0 to $10\nu_0$, $10\nu_0$ to $100\nu_0$, and $100\nu_0$ to $1000\,\nu_0$. (c) Find j for the same three frequency bands for $\alpha = -2$. (d) What is the integrated emissivity j if $\nu_2 \to \infty$ for the two cases $\alpha = -1$ and -2? Note that your conclusions also apply to the specific intensity $I(\nu)$; see (63). [Ans. $-$; $AB^2 \ln 10$; $-$; ∞, AB^3/ν]

Problem 8.52. (a) Carry out the approximate derivation of $j_\nu(\nu)$ (59). Follow the text but maintain the proportionality constants throughout the derivation. Start with the expression (42) for $n(U)$, and use the expression (27) for dU/dt as well as (13) for ν_c (rewrite the latter in terms of U, e, B, m and c). Assume the radiator is an electron, all pitch angles are $\phi = \pi/2$, and $\beta \approx 1$. To simplify the algebra, temporarily assign the symbols $k_1, k_2,$ and k_3 to the fixed algebraic and numerical coefficients (excluding variables U and B) in (13), (27), and (42) respectively, but substitute back the original values and symbols in the final result. Let B be constant throughout the emitting region. (b) What value of $a(p)$ in (59) would give agreement with your result? (c) Demonstrate that both sides of (59) have the same units. [Ans. $-$; ~ 0.1; $-$]

Problem 8.53. Consider the Crab nebula at the *x-ray* frequency, $\nu_{18} = 1 \times 10^{18}$ Hz (i.e., 4 keV). (a) Use the flux density at the earth obtained from Fig. 2 ($S_{18} = 1.2 \times 10^{-29}$ W m^{-2} Hz^{-1}) to find estimates of the specific intensity I_{18} (W m^{-2} Hz^{-1} sr^{-1}) and the volume emissivity

$j_{\nu,18}$ (W m^{-3} Hz^{-1}) at $\nu = 10^{18}$ Hz. Assume the emissivity is uniform throughout the nebula and that the thickness of the nebula along the line of sight equals its physical diameter (in meters). The distance from the earth to the Crab is $D \approx 6000$ LY and its angular diameter is $\sim 5'$. (b) The measured logarithmic slope of $S(\nu)$ in this region is $\alpha = -1.1$. Write the expression describing the spectrum of the volume emissivity $j_\nu(\nu)$ in terms of ν and numerical values. (c) Use the result for $j_{\nu,18}$ together with the theoretical expression for $j_\nu(\nu)$ (62) to find an expression for the particle number density spectrum $n(U)$ (electrons m^{-3} J^{-1}) in the form of (42) with all constants specified. Let $B = 5 \times 10^{-8}$ T, assume a one-to-one correlation between U and ν according to (14), let U_0 be the electron energy corresponding to $\nu_c = 10^{18}$ Hz, set $a(p) = 0.075$, and solve (62) for n_0. (d) Integrate $n(U)$ over the band of energies corresponding to the one-decade band of frequencies centered logarithmically on 10^{18} Hz to find the electron number density n_e emitting in that band. Again use (14) for energy-frequency conversion. Compare your answer to the electron density $n_e \approx 3 \times 10^{-7}$ m^{-3} you may have tabulated in part (b) of Prob. 46 for x-ray emitting particles emitting into one decade of frequency at 10^{18} Hz. [Ans. $\sim 10^{-23}$ W m^{-2} Hz^{-1} sr^{-1}, $\sim 10^{-39}$ W m^{-3} Hz^{-1}; $\sim 10^{-39} (\nu/\nu_{18})^{-1.1}$ W m^{-3} Hz^{-1}; $\sim 0.1(U/(2 \times 10^{-6}$ J$))^{-3.2}$ m^{-3} J^{-1}; $\sim 10^{-7}$ m^{-3}]

Problem 8.54. *Galactic radio background.* The particle number specific-intensity (spectrum) of cosmic ray electrons at the top of the atmosphere is well described from ~ 2 GeV to ~ 30 GeV by the power law

$$J(U) = 8.1 \times 10^{11} \left(\frac{U}{1 \text{ GeV}} \right)^{-2.65} \text{s}^{-1} \text{ m}^{-2} \text{ J}^{-1} \text{ sr}^{-1}.$$

The values of J fall below the power law at < 2 GeV owing to modulation by the solar wind. However, one may presume that, in interstellar space, the spectrum follows this power law even at energies below 1 GeV. Use this power law to find the expected radio-wave spectrum of the resulting synchrotron radiation with the interstellar magnetic field B as an unknown parameter. Then compare your radio-wave (photon) spectrum with the observed radio-wave spectrum to find the magnetic-field magnitude B. Proceed as follows: (a) From $J(U)$, find the particle number-density spectrum $n(U)$ (m^{-3} J^{-1}) of the electrons and evaluate it at 1 GeV. Hint: use the simple energy-independent conversion between J and n discussed in the text, before (42), for $v \approx c$. (b) Find the logarithmic slope α of the synchrotron photon spectrum. (c) Use (62) to find $j_\nu(\nu, B)$ and evaluate at $\nu = 10^8$ Hz for $a(p) = 0.08$. (d) Assume that the line of sight remains in the galactic disk in directions not directly in the plane nor toward the galactic center (to avoid discrete sources) and is of typical length $\sim 10^4$ LY ($\sim 10^{20}$ m) comparable to the 25 000 LY distance to the galactic center. Find the expected specific intensity spectrum of radio photons $I(\nu, B)$ (W m^{-2} Hz^{-1} sr^{-1}). Use (63) and be sure to include the dependence of I on both B and ν. Use 10^8 Hz as the reference frequency. (e) The observed spectrum of radio emission in the region $10^8 - 10^9$ Hz is

$$I(\nu) = 2.5 \times 10^{-21} \left(\frac{\nu}{10^8 \text{ Hz}} \right)^{-0.6} \text{W m}^{-2} \text{ Hz}^{-1} \text{ sr}^{-1}.$$

Compare the exponent in this expression with that calculated in (b) and comment. (f) Equate the observed and calculated specific intensities, and solve for the magnetic field B, for both 10^8 Hz and 10^9 Hz. Should the answers differ? (g) Find, from (14), the energies of the electrons that give rise to the photons at 10^8 Hz and 10^9 Hz as a check that we used the proper part of the electron spectrum. [Ans. $B \approx 0.6$ nT; about right for the interstellar medium.]

8.6   *Coherent curvature radiation*

Problem 8.61. Consider cyclotron and synchrotron radiation in a pulsar magnetic field of 10^8 T. (a) What frequency radiation would you expect from an electron spiraling *about* the field lines ($\phi = \pi/2$) for nonrelativistic electrons ($E \ll mc^2$)? Give the approximate energy of the emitted photon energy $h\nu$ in eV and the frequency band (radio, etc.). (b) Repeat (a) for relativistic electrons of energy 10 MeV. (c) What is the approximate maximum electron energy one might find orbiting at pitch angle $\phi = \pi/2$ in this high magnetic field according to our classical (nonquantum) development? Assume $\gamma \gg 1$. Comment on your answer. [Ans. $\sim$10 keV; $\sim$5 MeV; $\sim$15 MeV]

Problem 8.62. Consider curvature radiation wherein electrons follow the magnetic field lines with pitch angle $\theta \approx 0$. Let the field lines have radius of curvature equal to the neutron star radius, or $\rho = 10^4$ m. (a) At what electron energy does the emitted photon energy become equal to the electron energy according to an extrapolation of our classical theory – again, for $\gamma \gg 1$? (b) What is the frequency corresponding to the orbital period for a relativistic particle in a circular path of radius $R = \rho = 10^4$ m? (c) What is the critical photon frequency and associated photon energy emitted as curvature radiation by a 10^{12}-eV electron? (d) Find the energy of an electron that would emit curvature radiation as radio waves of (critical frequency) 500 MHz. [Ans. $\sim$$10^{14}$ eV; $\sim$5000 Hz; $\sim$250 MeV; $\sim$20 MeV]

Problem 8.63. Find the dependence of the volume emissivity $j_\nu(\nu)$ on $\nu, \rho, n_0, p,$ and N for (curvature) radiation from relativistic electrons on a track of curvature ρ (m). The energy-loss rate dU/dt is given in (73) and the number-density spectrum of the electrons is $n(U) = n_0 \cdot (U/U_0)^p$ (m^{-3} J^{-1}), where n is the total number of electrons per (m^3 J) in the observer's frame of reference. The electrons move along the field lines in bunches of N particles each. Follow the logic and assumptions used in the similar development for synchrotron radiation (42) to (55). As therein, you need not keep track of the constant coefficients, but do retain the five variables specified above. [Ans. $j_\nu(\nu) \propto N n_0 \, \rho^{(p-1)/3} \, \nu^{(p+2)/3}$]

9

Compton scattering

┌───┐

What we learn in this chapter

The **normal Compton effect** involves the collision of a photon with a nearly stationary electron. In the inverse process, a **high-energy electron** gives energy to a photon. This process, known as **inverse Compton (IC) scattering**, is important in the **jets of active galactic nuclei** and in **clusters of galaxies**.

Momentum and energy conservation yield the energy and scattered angle of the electron in the normal effect. The increase of the photon energy in the IC process follows from the application of the normal effect in the **rest frame of the energetic incident electron**. The final result is that a relativistic electron with Lorentz factor $\gamma = U/mc^2$ will **increase the photon energy** by a factor of $\sim\gamma^2$. In jets, for example, this can propel x-ray photons up to extreme **gamma-ray energies** detected by **TeV gamma-ray astronomers**. The modification of photon spectra due to single or multiple IC scatters is called **Comptonization**.

Energetic electrons in a nebula containing magnetic fields will radiate by the synchrotron process. The synchrotron photons may then interact with the energetic electrons that created them via the IC process and thus be boosted to extremely high energies. This is known as the **synchrotron self-Compton (SSC) process**. The **spectral energy distributions (SEDs)** of the **Crab nebula** and of **blazars** exhibit pronounced peaks attributed to IC scattering.

Thermal electrons in **clusters of galaxies** scatter the **cosmic microwave background (CMB)**, causing a shift of the blackbody CMB spectrum to slightly higher frequencies, which is known as the **Sunyaev–Zeldovich (S-Z) effect**. Detection of this in the **radio** band together with supplementary **x-ray spectra** of the thermal bremsstrahlung radiation from the cluster plasma yields the distance to the cluster, which, with the redshift, provides an independent measure of the **Hubble constant**. The **peculiar motions of clusters** can also be obtained from such studies. This pertains to the evolution of the **large-scale structure of the universe**.

└───┘

9.1 Introduction

An ejection from the core of an active galactic nucleus (AGN) could well consist of a cloud of relativistic electrons and magnetic fields that radiates by synchrotron radiation. If the cloud is sufficiently compact, the radiating electrons will begin to interact with the synchrotron

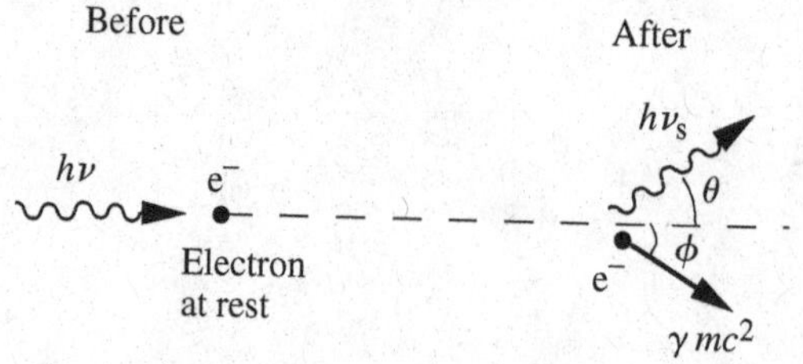

Figure 9.1. Normal Compton scattering. A photon scatters off of a stationary free (unbound) electron, giving part of its energy to the electron. The energy of the scattered photon is a function of the photon scattering angle θ.

photons in a process called *inverse Compton* (IC) *scattering*. The energetic electrons, on average, boost the photon energies by large amounts.

This process is the inverse of the normal Compton effect in which a photon gives energy to an (assumed) stationary electron. The amount of energy extracted from the electron in an IC interaction is much greater than that extracted in the emission of a synchrotron photon. In this manner, photons at radio frequencies can be boosted up to optical frequencies and beyond, or x rays can be boosted up to TeV gamma-ray energies.

The dynamics of Compton collisions are presented here. Because the calculation of the IC scattering process makes use of the normal effect, both will be derived. The synchrotron self-Compton (SSC) process, the Compton limit, and the Comptonization of a photon spectrum will be discussed qualitatively. Finally, we will present the Sunyaev–Zeldovich effect in which photons of the cosmic microwave background (CMB) are scattered by electrons of the plasmas in clusters of galaxies.

9.2   Normal Compton scattering

A photon interacts with a stationary electron and gives it some of its energy. We derive the increased wavelength of the scattered photon as a function of its angle of scatter.

Compton wavelength

Consider the collision of a photon with a stationary, free electron shown in Fig. 1. The photon loses energy as it scatters into angle θ, and this energy is given to the electron. The famous Compton relation between the incoming and scattered wavelengths, λ and λ_s, respectively, follows from energy and momentum conservation,

$$\lambda_s - \lambda = \frac{h}{mc}(1 - \cos \theta), \qquad \text{(Wavelength shift; Compton effect; m)} \qquad (9.1)$$

where m is the electron mass. This result is relativistically correct; it is derived in this section.

The quantity h/mc is known as the *Compton wavelength* λ_C; its value is

$$\lambda_C = h/mc = 2.43 \times 10^{-12} \text{ m} = 2.43 \text{ pm} \qquad \text{(Compton wavelength)} \qquad (9.2)$$
$$\rightarrow 1.23 \times 10^{20} \text{ Hz} \rightarrow 0.511 \text{ MeV},$$

where we use the relations $\lambda \nu = c$ and $E(\text{eV}) = h\nu/e$ to obtain the equivalent frequency and photon energy. The equivalent energy turns out to be the rest energy of the electron, $mc^2 = 0.511$ MeV.

According to (1), the wavelength shift $\lambda_s - \lambda$ at any given angle θ is of order 10^{-12} m and independent of the incident wavelength. Compare this with the much larger visible wavelengths, which are on the order of 600 nm $= 0.6 \times 10^{-6}$ m. The fractional shift, $(\lambda_s - \lambda)/\lambda$, becomes substantial only if the incoming wavelength λ is so short as to be comparable to 10^{-12} m, at which point the photon energy approaches that of the electron rest energy; see (2).

X-ray photons were used by Arthur Compton in 1922 to verify this effect. X rays produce a substantial fractional wavelength shift and, if sufficiently energetic, interact with electrons in a target material as if the electrons were free (unbound). The electrons are assumed to suffer negligible fractional ionization energy loss as they are ejected from the material. In the derivation of (1), the electron is treated as being completely free.

This was a landmark experiment, for it demonstrated the corpuscular nature of photons. The x-ray photons and the recoil electrons were found to obey energy and momentum conservation just as particles do. The observed increase of wavelength in such a scatter is unintelligible in the context of classical wave theory.

Momentum and energy conservation

The Compton wavelength shift (1) is derived from energy and momentum conservation with both the electron and the photon treated as particles. Two equations are required for the momentum: one for the longitudinal component and the other for the transverse component. The initial energy of the electron is its rest energy mc^2. Its recoil energy is γmc^2, and its momentum is $p = \gamma \beta mc = \gamma m v$ (7.14), where v is its recoil speed, m is its mass, $\beta \equiv v/c$, and $\gamma = (1 - \beta^2)^{-1/2}$. The energy and momentum of the incoming photon are, respectively, $h\nu$ and $h\nu/c$ (7.22). For the scattered photon, the energy and momentum are $h\nu_s$ and $h\nu_s/c$.

The three conservation equations follow from these expressions and Fig. 1:

$$h\nu + mc^2 = h\nu_s + \gamma mc^2 \qquad \text{(Energy conservation)} \qquad (9.3)$$

$$\frac{h\nu}{c} = \frac{h\nu_s}{c}\cos\theta + \gamma\beta mc\cos\phi \qquad \text{(Longitudinal momentum)} \qquad (9.4)$$

$$0 = \frac{h\nu_s}{c}\sin\theta - \gamma\beta mc\sin\phi. \qquad \text{(Transverse momentum)} \qquad (9.5)$$

The initial (before the interaction) momenta and energy are on the left sides. The longitudinal and transverse components of momentum are defined relative to the direction of the incident photon. The transverse component of momentum before (and after) the interaction is thus zero.

The expression (1) we wish to obtain describes the wavelength shift in terms of the scattering angle θ of the photon. In the following, we will eliminate ϕ from (4) and (5) and then use (3) together with the definition of γ to eliminate γ and β, the parameters that describe the electron energy and momenta. Before reading further, try it yourself (Prob. 21).

Solve (4) for $\cos\phi$ and (5) for $\sin\phi$ and substitute into the identity

$$\cos^2\phi + \sin^2\phi \equiv 1 \qquad (9.6)$$

to obtain

$$h^2(\nu^2 + \nu_s^2 - 2\nu\nu_s\cos\theta) = (\gamma\beta mc^2)^2. \qquad (9.7)$$

Square the energy equation (3) with the term γmc^2 isolated on the right side:

$$h^2(\nu^2 - 2\nu\nu_s + \nu_s^2) + 2hmc^2(\nu - \nu_s) + m^2c^4 = (\gamma mc^2)^2. \tag{9.8}$$

Subtract (7) from (8) to find

$$-2h^2\nu\nu_s(1 - \cos\theta) + 2hmc^2(\nu - \nu_s) + m^2c^4 = m^2c^4\gamma^2(1 - \beta^2). \tag{9.9}$$

The definition of γ gives $\gamma^2(1 - \beta^2) = 1$, and so

$$\frac{1 - \cos\theta}{mc^2} = \frac{\nu - \nu_s}{h\nu\nu_s} \tag{9.10}$$

$$\frac{1}{h\nu_s} - \frac{1}{h\nu} = \frac{1 - \cos\theta}{mc^2}, \tag{9.11}$$

which, with $\nu = c/\lambda$, becomes identical to (1). Our derivation of (1) is now complete.

Scattered frequency

The terms of (11) can be rearranged to yield a convenient form

$$\Rightarrow \quad h\nu_s = \frac{h\nu}{1 + \frac{h\nu}{mc^2}(1 - \cos\theta)}, \qquad \text{(Compton scattering)} \tag{9.12}$$

where ν and ν_s are the incident and scattered frequencies, respectively, and θ is the photon scattering angle. This form shows explicitly how the scattered photon energy is shifted significantly as the incident photon energy $h\nu$ becomes comparable to the rest energy mc^2 of the electron.

At much lower energies, the photon is scattered without significant reduction in energy. At a given incident photon energy, the decrease in energy is greatest when the photon is back scattered ($\theta \approx \pi$ radians). At very high photon energies ($h\nu \gg mc^2$), a back-scattered photon has energy $mc^2/2$ irrespective of the incoming photon's energy (Prob. 21).

9.3　Inverse Compton scattering

In a plasma of relativistic electrons emitting synchrotron radiation, the electrons typically have energies substantially greater than those of the synchrotron photons (Table 8.1). In this case, when the electrons and photons collide, the electrons typically *give* energy to the photon rather than vice versa, which accounts for the name *inverse Compton* (IC) process.

The Compton formula (12) was conditioned on the electron's initially being at rest. This formula can be applied in the inverse case if we first transform to the frame of reference in which the electron is at rest. The collision is then viewed in this rest frame and the scattered photon energy determined. Transformation back to the laboratory or observer frame yields the desired final energy, or frequency, of the scattered photon.

Here we restrict our development to the head-on collision of the electron and photon in which the photon is backscattered in the electron's frame of reference. Doing so gives the maximum possible energy to the scattered photon. See Prob. 31 for the overtaking case. We further choose to develop only the case in which the electron is highly relativistic ($v \approx c$).

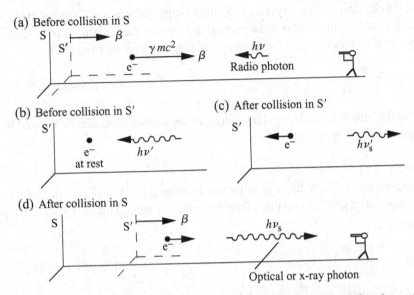

Figure 9.2. Inverse Compton effect for head-on interaction. (a) S frame before the scatter. A highly relativistic electron moves to the right and is about to collide with a left-moving, low-energy (radio) photon of energy $h\nu$. The S′ frame moves with the electron. (b) S′ frame before scatter. The electron is at rest, and so the scatter becomes a normal Compton scatter. The incident photon in S′ has been Doppler shifted to higher energy $h\nu'$ because of the motion of this frame relative to S. (c) S′ frame after scatter. The photon, having been backscattered with energy $h\nu'_s$, moves to the right with slightly less energy. (d) S frame after scatter. The scattered photon has a much higher energy $h\nu_s \gg h\nu'_s$ owing to the transformation back to frame S.

We also present, with plausibility arguments, a general expression for the energy loss rate of an electron in the presence of an isotropic distribution of photons.

Photon energy increase

Consider Fig. 2a, which shows the situation before a head-on collision. An energetic electron moves to the right in the laboratory frame of reference S with speed parameter $\beta = v/c$ and energy γmc^2, and a photon of (relatively low) energy $h\nu$ moves to the left.

Rest frame of electron

A moving frame S′ is defined in which the electron is at rest; this frame thus also moves to the right relative to frame S with speed parameter β. An observer in S′ finds the electron to be at rest and also finds the photon to be Doppler shifted to a higher energy $h\nu'$ because the photon velocity is opposed to that of S′ (Fig. 2b). The relativistic Doppler shift (7.38) gives

$$h\nu' = \left(\frac{1+\beta}{1-\beta}\right)^{1/2} h\nu. \qquad \text{(First Doppler shift)} \qquad (9.13)$$

Note that we are now using β and γ to describe the electron *before* the collision. In the calculation above for the normal Compton effect (Section 2), they describe the electron *after* the collision.

Because the electron is initially at rest in S′, the Compton scatter (in S′) obeys the Compton formula (12). Consider the case of back scattering in which the photon reverses direction in S′ ($\theta' = \pi$), as shown in Fig. 2c. The scattered energy $h\nu_s'$ in S′ is, from (12),

$$h\nu_s' = \frac{h\nu'}{1 + \frac{2h\nu'}{mc^2}}.$$

(Compton scatter in S′)   (9.14)

This is normal Compton scattering. The energy of the scattered photon is reduced because energy is given to the electron.

Laboratory frame

The scattered photon $h\nu_s'$ in S′ has a velocity that is in the same direction as that of S′, and so the reverse transformation back to the S frame results in a second upward shift in frequency to $h\nu_s$ (Fig. 2d):

$$h\nu_s = \left(\frac{1+\beta}{1-\beta}\right)^{1/2} h\nu_s'.$$

(Second Doppler shift)   (9.15)

In typical applications, the two upward Doppler shifts are much larger than the modest decrease in (14). The two Doppler shifts are both upward in frequency because the photon changed direction in the interaction.

The scattered energy $h\nu_s$ in S is the desired quantity except that it should be expressed in terms of the original photon energy $h\nu$ in frame S. We thus substitute $h\nu'$ (13) into (14) and the result $h\nu_s'$ into (15). In doing so, we assume the initial electrons are highly relativistic, $\beta \approx 1$ or, equivalently, $\gamma \gg 1$, and so we can use the relation

$$\left(\frac{1+\beta}{1-\beta}\right)^{1/2} \xrightarrow[\beta \approx 1]{} 2\gamma,$$

($\beta = v/c \approx 1$; or $\gamma \gg 1$)   (9.16)

which follows from multiplication of the left side by unity in the form $(1+\beta)^{1/2}/(1+\beta)^{1/2}$ and the definition of the Lorentz factor $\gamma = (1-\beta^2)^{-1/2}$. The result of the substitutions is

➡ $$h\nu_s \approx 4\gamma^2 h\nu \left(1 + \frac{4\gamma h\nu}{mc^2}\right)^{-1},$$

(Single IC back scatter; $\gamma \gg 1$)   (9.17)

where γ is the Lorentz factor of the high-energy electron that collides with a photon of energy $h\nu$ and mc^2 is the rest energy of the electron.

In the application of (17), the condition $4\gamma h\nu \ll mc^2$ is often satisfied. In this case, the energy of the scattered photon may be approximated as

➡ $$h\nu_s \approx 4\gamma^2 h\nu = 4\left(\frac{U}{mc^2}\right)^2 h\nu,$$

(Single head-on scatter; $4\gamma h\nu \ll mc^2$; $\gamma \gg 1$)   (9.18)

where γ has been expressed as the ratio of the total initial (rest + kinetic) energy of the electron U to the rest energy mc^2 (7.17). The expression is valid if the electron is highly relativistic ($\gamma \gg 1$) and if $4\gamma h\nu \ll mc^2$ (see (17)). Note that $2\gamma h\nu$ is the photon's incident energy in the moving (S′) frame; see (13) and (16).

The notable feature of (18) is that the energy of the photon is increased by a factor of $\sim\gamma^2$ in the interaction. Each factor of γ can be attributed to one of the Doppler-shift transformations.

Average over directions

In a real source, the electron will collide with many photons with various directions of travel and "impact parameters." Thus, not all collisions will be head-on back scatters; many other scattering angles in S' will occur and must be accounted for.

Collisions in which the photon overtakes the electron (in S) will result in a reduced photon energy, but, on average, the energy gain of the photons will be positive because there are relatively fewer overtaking than head-on collisions. Think of driving on a highway; one encounters more cars coming in the opposite direction than in the same direction. Also, special relativity tells us that, in the electron frame, the photons will appear to be concentrated in the oncoming direction (extreme aberration; Figs. 7.5b and 7.9a. In the latter, note that frame S moves to the left relative to S').

A proper calculation for an isotropic distribution of incident photon directions in the observer frame yields an average scattered energy that is one-third of that given in (18),

$$\Rightarrow \quad h\nu_{s,iso} = \frac{4}{3}\gamma^2 h\nu, \quad \text{(IC; average scattered energy; } 4\gamma h\nu \ll mc^2; \gamma \gg 1) \quad (9.19)$$

where $h\nu_{s,iso}$ is the final photon energy averaged over all angles, $\gamma = U/mc^2$ is the factor that describes the initial energy of the electron, and $h\nu$ is the initial photon energy.

The increase of photon energy can be a tremendous factor. For example, consider electrons of energy 10 GeV (10^{10} eV) that one might find in the Crab nebula (Table 8.1). Their Lorentz factor is $\gamma = U/mc^2 = 2 \times 10^4$, and $\gamma^2 = 4 \times 10^8$. Thus, a millimeter-wavelength radio photon of frequency 300 GHz will be shifted according to (19), on average, up to frequency $\nu_s = (4/3) \times 4 \times 10^8 \times 3 \times 10^{11} = 1.6 \times 10^{20}$ Hz, which is well into the gamma-ray band ($h\nu_s = 660$ keV). The electron involved in such a collision provides the energy gained by the photon.

Note that, in the foregoing example, the electron loses only 0.007% of its energy. If electron-photon IC interactions are frequent, however, this process can rapidly drain energy from the electrons. The interaction rate depends, of course, on the photon number density and the cross section.

Rate of electron energy loss

The IC formalism can also provide an approximate rate of energy loss dU/dt for an electron undergoing multiple IC scatters.

Cross section

The Thomson cross section, $\sigma_T = 6.65 \times 10^{-29}$ m^2 (2.49), is applicable to the Compton interaction if the energy of the photon in the electron rest frame S' is substantially (factor of 10 or more) less than the electron rest energy, $2\gamma h\nu \ll mc^2$. At higher photon energies, the cross section decreases slowly and is known as the Compton cross section (AM, Chapter 10). The Thomson cross section is appropriate for many astrophysical situations, and it is used here.

Single electron and many photons

Consider an electron traveling at speed $v = \beta c$ through a sea of photons with density n_{ph} photons/m^3 in frame S and with a distribution of energies $h\nu$. To calculate the electron rate of energy loss dU/dt properly, one should calculate the power in the scattered photons in the rest frame of the electron for an arbitrary incident photon direction, integrating over all photon energies. The next step would be to transform the power back to the observer frame and convert all variables to observer values. One would then average over incident photon directions and subtract the energies of the initial interacting photons. The negative of the result would be the desired electron-energy loss rate.

Rather than go through this, we offer an abbreviated argument that happens to give the correct result for relativistic electrons. Suppose that, in the observer's frame, the flux arriving at the speeding electron is $n_{ph}c$ (photons m^{-2} s^{-1}) and that the cross section for the scattering interaction is σ_T, the Thomson cross section. (Neither of these statements is true for substantial electron velocities.) The number of interactions per second would then be $n_{ph}c\sigma_T$.

In each of these hypothetical collisions, the electron would lose an amount of energy comparable to that gained by the photon which, for small initial photon energy, is about equal to $h\nu_{s,iso}$ (19). Multiplication of this energy by the number of collisions per second would yield the approximate energy lost per second by the electron, dU/dt,

$$\left(\frac{dU}{dt}\right)_{IC} \approx -\sigma_T n_{ph} c(h\nu_{s,iso}) = -\frac{4}{3}\sigma_T c\gamma^2 n_{ph} h\nu_{av}, \tag{9.20}$$

where we have invoked (19) to express the energy-loss rate in terms of the initial photon energy $h\nu$. We further generalize the expression to include a distribution of initial photon energies with average (over energy) $h\nu_{av}$.

The energy density of photons with which the electrons are colliding is $u_{ph} \equiv n_{ph}h\nu_{av}$, giving

➡ $$\left(\frac{dU}{dt}\right)_{IC} = -\frac{4}{3}\sigma_T c\beta^2\gamma^2 u_{ph}, \quad \text{(Rate of energy loss by electron due to inverse}$$
$$\text{Compton scattering; } \gamma h\nu \ll mc^2; \text{ W)} \tag{9.21}$$

where we have inserted, without argument, the electron speed factor $\beta^2 = v^2/c^2$ to obtain the more general expression that includes the nonrelativistic case. This turns out to be the result from a proper calculation. It even takes into account the contribution of the initial photon energy, $h\nu$, to the final scattered energy; the electron does not have to provide all of it.

The expression (21) is appropriate for an electron passing through photons with an isotropic distribution of velocities. It is generally valid as long the photon energy in the electron frame, $\sim\gamma h\nu$, is substantially less than mc^2.

Substitute values for σ_T and c into (21):

$$\left(\frac{dU}{dt}\right)_{IC} = -2.66 \times 10^{-20} \beta^2\gamma^2 u_{ph} \tag{9.22}$$
$$\text{(W)} \qquad\qquad\qquad\qquad\qquad (\text{J/m}^3).$$

These expressions refer to the energy loss by a single electron immersed in a sea of photons of energy density u_{ph}.

Volume emissivity

Consider a population of relativistic electrons of density n_e (m^{-3}), each of energy $U = \gamma mc^2$ with $\gamma \gg 1$. Let them be immersed in a sea of photons with energy density u_{ph} and photon energies negligible compared with the IC upscattered energies ($h\nu_{\rm av} \ll h\nu_{\rm s,iso}$). In this case, the rate of scattered energy emerging from a single electron approximates the negative of the electron energy loss rate (21). The radiation emerging from unit volume is thus the negative of (21) multiplied by the electron number density $n_{\rm e}$:

$$j = \frac{4}{3}\sigma_{\rm T}\, c\, \beta^2 \gamma^2 \, n_{\rm e}\, u_{\rm ph} \qquad \begin{array}{l} \text{(Integrated volume emissivity for IC;} \\ \gamma h\nu \ll mc^2; \ h\nu_{\rm av} \ll h\nu_{\rm s,iso}; \ \text{W/m}^3) \end{array} \qquad (9.23)$$

This is expressed as the spectral volume emissivity j (W/m^3) integrated over all emitted frequencies, $j = \int j_\nu {\rm d}\nu$. For a realistic system, one would also integrate over the spread of electron energies, γmc^2 (Prob. 33).

Comptonization

A population of photons that encounters a region containing free electrons will find its spectrum modified as a result of IC scatters given sufficient optical depth. If the electrons on average are more energetic than the photons, the photons will, on average, be scattered to higher energies. If, on the other hand, the electrons are less energetic, the photons will be scattered to lower energies. The modification of the photon spectrum by Compton scatters is called *Comptonization*. We mention two examples – one that is now deemed to be overly simplistic and the other a powerful tool for cosmological research.

Note that, in these two cases, the energetic electrons may not be relativistic (i.e., they may have kinetic energies substantially less than the rest energy $mc^2 = 511$ keV). Our limiting expressions above for $\gamma \gg 1$, (19) for example, may therefore not be applicable. More general expressions such as (21)–(23) must be used.

Black-hole binaries

The x-ray spectrum of an accreting black-hole binary system typically exhibits a thermal spectrum with kT to a few keV and a high-energy power-law tail that extends to 50 or 100 keV (Fig. 3). This tail has traditionally been attributed to IC scattering. Older models suggested that thermal radiation from the accretion disk around the black hole is the source of the photons. A corona of energetic electrons with a thermal distribution of $kT \approx 100$ keV overlying the accretion disk would scatter the photons to higher energies, thus creating the spectral tail.

This scenario is now known to be too simplistic. The hard tails in the spectrum appear to be related to jets emanating from the black hole; these jets have been seen by radio astronomers using high-resolution very long baseline interferometry (VLBI) (Fig. 7.8). There is also a "steep power-law" component in the spectrum that extends up to gamma-ray energies to 500 keV and more. This component sometimes exhibits rapid, quasi-periodic oscillations of intensity with frequencies extending up to ~450 Hz.

The sources can be highly variable on time scales of minutes to months. They will reside in one of several "states" in which one or another of the spectral components (e.g., thermal or power law) is dominant and then move to a state with another component dominant. Two such states are shown in Fig. 3. The accretion disk that feeds the matter into the black hole is continuously, or episodically, being replenished by a normal companion star (Section 4.4). The different states may be related to the rate at which matter is being deposited into the

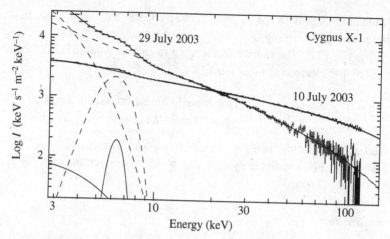

Figure 9.3. Two x-ray spectra of the black-hole binary Cygnus X-1 on a log-log plot illustrating the extreme variability of the source. In the 29 July 2003 "soft" spectrum, the thermal (modified blackbody) peak is quite pronounced (hump in upper curve), whereas the power-law extending to high energies is much more pronounced in the 10 July 2003 "hard" spectrum. The spectra are modeled with the components shown as dashed (29 July) and solid (10 July) lines. [J. Wilms *et al.*, *A&A* **447**, 245 (2006)]

accretion disk, to the amount of matter in the accretion disk, or to the manner in which the matter moves into the black hole.

The processes underlying all this activity are not yet understood. Studies in all wavebands that monitor the spectral and temporal variations guide the modeling. Inverse Compton scattering, including the SSC process (Section 4), may well play an important role.

Clusters of galaxies

A much better understood example of Comptonization, the Sunyaev–Zeldovich (S-Z) effect, is discussed below (Section 5). In this case, the hot plasmas in clusters of galaxies contain electrons that scatter the extremely low-energy photons of the 2.73-K cosmic microwave background (CMB). The photons gain energy, and this is revealed as a distortion of the CMB blackbody spectrum.

9.4   Synchrotron self-Compton (SSC) emission

Relativistic electrons in the presence of a magnetic field will surely emit synchrotron radiation at some level. If the radiation energy density of synchrotron photons within the source region is sufficiently intense, the photons will undergo IC scattering by the very same electrons that emitted them in the first place. Any such scatters will take place, of course, before the synchrotron photon leaves the source region. This is the *synchrotron self-Compton* (SSC) process (Fig. 4).

The two mechanisms underlying the SSC process, synchrotron radiation and inverse Compton scattering, have already been developed. Here we compare the electron energy loss rates for synchrotron and IC radiation, introduce the "Compton limit," and present two examples of SSC: the Crab nebula and blazars.

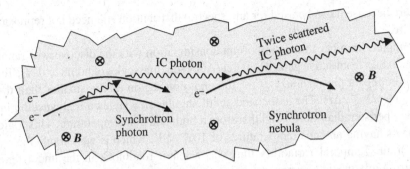

Figure 9.4. Synchrotron self-Compton (SSC) process in a nebula with magnetic field **B** pointed into the page. Synchrotron photons are emitted by energetic electrons, resulting in a high-radiation energy density in the nebula. The photons then undergo inverse Compton (IC) scattering to high energies by the same electrons that give rise to the synchrotron photons. If the probability of a second IC scatter (as shown) becomes significant, the electron energy is rapidly depleted.

Relative energy loss rates

An electron in a synchrotron nebula is subject to energy loss from both synchrotron radiation and IC scatters. The rates of such losses, $(dU/dt)_{IC}$ and $(dU/dt)_{syn}$, are easily compared. The expression for IC (21) is identical in form to that for synchrotron radiation (8.34) except for the nature of the energy density with which the energetic electron interacts. One can thus state, independent of electron energy, but still with the requirement that the photon energy in the electron frame be substantially less than mc^2, that

$$\frac{(dU/dt)_{IC}}{(dU/dt)_{syn}} = \frac{u_{ph}}{u_B}. \qquad \text{(Relative energy-loss rates; } \gamma h\nu \ll mc^2) \qquad (9.24)$$

The relative energy loss by electrons through the two processes depends solely on the relative energy densities of the photons and magnetic fields – the target material in each case. Inverse Compton losses are important if the ratio u_{ph}/u_B approaches order unity (e.g., $u_{ph}/u_B \gtrsim 0.1$).

Compton limit

Inverse Compton scattering can remove large amounts of energy from a typical electron – especially if *second-order scattering*, in which a given photon experiences two IC scatterings before leaving the nebula (Fig. 4), becomes important. This can occur in compact sources when radiation energy densities become comparable to magnetic energy densities. The photon is thus boosted by a second factor of γ^2, which can give it a huge energy. This can expend most of an electron's energy (Prob. 42).

Under these conditions, if more energy is pumped into the electrons of the nebula, it will immediately be radiated away as IC radiation. One can not make the nebula "hotter." This is known as the *Compton limit* and sometimes as the *Compton catastrophe*. It can be shown (not in this text) to occur when the nebula exhibits an antenna temperature of $\sim10^{12}$ K.

Recall that the *antenna temperature* is inferred by radio astronomers from a measured specific intensity at some frequency ν under the assumption of a blackbody spectrum in the Rayleigh-Jeans approximation (Fig. 6.2b). We sometimes think of it as an *apparent* temperature because it can be used for any spectral distribution and need not represent a true temperature.

The Compton limit became an important consideration with the discovery of rapid variations of radio flux from active galactic nuclei (AGN) on time scales of hours or days. Temporal variability implies a physical size D for the emission region that is smaller than the light-crossing time ($D \lesssim c\Delta t$). The associated small angular size yields a high specific intensity (energy flux per steradian-hertz) and therefore a high antenna temperature. This logic led to some sources having antenna temperatures of 10^{13-14} K, which is well above the $\sim 10^{12}$-K Compton limit. Temporal variability thus was a major roadblock in the understanding of radiation from relativistic plasmas.

This problem was resolved as it became apparent that jets of material ejected from AGN and other celestial objects are a common phenomenon. If a blob of emitting plasma is ejected at relativistic speeds toward an observer, the radiation from it will be substantially modified in the observer's frame of reference. The flux of photons will be emitted preferentially in the forward direction (headlight effect), and the measured specific intensity will be Doppler enhanced (7.78).

High intensity may also be due in part to coherent emission (Section 8.6). Also, the time scales of the intensity variability are dramatically reduced by the Doppler effect. The reduced light-crossing times suggest small angular sizes that further inflate the inferred antenna temperatures. The Compton limit is thus irrelevant to raw attenna temperatures not corrected for these effects.

The Compton-limit dilemma was mostly an issue in the days when AGN studies were confined primarily to radio and optical studies. Now we know that x-ray and gamma-ray luminosities often dominate the radio luminosity of AGN. This and the ubiquity of jets make modeling much more complicated than that leading to the Compton limit.

Inverse Compton peaks in SEDs

We present two examples of the SSC process, one galactic and one extragalactic. In each case the spectral energy distribution (SED), $\nu S(\nu)$, shows two pronounced maxima that are attributed, respectively, to synchrotron radiation (lower-frequency peak) and IC radiation (higher-frequency peak).

Crab nebula

The Crab nebula is a quite convincing example of IC scattering revealed to us through the recent advances in TeV astronomy. We know from the observed synchrotron radiation that the Crab nebula contains extremely high-energy electrons spiraling in magnetic fields of some tens of nanoteslas (Section 8.4 under "Crab nebula"). To produce gamma rays up to the observed ~ 40 MeV in such fields requires electrons of energies $U \approx 10^{15}$ eV (Table 8.1). With a Lorentz factor of $\gamma = U/mc^2 = 2 \times 10^9$ and $\gamma^2 = 4 \times 10^{18}$, these electrons would, according to (19), scatter the lowest energy (and most numerous) radio photons of 10^9 Hz up to $\sim 5 \times 10^{27}$ Hz or about 2×10^{13} eV ($= 20$ TeV).

This is the domain of TeV astronomy. Such radiation is detectable with current instrumentation. Telescopes observe the Cerenkov light released from atmospheric *extensive air showers* that are initiated by TeV photons. (We encountered TeV astronomy in Section 7.7; see discussion associated with Fig. 7.16.)

The spectral energy distribution (SED) of the photons emerging from the Crab nebula is shown in Fig. 8.2. This is a compilation of observational data from all wavebands, including the TeV results carried out with the HEGRA observatory in the Canary Islands. The dominant peak (center-left) is largely synchrotron radiation, whereas the prominent right-hand peak is attributed to IC scattering. Recall, from Section 8.2, that, in an SED plot, $\log(\nu S(\nu))$ versus $\log \nu$, a flat spectrum implies equal energy flux in each decade of frequency.

Modeling of the spectrum (dark lines in Fig. 8.2) as an SSC source indicates that IC-scattered synchrotron photons are the major contributor to the upper peak (solid line). The model best fits the data for a magnetic field of ~ 16 nT, which is comparable to the equipartition value of 50 nT we have been using.

Blazars

Active galactic nuclei come in many guises that are strongly determined by the view angle of the observer. If the observer lies in the plane of the accretion disk surrounding the massive black hole (Fig. 7.7), radiation from the central nuclear region will be partially or totally obscured at visible wavelengths. One sees instead radiation from local matter illuminated or heated by radiation from the active nucleus. Such objects are known as Seyfert 2 galaxies.

On the other hand, if the observer view is more or less normal to the accretion disk, the action close to the nucleus becomes visible. If the object has an active jet, the radiation can extend to gamma-ray energies that in some cases reach the TeV band. The radiation can be highly variable and polarized. In these latter cases, the observer is deemed to lie within the jet beam. Such objects are known as *blazars* or as BL Lacertae objects after the prototype object.

Blazars have SEDs that are typically two-peaked like the Crab SED (Fig. 8.2). Here, also, the peak at lower frequency is attributed to synchrotron radiation, and the one at higher frequency to IC scattering.

Figure 5 presents theoretical SEDs for two models of a blazar jet. A radiating plasma of electrons and photons moves at relativistic speed in a direction that is offset at some small angle from the observer. Photons are created and energized through the synchrotron and IC processes. In the observer frame, the photons are beamed and Doppler shifted according to the dictates of special relativity (Prob. 41). The resulting SED depends on the exact parameters chosen; the figure shows two extremes. The lower-energy case (LBL blazar) extends from the radio to the gamma-ray bands but is quiet in the TeV band. The high-energy case (HBL blazar), in contrast, reaches TeV energies but is quiet in the radio range.

Figure 6 shows the SED of data from several time periods from a real blazar, 3C454.3. The data show large temporal variations but generally exhibit the two peaks attributed to synchrotron and IC radiation. Diagnostics of the processes ongoing within the jet require repeated quasi-simultaneous observations over many wavebands so that the variations of the amplitude and shape of the SED as a function of frequency can be properly tracked. For example, in the SSC process, one would expect an IC enhancement to be preceded by a synchrotron enhancement.

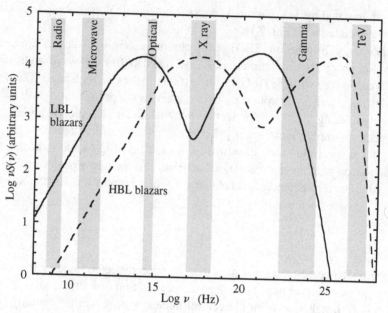

Figure 9.5. Theoretical spectral energy distributions (SED) for blazars across the broad range of frequencies where observations of blazars are carried out (shaded areas). The two curves exhibit extremes of possible spectra (high and low energy) and illustrate the importance of observations across many wavebands. In the model, the lower frequency peak arises from synchrotron photons, and the higher frequency peak from IC-scattered photons in an SSC source that is hurtling at relativistic speed toward the observer. In an SED plot, a flat spectrum represents constant energy flux per fixed log interval (e.g., a decade of frequency). [P. Giommi *et al.*, *A&A* **445**, 843 (2006); courtesy S. Colafrancesco]

9.5 Sunyaev–Zeldovich effect

Inverse Compton scattering plays a fundamental role in the interaction of two important characteristics of our universe. One is the cosmic microwave background radiation (CMB) that permeates the entire universe. It consists of a sea of very low energy photons with a blackbody spectrum of temperature $T_r = 2.73$ K and average energy $h\nu_{av} = 2.70\, kT_r = 6.4 \times 10^{-4}$ eV (6.32). The other characteristic is the existence of hot ionized gas (plasma) in the potential wells of clusters of galaxies.

The *Sunyaev–Zeldovich effect* is the distortion of the blackbody spectrum of the CMB owing to the IC interaction of the CMB photons with the energetic electrons of the cluster plasma. This effect, together with x-ray measurements of the cluster plasma, yields the distances to the clusters and thus allows determination of the Hubble constant. Surveys of many such clusters should yield the evolution of cluster numbers as the universe evolved.

Cluster scattering of CMB

Hot plasmas are known to exist in clusters of galaxies as evidenced by the thermal bremsstrahlung x-ray emission radiated by them. In a typical cluster, the particles might have kinetic energies of $kT_e \approx 5$ keV ($T_e \approx 10^8$ K). (Be careful to distinguish the CMB

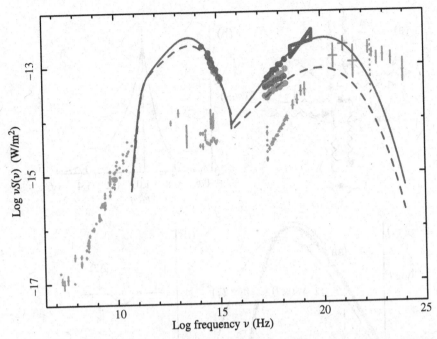

Figure 9.6. Spectral energy distribution (SED) of blazar 3C454.3 over many wavebands from many observatories. The two-peaked character of Fig. 5 is clearly evident in the data and in the simple one-zone model fits shown as solid and dashed lines. The filled circles signify recent quasi-simultaneous subsets of radio, optical, x-ray, and gamma-ray data, the latter three being mostly from the Swift gamma-ray burst observatory. The smaller points represent nonsimultaneous observations in the literature. The data show large intensity variations across most frequency bands. [P. Giommi et al., A&A, **456**, 911 (2006)]

radiation temperature T_r and the cluster electron temperature T_e.) An astronomer viewing the CMB sky at some frequency ν in the direction of such a cluster would measure a specific intensity somewhat lower or higher than that of the CMB. This is due to IC scattering of the CMB photons by the energetic electrons of the cluster plasma.

Note that the ∼5 keV electrons are much more energetic than the ∼10^{-3} eV photons. Nevertheless, they are not relativistic, because the electron kinetic energy E_k is much less than the rest energy, $E_k/mc^2 \approx kT_e/mc^2 \approx 0.01$; recall $mc^2 = 511$ keV. Hence, $\gamma = (mc^2 + E_k)/mc^2 = 1.01$, and, from (7.3), $\beta = v/c = 0.14$. The IC frequency-shift formulas provided above such as (18) are thus not applicable.

Average frequency increase

CMB photons approaching the observer and passing through a cluster may not scatter at all. However, if they do, they will be scattered out of the line of sight, whereas photons initially traveling in other directions may be scattered into the line of sight (Fig. 7a). From the isotropy of the CMB radiation, one can argue that the total number of photons arriving at the observer will be unchanged, but some of these will have undergone scattering. If the probability of any given photon being scattered in the cloud is small, as is the case for actual clusters, the proportion of scattered photons in the observed radiation will be small.

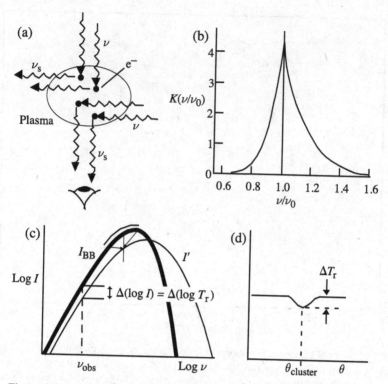

Figure 9.7. Sunyaev–Zeldovich effect. (a) Scattering of CMB photons in a cloud of plasma. The cloud does not affect the number of photons observed. In reality, only a small fraction of the observed CMB photons from the direction of a cluster have been scattered by cluster electrons. (b) Distribution of scattered photon energies for an initial monochromatic beam at $\nu = \nu_0$ and for $kT_e = 5.1$ keV. (c) Sketch of a blackbody spectrum (dark line) with the S-Z modified spectrum for $y = kT_e\tau/mc^2 = 0.15$ (light line); $T_r = 2.73$ K. The latter curve is broadened and shifted to the right. (d) Antenna temperature as a function of position on the sky for radio measurements in the Rayleigh–Jeans (low-frequency) portion of the spectrum. [(b) after Sunyaev & Zel'dovich, in ARAA **18**, 537 (1980), with permission; (c) after E. L. Wright, pvt. comm.; see http://www.astro.ucla.edu/~wright/SZ-spectrum.html; (d) S&Z, ibid, p. 551.]*

A head-on collision between a low-energy CMB photon and an energetic electron will result in the photon's gaining energy. An overtaking collision (a photon catching an electron) will cause the photon to lose energy. In this nonrelativistic case, these effects will cancel to first order in v/c. Examination of the Doppler shifts in these collisions, however, indicates that there is a net fractional photon frequency gain of order v^2/c^2; more energy is gained in head-on collisions than is lost in overtaking collisions (Prob. 51). Because the energy of a photon is $h\nu$, a fractional frequency increase equals the fractional energy increase. In a Maxwell–Boltzmann plasma of temperature T_e, the average kinetic energy of the electrons is $\langle mv^2/2\rangle_{av} = 3kT_e/2$, and so the fractional photon energy gain is of order $v^2/c^2 \approx 3kT_e/mc^2$.

A proper calculation must take into account the effect on monochromatic ($\nu = \nu_0$) photons of electron scatters averaged over all collision angles and electron speeds of a Maxwellian

---

* The modified curve of (c) in Fig. 7 is obtained by Wright from the exact formula (Eq. A8 of S&Z (1980)) rather than the approximate formula, Eq. A7. He points out that the latter expression is not appropriate for this large value of $y = 0.15$, but that it is often used for such plots (see Prob. 52). The approximation Eq. A7 is, however, appropriate for the small values of y characteristic of clusters of galaxies. (Wright, see link in caption)

distribution. The result, under the assumption that each photon is scattered just once in a plasma of $kT_e = 5.1$ keV, is a distribution of scattered-photon frequencies $K(\nu/\nu_0)$ that ranges up and down from ν_0 by as much as $\sim$50% (Fig. 7b). This is an asymmetric function in which the high-frequency wing is somewhat broader than the low-frequency wing.

The average fractional frequency shift of the distribution $K(\nu/\nu_0)$ turns out to be $(\Delta\nu/\nu)_{av} = 4kT_e/mc^2$. It is indeed of the order just discussed. For our x-ray–emitting plasma of $kT_e = 5.1$ keV, the average fractional photon frequency gain would thus be only about $4 \times 0.01 = 0.04$, or about 4%, if every photon were to be scattered once.

The probability of a scatter while a photon is in the cluster is, by definition, the optical depth τ of the cluster for $\tau \ll 1$. In the case of known clusters, only a small fraction, say $\sim$0.5%, of the detected CMB photons undergo a scatter; hence, the average fractional frequency shift will be reduced by the factor $\tau \approx 0.005$. We thus have, for nonrelativistic electrons,

$$\left(\frac{\Delta\nu}{\nu}\right)_{av} = \frac{4kT_e}{mc^2}\tau. \qquad \text{(Average fractional photon frequency}$$
$$\text{shift in Maxwellian electron gas)} \qquad (9.25)$$

For $\tau \approx 0.005$ and $kT_e/mc^2 = 0.01$, the average fractional frequency shift is only 2×10^{-4}. This upward shift of frequency is a net value; individual photons are shifted both upward and downward by much larger amounts.

Shifted spectrum

The number spectrum of the scattered photons is obtained by applying $K(\nu/\nu_0)$, the *kernel*, to each frequency bin $d\nu_0$, of the original spectrum $N(\nu_0)$ and summing (integrating) over all frequencies ν_0. Since the fraction of photons that are scattered is the optical depth τ (for $\tau \ll 1$), the factor τ must also be included in the integrand. Finally, the normalization of $K(\nu/\nu_0)$ must be addressed. A function of ν/ν_0 alone such as that in Fig. 7b (with unit area under the curve), will broaden in absolute frequency space as ν_0 is increased. (A 30% upward scatter from 1 Hz is 0.3 Hz, but from 2 Hz, it is 0.6 Hz.) Hence, to preserve normalization at unity, a factor $1/\nu_0$ must also be included.

The integral that gives the number spectrum of scattered photons is thus

$$N_{scatt}(\nu) = \int_0^\infty \tau N(\nu_0) K\left(\frac{\nu}{\nu_0}\right) \frac{d\nu_0}{\nu_0} \qquad (\text{s}^{-1}\ \text{m}^{-2}\ \text{Hz}^{-1}\ \text{sr}^{-1}) \qquad (9.26)$$

which can be evaluated for $N_{scatt}(\nu)$, given a known unscattered spectrum $N(\nu_0) = I(\nu_0)/h\nu_0$ and a known kernel. The latter is not given here; its shape depends on the "y" parameter, $y \equiv kT_e\tau/mc^2$.

The spectrum one observes is the sum of the scattered and unscattered number spectra where the latter is $(1 - \tau)N(\nu_0)$. For comparison to Fig. 7c, this combined spectrum must be converted to specific intensity $I'(\nu)$ (W m^{-2} Hz^{-1} sr^{-1}). In the figure, a 2.73-K (unscattered) blackbody spectrum is shown alongside the modified spectrum for $y = 0.15$ obtained from a proper calculation. This is an unrealistically large value of y; it is 1500 times greater than the expected values of $\lesssim 10^{-4}$.

The net effect is that the spectrum broadens slightly and shifts to the right by varying amounts; its shape is thus distorted. At a *fixed* frequency below the crossover point, the measured intensity is reduced, and above the crossover point, the intensity is increased (Fig. 7c). Further, in the low-frequency Rayleigh-Jeans regime, it can be shown that the modified curve approximates a power-law with the same logarithmic slope as the blackbody, i.e., 2.0. One can visualize the straight-line curve being shifted to the right parallel to itself, for small shifts.

Microwave observations of the CMB are most easily carried out in the low-frequency Rayleigh–Jeans region of the spectrum where $h\nu \ll kT_r$. Measurements of the CMB specific intensity in the direction of a cluster will thus show a lower intensity (or, equivalently, antenna temperature) than would a nearby portion of the sky offset from the cluster. The temperature profile for a hypothetical scan across a cluster is shown in Fig. 7d. Measurement of the CMB temperature as a function of position on the sky would thus exhibit antenna temperature dips in the directions of clusters that contain hot plasmas.

One might think that the modified CMB spectrum (Fig. 7c) is a thermal blackbody spectrum for a higher temperature because the peak intensity is shifted to the right, but it is not. The scattered spectrum has neither the shape nor the amplitude of a pure blackbody at the higher temperature. Recall from (6.18) that the integrated energy flux of blackbody radiation scales with temperature as T_r^4. Also the number density of blackbody photons (6.31) scales as T_r^3. We thus would expect a large increase of photon density and energy flux at a higher temperature if the radiation were blackbody, but this is not what we see in S-Z scattering (Fig. 7c). The total number of photons detected in a given time interval over the entire spectrum remains constant, and the total energy flux scales only to the extent that the individual photons gain energy (on average).

Intensity decrement

Now, let us quantitatively examine the effect of a spectrum shift to higher frequencies in the Rayleigh–Jeans region $h\nu \ll kT_r$. (For an arbitrary spectral shape, see Prob. 52.) The blackbody specific intensity (W m^{-2} Hz^{-1} sr^{-1}) is approximated in this region as a power law with logarithmic slope of 2 (6.8) by

$$I = CT_r\nu^2,$$

(Blackbody specific intensity; Rayleigh–Jeans approximation; W m^{-2} Hz^{-1} sr^{-1})   (9.27)

where C is a constant and T_r is the temperature of the CMB. Take this to be the spectrum before scattering.

If the spectrum is shifted parallel to itself on a log-log plot (as calculations show to be the case), the fractional frequency interval between the two curves does not change as a function of frequency – that is, $\varepsilon = \Delta\nu/\nu = (\nu' - \nu)/\nu = $ constant. Imagine, for simplicity, that each photon is scattered just once and boosted in frequency by this amount from ν to $\nu' = \nu(1 + \varepsilon)$. The photons in a given band move to the higher frequency, but the bandwidth they occupy also increases from $d\nu$ to the larger band $d\nu' = (1 + \varepsilon)\, d\nu$ at the higher frequency.

The number of photons in a given portion of the spectrum must be conserved when they are moved to higher frequency because scattering neither absorbs nor creates photons. Thus the number of photons at ν in the band $d\nu$ is equal to the number in the expanded band $d\nu'$ at the higher frequency ν':

$$N(\nu)\, d\nu = N'(\nu')\, d\nu'.$$

(Numbers conserved; s^{-1} m^{-2} sr^{-1})   (9.28)

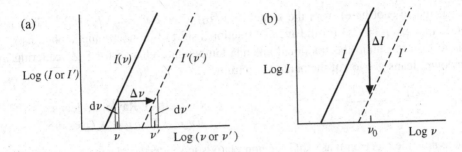

Figure 9.8. (a) Shift of a Rayleigh–Jeans power-law ν^2 spectrum by the S-Z effect. Photons shifted upward in frequency by a frequency-independent multiplicative factor have the same specific intensity, $I' = I$ (W m^{-2} Hz^{-1} sr^{-1}) at the higher frequency because the increased bandwidth $d\nu'$ (which lowers I) is offset by the increased photon energy $h\nu'$. (b) Change of specific intensity ΔI at a fixed frequency ν_0.

In terms of specific intensities, this may be written as

$$\frac{I(\nu)}{h\nu}d\nu = \frac{I'(\nu')}{h\nu'}\,d\nu'. \tag{9.29}$$

Solve for I' and apply the ratios $\nu'/\nu = 1 + \varepsilon$ and $d\nu/d\nu' = (1 + \varepsilon)^{-1}$ to find that

$$I'(\nu') = I(\nu). \tag{9.30}$$

The bandwidth and energy ($h\nu$) ratios cancel, and so the shifted specific-intensity spectrum has the same value at ν' as it had at ν before the shift (Fig. 8a). Thus, for our spectrum (27), the shifted spectrum is, from (30),

$$I'(\nu') = CT_r \left(\frac{\nu'}{1+\varepsilon}\right)^2. \tag{9.31}$$

In practice, we wish to compare the spectrum at one frequency in the cluster direction with that at the *same* frequency in an off-cluster direction (Fig. 7d). The unshifted (27) and shifted (31) spectra at the same frequency ν_0 are

$$I(\nu \to \nu_0) = CT_r\nu_0^2; \quad I'(\nu' \to \nu_0) = CT_r \left(\frac{\nu_0}{1+\varepsilon}\right)^2. \tag{9.32}$$

Comparison of these indicates that, at any fixed frequency ν_0 in the Rayleigh–Jeans regime, the intensity of the shifted spectrum is less than the unshifted spectrum (Fig. 8b). From (32) and by invoking $\varepsilon \equiv \Delta\nu/\nu \ll 1$, we find that the fractional decrease of I is

$$\frac{\Delta I}{I} = \frac{I' - I}{I} = \frac{1}{(1+\varepsilon)^2} - 1 \approx -2\varepsilon = -2\frac{\Delta\nu}{\nu}, \tag{9.33}$$

where $\Delta\nu/\nu$ is the fractional rightward shift of the spectrum in Fig. 8a. The factor of 2 in (33) arises directly from the logarithmic slope, $\alpha = 2$, of the unscattered spectrum.

Finally we can express the radiation (antenna) temperature decrement ΔT_r in terms of the temperature T_e of the electrons in the plasma. Set $\Delta I/I = \Delta T_r/T_r$ in (33) because the antenna temperature is proportional to the specific intensity in the Rayleigh–Jeans region (27). The average fractional frequency shift $\Delta\nu/\nu$ in the power-law region of the CMB spectrum may be calculated (not here) from the kernel $K(\nu/\nu_0)$ to be $\Delta\nu/\nu = kT_e\tau/mc^2$. (This is one-fourth that given in (25) for the average fractional shift of monochromatic photons.) Apply this to (33) to obtain the fractional antenna temperature change for S-Z scattering in the Rayleigh–Jeans portion of the photon spectrum,

$$\frac{\Delta T_r}{T_r} \approx -2\frac{kT_e}{mc^2}\tau, \qquad \text{(S-Z effect; Rayleigh–Jeans;} \\ h\nu \ll kT_r; \ h\nu \ll kT_e \ll mc^2) \qquad (9.34)$$

where m is the electron mass, and the nonrelativistic scattering plasma has temperature T_e, optical depth τ, and a Maxwell–Boltzmann distribution of speeds. This is the result we sought. See Prob. 52 for another approach to (34).

The optical depth may be written as (AM, Chapter 10)

$$\tau = \sigma_T \int_0^L n_e \, dx, \qquad \text{(Optical depth)} \qquad (9.35)$$

where $\sigma_T = 6.65 \times 10^{-29}$ m^2 is the Thomson cross section (2.49), which is applicable for nonrelativistic electrons and low-energy photons, n_e is the electron density (m^{-3}), and the integral is along the line of sight through the cluster plasma.

A typical cluster might have an average electron density of ~ 2500 m^{-3}, a (core) radius of $R_c \approx 1.0$ MLY $\approx 10^{22}$ m, and an electron temperature $kT_e \approx 5$ keV. The integration path through the diameter of the cluster would be two to three times the core radius. A typical optical depth is thus $\tau \approx 3\sigma_T(n_e)_{av}R_c$ or $\tau \approx 0.005$ as stated above before (25). The expected antenna temperature change, from (34), is then

$$\frac{\Delta T_r}{T_r} \approx -1 \times 10^{-4}, \qquad \text{(Typical antenna temperature change)} \qquad (9.36)$$

which yields a temperature change of $\Delta T_r \approx -0.3$ mK for the temperature $T_r = 2.7$ K of the CMB. This effect has been measured in dozens of clusters as of this writing.

The S-Z study of clusters is becoming an active research field as new microwave telescopes of greater sensitivity and improved angular resolution come on line. Interferometric temperature measurements of six clusters of galaxies are shown in Fig. 9. The white contours are temperature decrements relative to the CMB; the coolest is at the center.

Hubble constant

The S-Z effect in clusters of galaxies can be used to obtain the Hubble constant H_0. This cosmological parameter is the proportionality constant between the recession speed v of a galaxy, or cluster of galaxies, in the local universe and its distance r from the observer, or $v = H_0 r$ (AM, Chapter 9). The value of H_0 is a fundamental parameter in the evolution of the universe that determines, in part, whether it will expand forever or will eventually cease expanding and then collapse back to its initial hot, dense state. We encounter H_0 also in Section 6.3 (under "Hubble expansion"), 12.2 (after (12.1)), and 12.5 (under "Hubble constant").

A value of H_0 is obtained from a given galaxy only if one has independent measures of v and r. The former is readily obtained from its spectral redshift if one adopts the Doppler-shift interpretation of the redshift. (We return to this point below in our discussion of (41).) The traditional method of obtaining the distance to a celestial object depends on successive steps of a distance ladder, and these can introduce substantial errors. Thus astronomers have

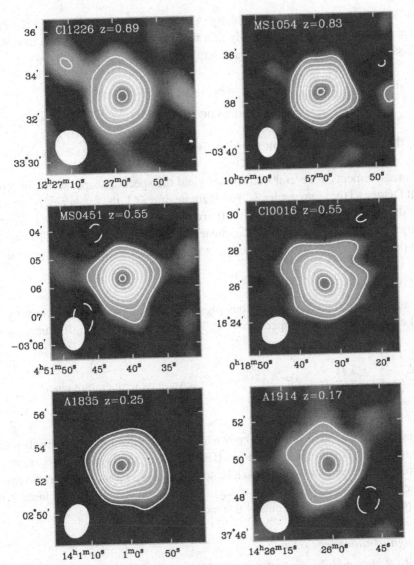

Figure 9.9. Interferometric images at 30 GHz of six clusters of galaxies at redshifts ranging from $z = 0.17$ to $z = 0.89$. The solid white contours indicate negative decrements to the CMB. The six clusters have comparable x-ray luminosities, suggesting similar values of n_e and R. The expected distance independence of the S-Z effect is apparent; the contours reach to similar depths despite the factor-of-five range in redshift. [BIMA and OVRA arrays; J. Carlstrom *et al.*, *ARAA* **40**, 643 (2002), with permission]

developed new independent methods of obtaining distances (AM, Chapter 9). One of these makes use of the S–Z effect (at radio frequencies) in conjunction with x-ray measurements to determine distances to clusters of galaxies.

Cluster x-ray intensity

For simplicity, consider a spherical cluster of radius R that contains a hydrogen plasma of uniform electron density n_e and temperature T_e throughout its volume. The plasma will radiate in x rays via thermal bremsstrahlung. The specific intensity I toward the center of the cluster is given in (5.43) for optically thin thermal bremsstrahlung radiation. We adapt it to our assumption of uniform density and temperature,

$$I_x(\nu_x, T_e) = \frac{C_1}{4\pi} \frac{g(\nu_x, T_e)}{T_e^{1/2}} \exp(-h\nu_x/kT_e)\, n_e^2 2R.$$

(Thermal bremsstrahlung; W m^{-2} Hz^{-1} sr^{-1})   (9.37)

Here g is the slowly varying Gaunt factor, ν_x is the frequency of the (x-ray) radiation, and the constant is $C_1 = 6.8 \times 10^{-51}$ J m^3 K$^{1/2}$.

X-ray measurements at several frequencies yield the spectral shape, which in turn yield the Gaunt factor and the temperature T_e. For example, if $g = 1$, the frequency ν_1 at which the spectrum drops to e^{-1} of its value at $\nu_x = 0$ gives $h\nu_1 = kT_e$ and $T_e = h\nu_1/k$. We neglect here the nonnegligible effect on I of the cluster's cosmological recession. The remaining unknown parameters in the specific-intensity function I_x, at some frequency ν_x, are n_e and R.

CMB decrement

Radio measurements of the CMB at various angular positions off and on the cluster yield the fractional change of antenna temperature (Fig. 7d) at the cluster center. It equals, from (34) and (35),

$$\frac{\Delta T_r}{T_r} = -2\frac{kT_e}{mc^2}\sigma_T\, n_e\, 2R.$$

(Antenna temperature change)   (9.38)

It, too, is a function of n_e and R.

The left sides of (37) and (38) are both known from measurements, and so the only unknown quantities in the two equations are n_e and R. Because different functions of n_e are involved, the equations are independent and thus can be solved for both n_e and R. The radio and x-ray measurements therefore yield absolute values of the electron density n_e and cluster radius R without a priori knowledge of the cluster distance.

Angular-diameter distance

The quantity R, thus derived, gives the physical size of the cluster. Independent imaging of the cluster in the radio or x-ray band yields the angular radius θ. If the cluster is spherical and the geometry of space is flat, the distance r to the cluster is

$$r = \frac{R}{\theta},$$

(Distance to spherical cluster)   (9.39)

where θ is in radians. In cosmology, this is known as the *angular-diameter distance*.

This is the distance needed, in conjunction with the speed v, to obtain the Hubble constant $H_0 = v/r$. The recession speed v is related to the fractional wavelength shift in the spectral lines from the cluster according to the Doppler-shift interpretation of the redshift as

$$\frac{v}{c} = \frac{\lambda - \lambda_0}{\lambda_0} \equiv z,$$

(Doppler shift; redshift)   (9.40)

which is applicable for nonrelativistic speeds $v \ll c$, and where z is the redshift parameter. Thus, from (39) and (40),

$$\Rightarrow \quad H_0 = \frac{v}{r} = \frac{cz}{R/\theta} = \frac{cz\theta}{R}. \quad \text{(Hubble constant from measurable quantities; s}^{-1}) \quad (9.41)$$

The parameters θ and z are directly measurable, whereas R is obtained from the microwave and x-ray measurements given in (37) and (38). The Hubble constant thus follows, in principle, from the described measurements. See Prob. 53 for a hypothetical example.

This illustrates the principle of the method. In practice, at cosmological distances at which $v \ll c$ approximations are not applicable, the x-ray intensity must be corrected for the cosmological redshift. Also, one must use the appropriate distance-redshift relation for the assumed geometry of the universe, $r = (c/H_0)f(z)$; see the discussion of the Hubble constant in Section 12.5 – particularly (12.48) and (12.51).

In general, the clusters will not have uniform electron density and temperature distributions; these quantities will vary with distance from the cluster center. In practice, to obtain a value of H_0 from a given cluster, one must make detailed measurements of the cluster with both radio waves and x rays as a function of angular position and frequency.

A value of the Hubble constant has been obtained from S-Z and x-ray (Chandra observatory) studies of 38 clusters out to redshifts $z \approx 0.9$ in a recent study. The result, for a currently favored set of cosmological parameters, is $H_0 = 23.6 \pm 3.5 \, \text{km s}^{-1} \, \text{MLY}^{-1}$, which is consistent with the value obtained from the Hubble Key Project that probes the nearby universe, $H_0 = 22.1 \pm 2.5 \, \text{km s}^{-1} \, \text{MLY}^{-1}$. (The nominal value of H_0 used in this text is $20 \, \text{km s}^{-1} \, \text{MLY}^{-1}$.) Sunyaev–Zeldovich studies of clusters of galaxies are becoming an important aspect of cosmological research.

Peculiar velocities of clusters

Galaxies and clusters of galaxies can and do have motions relative to the rest frame of the CMB at their respective locations in the expanding universe. In other words, they do not recede exactly as expected from the Hubble law. These motions relative to the Hubble flow are called *peculiar velocities*. On average, the plasma electrons in the cluster also have this velocity. The energies of the CMB photons that scatter from the electrons reflect this motion. This effect produces an additional slight distortion to the CMB spectrum, which is called the *kinetic S-Z effect*.

The kinetic effect turns out to be separable in the data from the *thermal* effect we have discussed here and is most apparent in the region where the thermal effect is zero (at the crossover point in Fig. 7c). Determinations of the peculiar velocities of clusters enable astronomers to map out the growth of large-scale structure in the universe. This topic is of fundamental importance, and the kinetic S-Z effect is a promising method for approaching it.

Nonthermal S-Z effect

The discussion of the S-Z effect in this chapter pertains to a thermal (nonrelativistic) distribution of electrons, the *thermal S-Z effect*. Many clusters exhibit radio halos that indicate the presence of electrons of GeV energies. Regions where there are relativistic electrons would also be expected to distort the CMB spectrum in recognizable ways – the so-called nonthermal

S-Z effect. With sufficient precision, the spectral shape could disentangle the several electron populations associated with a source. It is the thermal S-Z effect that is cosmologically most useful. For such studies, the nonthermal effects must be subtracted out.

Problems

9.2 Normal Compton scattering

Problem 9.21. (a) Solve the energy and momentum equations (3), (4), and (5) to obtain the Compton wavelength shift (1). (b) Demonstrate that a photon of wavelength $\lambda_C = h/mc$ has energy $E_v = mc^2$, where m is the electron mass. (c) In the normal Compton effect (electron at rest), find the energy of a backscattered photon ($\theta = \pi$) if the incoming photon has a large energy, $h\nu \gg mc^2$. What is the scattered energy if $\theta = \pi/2$ and if $\theta \to 0$, again for $h\nu \gg mc^2$? (d) In the backscattered case and still with $h\nu \gg mc^2$, doubling the incident photon energy results in no change in the scattered photon energy. What happens to the added energy? (e) Find an expression for the kinetic energy E of the scattered electron in terms of $h\nu, mc^2$, and θ. Evaluate it at $\theta = \pi, \pi/2$, and 0 and compare your answers with those of part (a). [Ans. $-; -; mc^2$ for $\theta = \pi/2; -; -$]

Problem 9.22. (a) Find values of the fractional energy remaining in a Compton-scattered photon, $h\nu_s/h\nu$, for five angles of scatter ($\theta = \pi/4, \pi/2, 3\pi/4$, and π) and for each of eight energies of the incoming photon ($h\nu = 0.01mc^2, 0.1mc^2, mc^2/2, mc^2, 2mc^2, 5mc^2, 10mc^2$, and $100\,mc^2$). Present your values on a single plot of $h\nu_s/h\nu$ versus θ. For each incoming energy $h\nu$, fit (roughly by hand) a curve to the five points for increasing θ. (Alternatively, create the plots with your computer.) (b) What additional insight does your set of curves give you? What fractional change in frequency, $(\nu_s - \nu)\nu$, would Compton have obtained at $\theta = \pi/2$ for 5-keV and 20-keV x rays? [Ans. $-; \sim -1\%, \sim -4\%$]

9.3 Inverse Compton scattering

Problem 9.31. Suppose a low-energy photon of frequency ν in a vacuum is moving in the *same* direction as a very energetic electron of speed v and mass m. (a) Under what condition could an interaction take place? (b) Find an expression for the final energy $h\nu_s$ of the scattered photon when there is an interaction. Assume the electron is initially relativistic, $\beta \approx 1$. Make sketches similar to those of Fig. 2 to illustrate your development. [Ans. $-; h\nu_s = 0.25\,h\nu/\gamma^2$ if $h\nu \ll \gamma mc^2$]

Problem 9.32. What is a typical energy of an IC-scattered photon when 2-GeV cosmic-ray electrons interact with the photons of the $T = 2.73$-K cosmic microwave background? The CMB background pervades the universe, and the cosmic rays pervade the Galaxy. What type of photons results from the scattering? Justify your use of the approximation (19). (Although such a background is observed, the IC process is not its principal origin.) [Ans. ~ 10 keV, x rays]

Problem 9.33. Find, from (23), an expression for the integrated (over frequency) volume emissivity j_{IC} (W/m^3) of IC radiation in a plasma of radiation energy density u_{ph} and relativistic electrons with a density-number spectrum $N_e(\gamma)d\gamma = N_1 \gamma^\alpha\,d\gamma$ (m^{-3}) between γ_{min} and γ_{max} and zero elsewhere, where N_1 and α are constants and γ is the electron's energy parameter, $\gamma = U/mc^2$. [Ans. $4\sigma_T c\beta^2 u_{ph} N_1(\gamma_{max}^{\alpha+3} - \gamma_{min}^{\alpha+3})/3(\alpha + 3)$]

9.4 Synchrotron self-Compton (SSC) emission

Problem 9.41. (a) Electrons of energy $U \approx 1$ GeV in a stationary knot of the jet of a compact AGN where the magnetic field is $B = 100$ nT produce synchrotron photons. What is the energy of the synchrotron photons that have the critical frequency (in joules and in electronvolts; refer to Chapter 8 and let $\phi = \pi/2$)? In what band are they? (b) These 1-GeV electrons also collide with the synchrotron photons, creating higher-energy photons through the IC process. What is the approximate energy (eV) of the scattered photons? What type of radiation is this? Justify your use of the approximation (19). (c) Repeat (a) and (b) for an electron energy of 10 GeV. (d) Suppose the knot emitting the IC radiation is being carried in a relativistic jet directed toward the observer. How fast, in terms of the Lorentz factor γ, must the knot of part (b) be moving to Doppler boost the IC photons of part (b) up to 0.2 TeV in order to make them accessible to TeV astronomers? Hint: see (7.70) and (7.71). What is the total energy (in electronvolts) of an electron at rest in the moving knot in the observer's frame of reference? Repeat for the IC photons of part (c). [Ans. ~10^{-4} eV; ~1 keV; ~5 MeV; ~150 TeV, ~15 GeV]

Problem 9.42. Consider a synchrotron self-Compton source at rest in the observer's frame. Again (see Prob. 41), the individual electrons have energy $U = \gamma mc^2 = 1$ GeV in a 100-nT field. The synchrotron photons thus have a typical (critical) frequency of $h\nu_c \approx 10^{-4}$ eV. (a) What is the angle-averaged photon energy of the IC radiation? Verify that the condition for using (19), $4\gamma h\nu \ll mc^2$, is satisfied. (b) Consider that one of the scattered photons encounters another electron (of the same energy γmc^2) before leaving the plasma and suffers a second head-on IC scattering with a 1-GeV electron. Can one again use the simplification (19)? (c) Find and evaluate an approximate expression for the energy of the twice-scattered photon in terms of γ, ν_c, and m. [Ans. –; no; ~0.9 GeV]

Problem 9.43. Consider a relativistic plasma radiating close to the Compton limit at $T \approx 10^{12}$ K. If an injector pumps more energy into the electrons, what happens to the energy? Under what hypothetical conditions might temperatures in excess of ~10^{12} K be observed?

9.5 Sunyaev–Zeldovich effect

Problem 9.51. Consider a Maxwellian plasma of temperature $kT_e = 5.0$ keV. (a) Find expressions for the Lorentz factor γ and speed parameter β for an electron that has kinetic energy $E_k = 3kT_e/2$ in terms of the ratio $r \equiv kT_e/mc^2$; use $r \ll 1$ to simplify. Evaluate for our 5-keV plasma; $mc^2 = 511$ keV. (b) Find an expression for the fractional frequency shift, $\Delta\nu/\nu$, of a CMB photon in a head-on collision with an electron of speed parameter β, as a function of β, and give its value for our electron. Note that the derivation in the text is not applicable because the electron is only mildly relativistic. Refer to Fig. 2, use (13), (14), and (15), and justify using $2h\nu' \ll mc^2$. (c) Repeat for an overtaking collision. (d) Find an expression for the net or average fractional frequency change $(\Delta\nu/\nu)_{av}$ of these two cases in terms of β and in terms of r. Give the numerical value for our case. Compare with (25). [Ans. ~0.15; ~0.4; ~ −0.3; $6r$]

Problem 9.52. The shift, $I' - I = \Delta I$, of any photon spectrum at a given frequency due to single scattering from a Maxwell–Boltzmann distribution of electrons is given by the formula (Sunyaev and Zeldovich, *Comments Ap. Sp. Sci.* **4**, 173 (1972))

$$I'(\nu) - I(\nu) = y\nu\left\{\frac{d}{d\nu}\left[\nu^4\frac{d}{d\nu}\left(\nu^{-3}I(\nu)\right)\right]\right\}; \quad y \equiv \frac{kT_e}{mc^2}\tau,$$

which is valid for $y \ll 1$ and $kT_e \gg h\nu$. (a) Evaluate the shift for the blackbody spectrum $I(\nu)$ (6.6) with a fixed radiation temperature T_r. Suggestion: define $a \equiv h/kT_r$, take derivatives, and then give your result in terms of $x \equiv a\nu = h\nu/kT_r$, y, and $I(\nu)$. (b) Find the limiting expressions for the Rayleigh–Jeans region ($x \ll 1$) and Wien ($x \gg 1$) zones. Compare the former with (34). (c) Create a log-log plot of $I(\nu)$ and $I'(\nu)$ with arbitrary scales, similar to that of Fig. 7c, for $y = 0.15$. Your result will differ from Fig. 7c. What is the likely source of this discrepancy? (d) Calculate the difference spectrum $I' - I$ for $y = 0.0005$ and $T = 2.73$ K, and plot $\Delta I = I' - I$ versus $\log \nu$. Label the axes of the plot quantitatively in SI units.

[Ans. $I' - I = yI \dfrac{x\,e^x}{e^x - 1} \left(x \dfrac{e^x + 1}{e^x - 1} - 4 \right)$; $-2yI$, $yIx(x - 4)$; $-$; $-$]

Problem 9.53. An idealized cluster of galaxies contains a plasma of protons and electrons that is uniform in density and temperature throughout its volume, which is perfectly spherical of unknown radius R. X-ray astronomers find the electron (equivalent) temperature to be $kT_e = 5.1$ keV from the shape of the spectrum, which indicates thermal bremsstrahlung emission. They also find the specific intensity at the center of the cluster on the flat part of the spectrum ($h\nu_x \ll kT_e$) to be $I_x = 1.5 \times 10^{-27}$ W m^{-2} Hz^{-1} sr^{-1} and the angular radius of the cluster to be $\theta = 1.5'$. Radio astronomers measure the maximum decrease in CMB temperature (toward the cluster center) to be $\Delta T_r/T_r = -3 \times 10^{-5}$. (a) Derive expressions and find numerical values for the electron density n_e (m^{-3}) and radius R (m) of the cluster in terms of the measured or known quantities σ_T, I_x, T_e, $f \equiv -\Delta T_r/T_r$, $\varepsilon \equiv kT_e/mc^2$, and the constant $A = C_1/4\pi$; see (37) and (38). Neglect effects of special relativity due to the recession velocity of the cluster. Express the numerical value of R in meters and MLY. Assume the Gaunt factor is unity. (b) Use the measured angular radius of the cluster to find the distance r from the earth to the cluster in LY. Adopt flat geometry. (c) The fractional wavelength change of spectral lines due to the recessional velocity of the cluster is $z \equiv \Delta\lambda/\lambda = 0.20$. What is the value of the Hubble constant $H = v/r$ in terms of z and r and numerically in units of (km s^{-1} MLY^{-1})? Hint: relate z and v. [Ans. $\sim 10^3$ m^{-3}, ~ 1 MLY; ~ 3 GLY; ~ 20 km s^{-1} MLY^{-1}]

10

Hydrogen spin-flip radiation

What we learn in this chapter

The gaseous matter between the stars, the **interstellar medium** (ISM), is a **complex mix** of gases (atomic and molecular), dust, radiation, magnetic fields, and cosmic rays. The dominant component is **hydrogen**. In its atomic form in the ISM, it emits radiation at the radio frequency of **1420 MHz**. The transition is between the two **hyperfine states** of the ground state. Their energy difference, and hence the frequency of the transition, arises from the **interaction of the magnetic moments** of the proton and electron constituting the atom.

The **orbital motions of clouds** of gas about the center of the Galaxy may be tracked through studies of the 1420-MHz radiation. Multiple clouds at a given galactic longitude are distinguished by their different **Doppler shifts**. Application of the **tangent-point method** yields a **rotation curve** (orbital speed versus galactocentric radius) for the Galaxy. The curve is quite flat, indicating large amounts of unseen mass (**dark matter**). This is typical of other galaxies also. A flat rotation curve is expected for a spherical (or spheroidal) mass distribution with density that decreases as r^{-2}. The **differential motion** of matter in the solar region of the Galaxy may be described with the **Oort constants**, which represent the **shear** and **vorticity** of the galactic material.

In the presence of an external magnetic field, the upper state of the hyperfine transition of hydrogen is split into three levels, which is known as the **Zeeman effect**. This permits radio astronomers to detect **magnetic fields in interstellar clouds**. Radiation from a background continuum source is absorbed by atoms in the cloud at two of the closely spaced Zeeman frequencies, one for left **circularly polarized radiation** and the other for right circular polarization. Detection of the line splitting is possible through **subtraction of the line profiles** obtained at the two polarizations. **Magnetic fields of 1–30 nT** are found for clouds of the ISM.

10.1 Introduction

The sun is one of more than 10^{11} stars in the (Milky Way) Galaxy, herein called simply the Galaxy with capital G to distinguish it from external galaxies (lower case). The stars in a galaxy occupy only a minuscule portion of its volume. The nearest star is 4.3 LY distant from the sun. Consider that the sun, of size $R_\odot = 7 \times 10^8$ m, occupies a cube of space of 4.3 LY

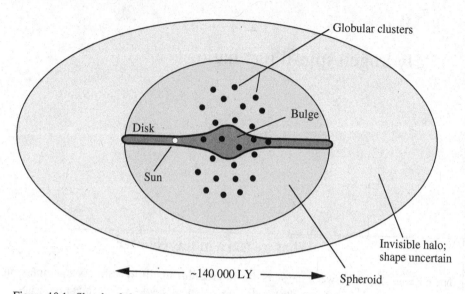

Figure 10.1. Sketch of the Galaxy showing the disk, the spheroid, globular clusters, and the invisible halo containing the dark gravitational matter.

$(4.1 \times 10^{16}$ m) on a side. The sun thus occupies only $\sim 10^{-23}$ the available volume, and this is typical of most stars in the Galaxy. The empty space between stars is occupied by tenuous gas, dust grains, energetic cosmic-ray particles, and radiation of all kinds, including starlight and microwave background radiation. This is known as the *interstellar medium* (ISM).

The initial objective of this chapter is to present some of the physics underlying the 1420-MHz radiation emitted by the most common element of the ISM, neutral hydrogen. We then describe how radio astronomers use this radiation to map the location and motions of hydrogen clouds throughout the Galaxy as well as in other galaxies. Finally, we consider how observations of the 1420-MHz line in absorption can reveal magnetic fields in interstellar clouds through the Zeeman effect. We begin with brief descriptions of the Galaxy and the ISM.

10.2   The Galaxy

Stellar content

A schematic of the Galaxy is shown in Fig. 1. In Fig. 2, we show photographs of two spiral galaxies of similar structure, M81 and M101. A typical *spiral galaxy* consists of a disk with spiral arms where star formation is still taking place and a large spheroidal region populated by older low-mass stars and gas. (A spheroid is formed by the rotation of an ellipse about one of its axes of symmetry). Some characteristics of the Galaxy are summarized in Table 1.

Figure 10.2. Two spiral galaxies, each subtending $\sim 0.5°$ with north up and east to the left. (a) Spiral galaxy M81 (NGC 3031), which is ~ 5 MLY distant from the sun and $\sim 35\,000$ LY in diameter. Our own Galaxy might appear like this to an astronomer in M81. (b) Spiral galaxy M101 (NGC 5457), a "face-on" spiral at a distance of ~ 17.6 MLY and $\sim 150\,000$ LY in diameter. It played a role in the 1920 Shapley–Curtis debate on the distances of spiral nebulae. [(a, b) Canada-France-Hawaii Telescope/ J.-C. Cuillandre/Coelum.]

(a)

(b)

Table 10.1. *Characteristics of the (Milky Way) Galaxy[a]*

Characteristic	Value
Visible radius	~80 000 LY
Center to sun distance	~25 000 LY
Scale height at solar radius[b]	400−4000 LY
Disk mass-to-light (V) ratio[c]	~5 $M_\odot/L_\odot$
Disk luminosity (V band)	$1.2 \times 10^{10} L_\odot$
Disk mass (dynamical)[d]	$6 \times 10^{10} M_\odot$
Luminosity (V band)	$1.4 \times 10^{10} L_\odot$
Mass out to radius 25 000 LY[e]	~$9.5 \times 10^{10} M_\odot$
Mass out to 114 000 LY[e]	~$4 \times 10^{11} M_\odot$
Mass out to 800 000 LY[e]	~$1.3 \times 10^{12} M_\odot$
Solar (LSR) rotational velocity about Galaxy center[f]	~220 km/s

[a] V. Trimble in *Allen's Astrophysical Quantities*, 4th Ed., ed. A. N. Cox, AIP Press 1999, p. 569; J. Binney and S. Tremaine, *Galactic Dynamics*, Princeton University Press, 1987, p. 17; F. Eisenhauer *et al.*, ApJ **597**, L121 (2003).
[b] Scale height depends on component. The neutral hydrogen scale height is ~400 LY. Cool gas and young stars are more concentrated in the plane than are older stars and hot gases.
[c] From measurements in the local solar region of stellar altitudes above and below the galactic plane and also the stellar vertical speeds and luminosities. See discussion in Binney and Tremaine, p. 201.
[d] Based on the previous two entries and thus includes luminous and nonluminous mass. See Binney and Tremaine, p. 90.
[e] From rotation data, and thus includes dark matter.
[f] Local standard of rest (LSR) is barycenter of stars in solar neighborhood.

The disk contains *population I* stars that include recently formed massive stars. The spheroid contains *population II* stars of greater ages from an earlier epoch of star formation. These stars are of relatively low mass because those that were more massive exhausted their nuclear fuel rapidly and have long since evolved to less visible white dwarfs, neutron stars, or black holes.

Small clusters containing 10^5 or 10^6 population II stars surround many galaxies. They are known as *globular clusters*. About 200 of them orbit the center of the Galaxy (Fig. 1).

The orbital motions of the stars and clouds about the galactic center can be tracked by means of optical and radio spectral line studies. Astronomers find that these objects are bound to the Galaxy by forces far exceeding the gravity that would be associated with the luminous (i.e., visible) stars and matter. This is true even if one includes the gravity due to the nonluminous remnants of stars (white dwarfs, neutron stars, and black holes) that we expect to be present. One must therefore postulate a huge invisible halo of *dark matter*, the nature of which is not known.

The ratio of disk mass to disk luminosity M/L_V is ~5 in solar units. (The mass-to-luminosity ratio for the sun in these units is unity.) This indicates that, on average, the matter in the disk is less luminous per kilogram than the sun. Nonluminous matter such as gas, stellar remnants (e.g., white dwarfs), and dark matter serve to raise the ratio. Also, the stellar matter is made up, on average, of stars somewhat less massive than the sun

(Prob. 21). These are significantly less luminous per unit mass; recall the scaling $L \propto M^3$ (Table 4.3).

At large radial distances, the mass-to-light ratios inferred from stellar motions (see Section 4) become very large (e.g., at 800 000 LY, it is $\sim$100). This is indicative of large amounts of dark matter.

Interstellar medium (ISM)

The *interstellar medium* (ISM) as described in this section inhabits the space between the stars of the (Milky Way) Galaxy. This is typical of other *spiral galaxies* as well. In contrast, *elliptical galaxies* do not have spiral arms and contain little gas and dust. Neutral hydrogen, the focus of this chapter, is the dominant component of the ISM.

Gases

Measurements of the gas content in the solar vicinity indicate that the mass density of gas is 2.8×10^{-21} kg/m^3. A cube 4.3 LY on a side would contain 2×10^{29} kg at this density, which is 10% of the solar mass. Thus, one might infer that roughly 10% of the mass in the Galaxy consists of gas. A more careful assessment confirms this result.

Although the interstellar gases are quite tenuous, the huge volumes lead to large masses of gas. If 10% of the entire galactic disk consisted of gas, the mass in gas would be $\sim$6 $\times$ 10^9 $M_\odot$. The gas composition in the disk is presumably approximated by the solar system abundances, which by mass are 71% hydrogen, $\sim$27% helium, and 2% heavier elements; see AM, Chapter 10. The percentages by *number* are quite different because of the differing masses of the several elements (e.g., 9% of the atoms are helium).

The gas is found in atomic and molecular forms, and it can be neutral or ionized; it has a complex mix of densities, temperatures, and ionization states. It appears as luminous, colorful H II (ionized hydrogen) regions and as supernova remnants. It is gradually, and at times drastically, modified by the evolution of stars. The heavy elements of the ISM are continually being enriched by the ejection of matter from stars by means of stellar winds and supernova explosions.

Neutral hydrogen

Most of the gaseous hydrogen in the ISM ($\sim$97%) is neutral, whereas a small percentage ($\sim$3%) is ionized. A substantial proportion (roughly half) is in the form of molecular hydrogen, H_2, mostly in dense molecular clouds.

The number density of atoms varies widely from place to place in the Galaxy, but at the radius of the sun, it is estimated to be

$$n_{gas} \approx 1 \times 10^6 \text{ atoms m}^{-3}. \tag{10.1}$$

The number density of neutral atomic hydrogen (H I) in diffuse regions can range up to 10^9 atoms/m^3 in so-called H I clouds, and to $>10^{13}$ atoms/m^3 in the cores of the densest molecular clouds. The kinetic temperature of the neutral hydrogen in the H I clouds is $T \approx 80$ K, but at some locations there is a "warm neutral" component of $T \approx 6000$ K.

Neutral hydrogen in the ISM makes itself evident to optical, ultraviolet, and radio astronomers through its absorption and emission spectra, which reveal the hydrogen's densities and temperatures. For example, the spectra of bright, ultraviolet-emitting stars will show an absorption line at 10.2 eV, the *Lyman α transition* (see Fig. 3 below), given the presence of neutral hydrogen between the star and the observer.

These ultraviolet observations must be made from space observatories because of atmospheric absorption. They probe only selected regions close to the sun because the background stars must be sufficiently close and luminous to provide adequate flux for the absorption measurements. The number of such stars is limited.

A more powerful probe of the cold neutral hydrogen in the ISM can be undertaken through the detection of radio waves emitted by hydrogen clouds at 1420 MHz (Section 3). This radiation is due to a spin-flip transition in the ground state of atomic hydrogen. Such radiation allows one to trace the locations and motions of H I clouds in the Galaxy.

Ionized hydrogen

The ionized hydrogen (H II) component is a tracer of high-energy processes because it takes 13.6 eV to ionize a hydrogen atom. Cosmic-ray protons, background ultraviolet (UV) light from hot stars, shock waves from supernovae, and intense UV emission from newly formed stars all serve to ionize hydrogen. The proportion of ionized atoms in a given region can vary substantially from <0.1% in cold dense regions to ~100% in extended, hot, low-density regions.

Ionized hydrogen is dramatically evident in H II nebulae, which contain young luminous stars. Such stars emit strongly in the ultraviolet. The gas clouds typically associated with regions of newly formed stars are illuminated by this radiation. The radiation ionizes the atoms, which, in turn, radiate during recombination.

Four components of the gaseous ISM

The several components of the diffuse ISM are usually denoted in terms of their temperature and ionization state. They are designated as "cold," "warm," or "hot" (C, W, or H) as determined by the temperature and as neutral or ionized (N or I) as dictated by the degree of ionization. The state one finds will vary with location. In one scenario, the warm neutral medium fills much of the space whereas the hot ionized medium is in bubbles created by supernova ejecta.

The ISM is usually described through a limited set of such states (e.g., the states or components given in Table 2) rather than as a smooth continuum of conditions. This is useful but is undoubtedly an oversimplification because of the variability of conditions from place to place in the Galaxy. The first three components listed in Table 2 are detected at radio wavelengths either in absorption, emission, or both. The WIM is detected as diffuse Hα (optical Balmer spectral line) emission and through the dispersion of radio waves, and the HIM is detected through thermal bremsstrahlung x-ray emission (Chapter 5) and UV absorption line studies.

If different regions are to coexist in the Galaxy as more or less static structures, the pressures in adjacent regions should be approximately equal. The thermal pressure P is due to the kinetic energies of the particles, namely, that given by the ideal gas law (3.38),

$$P = nkT, \hspace{4cm} \text{(N/m}^2\text{)} \hspace{1cm} (10.2)$$

Table 10.2. *Components of the diffuse ISM*[a,c]

	Component (state)	Temperature (K)	Average H density $n_H(m^{-3})$[b]	Fractional ionization n_e/n_H	Comment
CNM	Cold neutral medium	~80 K	~4×10^7	~10^{-3}	H I clouds of size ~3–30 LY and 1–1000 $M_\odot$
WNM	Warm neutral medium	~6000 K	~4×10^5	~0.1	Intercloud medium
WIM	Warm ionized medium	~8000 K	~3×10^5	~0.7	~30% of ISM mass
HIM	Hot ionized medium	~10^6 K	~2×10^3	~1.0	Sun resides in local hot bubble of HIM

[a] The values in this table must be treated with caution; in some cases they are not well known and in others the range of measured values is considerable. See the recent review by D. P. Cox, *ARAA* **43**, 337 (2005) for a more contemporary assessment.

[b] Incorporates all hydrogen atoms, whether neutral of ionized.

[c] References: L. Cowie and A. Songaila, *ARAA* **24**, 499 (1986); S. Kulkarni and C. Heiles in *Galactic and Extragalactic Radio Astronomy* 2nd Ed., eds. G. Verschuur and K. Kellermann, Springer-Verlag, 1988; C. McKee and J. Ostriker, *ApJ* **218**, 159 (1977).

where $n = \rho/m_{av}$ is the particle number density, k is the Boltzmann constant, T is the temperature, ρ is the mass density and m_{av} is the average particle mass. The pressure of cosmic rays, magnetic fields, and turbulent motion must be taken into account in proper models. The ratio of the thermal pressure to the Boltzmann constant in the galactic plane is generally taken to be about

$$\frac{P}{k} = nT \approx 4 \times 10^9 \text{ m}^{-3} \text{ K}. \tag{10.3}$$

The ratio P/k is quoted here in units that directly reflect the density and temperature. The pressure P/k varies somewhat with location and reaches below 10^9 m^{-3} K and above 10^{10} m^{-3} K in different portions of the Galaxy (Prob. 22).

In ionized regions, both electrons and protons must be included in the particle density n in (2). The values of n_H in Table 2 represent the number density of the hydrogen nuclei (protons) whether ionized or not. In a highly ionized region of mostly hydrogen, $n_e \approx n_H$ and $P/k = 2n_H T$. Finally, we caution that this quasi-static approach is highly simplistic. The ISM is an active turbulent environment owing to stellar winds, star formation, and supernovae.

Molecules

Molecules in the ISM are of major interest to us because a large part of the material of our habitat, not to mention our own bodies, consists of molecules. It turns out that the chemistry for the formation of large quantities of molecules takes place in the interstellar medium. Fully about half the mass of the Galaxy's interstellar medium is in the form of molecules.

A wide variety of molecular types, consisting of up to at least thirteen atoms, have been detected by radio astronomers since the discovery of hydroxyl (OH) in 1963. They include molecular hydrogen (H_2), carbon monoxide (CO), water (H_2O), ammonia (NH_3), and ethyl alcohol (CH_3CH_2OH). The latter consists of nine atoms and is found in alcoholic beverages. These molecules are located in interstellar clouds – both diffuse and dense – shocked regions, and circumstellar envelopes.

Table 10.3. *Energy densities in the ISM*[a]

Component	(J/m^3)	(MeV/m^3)
Radiation from stars	0.7×10^{-13}	0.4
Kinetic gas motion	0.5×10^{-13}	0.3
Cosmic microwave background	0.4×10^{-13}	0.3
Cosmic rays	1.6×10^{-13}	1.0
Magnetic fields	0.4×10^{-13}	0.2

[a] References: Allen, *Astrophysical Quantities*, 3rd ed., Athlone, 1976, p. 269; D. C. Backer in *Galactic and Extragalactic Radio Astronomy*, 2nd ed., eds. G. Verschuur and K. Kellermann, Springer-Verlag, 1988, p. 498. Also see J. S. Mathis in *AQ*, 4th Ed., ed. A. N. Cox (1999) for a brief overview of the ISM.

Most of these molecular species are found in *molecular clouds* that have sizes from 1 LY up to ~500 LY and number densities (including atoms) from 10^7 to 10^{13} m^{-3} (Prob. 23). Molecules are found at the sites of star formation and are often associated with cool but luminous M-giant stars with intense stellar winds. *Giant molecular clouds* have masses ranging up to $10^5 M_\odot$ and are the formation sites of massive stars. Within the clouds, molecules are protected from stellar ultraviolet radiation, which would otherwise dissociate them.

Dust, radiation, cosmic rays, and magnetic fields

Interstellar dust grains constitute $\gtrsim 1\%$ of the interstellar medium's mass. The dust is very effective at absorbing optical radiation, and so most of the Galaxy is not visible to optical astronomers (AM, Chapter 10). The dust is not uniformly distributed in the Galaxy but rather is lumped into interstellar clouds with the interstellar gases, which can be the locations of star formation. In fact, the dust helps bring about the collapse of clouds because energy must be released during a collapse. This will heat local dust, which can radiate through the emission of infrared (IR) radiation. The IR can escape the region because opacities for absorption are much less for infrared than for optical radiation.

Other components of the ISM include cosmic rays (energetic charged particles), magnetic fields, and radiation in several forms: x rays, starlight, and the cool ($T = 3$ K) cosmic microwave background (CMB) radiation that is the remnant of the hot early universe. These components have the approximate energy densities (J/m^3) given in Table 3. With the exception of the CMB, these values vary considerably from place to place in the Galaxy. Each of these components is of the order ~1 MeV/m^3.

10.3   Hyperfine transition at 1420 MHz

The most powerful probe of cold neutral hydrogen in the Galaxy, and also in other galaxies, is the 1420-MHz radiation emitted by hydrogen atoms. The transition arises from the *hyperfine splitting* of the ground state of hydrogen (Fig. 3).

The proton and the electron both exhibit angular momentum (*spin*) and a related *magnetic moment*. In the ground ($n = 1$) state of hydrogen, the electron spin can be either parallel or antiparallel to the spin of the proton. The magnetic moments interact to change the energies of the two states slightly (Fig. 3, lower right). The small energy difference between the

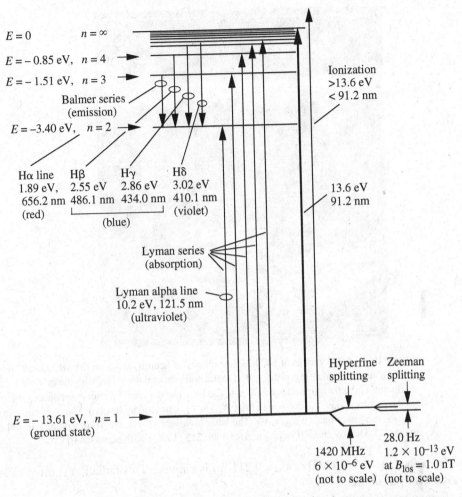

Figure 10.3. Energy levels of the hydrogen atom. Downward Balmer and upward Lyman transitions are shown. The minuscule hyperfine splitting of the ground state is shown with greatly exaggerated spacing together with the even finer Zeeman splitting of the upper hyperfine level. The energy difference between the upper and lower Zeeman states is given for a 1-nT line-of-sight component of magnetic field in a hydrogen cloud. [Adapted from H. Bradt, *Astronomy Methods*, Cambridge University Press, 2004, Fig. 10.1, with permission]

two states, $\Delta E = 5.874 \times 10^{-6}$ eV, leads to the $\nu = 1420$-MHz transition according to $\nu = \Delta E / h$ where h is the Planck constant.

The more stable of the two states is that in which the magnetic moments of the proton and electron are coaligned (parallel). The high-energy state occurs when they are antiparallel. Because the magnetic moment of an electron is opposed to its spin, the more stable state has opposed spins.

Collisions of atoms in the gas excite them to the higher state. The kinetic energies of the atoms are easily sufficient because, even in the coldest regions at temperatures of ~ 100 K, we have $E_k \approx kT \approx 10^{-2}$ eV, or about 1000 times the required excitation energy of 6×10^{-6} eV. A given atom in isolation would decay back to the lower state in ~ 11 million yr, but collisional deexcitation in the ISM substantially reduces this time. Upon deexcitation, a radio photon

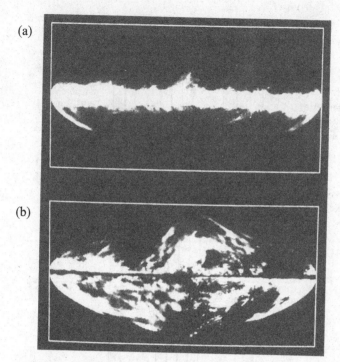

Figure 10.4. All-sky maps of 1420-MHz emission of neutral hydrogen from the Galaxy in galactic coordinates. The galactic equator is horizontal and centered vertically with the galactic center at its midpoint. Galactic longitude increases to the left. The degree of whiteness indicates the intensity of the radiation (i.e., the column density N_H (m^{-2}) of neutral hydrogen). The upper map (a) shows N_H and the lower map (b) $N_H \sin b$. The latter highlights deviations from a plane-parallel galactic model. [J. Dickey and F. Lockman, *ARAA* **28**, 215 (1990), with permission]

of frequency $\nu_0 = 1420$ MHz ($\lambda_0 = 0.211$ m) is emitted. This radiation is often called *21-centimeter radiation.*

Fortunately, both the Galaxy and the earth's atmosphere are quite transparent to radiation of this frequency. Because large numbers of atoms are decaying in the Galaxy, a substantial signal can be, and is, detected by radio astronomers. The 1945 prediction and 1951 discovery of this radiation are milestones in the history of radio astronomy.

The primary goal of this section will be the derivation of the energy difference of the hyperfine splitting, which in turn yields the famous 1420-MHz frequency.

Sky at 1420 MHz

Figure 4 shows two all-sky views of the 1420-MHz radiation in galactic coordinates. The upper map shows detected intensity (degree of whiteness) of the radiation as a function of galactic coordinates. To a large extent, the Galaxy is optically thin at 1420 MHz, and so the intensity in many directions is, to a first approximation, a measure of the column densities of neutral hydrogen, N_H (H I atoms per square meter), in those directions. The map shows a strong concentration of hydrogen along the galactic plane with some fingers of emission extending above and below it.

The lower map of Fig. 4 is constructed from the same data, but the intensity is multiplied by $\sin b$, where b is the galactic latitude; in other words this map is one of $N_H \sin b$. It

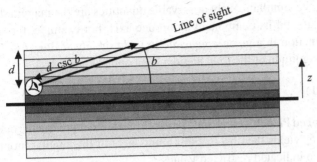

Figure 10.5. Parallel-plane model of the Galaxy with gas density varying as a function of the vertical (z) direction only. It is shown here as discrete layers, but it is more realistically a continuum of density decreasing with $|z|$. The column density along the line of sight from an observer, not necessarily on the galactic plane, varies as $N_H \propto \csc b = 1/\sin b$.

exaggerates the weaker intensities at the higher galactic latitudes relative to the lower latitudes and thus reveals what appear to be clouds, filaments, and sheets of hydrogen.

A simplified view of the three-dimensional structure of the Galaxy is to consider it as a parallel-plane system of layers of various H I densities, as shown in Fig. 5. The H I density varies only in the z direction, and it is uniform in the x, y directions. In the limit of very thin layers, the density distribution would decrease continuously with z.

Consider the observer (the "eye" in Fig. 5) to be at some height from the galactic plane and viewing in the direction of galactic latitude b. Let the line of sight extend completely through the layers of the Galaxy to "infinity." The amount of material (column density, atoms m^{-2}) along the line of sight in any one incremental layer of constant density is proportional to $1/\sin b$. Hence, the amount summed over all layers is also proportional to $1/\sin b$.

The measured intensities of 1420-MHz radiation (i.e., the column density of H I) would thus vary as $N_H \propto 1/\sin b$ if the effects of self-absorption, differing excitation rates, and so on were ignored. If we then plot the product $N_H \sin b$, the result would be independent of galactic latitude, and the lower map of Fig. 4 would show a completely uniform distribution.

The real map of $N_H \sin b$ (Fig. 4b) thus shows the deviations of hydrogen in our Galaxy from the plane-parallel model of Fig. 5. There is much structural information in this figure. The filaments extending from the plane were probably created by supernova outbursts in the Galaxy.

Quantization fundamentals

The derivation of the hyperfine energy difference springs from the quantization of angular momentum and its relation to the magnetic moment.

Angular momenta

The hydrogen atom in its ground state consists of an electron and proton each of spin 1/2. The ground state has zero *orbital* angular momentum. The *total* angular momentum F thus consists solely of the *spin* angular momenta of the electron S and the proton I,

$$F = S + I. \qquad \text{(Total angular momentum)} \qquad (10.4)$$

In quantum mechanics, the simultaneously observable quantities are the magnitude of the angular momentum's square and its component along one axis. For example, the operator $F \cdot F$ acts on the wave function in a given *eigenstate* to yield the *eigenvalue* $F(F + 1)\hbar^2$, which is the observable magnitude of the total angular momentum squared:

$$F \cdot F \Rightarrow F(F + 1)\hbar^2, \qquad \text{(Eigenstate and eigenvalue)} \qquad (10.5)$$

where $\hbar = h/2\pi$ is the reduced Planck's constant. We use the arrow to represent the operation.

Similarly, the operator F_z yields the observable z component of the angular momentum, $m_F \hbar$, where m_F takes on the indicated restricted values

$$F_z \Rightarrow m_F \hbar; \quad m_F = F, F - 1, F - 2, \ldots, -F. \qquad \text{(Angular momentum component)} \qquad (10.6)$$

For $F = 1$, the possible observable values of F_z are threefold, $F_z \Rightarrow -1\hbar, 0, +1\hbar$. For $F = 0$, there is a single value, $F_z \Rightarrow 0\hbar$. The $F = 1$ states are coincident in energy if there is no external magnetic field to separate them; they are *degenerate*. It is the energy difference between the $F = 0$ and $F = 1$ states that gives rise to the 1420-MHz radiation.

The spin angular momenta of the proton and electron are treated similarly. Thus, for the electron with "spin 1/2" ($S = 1/2$),

$$S \cdot S \Rightarrow S(S + 1)\hbar^2$$
$$S_z \Rightarrow m_S \hbar; \quad m_S = +1/2, -1/2. \qquad \text{(Electron eigenvalues)} \qquad (10.7)$$

The spin therefore has two states or alignments, spin "up" ($m_s = +1/2$) and spin "down" ($m_s = -1/2$). The proton is also "spin 1/2" ($I = 1/2$):

$$I \cdot I \Rightarrow I(I + 1)\hbar^2$$
$$I_z \Rightarrow m_I \hbar; \quad m_I = +1/2, -1/2. \qquad \text{(Proton eigenvalues)} \qquad (10.8)$$

One can visualize the angular momentum as a precessing vector, as shown in Fig. 6a,b for S and F. The allowed states are limited to those shown. The magnitudes and z projections are measurable, but the azimuthal angle is completely undetermined. The vector summation of $S + I$ for the $F = 0$ and $F = 1$ states could appear as in Fig. 6c,d. Note that S and I are more tightly coupled in the $F = 0$ state than in the $F = 1$ state. This leads to the asymmetry in the energies of the two states, (31) and (32), calculated below.

The electron and the proton each can have up (U) or down (D) spin. Thus, they can be combined in four ways (UU, UD, DD, DU). The combinations that are eigenstates of $F \cdot F$ and F_z are

$$
\begin{array}{cccc}
\text{UU} & (\text{UD} + \text{DU})/\sqrt{2} & \text{DD} & (\text{UD} - \text{DU})/\sqrt{2} \\
F = 1 & F = 1 & F = 1 & F = 0 \\
m_F = +1 & m_F = 0 & m_F = -1 & m_F = 0.
\end{array}
\qquad (10.9)
$$

It is not intuitively apparent from classical arguments why the $(\text{UD} + \text{DU})/\sqrt{2}$ combination would represent a total angular momentum $F = 1$, but it does.

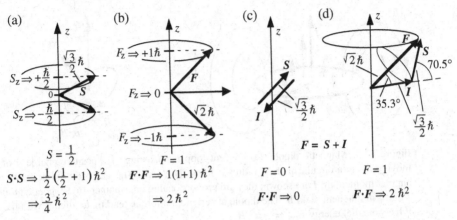

Figure 10.6. Quantum states of angular momentum. (a) The two states of spin-angular momentum S for spin 1/2 particle. The vector precesses around the z-axis with indeterminate azimuth (in a semiclassical picture). Its magnitude squared is $S(S+1)\hbar^2$, where $S = 1/2$ and $\hbar \equiv h/2\pi$. The observable z component of angular momentum S_z can take on the value $+\hbar/2$ or $-\hbar/2$. (b) The three states of angular momentum vector F for $F = 1$ ground-state hydrogen. (c,d) Coupling of the proton and electron spins I and S, respectively, for the $F = 0$ and $F = 1$ states of neutral hydrogen. The degree of coupling is controlled by the requirement that $F = S + I$. The coupling is stronger for the $F = 0$ state. (This classical visualization is less than perfect; see text.)

These combinations comprise four hyperfine states of the ground state of the hydrogen atom, one with $F = 0$ and three with $F = 1$. As noted, after (6), the three $F = 1$ levels are degenerate in the absence of an external magnetic field, and our interest lies in the difference in the potential energies of the $F = 0$ and $F = 1$ states (Figs. 6c,d).

To carry along a classical picture during our derivation, we will visualize the $F = 1$ state as UU or DD (parallel spins) and the $F = 0$ state as UD or DU (antiparallel spins) even though this is not strictly correct according to (9).

Magnetic moments

How is the magnetic moment related to the angular momentum? Consider first the classical case. Let a charge q with mass m (Fig.7a) orbit in a circle at radius r about the z-axis (unit vector $\hat{z}$) with angular velocity ω. The angular momentum is $J = mr^2\omega\hat{z}$. The magnetic moment μ is defined as the current in a loop times the area of the loop, $\mu = iA\hat{z}$ (C s^{-1} m^2). For one orbital revolution of period P, the current is $i \equiv \Delta q/\Delta t = q/P = q\omega/(2\pi)$ and $A = \pi r^2$; thus, $\mu = (q\omega r^2/2)\,\hat{z}$.

The quantities μ and J are both proportional to $r^2\omega$ and hence to each other, and they are directed parallel to each other for positive charge q (Fig. 7a). Consequently, the classical relation between magnetic moment and angular momentum for an orbiting charge is

$$\mu = \frac{q}{2m}J.$$

(Magnetic moment for orbital angular momentum; classical; J/T) (10.10)

Now consider a spinning sphere with uniform mass and charge distributions. It actually consists of many orbiting charge and mass elements, each obeying (10). Thus (10) is also valid for the entire sphere.

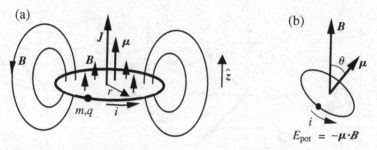

Figure 10.7. Magnetic dipole. (a) Current loop i consisting of a positive charge q of mass m moving along a circular path of radius r about the z-axis. The magnetic dipole moment $\boldsymbol{\mu}$ and angular momentum $\boldsymbol{J}$ are shown; they are proportional to one another. (b) Magnetic moment $\boldsymbol{\mu}$ in a uniform magnetic field $\boldsymbol{B}$. The dipole experiences a torque tending to align it with the $\boldsymbol{B}$ field. It has potential energy $E_{\text{pot}} = -\boldsymbol{\mu} \cdot \boldsymbol{B}$.

One might expect that the electron – a mathematical point charge with zero radius – would have zero magnetic moment and zero angular momentum. Nevertheless, it has finite values of both, but their ratio differs from that of (10). A multiplicative dimensionless correction (the *g factor*) is introduced to account for the difference. By convention, the g factor carries sign, and the symbol e is the *magnitude* of the electron charge. Thus, for the electron spin angular momentum $\boldsymbol{S}$, one can write, from (10),

$$\boldsymbol{\mu}_e = g_e \frac{e}{2\,m_e} \boldsymbol{S}, \qquad\qquad (g_e = -2.002\,32\ldots) \qquad (10.11)$$

where, from experiment, $g_e = -2.002\,32$. The magnetic moment of the electron is opposed to the direction of the spin $\boldsymbol{S}$ (as expected for an orbiting negative charge) and is about twice the magnitude expected for a uniformly charged sphere with the same angular momentum, (Prob. 32).

It is convenient to define the *Bohr magneton* as

$$\mu_B \equiv \frac{e\,\hbar}{2\,m_e} = 9.274\,01 \times 10^{-24}\ \text{J/T}, \qquad \text{(Bohr magneton)} \qquad (10.12)$$

and so we can write

➡ $$\boldsymbol{\mu}_e = g_e \frac{\mu_B}{\hbar} \boldsymbol{S}, \qquad \text{(Electron magnetic moment)} \qquad (10.13)$$

where $\boldsymbol{S}$ is the electron's spin angular momentum. The quantities $\boldsymbol{S} \cdot \boldsymbol{S}$ and S_z are angular momentum operators, and hence $\boldsymbol{\mu}_e \cdot \boldsymbol{\mu}_e$ and μ_z are also operators with eigenvalues that are quantized similarly to those of $\boldsymbol{S}$, as given in (7).

The proton consists of three quarks with fractional charges, and so its magnetic moment will also deviate from (10) – in this case by the factor g_p,

➡ $$\boldsymbol{\mu}_p = g_p \frac{\mu_N}{\hbar} \boldsymbol{I}, \qquad \text{(Proton magnetic moment; } g_p = +5.585\,69.) \qquad (10.14)$$

where $g_p = +5.586\ldots$, and μ_N is the *nuclear magneton*,

$$\mu_N \equiv \frac{e\,\hbar}{2\,m_p} = 5.050\,783 \times 10^{-27}\ \text{J/T}. \qquad \text{(Nuclear magneton)} \qquad (10.15)$$

The proton and electron have the same angular momentum ("spin 1/2"), but (12) through (15) tell us that the magnetic moment of the proton is a factor of $\sim$650 less in magnitude than that of the electron. From (10), we note that, classically, two uniformly charged spheres with

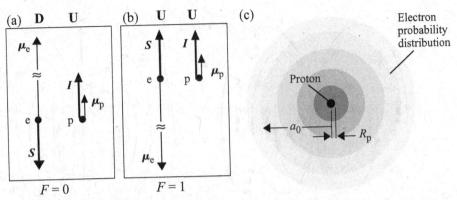

Figure 10.8. (a,b) Spin angular momenta of the electron S and proton I in the down-up and up-up configurations, which correspond (imperfectly) to the $F = 0$ and $F = 1$ states, respectively, together with the magnetic moments. The proton and electron angular momenta are equal in magnitude, but the electron magnetic moment is much larger than the proton moment. (c) Schematic representation of the exponential probability distribution of the electron cloud (wave function) of characteristic radius a_0. The electron has a very low probability, $\sim 10^{-15}$, of being within the tiny proton.

equal total charges and angular momenta but different masses would have different magnetic moments, with the more massive having the lesser. (A large mass at fixed $J = mr^2\omega$ implies a small ω or r – or both – and hence a small μ.)

The relative orientations and magnitudes of the magnetic moments and angular momenta for the $F = 0$ and $F = 1$ states are shown in Fig. 8a,b. This view is again flawed in that we take DU to be the $F = 0$ state and the UU state to be perfectly aligned; see (9) and Fig. 6d. Note the different magnitudes of the magnetic moments and also that parallel angular momenta imply antiparallel magnetic moments and vice versa.

Line splitting

The energies of the $F = 0$ and the $F = 1$ states differ owing to the interaction of the magnetic moments of the electron and proton. It is as if two little magnets were brought together. They have different potential energies if they are oriented parallel to each other or antiparallel; the energy difference also depends on their relative locations (e.g., end to end or side by side). We will calculate the potential energies of the $F = 0$ and $F = 1$ states separately.

Magnetic dipole in a magnetic field

A magnetic dipole will tend to align with an external magnetic field just as a compass needle aligns itself with the magnetic field of the earth. One can visualize the external field acting on the "current loop" that constitutes the dipole to provide the required torque (Fig. 7b). Integration of this torque over the appropriate angles yields a potential energy that is a function of the angle θ between the magnetic field B and magnetic moment μ vectors.

The lowest potential energy occurs when the vectors are aligned (approximately as shown in Fig. 7b). If the energy is defined to be zero at $\theta = 90°$, the potential energy E_{pot} turns out to be the negative dot product of the vectors μ and B,

➡ $$E_{pot} = -\mu \cdot B \quad \text{(Potential energy; dipole in a magnetic field; classical)} \quad (10.16)$$

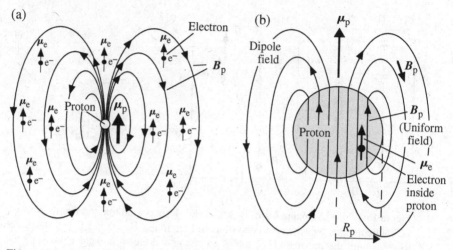

Figure 10.9. Calculation of the interaction energy between the electron magnetic moment and the magnetic field of the proton. The two magnetic moments, $\boldsymbol{\mu}_e$ and $\boldsymbol{\mu}_p$, are shown parallel to each other, which is the $F = 0$ case in our imperfect analogy. (a) The electron magnetic moment $\boldsymbol{\mu}_e$ (light vertical arrows) is shown at several different positions in the proton's magnetic field. The average interaction energy turns out to be zero. (b) Blowup of central region including the interior of the proton. The field $\boldsymbol{B}_p$ is taken to be uniform inside the proton and parallel to $\boldsymbol{\mu}_p$. The overlap of the electron wave function with the proton, the *contact term*, gives a negative energy shift.

or

$$E_{\text{pot}} = -|\boldsymbol{\mu}|\,|\boldsymbol{B}|\cos\theta. \tag{10.17}$$

In this expression, the energy is lowest when the vectors are aligned ($\theta = 0$). In the calculation of the energy difference between the $F = 0$ and $F = 1$ states, it is the proton that provides the "external" magnetic field.

Three interaction terms

The energy difference between the hydrogen hyperfine states is formally calculated from the interaction terms of the several magnetic moments and fields of the problem. There turn out to be three terms in the energy operator (the *Hamiltonian*) that give the interaction energy. The first term is the interaction of the proton magnetic moment with the magnetic field associated with the orbital angular momentum of the electron. Because the latter is zero (ground state), this part of the interaction energy is zero.

The second term is the interaction of the proton and electron magnetic moments, or more precisely, the interaction of the electron magnetic moment with the magnetic field of the proton over the large region of space where they overlap. The electron has a finite probability of being any place over the large region occupied by its wave function out to, and beyond, the Bohr orbit at radius a_0,

$$a_0 = 5.291\,772 \times 10^{-11}\ \text{m}, \tag{10.18}$$

as illustrated in Fig. 8c and by the many little arrows in Fig. 9a. This region is much larger than that of the proton itself, which has radius of only $\sim 10^{-15}$ m. Nevertheless, the magnetic

Table 10.4. *Hyperfine splitting (ground-state hydrogen)*

System state	$F = 0$	$F = 1$
Angular momentum	Antiparallel[a]	Parallel[a]
Magnetic moment	Parallel[a]	Antiparallel[a]
Potential energy	Low	High

[a] Relative orientations of the vectors for the electron and the proton in our classical visualization. See (9) and associated discussion.

field of the proton $\boldsymbol{B}_\mathrm{p}$ extends over this region, as illustrated by the oval field lines in Fig. 9a. As drawn, the two magnetic moments $\boldsymbol{\mu}_\mathrm{e}$ and $\boldsymbol{\mu}_\mathrm{p}$ are parallel, but the direction of $\boldsymbol{B}_\mathrm{p}$ varies with position.

It is apparent from Fig. 9a that, in regions of strong magnetic field (near the poles of the proton), the vectors $\boldsymbol{B}_\mathrm{p}$ and $\boldsymbol{\mu}_\mathrm{e}$ are parallel but that they are antiparallel over much larger regions of weak fields. It turns out that integration of $-\boldsymbol{\mu}_\mathrm{e} \cdot \boldsymbol{B}_\mathrm{p}$ over the entire overlap region, but excluding the interior of the proton, yields zero net potential energy, and so this term also makes no contribution.

The third term is the so-called contact term, which does have a net effect. This occurs when the electron is directly superposed on the proton. The contact term arises in general for two pointlike dipoles. For our case, we may gain some insight into it by considering the magnetic field inside the proton. This field may be approximated as uniform (Fig. 9b). This portion of the proton's field was ignored in the integration just described (Fig. 9a) even though the "$\boldsymbol{\mu}_\mathrm{e}$-up" electron in our example has a finite probability for being inside the proton. Because $\boldsymbol{B}_\mathrm{p}$ is everywhere up inside the proton, this entire region contributes a negative (low) potential energy.

There is no offsetting high-potential region, and so the net effect of the contact term is a lower potential for this case of parallel magnetic moments (or antiparallel spins) shown in Fig. 9b. If, in contrast, $\boldsymbol{\mu}_\mathrm{e}$ is down, the magnetic moments are antiparallel inside the proton, and the contact term contributes a positive potential energy shift. These two states are equivalent (in our imperfect analogy) to the $F = 0$ and $F = 1$ eigenstates, respectively, as summarized in Table 4.

Overlap of electron wave function with a proton

How large is the energy difference between these two states? Naively, one might expect it to be approximately $2\mu_\mathrm{e}B_\mathrm{p}$. In fact, this is a huge overestimate because the pointlike electron has a very low probability of being inside the proton. The characteristic radius of the electron wave function (18) is $a_0 \approx 10^{-10}$ m (shaded regions in Fig. 8c), whereas the proton radius is only $R_\mathrm{p} \approx 10^{-15}$ m. The probability $\mathcal{P}$ of finding the electron inside the small proton is roughly the ratio of the volumes $\mathcal{P} \approx R_\mathrm{p}^3/a_0^3 \approx 10^{-15}$. The energy shift due to the electron's presence inside the proton is thus the difference of the interaction energies $-\boldsymbol{\mu}_\mathrm{e} \cdot \boldsymbol{B}_\mathrm{p}$ (16) for the $F = 0$ and $F = 1$ eigenstates reduced by the factor $\sim R_\mathrm{p}^3/a_0^3$.

The probability $\mathcal{P}$ is obtained formally from an integration of the electron's wave function squared over the volume occupied by the proton. The wave function $\psi(r)$ for the ground state of the hydrogen atom is derived from the Schrödinger wave function; it turns out to be

$$\psi(r) = \frac{1}{\sqrt{4\pi}} \frac{2}{a_0^{3/2}} \exp(-r/a_0).$$ (Wave function; ground state)   (10.19)

The probability per unit volume of finding the electron at some location is the product of the wave function ψ and its complex conjugate ψ^*, which in this case is equal to the function itself. Multiplication by the volume of a radial shell $4\pi r^2\, dr$ yields the probability of finding the electron in the shell.

The total probability $\mathcal{P}$ of finding the electron within the radius R_p is therefore

$$\mathcal{P}(r \leq R_p) = \int_0^{R_p} \psi^*(r)\psi(r)\, d^3 r = \int_0^{R_p} \frac{1}{4\pi} \frac{4}{a_0^3} \exp\left(\frac{-2r}{a_0}\right) 4\pi r^2\, dr.$$ (10.20)

Because $R_p \ll a_0$, the exponential may be approximated as unity, $\exp(-2r/a_0) \to 1$,

➡ $$\mathcal{P}(r \leq R_p) = \int_0^{R_p} \frac{4}{a_0^3} r^2 dr = \frac{4}{3} \frac{R_p^3}{a_0^3}.$$ (Probability of finding electron inside proton)   (10.21)

This better calculation, also an approximation, yields our previous estimate of $\mathcal{P}$ but with an additional coefficient of 4/3. (See Prob. 34 for another level of approximation.)

Magnetic field inside the proton

The magnetic field inside the proton $\boldsymbol{B}_p$ is also needed for the calculation of the $-\boldsymbol{\mu}_e \cdot \boldsymbol{B}_p$ energy shift. A simple approximation is to assume that the proton is a uniformly magnetized sphere of sufficient strength to give the observed magnetic dipole moment $\boldsymbol{\mu}_p$ (Fig. 9b). Under this assumption, the relation between these parameters is

$$\boldsymbol{B}_p = \frac{\mu_0}{2\pi R_p^3} \boldsymbol{\mu}_p,$$ (Uniformly charged magnetized sphere)   (10.22)

where μ_0 is the permeability constant for a vacuum, *not* a magnetic moment, defined as

$$\mu_0 = 4\pi \times 10^{-7}\, \mathrm{T\,m\,A^{-1}}.$$ (10.23)

The magnetic field (22) happens to be exactly the magnetic field at the center of a circular loop of current i that gives rise to a magnetic moment $\boldsymbol{\mu}_p$. The Biot–Savart law readily yields this result (Prob. 33).

Substitute the expression (14) for $\boldsymbol{\mu}_p$ into (22):

➡ $$\boldsymbol{B}_p = \frac{\mu_0}{2\pi R_p^3} \frac{g_p \mu_N}{\hbar} I.$$ (Approximate magnetic field inside a proton)   (10.24)

This is the magnetic field inside the proton we need to calculate the $-\boldsymbol{\mu}_e \cdot \boldsymbol{B}_p$ interaction.

Spin-spin coupling

The desired potential energy E_{pot} is the eigenvalue of the operator $-\boldsymbol{\mu}_e \cdot \boldsymbol{B}_p$ (16) applied to the state in question ($F = 0$ or $F = 1$) reduced by the probability $\mathcal{P}$ that the electron is found inside the proton (21):

$$-\mathcal{P}\boldsymbol{\mu}_e \cdot \boldsymbol{B}_p = -\frac{4}{3}\frac{R_p^3}{a_0^3}\boldsymbol{\mu}_e \cdot \boldsymbol{B}_p \Rightarrow E_{pot}. \tag{10.25}$$

Substitute for $\boldsymbol{\mu}_e$ from (13) and for $\boldsymbol{B}_p$ from (24) to obtain

$$-\frac{4}{3}\frac{\mu_0}{2\pi}\frac{g_e g_p \mu_B \mu_N}{a_0^3}\frac{\boldsymbol{S} \cdot \boldsymbol{I}}{\hbar^2} \Rightarrow E_{pot}. \tag{10.26}$$

Note that the size of the proton, R_p, drops out; the contact term does not require a finite size for either particle.

Substitute into (26) numerical values for the factors g_e (11), μ_B (12), g_p (14), μ_N (15), a_0 (18), and μ_0 (23) together with values of e, m_e, m_p, and $\hbar = h/2\pi$:

$$\Rightarrow \quad 9.427\,62 \times 10^{-25}\,\frac{\boldsymbol{S} \cdot \boldsymbol{I}}{\hbar^2} \Rightarrow E_{pot}. \qquad \text{(J)} \tag{10.27}$$

The interaction energy of the magnetic moments thus reduces to the dot product of the spin operators for the electron and proton, which is known as *spin-spin coupling*. We will apply the operator $\boldsymbol{S} \cdot \boldsymbol{I}$ to the wave function to obtain the potential energy.

The dot product $\boldsymbol{S} \cdot \boldsymbol{I}$ can be expanded by taking the dot product of $\boldsymbol{F}$ (4) with itself,

$$\boldsymbol{F} \cdot \boldsymbol{F} = (\boldsymbol{S} + \boldsymbol{I}) \cdot (\boldsymbol{S} + \boldsymbol{I}) = \boldsymbol{S} \cdot \boldsymbol{S} + \boldsymbol{S} \cdot \boldsymbol{I} + \boldsymbol{I} \cdot \boldsymbol{S} + \boldsymbol{I} \cdot \boldsymbol{I} \tag{10.28}$$
$$= \boldsymbol{S} \cdot \boldsymbol{S} + 2\boldsymbol{S} \cdot \boldsymbol{I} + \boldsymbol{I} \cdot \boldsymbol{I}$$

$$\boldsymbol{S} \cdot \boldsymbol{I} = \frac{1}{2}(\boldsymbol{F} \cdot \boldsymbol{F} - \boldsymbol{S} \cdot \boldsymbol{S} - \boldsymbol{I} \cdot \boldsymbol{I}), \tag{10.29}$$

and, from (5), (7), and (8),

$$\boldsymbol{S} \cdot \boldsymbol{I} \Rightarrow \frac{1}{2}\hbar^2[F(F + 1) - S(S + 1) - I(I + 1)]. \tag{10.30}$$

The electron and proton spin quantum numbers are $S = 1/2$ and $I = 1/2$. The interaction operator for the $F = 0$ state gives the following eigenvalue:

$$\boldsymbol{S} \cdot \boldsymbol{I} \Rightarrow \frac{1}{2}\hbar^2\left[0 - \frac{3}{4} - \frac{3}{4}\right] = -\frac{3}{4}\hbar^2. \qquad (F = 0) \tag{10.31}$$

For the $F = 1$ state,

$$\boldsymbol{S} \cdot \boldsymbol{I} \Rightarrow \frac{1}{2}\hbar^2\left[2 - \frac{3}{4} - \frac{3}{4}\right] = +\frac{1}{4}\hbar^2. \qquad (F = 1) \tag{10.32}$$

Note that the $F = 1$ state has a positive potential energy as expected (Table 4); the magnetic moments are antiparallel. The $F = 0$ state is negative as expected, but its magnitude is three times greater. Recall our mention of the tighter coupling in this state (Fig. 6c).

Energy difference

Finally, apply (31) and (32) separately to (27) to obtain the eigenvalue E_{pot} for each state $F = 0$ and 1, and then take the difference to obtain the energy difference between the two states,

$$\Delta E_{pot} = E_{pot,\, F=1} - E_{pot,\, F=0} = 9.427\,62 \times 10^{-25} \left[\frac{1}{4} - \left(-\frac{3}{4} \right) \right] \quad (J) \quad (10.33)$$

$$\Delta E_{pot} = 9.427\,62 \times 10^{-25}\,J \qquad \text{(Energy of hyperfine splitting)} \quad (10.34)$$
$$= 5.9 \times 10^{-6}\,eV.$$

This is the separation in energy units of the two hyperfine states illustrated in the lower right corner of Fig. 3. The associated frequency is

$$\nu = \frac{\Delta E_{pot}}{h} = \frac{9.427\,62 \times 10^{-25}}{6.626\,07 \times 10^{-34}} = 1422.8\,\text{MHz}. \quad (10.35)$$

This is the desired answer according to our (approximate) calculation.

The precise result from a correct quantum electrodynamics calculation and from measurements is

$$\Delta E = h\nu = 9.412 \times 10^{-25}\,J, \quad (10.36)$$

and, with greater precision,

➡ $$\nu = 1\,420\,405\,752\,\text{Hz} = 1420.4\,\text{MHz}. \quad (10.37)$$

This is the desired result that is of such great importance. Our approximate calculation is therefore correct to 0.2%.

10.4   Rotation of the Galaxy

The flattened-disk shape and spiral structure of our Galaxy (Fig. 1) and others strongly suggest that the stars and matter rotate about the axis of symmetry. If, as is believed, galaxies form from infalling gas, the angular momentum of the gas will tend to prevent the collapse in directions perpendicular to the net angular momentum vector. In contrast, gas falling in along the rotation axis is not so impeded. The result is a disklike shape such as that of an accretion disk, the planetary system of the solar system, or the rings around Saturn.

Spectral Doppler studies in the optical and at 1420 MHz in the radio clearly show that the stars and gaseous matter in spiral galaxies rotate about the center of the galaxy – probably in nearly circular orbits, as illustrated in Fig. 10a. When viewed edge-on, more or less, the spectral lines from one side of the disk are shifted upward in frequency, whereas those on the other side are shifted downward, all relative to the center. Doppler studies of gas and stars within the Galaxy clearly show that it is no exception to this rotational motion.

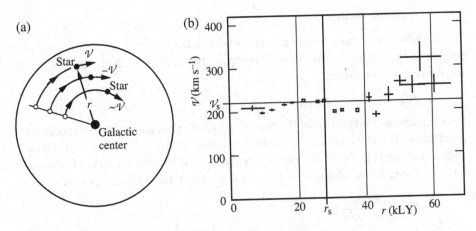

Figure 10.10. (a) Differential rotation of gas and stars in the Galaxy viewed from above the plane with rotation speeds $\mathcal{V}$ approximately independent of radius r from the galactic center. It thus takes longer for stars in outer orbits to orbit the center. (b) Rotation curve $\mathcal{V}(r)$ of the Galaxy for the IAU values of distance of the sun from the center of the Galaxy ($r_s = 8.5$ kpc $= 27.7$ kLY) and the solar speed, ($\mathcal{V}_s = 220$ km/s). Overall, the curve is quite flat, indicating substantial mass out to radii well beyond the sun. The data for distances within the solar radius are obtained from radio studies of atomic hydrogen; at greater radii, data are obtained from radio and optical studies of stars and also of H II regions with their accompanying stellar associations. [(b) Adapted from M. Fich and S. Tremaine, *ARAA* **29**, 409 (1991), with permission]

The velocity of the sun relative to distant galaxies is obtained from studies of spectral lines of nearby galaxies and globular clusters. It has been found that the sun and its immediate neighbors (the *local standard of rest*, LSR) are moving ~220 km/s in the direction of the constellation Cygnus. This is about 90° from the galactic center in the constellation Sagittarius. The sun's motion is thus consistent with rotation about the galactic center. This velocity allows us to estimate the time it takes the sun to circle the galactic center. Given the ~25 000 LY distance to the center, the period is 2.1×10^8 years. Compare this with the ~10^{10}-yr age of the Galaxy.

The motion of a given star in a galaxy is governed by the net gravitational force on it. This force arises from all other matter in the galaxy. The *rotation curve*, a plot of rotational speed $\mathcal{V}$ versus distance from the galactic center, is an indicator of the mass distribution within the galaxy. The rotation curves of most galaxies, including our own, indicate that large quantities of *dark matter* are associated with them.

A rotation curve for the (Milky Way) Galaxy is shown in Fig. 10b. Note that it is quite flat, indicating that the speed $\mathcal{V}(r)$ does not vary much with radius. This implies that the outer stars fall behind the inner stars as illustrated in Fig. 10a; the Galaxy does not rotate as a rigid body. This slippage is known as *differential rotation*.

It is possible to construct the rotation curve of the Galaxy, in part, from the intensities and Doppler shifts of the 1420-MHz radiation emitted from hydrogen in or near the galactic plane. In this section, we discuss (*i*) differential rotation and its relation to the mass of the Galaxy, (*ii*) the tangent-point method for determining the rotation curve and subsequently the distribution of hydrogen clouds in the Galaxy, and finally (*iii*) the Oort constants, which relate the measurements of star motions in the vicinity of the sun to the differential rotation of our region of the Galaxy.

Galactic models

The distribution of mass within a galaxy dictates the shape of the rotation curve. Here we present several simplified models and compare the rotational velocity distribution $\mathcal{V}(r)$ with the flat distribution typically observed.

Pointlike central mass

Let the preponderant portion of the mass of a hypothetical galaxy be incorporated in a central point mass. The stars and gas that orbit it are thus influenced primarily by the central mass as are the planets in the solar system. For circular orbits with azimuthal linear velocity $\mathcal{V}$ at radius r, the radial acceleration is $\mathcal{V}^2/r$, and so the radial component of the force is

$$F_{\mathrm{r}} = -\frac{m\mathcal{V}(r)^2}{r}. \tag{10.38}$$

The radial dependence of $\mathcal{V}$ is indicated to emphasize that this velocity in general will depend on the distance from the galactic center. Invoke the Newtonian gravitational force law to obtain $\mathcal{V}(r)$ for circular Keplerian orbits as follows:

$$\frac{-GMm}{r^2} = \frac{-m\mathcal{V}^2}{r} \qquad \text{(Circular motion; point mass)} \tag{10.39}$$

$$\Rightarrow \quad \mathcal{V}(r) = \left(\frac{GM}{r}\right)^{1/2} \propto r^{-1/2}. \qquad \text{(Keplerian orbital speed)} \tag{10.40}$$

The orbital velocity $\mathcal{V}(r)$ of the orbiting stars decreases with their distance (radius) from the center unlike the result for the Galaxy (Fig. 10b). This result remains valid if the central object has a significant size as long as it is spherically symmetric and entirely within the orbit of the innermost orbiting object. This is nothing more than the standard satellite problem. The radial dependencies of the force F_{r} and $\mathcal{V}$ for a point central mass are shown in Fig. 11b.

Galactic mass

An estimate of the Galaxy's mass that is within the solar orbit may be obtained from (40) given the solar speed $\mathcal{V}_{\mathrm{s}} \approx 220$ km/s and radius $\sim 25\,000$ LY (Table 1),

$$M_{\mathrm{Gal}} = 8.6 \times 10^{10}\ M_{\odot}, \qquad \text{(Galactic mass within solar orbit)} \tag{10.41}$$

in approximate agreement with the value in Table 1. This result does not include mass at galactic radii beyond the sun. Also, it is based on an unrealistic spherical distribution for the mass within the solar orbit. (The forces from a spherically symmetric mass distribution follow from Gauss's law.)

Spherical and spheroidal distributions

The forces and speeds of orbiting material in a galaxy are more easily calculated if oblate spheroids are adopted as approximations of the true mass distribution and consideration is restricted to equatorial circular orbits. An ellipse rotated about its minor axis generates an oblate spheroid suggestive of a discus, which is somewhat akin to the shape of a spiral galaxy. Rotation about the major axis generates a prolate spheroid, which is reminiscent of a football (American style).

A spheroid can be made up of spheroidal shells, each of uniform mass density, that all have the same shape (eccentricity) as the outer surface. The shells will differ only in their

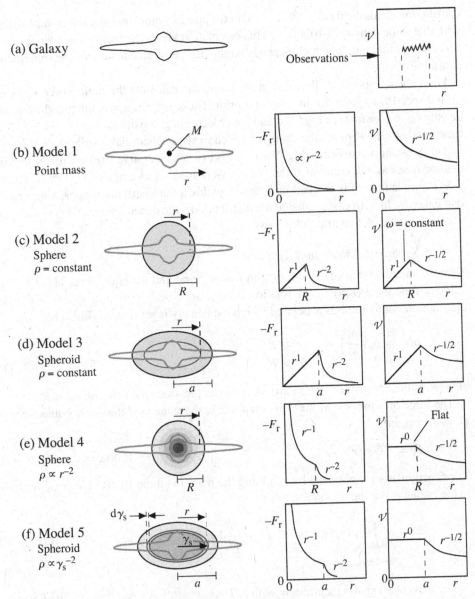

Figure 10.11. Models of galactic rotation for various distributions of mass in the Galaxy showing (left) sketches of the assumed distributions, (center) plots of $-F_r$ versus r for stars in the equatorial plane, and (right) plots of $\mathcal{V}(r)$, the speed of a star in a circular orbit in the equatorial plane. Radial dependencies are indicated.

sizes (semimajor and semiminor axes). Each such shell can be designated with the radial parameter γ_s (Fig. 11f). The value of γ_s thus specifies one such shell. We set it to be equal to r on the equatorial plane, $\gamma_s = r$.

It turns out that a circular orbit in the equatorial plane of a spheroidal mass distribution with a given distribution of density $\rho(\gamma_s)$ has velocity function $\mathcal{V}(r)$ characterized by the same radial dependence on r as it would have for a spherical mass distribution with the identical density function $\rho(r)$. Thus, calculation of $\mathcal{V}(r)$ for some chosen *spherical* density distribution yields the form of $V(\gamma_s = r)$ for the possibly more realistic *spheroidal* distribution with that same density distribution. This calculation yields the radial dependence of $V(r)$ on r, but not its magnitude, which depends on the degree of eccentricity of the spheroidal mass distribution.

For a chosen spherical distribution $\rho(r)$, one can calculate the radial force $F_r(r)$ and the orbital speed $V(r)$ for a star in a circular orbit. These functions are illustrated for a constant density ($\rho = $ constant) in Fig. 11c and for a decreasing density ($\rho \propto r^{-2}$), in Fig. 11e. The functional shapes $V(r)$ are the same for the equivalent spheroidal distributions (Fig. 11d,f).

The constant-density case (Fig. 11c,d) leads to a velocity curve $V(r)$ within the distribution that increases as r^1 exactly as if the Galaxy were rotating as a rigid body (Prob. 42c,d). The $\rho \propto r^{-2}$ case (11e,f) is worked out here. It yields a flat distribution approximating that of real galaxies. In both cases, outside the distributions, the dependence is $V(r) \propto r^{-1/2}$, which is characteristic of a central point mass.

Spherical distribution with $\rho \propto r^{-2}$

Take the spherical mass distribution with $\rho \propto r^{-2}$ and find the radial force law $F_r(r)$ and the velocity function $V(r)$ (i.e., the rotation curve). Specifically, adopt a spherical r^{-2} density distribution out to a radius R beyond which the density is set to zero (Fig. 11e):

$$\rho(r) = \rho_0 \left(\frac{r_0}{r}\right)^2 \quad (r < R)$$
$$\rho(r) = 0 \quad (r > R). \tag{10.42}$$

Here ρ_0 is the density at some radius r_0; for example, one could choose $r_0 = R$.

The force on a particle of mass m at radius r is the same as if the interior mass $\mathfrak{M}(r)$ were all at the origin:

$$F_r = -G\mathfrak{M}(r)m/r^2. \qquad \text{(Force equation)} \tag{10.43}$$

The function $\mathfrak{M}(r)$ is obtained by summing the mass in all the shells of volume $4\pi r^2\,dr$ out to the radius r. For the region $r < R$,

$$\mathfrak{M}(r) = \int_0^r \rho(r)4\pi r^2 dr = \int_0^r \rho_0 \frac{r_0^2}{r^2}4\pi r^2 dr. \qquad (r < R) \tag{10.44}$$
$$= 4\pi\rho_0 r_0^2 r$$

The mass $\mathfrak{M}(r)$ grows indefinitely with r. Thus, a cutoff at some finite radius R is required to avoid an infinite mass. Substitute (44) into the force equation (43) to obtain the radial force within the mass distribution ($r < R$):

$$F_r = -\frac{G\mathfrak{M}(r)m}{r^2} = -\frac{4\pi Gm\rho_0 r_0^2}{r} \propto -r^{-1}. \tag{10.45}$$

The force is attractive and decreases in magnitude with radius more slowly than the r^{-2} dependence for a point mass. For circular motion, one has $F_r = -mV^2/r$. Use (45) for F_r and solve for $V(r)$:

$$\Rightarrow \quad V = \left(-\frac{F_r}{m}r\right)^{1/2} = (4\pi G\rho_0)^{1/2} r_0 \propto r^0 = \text{constant.} \quad (r < R; \rho \propto r^{-2}) \tag{10.46}$$

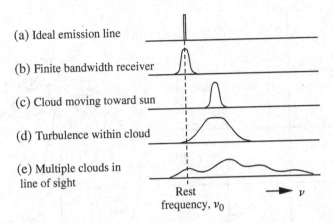

(a) Ideal emission line

(b) Finite bandwidth receiver

(c) Cloud moving toward sun

(d) Turbulence within cloud

(e) Multiple clouds in
line of sight

Rest
frequency, ν_0

ν

Figure 10.12. Hypothetical spectral-line profiles of specific intensity versus frequency showing (a) the ideal line and (b)–(e) the effect of four widening mechanisms. [Adapted from J. D. Kraus, *Radio Astronomy*, 2nd ed., Cygnus-Quasar Books, 1986, p. 8-91, with permission]

Thus, as anticipated, the r^{-2} spherical mass distribution yields a rotation curve $\mathcal{V}(r)$ that is flat for orbits within the distribution (Fig. 11e), which is similar to the actual observed rotation curve of the Galaxy and many other galaxies. Given the similar solutions for spheres and spheroids, this flat shape is also valid for the spheroidal distribution $\rho(\gamma_s) \propto \gamma_s^{-2}$ (Fig. 11f), though, as stated, its magnitude will depend on the eccentricity of the spheroid. The spheroidal shape, of course, is only an idealization of the possible shape of a galactic dark-matter halo.

Tangent-point method

How is the rotation curve $\mathcal{V}(r)$ for the Galaxy obtained in practice? Astronomers measure line-of-sight Doppler velocities v_r and are able to deduce from them locations and speeds (relative to the galactic center) of stars and gas in the Galaxy. Optical astronomers use spectral lines from stars and radio astronomers use the 1420-MHz emission line from clouds of neutral hydrogen. The *tangent-point method* presented here makes use of Doppler data at 1420 MHz to determine the rotation curve for radii within the sun's orbit.

Hydrogen profiles

A hypothetical, narrow 1420-MHz spectral line is broadened through several effects (Fig. 12). The most pertinent to our discussion is due to the contributions of several clouds along the line of sight, each with a different radial velocity relative to the observer (Fig. 12e). Our objective is to interpret such profiles in terms of differential rotation of the Galaxy.

Actual profiles from an early set of measurements are sketched in Fig. 13a for differing galactic longitudes ℓ. This longitude is defined in Fig. 13b, a sketch of the Galaxy viewed from the north galactic pole, as being measured from the sun in the counterclockwise direction with $\ell = 0$ in the direction of the galactic center.

The vertical scale of Fig. 13a is specific intensity $I(\nu)$ given as the equivalent (blackbody) antenna temperature T (6.8). In these profiles, all the detected photons are presumed to have been emitted by hydrogen atoms at $\nu_0 = 1420$ MHz in the frame of reference of the emitting

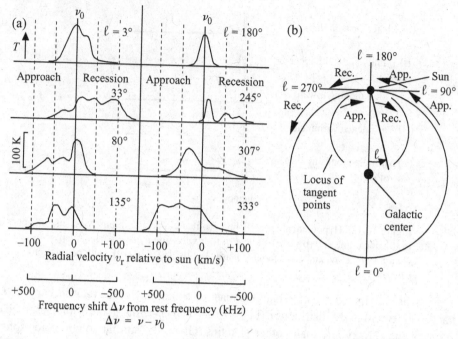

Figure 10.13. (a) Sketches of actual profiles (antenna temperature versus frequency) of H I emission measured at different galactic longitudes in the galactic plane. (b) View of the Galaxy from the north galactic pole in the rotating frame of reference in which the sun and galactic center are both at rest. In this frame, the gas above the sun appears to be moving to the left and that below the sun to the right. The locus of tangent points are the regions sampled to construct a rotation curve. [(a) J. D. Kraus, *ibid.*; adapted from F. Kerr and G. Westerhout in *Galactic Structure*, A. Blaauw and M. Schmidt, eds., U. Chicago, 1965, p. 178. (b) Adapted from J. D. Kraus *ibid.*, p. 8-92, with permission]

atom. In the frame of reference of an observer at the sun, different clouds with different line-of-sight velocities (relative to the sun) will exhibit different Doppler shifts. That is, each cloud appears as a distinct bump at its own particular frequency in the profiles of antenna temperature versus frequency.

The horizontal axis for these profiles (Fig. 13a) is labeled with line-of-sight velocities, v_r, and also with the directly measured *frequency shifts* Δv (lower axis). The frequency shift is referenced to the rest frequency v_0 of the hydrogen atom,

$$\Delta v = v - v_0; \quad v_0 = 1420\,\text{MHz}, \tag{10.47}$$

where v is the observed frequency. This is related to the line-of-sight velocity v_r, which is also called the *radial velocity*, by the classical Doppler shift formula (7.35):

$$\frac{\Delta v}{v} \equiv \frac{v - v_0}{v_0} = -\frac{v_r}{c}. \tag{10.48}$$

Note the sign convention, which is universal in astronomy: *receding* clouds (which exhibit negative frequency shifts) are assigned *positive* velocities. We invoke only the classical Doppler effect here because the velocities are nonrelativistic; 220 km/s is about $10^{-3}\,c$.

Astronomers often quote velocities (km s^{-1}) when they really have measured frequency shifts. The ratio v/c is a physical expression of the fractional frequency shift. It is the same for different spectral lines (of the same or differing kinds of atoms) emanating from the same cloud with a common velocity. This allows one to directly compare velocities obtained from different atomic spectral lines.

Working model of galactic rotation

The profiles of Fig. 13a for the Galaxy exhibit strong positive radial velocities at galactic longitudes $\ell = 33°$ and $\ell = 245°$ relative to the sun (i.e., adjusted for the earth's motion about the sun); these clouds are receding from the sun. At other galactic longitudes ($\ell = 135°, 307°, 333°$) the velocities are mostly negative and hence mostly approaching the sun.

These gross features can be understood in terms of a simple model illustrated in Fig. 10a wherein (*i*) the gas and stars (including the sun) move clockwise in concentric circles about the galactic center as viewed from the north galactic pole, and (*ii*) the angular velocity of rotation, ω (rad/s), decreases as radius r increases. That is, the outer material lags the inner material. This is the case if, for example, the speed is approximately the same at all radii as drawn in Fig. 10a.

These gas motions from the perspective of an observer at the sun are shown in Fig. 13b. In the *rotating frame* of reference, the sun and the galactic center are at rest. As indicated by the arrows, the gas outside (above) the sun's radius moves (in the sun's frame of reference) to the left, and that inside (below) moves to the right in accord with our model in which angular velocity decreases with radius. Thus, to the upper right and lower left of the sun, the gas is approaching ("App.") the sun, and to the upper left and lower right, it is receding ("Rec.") from the sun. This is in agreement with the broad features of the measured profiles shown in Fig. 13a.

The profiles are thus consistent with a simple picture of galactic rotation; namely, the gas moves in circular paths with angular velocity $\omega(r)$ that decreases with increasing distance r from the galactic center. This same picture is obtained from the study of galaxies outside our own. If, as indicated in Fig. 10b, the rotation of matter in the Galaxy in terms of azimuthal velocity $V(r)$ is relatively constant, then $\omega(r)$ decreases with r:

$$\omega(r) = \frac{V(r)}{r} \propto r^{-1}. \tag{10.49}$$

If the gas clouds in the Galaxy move with the stars in their vicinities (a reasonable assumption), this analysis tells us about the organized internal motions of the Galaxy as a whole.

Geometry

One can use this axisymmetric model together with the data (Fig. 13a) to determine the *rotation curve*, $\omega(r)$, or equivalently $V(r)$. The geometry of Fig. 14a shows the motion seen by an observer in a fixed inertial frame of reference (not the rotating frame of Fig. 13b). The sun rotates at angular velocity ω_s at radius r_s with velocity $V_s = \omega_s r_s$. If the antenna on the sun is directed toward the galactic plane at galactic longitude ℓ, the line-of-sight component of the velocity of the sun is $\omega_s r_s \sin \ell$.

A cloud at point P on the line of sight at a position specified by the angle δ (Fig. 14a) rotates with angular velocity ω on a circle at radius r. The speed of cloud P along the inner circle is thus $V = \omega r$, or, more explicitly, $V(r) = \omega(r)r$. Its component along the line of sight

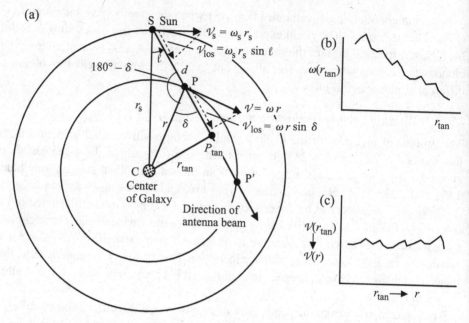

Figure 10.14. (a) Geometry for the determination of the rotation curve $\mathcal{V}(r)$. The Galaxy is viewed from the galactic north pole from an inertial frame of reference; both the sun S and a cloud at point P rotate clockwise about the galactic center. (b) Sketch of a hypothetical rotation curve $\omega(r)$ derived from tangent point measurements in which the outer parts of the galaxy rotate with smaller angular velocity ω than do the inner parts. (c) Sketch of the rotation curve of (b) redrawn in terms of the azimuthal velocity $\mathcal{V}(r) = \omega(r)\,r$. [(a) Adapted from J. D. Kraus, *ibid.*, p. 8-92, with permission]

is $\omega r \sin\delta$. According to an observer on the sun, the *apparent* radial line-of-sight velocity v_r of the cloud at P is the difference of the radial velocities of the cloud and the sun,

$$v_r = \omega r \sin\delta - \omega_s r_s \sin\ell.$$
(10.50)

Invoke the law of sines for the triangle SCP:

$$\frac{r}{r_s} = \frac{\sin\ell}{\sin(180° - \delta)} = \frac{\sin\ell}{\sin\delta}$$
(10.51)

or

$$r \sin\delta = r_s \sin\ell,$$
(10.52)

which is directly self-evident from the geometry. Substitute into (50) to obtain

➡ $$v_r = (\omega - \omega_s)r_s \sin\ell.$$ (Relative line-of-sight velocity from sun)   (10.53)

Measurement of the apparent radial velocity v_r of a cloud at longitude ℓ thus yields the desired angular velocity ω of that cloud given only knowledge of the solar values r_s and $\omega_s = \mathcal{V}_s/r_s$ (Table 1).

Rotation curve

The desired rotation curve $V(r)$ or $\omega(r)$ requires knowledge not only of ω but also of the distance r from the galactic center to each of the clouds used to construct the curve. Along any given line of sight, this distance can be found for a cloud at the tangent point P_{tan} (Fig. 14a). Measurements of ω and r at several tangent points (i.e., at several longitudes ℓ) and the assumption of axial symmetry allow one to construct the rotation curve $\omega(r)$.

The largest angular velocity along the line of sight SPP′ will arise from a cloud at the tangent point P_{tan} because there the line of sight is closest to the galactic center. According to our simple model, the angular velocity increases toward lesser radii. In turn, this gives rise to the largest v_r; see also (53). Thus, in any given profile, the emission with the greatest v_r would arise from the vicinity of the tangent point for that particular line of sight if there were sufficient emitting atoms at P_{tan}.

The measured magnitude of $v_{r,max}$ along a given line of sight (i.e., in a given profile) thus yields the value of ω of the gas at the tangent point radius, r_{tan}. Solve (53):

$$\omega(r_{tan}) = \frac{v_{r,max}(\ell)}{r_s \sin \ell} + \omega_s; \quad r_{tan} = r_s \sin \ell. \tag{10.54}$$

The geometry also provides, in (54), the galactocentric radius r_{tan} associated with $\omega(r_{tan})$.

In this manner, a value of $v_{r,max}$ provides a value of ω at a known radius; this is one point on the rotation curve, a plot of $\omega(r_{tan})$ versus r_{tan} (Fig. 14b). Measurements at other galactic longitudes ℓ provide additional points at other values of r_{tan}, and hence the complete curve can be constructed for orbits interior to the sun. This curve represents the measured values at a series of tangent points – one for each measured ℓ. The locus of the tangent points in the Galaxy are the arcs in Fig. 13b.

The plot $\omega(r_{tan})$ can be redrawn as $V(r_{tan})$. Use the relation between linear and angular velocities, $V(r_{tan}) = \omega(r_{tan})r_{tan}$, to rewrite (54):

$$V(r_{tan}) = v_{r,max}(\ell) + \omega_s r_{tan}; \quad r_{tan} = r_s \sin \ell. \quad \text{(Rotation curve)} \tag{10.55}$$

Because we have multiplied by r_{tan}, the slope of $V(r_{tan})$ is greater than that of $\omega(r_{tan})$ (Fig. 14c). The plot $V(r_{tan})$ thus also derives from the measured maximum Doppler shift, $v_{r,max}$, along various lines of sight ℓ.

This rotation curve will be a general description of the rotation elsewhere in the Galaxy because, in our axisymmetric model, ω and V are postulated to be the same all around the circle at radius r. We thus substitute r for r_{tan} in the axis labels of Fig. 14c. The two plots (Fig. 14b,c) specify to what extent the outer parts of the Galaxy slip behind the inner parts as they rotate.

The tangent-point method does not work for galactic longitudes $\ell > 90°$; there is no peak recession speed along the line of sight. The distance from the sun to a tracer object must be independently determined. This, together with ℓ and r_s, yields the distance r of the object from the galactic center, and the measured v_r of the object yields ω; the expression (53) remains valid for these positions. Bright stars and new star associations are two types of objects used by optical astronomers for tracer objects. The distances are determined, respectively, by spectral classification and main-sequence fitting.

The rotation curve for the Galaxy in Fig. 10b within the radius of the sun (~25 kLY) is constructed from H I emission data with the tangent-point method. The curve is approximately flat, which is indicative of large amounts of matter reaching out to the solar radius and beyond; otherwise, it would be decreasing, as in the models of Fig. 11 for $r > R$. Caution is required here; the slope of the V versus r curve will change if a different value of V_s or r_s is adopted. This follows from $\omega_s = V_s/r_s$ and (55); see Prob. 45.

The rotation curve of Fig. 10b shows average values at each radius. It therefore suppresses relatively large point-to-point fluctuations and significant irregularities in the rotation curves. Galaxies are known to have large-scale distortions owing to spiral arms and close encounters or collisions with other galaxies. Smaller-scale irregularities are also observed. The detailed rotational data indicate that our Galaxy is no exception.

Construction of a hydrogen-cloud map

Heretofore we have considered only the locations of the fastest moving clouds, and they were assumed to lie at the tangent points. The location (along the line of sight) of *all* the clouds that appear as peaks in a given profile may be determined from the rotation curve. The measured radial velocity v_r of each peak is associated uniquely with a value of $\omega(r)$ through (53). The rotation curve then yields the galactic radius r if axisymmetric rotation is assumed. This locates the cloud along the line of sight but with a twofold ambiguity.

Clouds at points P and P' in Fig. 14a would both show the same v_r. Additional information can help determine which of two such locations is correct. For example, the angular thickness in galactic latitude of the more distant hydrogen cloud would be less than that of the closer one if their actual linear thicknesses were similar.

The locations of many such clouds make it possible to construct a map of neutral hydrogen such as that of Fig. 15. The map shows that the distribution of the gas in the Galaxy is quite complex; the axisymmetric model is clearly only an approximation of the true situation.

Summary

Let us summarize the steps taken to obtain a map of hydrogen gas in the Galaxy with the tangent-point method.

(i) In the spectral data along one line of sight, find the cloud with the maximum radial velocity $v_{r,max}$ and associate it with the tangent point.

(ii) Calculate the angular velocity $\omega(r_{tan})$ about the galactic center at the tangent point from $v_{r,max}$ with (53).

(iii) Repeat (i) and (ii) at many different galactic longitudes to obtain the rotation curve $\omega(r_{tan})$ versus r.

(iv) For each peak in the intensity profiles, use the measured v_r to find ω through (53). Then use the rotation curve from (iii) to deduce the distance to the emitting cloud, hence locating it in the Galaxy. Construct a map of cloud locations and luminosities in the Galaxy

(v) For points outside the solar radius, use independent distance measures and the measured line-of-sight velocities to determine values of $\omega(r)$.

Flat rotation curves and dark matter

We have seen that if the mass in a galaxy is all within the radius in question, $V(r)$ would decrease approximately as $r^{-1/2}$. Astronomers have yet to see the velocity begin to decrease

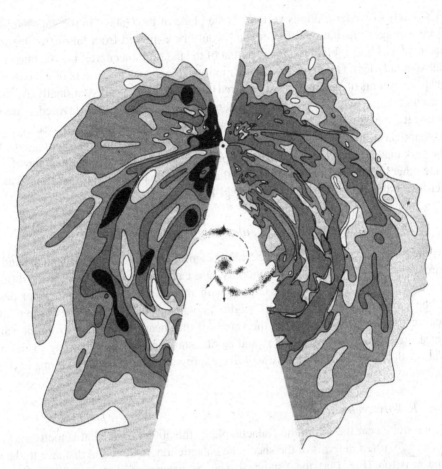

Figure 10.15. Distribution of neutral hydrogen in the galactic plane inferred from early 1420-MHz observations. The sun is the dot at the upper center and the galactic center is at the figure center. [D. Mihalas and J. Binney, *Galactic Astronomy*, Freeman, 1981, p. 528; outer parts from F. Kerr and E. Westerhout, *ibid.*, Plate II, p. 172; inner parts from M. Schmidt, *BAN* **13**, 247 (1957) and from G. Rougoor and J. Oort, *Proc. Nat. Acad. Sci. USA*, **46**, 1 (1960)]

in this manner in our Galaxy and many others. Studies of large-radii orbits ($>100\,000$ LY) of globular clusters also indicate large amounts of unseen matter in the Galaxy. As noted (Table 1), a total mass of the Galaxy of $\sim 10^{12}\,M_\odot$ can be deduced from the data. The actual number depends on the radius at which the speed starts decreasing, *if* it does.

One can tabulate the amount of known matter in a galaxy. Luminous matter is that which is directly visible. It could be hydrogen gas observed through its radio emission, neutron stars observable through radio or x-ray emission, normal stars seen in visible light, and so on. In addition, one must include objects of known types that may be obscured (e.g., stars behind the dust of the Milky Way) or are currently nonluminous (e.g., nonradiating or nonaccreting black holes, neutron stars, and white dwarfs). The estimates are legitimate statistical extrapolations from known phenomena.

The result is that matter of known types clearly falls short by a factor of as much as 10 or more of the mass inferred from galactic rotation studies. This is an important facet of the evidence for substantial amounts of nonluminous or dark matter in the Galaxy.

The study of stellar motions vertical to the plane of the Galaxy in the solar vicinity also reflects stronger gravitational forces than would be expected from luminous matter in the galactic disk. This problem is similar to that of the flat rotation curves, but the mass involved is substantially less. On larger scales, the motions of galaxies in clusters of galaxies indicate that large amounts of dark matter are required to bind the clusters gravitationally (Section 2.4). On a much larger scale, even greater amounts are required to satisfy favored cosmological models. It appears that dark matter pervades our universe on many distance scales.

Dynamical studies of disks show that instabilities would disrupt the rotating galactic disk if the dark matter needed to solve the rotation-curve problem resided in the plane of the disk. On the other hand, these analyses indicate that the dark matter could be distributed as a large spheroidal halo such as that sketched in Fig. 1.

Differential rotation in the solar neighborhood

Early measures of differential galactic rotation came from the study, in the optical band, of the apparent velocity of stars in the solar vicinity. One can quantitatively describe the differential rotation at any one location in the Galaxy with *shear* and *vorticity*. The former describes the slippage of material relative to its neighbors, whereas the latter describes the rotation of a local region. It is possible to find the values of the shear and vorticity from the values of the measured velocity components v_r and v_θ of a star near the sun according to an observer moving with the sun. To do this, we will find expressions for v_r and v_θ in terms of the shear and vorticity.

Relative velocities

For a star quite near the sun in the Galactic plane, the observable radial velocity v_r, given in (50), can be written in terms of the shear, the galactic longitude ℓ, and distance to the star d. Define, as before (Fig. 14a), the distance r to be the distance of a local star (Point P) from the galactic center. Require the star (Point P) to be close to the sun, $d \ll r_s$, and in the galactic plane, but it may lie in any azimuthal direction specified by ℓ.

The desired relation is obtained by expanding the angular velocity ω about its value ω_s at the position of the sun $r = r_s$:

$$\omega(r) = \omega_s + \left(\frac{d\omega}{dr}\right)_{r_s}(r - r_s) + \cdots. \tag{10.56}$$

Substitute this into v_r (50), evaluate the derivative for $\omega(r) = V(r)/r$, eliminate δ with (52), and prove from the geometry that $r - r_s \approx -d(\cos \ell)$ for $d \ll r_s$ to find (Prob. 46a) the desired relation – namely, v_r as a function of d, ℓ, V_s, r_s, and $V(r)$:

➡ $$v_r = Ad\sin 2\ell; \quad A = \frac{1}{2}\left[\frac{V_s}{r_s} - \left(\frac{dV}{dr}\right)_{r_s}\right], \quad \text{(Radial velocity of star in terms of}$$
$$\text{Oort constant } A; d \ll r_s) \tag{10.57}$$

where the parameters describing the rotation in the nonrotating frame of the galactic center have been collected into the coefficient A. They are the radial position and velocity of the sun, r_s and V_s, and the change of V with radius in the solar vicinity. The expression (57) gives the apparent radial velocity v_r (according to a solar observer) of a nearby star at distance d and galactic longitude ℓ if the Galaxy's local rotation is characterized by the parameter A.

Similarly, one can obtain an expression for v_θ, the star's velocity normal to the line of sight, again according to the observer traveling with the sun; thus, $v_\theta = V_{\text{star},\perp} - V_{\text{s},\perp}$, where $V_{\text{star},\perp}$ and $V_{\text{s},\perp}$ are the velocity components normal to the line of sight to the star. Show, from geometry similar to that of Fig. 14a, that $v_\theta = [\omega(r)(r_s \cos \ell - d)] - \omega_s r_s \cos \ell$. Then, as we did for v_r, expand ω about its value at the position $r = r_s$ (56) to find (Prob. 46b)

$$\mathrel{\rightarrow} \quad v_\theta = d(A \cos 2\ell + B); \quad B = -\frac{1}{2}\left[\frac{V_s}{r_s} + \left(\frac{dV}{dr}\right)_{r_s}\right], \quad \begin{array}{l}\text{(Azimuthal velocity;}\\ d \ll r_s) \end{array} \quad (10.58)$$

where the coefficient A is the same as in (57) and where B, similarly, collects rotation parameters. (Do not confuse B with the vector magnetic field $\mathbf{B}$ and its component B_z used elsewhere in this chapter.)

Keep in mind that v_r and v_θ are the velocities measured by the observer who travels with the sun. This frame of reference translates to the right in Fig. 14a with the 220-km/s velocity of the sun relative to the galactic center. In contrast, the symbols V_{star} and V_s are the speeds of the star and the sun in a nonrotating frame at rest with respect to the galactic center.

Oort constants

The constants A and B are known as the *Oort constants* after Jan Oort, who first formulated these expressions in 1927. Measurements of the radial velocity v_r (m/s) and the proper motion $\mu = v_\theta/d$ (radians per second) of a given nearby star, or a group of stars, directly provide, through (57) and (58), values of A and B if the distance d and galactic longitude ℓ are known. The proper motion μ is the angular motion on the celestial sphere relative to background galaxies (see AM, Chapter 4). (Do not confuse μ with the vector magnetic moment $\boldsymbol{\mu}$ and its z component μ_z used elsewhere in this chapter.)

The first term in the brackets of each constant in (57) and (58) is the angular velocity of the sun $\omega_s = V_s/r_s$ about the galactic center, and the second term is the rate at which the rotational velocity V changes with galactocentric radius at the radial position of the sun. These are the two physical parameters one wishes to determine. They can be extracted as follows:

$$\omega_s = \frac{V_s}{r_s} = A - B; \quad \left(\frac{dV}{dr}\right)_{r_s} = -(A + B). \quad \text{(Local galactic quantities)} \quad (10.59)$$

In practice, of course, one would measure numerous stars to average properly over fluctuations of stellar velocities from the general galactic rotation.

Shear and vorticity

The constants A and B describe, respectively, the local *shear* and *vorticity* of the medium. To gain some insight into these quantities, consider a limiting example such as a hypothetical galaxy that rotates as a rigid body with angular velocity ω_s in the clockwise direction; that is, $V = (V_s/r_s)r$. Take the derivatives in (57) and (58) to find that $A = 0$ and $B = -\omega_s$.

The former constant ($A = 0$) indicates that, locally, there is no shear (i.e., no slippage) between layers of the medium, which indeed is the case for the rigid galaxy. The latter constant ($B = -\omega_s$) tells us that the vorticity (i.e., the local angular velocity) of the material is of magnitude $\omega_s = V_s/r_s$. In inertial space, there is a vorticity in any local region of our rigid galaxy because each portion of the disk rotates with angular velocity ω_s. The vorticity is simply due to this rigid rotation. (The minus sign for B is a consequence of the chosen definitions.)

Substitute these values of A and B into the expressions (57) and (58) to find that the measured velocities of a nearby star would be $v_r = 0$ and $v_\theta = -\omega_s d$. A nearby star would remain at a fixed distance from the sun on a rigid galaxy and hence would exhibit no radial motion according to the solar observer. The same star, though, would appear to rotate through the distant galaxies at the angular velocity $-\omega_s$ of the GC–sun system, where GC is the galactic center. The direction of motion would be negative, which is clockwise in Fig. 14a; note that positive ℓ is defined counterclockwise and that $\omega_s = V_s/r_s$ is a positive quantity.

The expressions for v_r (57) and v_θ (58) indicate that, in the presence of differential rotation, the two velocities should cycle through two complete cycles of the sine (or cosine) function as galactic longitude ℓ is varied through 360° for fixed distance d. This double oscillation is clear in the radio data of Fig. 13a. It was this feature in early optical data from nearby stars that first established the existence of differential rotation.

Finally, one can demonstrate (Prob. 46c) that the Oort constants can be written in terms of angular velocity ω instead of linear velocities V,

$$A = -\left(\frac{r}{2}\frac{d\omega}{dr}\right)_{r_s} ; \quad B = -\left[\frac{1}{2r}\frac{d(r^2\omega)}{dr}\right]_{r_s}, \tag{10.60}$$

which may give one more insight into the meaning of shear A and vorticity B. Shear exists if ω varies with radius. Vorticity exists if $r^2\omega$ changes with radius, where $r^2\omega$ is the angular momentum (about the galactic center) per unit mass or the *specific angular momentum* (Prob. 48). In the rigid-body case, ω does not vary with radius but r and hence $r^2\omega$ do; thus, the shear is zero and the vorticity is finite, as we found.

Centers of galaxies

The study of our own galaxy's structure is difficult because we are embedded within it. Current rotation studies focus mostly on determining the rotation curves of other galaxies. This has become a major industry entailing studies of large samples of galaxies of each of the known types. A primary goal is the study of mass distributions and thereby the distribution of dark matter.

A major thrust of current research is the study of the cores of galaxies to obtain the mass distributions very close to the center. The extreme luminosities and rapid temporal x-ray variability of the cores of some galaxies – and in particular Seyfert galaxies and quasars – which are generically known as active galactic nuclei (AGN), have long suggested that they harbor massive black holes with masses of order 10^6 to $10^8 \, M_\odot$.

The advent of powerful telescopes in all bands with excellent angular and spectral resolution makes it possible to probe very near the central regions – especially of our own galactic center. Infrared astronomers observe broadened spectral lines at the center, indicating high orbiting velocities in its proximity which, in turn, indicates a very large mass (or masses) in this region.

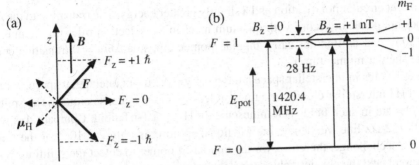

Figure 10.16. Zeeman absorption by a hydrogen atom in an external magnetic field B. The z-axis is taken to be in the direction of B. (a) The three allowed projections ($m_F = +1, 0, -1$) of the angular momentum for the $F = 1$ state. The atomic magnetic moment μ_H, which is directed opposite to the angular momentum, interacts with the magnetic field to perturb the energy state. (b) Energy levels showing (left) the 1420-MHz transition for $B = 0$ and (right) the Zeeman splitting of the upper energy level into the three sublevels for $B_z > 0$. The two allowed transitions to the upper states are shown; they differ in frequency by 28 Hz for $B_z = 1$ nT.

More dramatically, individual stars at the center of the Galaxy have been imaged over several years with sufficient resolution to track their orbits directly. These studies indicate an extreme mass density at the center, which most likely is a black hole of mass $\sim 4 \times 10^6\ M_\odot$ (Section 1.7 and Fig. 1.16).

10.5 Zeeman absorption at 1420 MHz

The physics of the hyperfine transition in hydrogen may be extended to describe the *Zeeman effect*, the splitting of atomic energy levels in the presence of an external magnetic field. The upper level of the hyperfine transition exhibits this splitting as a slight broadening of the 1420-MHz line when it is observed in absorption against a bright background source. This effect is used by radio astronomers to find the strengths of magnetic fields in interstellar clouds.

Zeeman effect

The upper energy level of the 1420-MHz hyperfine transition of hydrogen is the $F = 1$ state. Its angular momentum has three possible projections onto a given axis (Fig. 16a). Following (6), the three states are

$$F_z \Rightarrow m_F \hbar; \quad m_F = -1, 0, +1. \tag{10.61}$$

The quantum number m_F can take on any of the three indicated values. Normally the three states have the same energy (i.e., they are degenerate), but if the atoms are immersed in an external magnetic field, the three states take on slightly different energies (Fig. 16b). This is a consequence of the interaction of the magnetic moment of the hydrogen atom with the external field.

An absorption transition from $F = 0$ to $F = 1$ in the presence of an external magnetic field will excite an atom up to either the $m_F = -1$ or the $m_F = +1$ state. A transition to the $m_F = -1$ state requires that the absorbed photon be of left circular polarization to conserve angular momentum and also of the correct energy, whereas transition to the $m_F = +1$ state requires right circular polarization and a slightly greater energy. The transition to the $m_F = 0$ state does not occur in accord with quantum mechanical selection rules that require $\Delta F = \pm 1$ and $\Delta m_F = \pm 1$. (Photons carry angular momentum, so a $\Delta m_F = 0$ transition could not conserve angular momentum.)

Magnetic fields in interstellar space are typically ~ 0.3 nT in interstellar space, ~ 1 to 2 nT in typical H I interstellar clouds, and up to at least ~ 30 nT in dense clouds. The splitting of the $F = 1$ state in such fields is a minuscule 28 Hz in a 1-nT field. Compare this with the 1 420 405 752-Hz line frequency; the fractional splitting is only 2×10^{-8} of the transition energy. It is a challenge, therefore, for the radio astronomer to detect the splitting.

We present next the physics behind the Zeeman splitting and its quantitative relation to the magnetic field. We then discuss how a radio astronomer makes use of absorption data with polarizing antenna feeds to determine the values of magnetic fields in interstellar clouds.

Energetics

The magnetic moment of the hydrogen atom is dominated by the much larger moment of the electron (Fig. 8a,b). The atom thus has a net magnetic moment even if the proton and electron moments are opposed.

In the $F = 0$ state, the total angular momentum is zero. Stated otherwise, the atom has no preferred orientation. The *measurable* magnetic moment is thus zero. Think of the system rotating randomly about all axes so that the moment averages to zero. There is therefore no measurable interaction of the atom in this state with an external magnetic field.

In the $F = 1$ state, the magnetic moment, being directed opposite to the angular momentum, is also constrained to three states (Fig. 16a). When $m_F = +1$, the angular momentum $\boldsymbol{F}$ of the atom is roughly in the direction of the magnetic field $\boldsymbol{B}$, and the magnetic moment $\boldsymbol{\mu}$ is opposed to it. Consequently, this is the state with the highest potential energy according to the interaction energy $E_{\text{pot}} = -\boldsymbol{\mu} \cdot \boldsymbol{B}$ (16). Conversely, the $m_F = -1$ state has the lowest potential energy.

The magnitude of the splitting in energy units follows directly from our development in Section 3. Here, the magnetic field is the external field and not the field of the proton as before, and the magnetic moment is the magnetic moment $\boldsymbol{\mu}_H$ of the entire atom in the $F = 1$ state rather than that of the electron $\boldsymbol{\mu}_e$. The interaction energy is thus, from (16),

$$E = -\boldsymbol{\mu}_H \cdot \boldsymbol{B} = -\mu_{H,z} B_z, \qquad \text{(Potential energy)} \qquad (10.62)$$

where B_z has the magnitude of $\boldsymbol{B}$ because the z-axis is defined to be in the direction of $\boldsymbol{B}$.

Because $\boldsymbol{\mu}_H$ arises almost totally from the electron, we take $\mu_{H,z} \approx \mu_{e,z}$. The eigenvalue of $\mu_{e,z}$, in terms the electron g factor, the Bohr magneton, and the electron spin quantum number m_S is, from (13) and (7),

$$\mu_{e,z} \Rightarrow g_e \mu_B m_S. \qquad (10.63)$$

This yields an interaction energy of, from (62),

$$E = -g_e \mu_B m_S B_z, \qquad \text{(Interaction energy)} \qquad (10.64)$$

where the allowed values of m_S are $+1/2$ and $-1/2$. We associate these two values with the $m_F = +1$ and -1 states, respectively. This is consistent with our imperfect association of parallel S and I vectors with the $F = 1$ state; see Fig. 8b and discussion thereafter. In this view, when S is aligned with (or opposed to) the external magnetic field direction, F must be also; see Fig. 6c. Because $g_e \approx -2.0$ and $m_S = \pm 1/2$, the interaction energies (64) for the two m_F states are

$$E \approx \mu_B B_z; \qquad m_F = +1$$
$$E \approx -\mu_B B_z; \qquad m_F = -1. \tag{10.65}$$

The quantity measured by radio astronomers is the separation of the $+1$ and -1 levels. The magnitude of the effect is therefore approximately

$$\Delta E = E(m_F = +1) - E(m_F = -1). \tag{10.66}$$

From (65),

$$\Rightarrow \qquad \Delta E \approx 2\mu_B B_z = \frac{e\hbar}{m_e} B_z, \qquad \text{(Zeeman splitting; J)} \tag{10.67}$$

where B_z is the magnitude of the external magnetic field and the Bohr magneton μ_B is taken from (12).

Angular momentum and polarization

A crucial feature of the absorption transition from the $F = 0$ ground state to the excited $F = 1$ state is the change of the atom's angular momentum. The atom gains $1\hbar$ in going from $F = 0$ ($m_F = 0$) to the $m_F = +1$ (sub)state of $F = 1$ and loses $1\hbar$ unit in going from $F = 0$ to the $m_F = -1$ (sub)state of $F = 1$. The absorbed photon must (and does) provide the required angular momentum.

Photons carry one unit of angular momentum J, which is realized by their circular polarization (see Section 11.4 for a description of circular polarization). This angular momentum can be projected with or against the direction of travel ($J_z \Rightarrow +1\hbar$ or $J_z \Rightarrow -1\hbar$, respectively). In the former case, the polarization is described as *right circular polarization* (RCP). In the latter, it is *left circular polarization* (LCP). This is the convention of the Institute of Radio Engineers (IRE). The "classical" definition is the opposite. Figure 11.9 illustrates the electric field variation of RCP and LCP (IRE convention).

A continuum source, an interstellar cloud, and observers are illustrated in Fig. 17. The magnetic field in the cloud is directed toward the observer in the $+z$ direction. An incoming photon with angular momentum directed to the right (RCP) with exactly the energy $h\nu_R$ can raise an atom from the $F = 0$ ($m_F = 0$) to the $F = 1$ ($m_F = +1$) state. The atom will then have its angular momentum directed toward the observer.

In such absorption events, the photons are lost and the observer would note a deficit of RCP photons at ν_R. Observations with an RCP antenna feed over a spectral band that includes ν_R would therefore exhibit an absorption line in the background continuum source at frequency ν_R.

Similarly, observations with an LCP feed will yield an absorption line at the lower frequency ν_L (Fig. 17). It is an important aid in the detection of this effect that the two absorption lines can be detected independently with different antenna feeds.

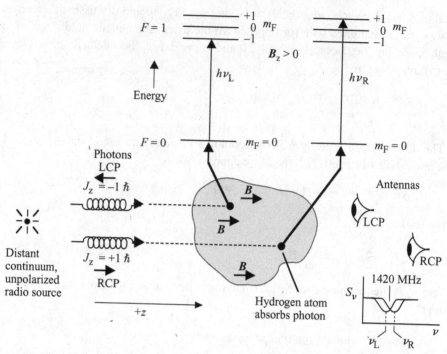

Figure 10.17. Role of polarization and angular momentum in absorption of 1420-MHz photons by hydrogen atoms in a cloud. Photons with the correct angular momentum, $J_z \Rightarrow -1\hbar$ (left circularly polarized, LCP), and correct energy $h\nu_L$ are absorbed at frequency ν_L by atoms in the ground state. Right circularly polarized (RCP) photons are absorbed at a slightly higher frequency. The observer detects absorption lines with LCP and RCP antenna feeds at the two slightly different frequencies.

Frequency difference

The difference of these two frequencies follows directly from the energy difference, $h\nu_R$ and $h\nu_L$, expressed as

$$\Delta E = h\nu_R - h\nu_L. \tag{10.68}$$

Equate (68) to (67) to yield for $B_z = 1\,\text{nT}$

$$\nu_R - \nu_L = \frac{eB_z}{2\pi m_e} \underset{B_z=1\,\text{nT}}{\rightarrow} 27.95\ \text{Hz}, \tag{10.69}$$

where B_z is the parallel (line-of-sight) component of $\mathbf{B}$. We thus can write

$$\blacktriangleright \qquad \nu_R - \nu_L = \frac{eB_z}{2\pi m_e} = 27.95 \left(\frac{B_z}{1.0\,\text{nT}} \right)\ \text{Hz}, \qquad \text{(Zeeman splitting)} \tag{10.70}$$

where e and m_e are the charge and mass of the electron, respectively, and where B_z is taken to be positive in the direction of photon travel (i.e., toward the observer). The astronomical convention is the reverse of this; the positive direction of the line-of-sight component B_{los} of $\mathbf{B}$ is in the direction away from the observer. Thus, $B_{los} = -B_z$.

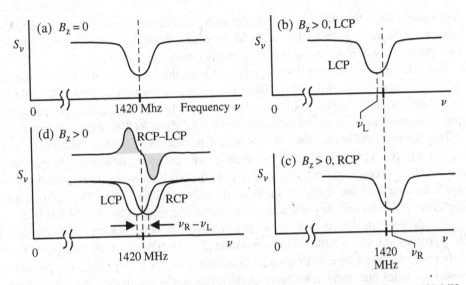

Figure 10.18. Hypothetical absorption-line spectra illustrating Zeeman splitting at 1420 MHz. (a) Absorption at 1420 MHz, for either LCP or RCP radiation, when there is no magnetic field B_z parallel to the propagation direction in the absorbing cloud. (b,c) Absorption line in presence of a measurable B_z with LCP and RCP antenna feeds. The magnetic field is taken to be in the propagation direction. (d) Difference of the RCP and LCP signals (shaded). Here, the separation of the LCP and RCP lines is greatly exaggerated. The difference signal is most pronounced in the wings (outer edges) of the line.

A magnetic field at right angles to the line of sight will give rise to a triplet of *linearly* polarized lines. In this case, the central line is polarized parallel to the magnetic field and is unshifted in frequency. The outer two lines are polarized perpendicular to the field and are shifted left and right by the same amounts as ν_R and ν_L for a radial B field. Because both of the latter have the same polarization, it is effectively impossible to distinguish them for typical astrophysical conditions (Prob. 53).

The frequency difference (69) is easy to recall in that it equals the *cyclotron frequency* qB/m (8.3) for the z component of B if $q = e$. (The 2π in the denominator converts from radians/s to cycles/s.) This is the frequency with which a particle of charge e and mass m rotates around a tube of magnetic field lines; it is independent of the particle energy for nonrelativistic particles.

Detection of Zeeman splitting

To detect this "splitting" of the 1420-MHz line, a radio astronomer would (*i*) install a circular polarization feed (e.g., for LCP radiation) at the focus of the antenna, (*ii*) adjust the receiver to about 1420 MHz, (*iii*) point the antenna to a distant unpolarized radio source with a continuum spectrum, and (*iv*) measure the intensity at all frequencies in the region of 1420 MHz. The spectrum (flux versus frequency) thus measured should show an absorption line at 1420 MHz owing to the intervening hydrogen atoms (e.g., in a cloud).

If the magnetic field in the cloud is too small to yield a detectable effect ($B_z \approx 0$), the absorption-line frequency would be at the rest frequency ν_0 of the 1420-MHz line (Fig. 18a).

(We assume here, that there is no Doppler shift from cloud or observer motion.) If, on the other hand, there is a detectable net field directed toward the observer ($B_z > 0$) the LCP absorption line would appear at the lowered frequency $\nu_L < \nu_0$ (Fig. 18b).

The entire process would then be repeated with the RCP feed installed in the antenna and the value of ν_R recorded (Fig. 18c). The values of the two frequencies would then be substituted into (70), which would be solved for B_z. Modern radio telescopes permit the frequent electronic switching back and forth between LCP and RCP to reduce systematic effects.

The frequency difference for a typical interstellar magnetic field is a minuscule 28 Hz for $B_z = 1$ nT (70). This is very much less than the instrumental line width. The measurements are therefore exceedingly difficult in practice. If both lines were detected together with a single antenna feed, the spectrum would be the sum of the two lines. The splitting would be apparent only as an infinitesimal and undetectable widening of the 1420-MHz line.

With the separate detection of the two polarizations, however, one can subtract the LCP and RCP profiles to find a unique difference profile (Fig. 18d) from which the desired frequency difference can be obtained (Prob. 52). Sensitive searches for splitting are possible if the observer carries out many scans in frequency, repeatedly alternating the two polarizations, and follows this with a careful subtraction of the profiles of the two spectral lines.

Cloud magnetic fields

Zeeman measurements can yield the magnetic field in a given cloud along the line of sight. A 1420-MHz absorption feature due to a given cloud will exhibit the same Doppler shift as emission from the same cloud. As in emission profiles (Fig. 13a), the shifts in absorption frequencies from cloud to cloud can exceed the line widths and hence can be distinguished from one another (Probs. 51 and 54). Measurement of the Zeeman frequency difference for each such absorption line can yield a value of B_z for each of the associated clouds.

Recall that the absorption lines are measured in the direction of an intense background continuum source. Emission lines from an intervening cloud such as those used in the Galaxy-rotation studies (Section 4) would not be detectable along this exact line of sight but would be detected in a direction displaced from it by a few beamwidths. See AM, Chapter 11 for a discussion of spectral line formation.

Detection of Zeeman splitting is most successful when the absorption line is narrow and deep because these characteristics optimize the detection of the asymmetry in the subtracted RCP and LCP profiles. Such lines preferentially come from cold H I clouds rather than the intercloud medium. In contrast, Faraday rotation (Chapter 11) samples the magnetic fields in the ionized warmer regions of the interstellar medium.

Positive detections of Zeeman splitting are now being obtained. A cloud in front of the Crab nebula (which provides the continuum radiation) yields a magnetic field of 0.35 nT. Clumps in front of another bright supernova remnant, Cas A, yield values ranging from 2 to 4 nT. A measurement of the field in an interstellar shock (the Eridanus shell) yields 0.7 nT. In atypical clouds near the star formation region of Orion, larger values of 5 and 7 nT have been measured.

High angular resolution studies in the directions of star-forming regions W3, M17, and W49A have yielded values up to 30 nT. Figure 19 shows the results from the direction of the nebula W49A with field strengths reaching beyond 15 nT. Strong continuum emission from regions of ionized hydrogen (H II regions) in W49A provide the flux for the absorption

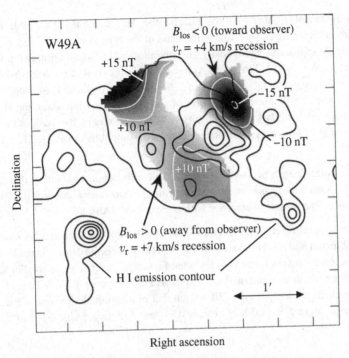

Figure 10.19. Radio measures of magnetic fields in the star formation region W49A. The black contours signify continuum emission at 1420 MHz with an angular resolution of 10″. The two gray-scale areas are regions where Zeeman splitting in absorption is detected with angular resolution of 40″. The darker the region, the greater the splitting and thus the stronger the line-of-sight magnetic field B_{los}. The labeled white contours give the strength of B_{los}; positive values indicate B_{los} is directed away from the observer, which is the astronomical convention. The magnetic field points in opposite directions in the two regions. The Doppler shift of the 1420-MHz absorption line indicates the two absorbing regions are receding at different velocities; one is at +4 km/s and the other at +7 km/s. Fields in excess of 15 nT are detected. With improved angular resolution, higher fields up to 30 nT are detected in localized regions. [C. Brogan and T. Troland, *ApJ* **550**, 799 (2001)]

measures. The detected Zeeman splitting could occur anywhere along the line of sight to W49A, but it is attributed to the locale of W49A.

Galactic magnetic fields are also studied extensively with other methods. Synchrotron radiation (Chapter 8) at radio, optical, and x-ray wavelengths reveals high-energy electrons in the presence of magnetic fields. Radio waves from polarized background sources passing through ionized plasmas with magnetic fields experience detectable Faraday rotation (Chapter 11). Finally, polarization of optical light from background stars is due to the alignment of interstellar grains by galactic magnetic fields (AM, Chapter 10).

Problems

10.2 The Galaxy

Problem 10.21. (a) What is the approximate absolute visual magnitude of the Galaxy, $M_{V,G}$? Neglect the effect of interstellar absorption. Use the Galaxy V–band total luminosity given in Table 1, $L_V = 1.4 \times 10^{10} \, L_\odot$, and let $L_\odot \approx L_{V,\odot}$. For the sun, $M_{V,\odot} = 4.83$. (b) Suppose that the mass-to-(visual)-light ratio of the galactic disk $M/L_V = 5M_\odot/L_\odot$ (Table 1) is due solely to luminous stars. Roughly estimate the mass of the typical star in the disk in units of solar mass. Use the approximate luminosity–mass relation for hydrogen-burning ("pp chain") main-sequence stars, $L \propto M^3$ (Table 4.3). Assume $L \approx L_V$. (c) If all stars in the disk had this mass, how many stars would the disk contain? Use disk mass in Table 1. (d) Suppose now, more realistically, that only about one-quarter of the mass is in visible stars, the rest being in dust, compact stars, and dark matter. In this case, what would be the mass of the typical star and the number of stars in the disk under the same assumptions? [Ans. $\sim$–21; $\sim1/2\,M_\odot$; $\sim10^{11}$; $\sim10^{11}$]

Problem 10.22. Calculate the gas pressure for each of the components of the interstellar medium listed in Table 2. Assume the gas is solely hydrogen and properly take account of the fractional ionization. How do the pressures compare with one another? [Ans. all $\sim10^{-13}$ N/m^2]

Problem 10.23. An approximately spherical molecular cloud of 100-LY diameter contains a uniform density of matter and has a total mass of $10^6\,M_\odot$. (a) What is the density in nucleons/m^3? (Protons and neutrons are nucleons.) (b) By what factors is this greater or smaller than the number densities of the intercloud medium and the H I clouds given in Table 2? (c) By what factor does the cloud nucleon density differ from that of a manmade laboratory vacuum of 10^{-6} torr at temperature $T = 300$ K? (1 torr $= 133$ pascal (N/m^2). [Ans. $\sim10^9$ m^{-3}; $\sim10^4$, $\sim10^2$; $\sim10^{-9}$]

10.3   *Hyperfine transition at 1420 MHz*

Problem 10.31. (a) Show that the units of magnetic moment are C m^2 s^{-1}. (b) Demonstrate that this is equivalent to J/T (joules per tesla). Hint: use the magnetic force law, $F = i(d\boldsymbol{l} \times \boldsymbol{B})$, to relate magnetic field to current.

Problem 10.32. (a) Consider a spinning sphere uniformly charged throughout its volume with total charge q. Argue that its magnetic moment is given by (10) if the mass m is also distributed uniformly throughout the volume. (b) Find simple (even if artificial) arrangements of charge and mass that would give ratios of magnetic moment to angular momentum that are twice, and 5.6 times, those given by (10), like the electron and proton, respectively.

Problem 10.33. Use the Biot–Savart law quoted here,

$$d\boldsymbol{B} = \frac{\mu_0 i}{4\pi} \frac{d\boldsymbol{l} \times \boldsymbol{r}}{r^3},$$

to estimate the field inside a proton of radius R_p and magnetic moment μ_p. Assume the magnetic moment is due to a single current loop of radius R_p. Calculate the field in the center of the loop to obtain the expression (22).

Problem 10.34. Calculate the probability that the electron is inside the proton as in (21) but retain the first-order term (linear in r) of the exponential. Does this serve to bring the frequency of the hyperfine line result obtained by us to agree with the actual value of 1420.4 MHz? Adopt $R_p \approx 1.4 \times 10^{-15}$ m. [Ans. no]

10.4   *Rotation of the Galaxy*

Problem 10.41. Given the velocity of the sun in its rotation about the center of the Galaxy, $\sim$220 km/s, and the distance to the galactic center, 25 000 LY, find (a) the period of rotation and (b) a rough estimate of the mass of the Galaxy interior to the sun's orbit. Assume the sun is in a circular Kepler orbit about a spherical mass distribution. (c) Calculate the speed of the earth in its orbit about the sun and compare this with the speed of the sun in its orbit. Assume a circular orbit of radius $r_E = 1$ AU $= 1.496 \times 10^{11}$ m. [Ans. $\sim 10^8$ yr; $\sim 10^{11}$ $M_\odot$; ~ 30 km/s]

Problem 10.42. Consider a spherical mass distribution (a galaxy) of radius R with mass density $\rho(r) = \rho_0(r_0/r)$ out to radius R and $\rho = 0$ at $r > R$. (a) Find the included mass function $\mathfrak{M}(r)$ and the total mass $M = \mathfrak{M}(R)$ of the galaxy with the latter in terms of r_0, ρ, and R. (b) For this galaxy, find the expression for the radial force $F(r)$ and the velocity $\mathcal{V}(r)$ for a test particle (star) of mass m in a circular orbit with $r < R$. Find the expressions for $F(r)$ and $\mathcal{V}(r)$ outside the distribution. Make qualitative sketches of $F(r)$ and $\mathcal{V}(r)$. (c) Repeat (a) for a different spherical distribution – namely constant density $\rho = \rho_0$ out to radius R and $\rho = 0$ thereafter. (d) Repeat (b) for this constant-density distribution. Compare your results with the rotation curves in Fig. 11. [Ans. $M \propto \rho_0 r_0 R^2$; $\mathcal{V}(< R) \propto r^{1/2}$, $\mathcal{V}(> R) \propto r^{-1/2}$; $M \propto \rho_0 R^3$; –]

Problem 10.43. Consider a spherical mass distribution (a galaxy) of radius R with mass density $\rho(r) = \rho_0[1 - (r/R)]$ for $r < R$ and $\rho(r) = 0$ for $r > R$. (a) What are the included mass and total mass of the galaxy? (b) Find expressions for the force $F(r)$ and the circular velocity $\mathcal{V}(r)$ for a test particle (star) of mass m within the mass distribution ($r < R$) at $r = R$ and at $r > R$. (c) Qualitatively sketch the functions $F(r)$ and $\mathcal{V}(r)^2$. It is helpful to solve for the positions of zero slope and to examine the continuity of values and slopes at $r = R$. [Ans. $M \propto \rho_0 R^3$; $\mathcal{V}^2 = \pi G\rho_0 R^2/3$ at $r = R$; –]

Problem 10.44. (a) What is the approximate specific intensity I (W m^{-2} Hz^{-1} sr^{-1}) of the higher peaks in Fig. 13a? Assume the baseline is at $T = 0$ K. Refer to Chapter 6 if necessary. (b) Confirm that the relation between Doppler velocity and frequency shift on the dual horizontal axes is correctly given in the figure. [Ans. $I \approx 10^{-19}$ W m^{-2} Hz^{-1} sr^{-1}; –]

Problem 10.45. (a) Suppose that our Galaxy rotates as a rigid body. Demonstrate how you would use the tangent-point method to determine this fact. (b) For a nonrigid galaxy, show analytically how the slope of the $\mathcal{V}$ versus r curve at some galactic longitude ℓ depends on the adopted value for the speed of the sun $\mathcal{V}_s$ about the galactic center and on the distance r_s of the sun from the galactic center. Hint: take the derivative of (55) with respect to r_{tan} and express the result in terms of $v_{r,max}$, $\mathcal{V}_s(= \omega_s r_s)$, $r_s(= r_{tan}/\sin \ell)$, and ℓ. (c) If the rotation curve $\mathcal{V}$ versus r is essentially flat at some ℓ, what is the net effect on the slope of $\mathcal{V}(r)$ of changing from the 1964 IAU values of $\mathcal{V}_s = 250$ km/s, $r_s = 10$ kpc to the 1985 values $\mathcal{V}_s = 220$ km/s, $r_s = 8.5$ kpc?

$$[\text{Ans.}-; \quad \frac{d\mathcal{V}}{dr_{tan}} = \frac{1}{r_s}\left[\frac{dv_{r,max}}{d(\sin \ell)} + \mathcal{V}_s\right]; \sim 1 \text{ km s}^{-1} \text{ kLY}^{-1}]$$

Problem 10.46. Follow the suggestions in the text to obtain the expressions for (a) the radial velocity (57) and (b) the azimuthal velocity (58) and hence the expressions for the Oort constants A and B. (c) Demonstrate that the Oort constants in terms of ω, as given in (60), yield the forms in terms of linear velocities (57) and (58) if the relation $\omega = \mathcal{V}/r$ is applied.

Problem 10.47. (a) Find the two Oort constants in terms of ω_s for a star at distance $d \ll r_s$ from the sun if the rotation curve is flat; that is, $\mathcal{V}(r) = \mathcal{V}_s$. (b) For this case, what are the relative velocity components of a star or cloud, v_r and v_θ, in terms of ω_s, d, and ℓ? (c) For this flat-rotation case, find, from elementary geometrical considerations, the approximate time it would take a star cluster of diameter 20 LY at the solar radius (25 kLY) to be disrupted by the shear; $\mathcal{V}_s = 220$ km s^{-1}. [Ans. $A = -B$; $v_r = (\omega_s d \sin 2\ell)/2$; few $\times 10^7$ yr]

Problem 10.48. (a) What rotation function $\mathcal{V}(r)$ for the Galaxy would lead to zero vorticity ($B = 0$) near the sun given a finite solar velocity $\mathcal{V}_s$? Suggestion: try a power law, $\mathcal{V}(r) = \mathcal{V}_s(r_s/r)^n$, and solve for n. (b) Describe in simple physical terms why this variation is considered to have "zero vorticity" in two ways. First, consider (60). Second, consider the motion of a star at $\ell = 0$ relative to the sun (i.e., evaluate v_r and v_θ). [Ans. $n \approx 1$; –]

Problem 10.49. (a) How long a period must elapse between photographs of the face on spiral galaxy M101 (Fig. 2b) for the rotation of the stars about the center to become evident (e.g., through astrometric measurements)? M101 is about 150,000 LY in diameter and 17.6 MLY distant. Assume that the outer stars rotate with a period of about 1×10^8 years and that a $0.01''$ motion of a star would be detectable. (b) Proper motions of $\sim 0.02''$/yr were erroneously reported in M101 in about 1920. At the quoted distance of 17.6 MLY, how fast would the stars be moving compared with the speed of light? What might this have suggested to astronomers about the distance to M101? [Ans. ~ 200 yr; $\sim 2c$]

10.5   Zeeman absorption at 1420 MHz

Problem 10.51. (a) How big is the Zeeman frequency difference for (*i*) a typical interstellar field, $B \approx 0.3$ nT, directed toward the observer; (*ii*) the field in an interstellar cloud, 2 nT; and (*iii*) the earth's magnetic field, $B \sim 50$ μT? For each case, give the ratio $\Delta\nu/\nu$ of the frequency difference to the basic frequency (1420 MHz). (b) What is the line-of-sight velocity that corresponds to the frequency shift for $B = 2$ nT? How does this compare with the ~ 10 km/s features seen in line-of-sight-velocities of interstellar clouds at 1420 MHz (e.g., in Fig. 13a)? (c) What would you change in Fig. 17 if the magnetic field in the cloud were directed to the left rather than to the right? Explain your answer. [Ans. (*i*) ratio $\sim 10^{-8}$; ~ 10 m/s; –]

Problem 10.52. Let the line profiles for LCP and RCP in a 1420-MHz absorption observation be rectangular, of width W, and of depth D. A magnetic field separates the lines by $\Delta\nu \ll W$, where $\nu_R > \nu_L$. (a) Sketch the appearance of the difference of the LCP and RCP line profiles (as in Fig. 18d) and label the dimensions in terms of given parameters. (b) Given such a result, an observer measures the (positive) area A_+ under the positive part of the difference curve and the (negative) area A_- under the negative part. What is the desired frequency shift, $\Delta\nu = \nu_R - \nu_L$, in terms of A_+, A_-, D, and W? [Ans. –; $(A_+ - A_-)/2D$]

Problem 10.53. Consider the polarization characteristics of spectral lines exhibiting Zeeman splitting when observed *perpendicularly* to the interstellar magnetic field. The polarization characteristics for this case are described in the text after (70). Again, astronomers can measure each polarization separately. (a) Make qualitative sketches with quasi-Gaussian profiles showing your expectation of the appearance of the profiles anticipated from each polarization measurement (feed normal and perpendicular to B) for both strong and weak magnetic fields. For the weak field case, sketch your estimation of the appearance of the difference profile of the two polarizations. (It is helpful to use one-half the measured profile amplitude for the perpendicular polarization.) Discuss the relative difficulties of determining the B field in each case. (b) Repeat part (a) (weak field case only) for the rectangular line profiles of Prob. 52. Make careful expanded plots of the measured profiles and also a difference plot. Compare with the B parallel case (Prob. 52). Discuss the relative difficulties of measuring B fields parallel and normal to the line of sight. You may need to go back and change your sketches in (a).

Problem 10.54. (a) Argue that an *absorption* line will exhibit a shift in frequency that reflects the line-of-sight velocity of the absorbing atoms rather than that of the background continuum source. (b) Clarify with a sketch of an idealized spectral profile the argument that Zeeman splitting in principle can, under proper conditions, yield the magnetic fields for each distinct cloud along the line of sight. What is the required condition?

11

Dispersion and Faraday rotation

What we learn in this chapter

Radio waves traversing the **interstellar medium** (ISM) reveal a great deal about the medium because the waves are modified by **plasmas** and **magnetic fields** during their transit through the Galaxy. **Dispersion** is the variation with frequency of the **group velocity** of a radiation pulse. The measured spread of arrival times is directly related to the **dispersion measure** (DM), which is the integral of the electron density n_e along the line of sight.

Faraday rotation is the rotational displacement of the electric vector of **linearly polarized (LP) radiation** about the propagation direction. The measured quantity, the **rotation measure** (RM), is the integral along the propagation path of the product $n_e B_z$, where B_z is the component of the galactic magnetic field $\boldsymbol{B}$ parallel to the propagation direction.

The relations between the interstellar quantities and these two phenomena follow from **Maxwell's equations** as they apply to a dilute plasma. Their solution leads to a frequency-dependent **phase velocity** in terms of the **dielectric constant** of the medium. The square root of the latter is the frequency-dependent **index of refraction**. This in turn leads to the **dispersion relation** – a relation between the index of refraction, frequency, and wavelength. **Faraday rotation** follows from the determination of the indices of refraction for left and right **circularly polarized radiation**, which are slightly different in the presence of a magnetic field with a line-of-sight component B_z.

For each of these cases, the required dielectric constant is obtained by considering, with the aid of **Newton's second law**, the effect of an oscillating electric field on a free electron. The derived electron displacements yield the **vector polarization P** of the medium. This in turn is directly related to the dielectric constant and index of refraction.

The radiation from **radio pulsars** and **polarized extragalactic radio sources** serves to probe the ISM through these processes. Measurements of dispersion and polarization rotation yield locations of ionized plasmas and magnetic fields in the Galaxy. These results complement those of the neutral hydrogen studies discussed in Chapter 10.

11.1 Introduction*

The electrons and magnetic fields of the ionized interstellar medium (ISM; Section 10.2) serve to alter the character of radio waves propagating through it. The effect of the electrons is to slow the propagation speed of the low-frequency components of a signal relative to that of the higher frequencies. This phenomenon is known as *dispersion*.

The phenomenon is observed through the detection of pulses from radio pulsars. Dispersion depends on the integrated density of free electrons along the line of sight over the distance the electromagnetic (EM) wave travels. The degree of dispersion is thus a measure of the column density of free electrons along the line of sight to the pulsar emitting the pulses.

Magnetic fields in the ISM together with the free electrons of a plasma lead to *Faraday rotation* of electromagnetic waves. A linearly polarized (LP) EM wave traveling through such a medium will experience a gradual rotation of the plane of polarization. The amount of this rotation provides a measure, summed along the line of sight, of the product of the line-of-sight component of the interstellar magnetic field B_z and the free-electron density n_e. The direction of rotation provides the sign of the average, $\langle n_e B_z \rangle_{av}$.

Radiation from supernova remnants and radio pulsars is typically polarized. Such a source in the Galaxy is thus able to serve as a probe of the magnetic field in the ISM given knowledge of the electron density along the line of sight. Radiation from pulsars is both pulsed and polarized, and so they provide measurements of both dispersion and Faraday rotation. Together these two quantities yield quantitative measures of the n_e and B_z averaged along the line of sight to the pulsar.

As a reference, a typical ion density in the ISM can be taken to be that of the warm neutral medium (WNM; Table 10.2), namely, $\sim 4 \times 10^5$ m^{-3}. As illustrated in Section 10.2, however, the density variations from location to location in the Galaxy can be huge (many factors of 10). A typical magnetic field may be taken to be ~ 0.3 nT, but again the values and the directions of the fields vary greatly from place to place.

In this chapter, the quantitative relations between the physical quantities (n_e and B_z) and the measured frequency dispersion and polarization rotation angle are derived. In essence, the derivations consist of the calculation of indices of refraction of an EM wave in a plasma with and without a magnetic field. We begin by reviewing the concept of index of refraction, which is approached through Maxwell's equations.

11.2 Maxwell's equations and the index of refraction

The derivation of the index of refraction from Maxwell's equations and its relation to the "polarization" of a plasma are presented. The goal is to give the reader a physical feeling for some of the theoretical underpinnings of dispersion and Faraday rotation.

---

* In this chapter, be careful to distinguish the following quantities: (*i*) the polarization vector P of a medium and the polarization of an electromagnetic (EM) wave indicated by its type (LP, RCP, and LCP for linear, right circular, and left circular polarizations, respectively), (*ii*) the vector dipole moment μ and the scalar magnetic permeabilities μ and μ_0, and (*iii*) the oscillating magnetic field B (Times font; component B_z) of an EM wave and the static magnetic field B (Helvetica; component B_z) of the ISM. Other abbreviations used are DM and RM for dispersion and rotation measures, respectively. We use η for the index of refraction rather than n to avoid confusion with particle and photon densities.

Table 11.1. *Maxwell's equations*

Vector form in medium[a]	Component form in vacuum[b]	Name
$\boldsymbol{\nabla} \cdot (\kappa_e \boldsymbol{E}) = \dfrac{\rho}{\varepsilon_0}$	$\dfrac{\partial E_x}{\partial x} = 0$	Gauss's law
$\boldsymbol{\nabla} \times \boldsymbol{E} = -\dfrac{\partial \boldsymbol{B}}{\partial t}$	$\dfrac{\partial E_x}{\partial z} = -\dfrac{\partial B_y}{\partial t}$	Faraday's law
$\boldsymbol{\nabla} \times \left(\dfrac{\boldsymbol{B}}{\kappa_m}\right) = \mu_0 \varepsilon_0 \kappa_e \dfrac{\partial \boldsymbol{E}}{\partial t} + \mu_0 \sigma \boldsymbol{E}$	$\dfrac{\partial B_y}{\partial z} = -\mu_0 \varepsilon_0 \dfrac{\partial E_x}{\partial t}$	Ampere's law (modified)
$\boldsymbol{\nabla} \cdot \boldsymbol{B} = 0$	$\dfrac{\partial B_y}{\partial y} = 0$	Gauss's law for magnetic field

[a] For a linear isotropic medium characterized by ρ, σ, κ_m, and κ_e.
[b] For $\rho = \sigma = 0$, $\kappa_m = \kappa_e = 1$, $E_y = E_z = B_x = B_z = 0$.

The equations

The differential EM equations of Maxwell are given in Table 1 for certain restricted conditions – namely, that the medium be isotropic and linear in its response to electric and magnetic disturbances. The equations describe the behavior of the electric $\boldsymbol{E}(x, y, z, t)$ (V/m) and magnetic $\boldsymbol{B}(x, y, z, t)$ (T) vectors that constitute the traveling waves we know as EM radiation.

The vector equations show that the values of $\boldsymbol{E}$ and $\boldsymbol{B}$ at some position x, y, z and time t depend on each other and on several parameters that describe the medium through which the wave travels. These include the local *charge density* ρ (C/m^3), the *conductivity* σ (ohm^{-1} m^{-1} = A^2 s^3 m^{-3} kg^{-1}), and the susceptibility of the medium to magnetic and electric fields through the (magnetic) *relative permeability* κ_m and the (electric) *dielectric constant* κ_e. The latter two are dimensionless quantities defined, respectively, as the *permeability* μ and *permittivity* ε of the medium referenced to the vacuum values μ_0 and ε_0:

$$\kappa_m \equiv \frac{\mu}{\mu_0}; \quad \kappa_e \equiv \frac{\varepsilon}{\varepsilon_0}; \quad \boldsymbol{J} = \sigma \boldsymbol{E}. \tag{11.1}$$

The conductivity σ relates the current density $\boldsymbol{J}$ (A/m^2) to the applied electric field $\boldsymbol{E}$ (V/m). The expression $\boldsymbol{J} = \sigma \boldsymbol{E}$ is a restatement of Ohm's law but at a point in the medium rather than for a macroscopic sample.

The elegant equations of Table 1 are obtained from tabletop experiments. The goal, in general, is to solve the equations for the position and time variation of $\boldsymbol{E}$ and $\boldsymbol{B}$ for given situations. The last column of Table 1 gives the "laws" the equations represent.

Vacuum solution

It is possible to demonstrate that a solution of Maxwell's equations in a vacuum is a plane-parallel wave traveling in some direction with $\boldsymbol{E}$ and $\boldsymbol{B}$ both normal to this direction and also normal to each other and with their amplitudes having a constant ratio at all locations. In addition the cross product of the $\boldsymbol{E}$ and $\boldsymbol{B}$ yields the propagation direction; hence, if we specify the propagation to be in the $+z$ direction and $\boldsymbol{E}$ to be directed in the $+x$ direction, then $\boldsymbol{B}$ will be in the $+y$ direction (see Fig. 1).

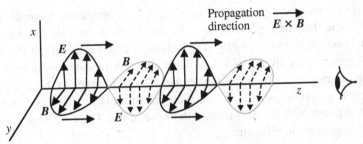

Figure 11.1. Linearly polarized electromagnetic (EM) wave at a fixed time. The wave propagates down the $+z$-axis with the oscillating E vector polarized in the x direction and B in the y direction. At any instant, the values of E_x and B_y are constant everywhere on any x–y plane. The values of E_x and B_y are in a fixed ratio at all positions and times.

Wave equations

We now demonstrate that this traveling wave is a solution of Maxwell's equations in a vacuum. Write the vector equations of Table 1 for a vacuum by setting $\kappa_m = 1$, $\kappa_e = 1$, $\sigma = 0$, and $\rho = 0$, and then write them in component form for the case just cited in which E_x and B_y are the sole components of E and B and all other components equal zero. This yields the expressions in the second column of Table 1.

Next, find, from these component expressions an equation solely in E_x. Operate on both sides of Faraday's law (row 2, column 2 of Table 1) with the partial derivative $\partial/\partial z$ and on both sides of Ampere's law with $\partial/\partial t$, and then eliminate the $\partial^2/\partial z\,\partial t$ term to find (Prob. 22)

$$\frac{\partial^2 E_x(z,t)}{\partial z^2} = \mu_0 \varepsilon_0 \frac{\partial^2 E_x(z,t)}{\partial t^2}; \quad \frac{\partial^2 B_y(z,t)}{\partial z^2} = \mu_0 \varepsilon_0 \frac{\partial^2 B_y(z,t)}{\partial t^2}.$$

(Wave equations in vacuum)   (11.2)

Similar treatment yields an identical equation for B_y. These are called *wave equations* because a propagating wave solution satisfies them.

One can readily confirm, by substitution into (2), that any function f of the form $E_x = f(\omega t \pm kz)$ is a solution of the wave equation in E_x, but only if the ratio ω/k is equal to $(\mu_0 \varepsilon_0)^{-1/2}$ (Prob. 22). Here, ω and k are, respectively, the *angular frequency* $\omega = 2\pi\nu$ and *wave number* $k = 2\pi/\lambda$, where ν and λ are, respectively, the frequency and wavelength of the wave. Similarly the same function, $B_y = f(\omega t \pm kz)$, satisfies the B_y equation with the same restriction, $\omega/k = (\mu_0 \varepsilon_0)^{-1/2}$. It is common to adopt a sinusoidal variation for the function f; the solutions to (2) thus become

$$E_x = E_0 \cos(\omega t - kz); \quad B_y = \frac{E_0}{c} \cos(\omega t - kz).$$

(11.3)

We set the amplitude of B_y to be $1/c$ the amplitude of E_x in (3). This follows directly from Faraday's law in component form (Table 1). Substitute E_x (3) into it, solve for B_y by differentiation and integration, and then apply the equivalence $c = (\mu_0 \varepsilon_0)^{-1/2}$, which we justify below (5). The ratio $B_y/E_x = 1/c$ is the same at all z and t according to (3). The ratio $1/c$ is an artifact of SI units; the ratio of amplitudes is unity in cgs units. It turns out that, in any units, the energy carried by the B field is equal to that carried by the E field.

The expressions (3) describe a plane wave of E_x and B_y propagating down the positive z-axis. To see this, note that the argument, $\omega t - kz$, can be held constant as t increases if z continuously increases at the correct rate. The values of z and t determine the value of E_x through (3). A given value of E_x will thus be found at increasing positions z as time progresses. In this manner, every point on the cosine curve (3) moves to the right at some *phase velocity* v_p; it is a traveling wave. The same arguments hold for B_y, which propagates to the right always in phase with E_x. This then, is the wave of Fig. 1 that we have now shown satisfies Maxwell's equations in a vacuum.

Phase velocity

The phase velocity $v_p = \Delta z/\Delta t$ of a wave is the speed at which a feature, such as a wave peak, moves in the propagation direction – the z-axis in our case. It is readily obtained by requiring that E_x be constant, or equivalently, that the argument of (3), $\omega t - kz$, be constant. Set the differential to zero, $\omega \Delta t - k\Delta z = 0$, to find that

$$v_p = \frac{\Delta z}{\Delta t} = \frac{\omega}{k}. \tag{11.4}$$

See an elementary example in Prob. 21. The definitions of ω and k give us $\omega/k = \lambda \nu$; hence, we have the familiar expression for the speed of a wave, $v_p = \lambda \nu$.

As stated, the expressions (3) satisfy Maxwell's equations for a vacuum (2) only if $\omega/k = (\mu_0 \varepsilon_0)^{-1/2}$. The experimental values of μ_0 and ε_0 yield

$$\Rightarrow \qquad v_p = \frac{\omega}{k} = (\varepsilon_0 \mu_0)^{-1/2} = 2.998 \times 10^8 \text{ m/s} = c, \quad \text{(Phase velocity in vacuum)} \tag{11.5}$$

where $\mu_0 = 4\pi \times 10^{-7}$ kg m s^{-2} A^{-2} and $\varepsilon_0 = 1/\mu_0 c^2 = 8.854 \times 10^{-12}$ A^2 s^4 kg^{-1} m^{-3}. The wave speed of our solution (3) is exactly the measured speed of light in a vacuum. This agreement established the electromagnetic nature of light and is the great triumph of Maxwell's theory.

The derived phase velocity of EM radiation in a vacuum is independent of the frequency ω of the radiation. Because an arbitrary waveform can be constructed of many sinusoidal waves of different frequencies, any waveform will propagate through a vacuum without distortion. As will be demonstrated next, this is not the case for waves propagating through a plasma. In that situation, the phase velocity is a function of frequency; hence, waveforms will vary with position as they propagate through the medium.

EM waves in a dilute plasma

The presence of matter dramatically changes the behavior of EM radiation. We start again with the complete vector form of Maxwell's equations (Table 1). Recall that, as written, they apply to a medium that is isotropic and linear in its response to electric and magnetic fields. We make no attempt to explore the various possible solutions of Maxwell's equations for different conditions. A dilute, ionized plasma (with or without a weak static magnetic field) is the case of interest to us.

Wave solution

In general, the conductivity of a plasma can be high and must be taken into account. The *conductivity* (or its inverse, the *resistivity*) arises from the collisions of electrons with atoms.

Table 11.2. *Maxwell's equations for dilute nonferromagnetic plasma*

Vector form in medium[a]	Component form[b]	Name
$\nabla \cdot (\kappa_e E) = 0$	$\dfrac{\partial(\kappa_e E_x)}{\partial x} = 0$	Gauss's law
$\nabla \times E = -\dfrac{\partial B}{\partial t}$	$\dfrac{\partial E_x}{\partial z} = -\dfrac{\partial B_y}{\partial t}$	Faraday's law
$\nabla \times B = \mu_0 \varepsilon_0 \kappa_e \dfrac{\partial E}{\partial t}$	$\dfrac{\partial B_y}{\partial z} = -\mu_0 \varepsilon_0 \kappa_e \dfrac{\partial E_x}{\partial t}$	Ampere's law (modified)
$\nabla \cdot B = 0$	$\dfrac{\partial B_y}{\partial y} = 0$	Gauss's law for magnetic field

[a] For a linear isotropic medium characterized by κ_e with $\rho = \sigma = 0$, $\kappa_m = 1$. See Table 1.
[b] For $E_y = E_z = B_x = B_z = 0$.

In the low density of plasmas (compared with solids), radiation wiggles the electrons at such high frequencies that the electrons do not experience a general migration in their oscillatory motion; hence, there is no general current flow. It is thus appropriate for us to set $J = \sigma E$ to zero in the third Maxwell equation for our low-density, high-frequency case. (In a proper analysis, σE is not set to zero, and σ turns out to be finite but totally imaginary.)

In addition, the usual plasma will not be ferromagnetic, and thus $\kappa_m \approx 1$. Also, because the plasma is neutral, the charge density is zero, $\rho = 0$. The four Maxwell's equations of Table 1 may thus be written as presented in Table 2.

It turns out that, as before, any function of $\omega t - kz$ will satisfy these equations; see (8) below and Prob. 23. Following the vacuum case, we again try a plane wave solution wherein E is directed in the x direction and travels in the z direction (Fig. 1),

$$E = E_x \hat{i} = E_0 \cos(\omega t - kz)\hat{i}, \tag{11.6}$$

where $\hat{i}$ is a unit vector in the x direction. Application of $E_x = E_0 \cos(\omega t - kz)$ to Faraday's law in component form yields B_y (Prob. 23),

$$B = B_y \hat{j} = E_0 \frac{k}{\omega} \cos(\omega t - kz)\hat{j}, \tag{11.7}$$

where $\hat{j}$ is a unit vector in the y direction. Use Faraday's and Ampere's laws to construct wave equations as before. Comparison of Tables 1 and 2 indicates that they will be identical to the vacuum expressions except that ε_0 becomes $\varepsilon_0 \kappa_e$ as follows:

$$\frac{\partial^2 E_x(z, t)}{\partial z^2} = \mu_0 \varepsilon_0 \kappa_e \frac{\partial^2 E_x(z, t)}{\partial t^2}; \quad \frac{\partial^2 B_y(z, t)}{\partial z^2} = \mu_0 \varepsilon_0 \kappa_e \frac{\partial^2 B_y(z, t)}{\partial t^2}.$$
$$\text{(Wave equation)} \tag{11.8}$$

Phase velocity

Apply the trial solution E_x (6) or B_y (7) to find (Prob. 23) that it indeed satisfies the wave equation, but only if

$$\frac{\omega}{k} = (\mu_0 \varepsilon_0 \kappa_e)^{-1/2}. \tag{11.9}$$

The ratio ω/k is the phase velocity for the wave; recall that $v_p = \lambda \nu = \omega/k$. In this case, it is no longer equal to $(\mu_0 \varepsilon_0)^{-1/2} = c$ but is rather

➡ $$v_p = \frac{\omega}{k} = \frac{c}{\kappa_e^{1/2}}.$$ (Phase velocity; plasma at high frequency)   (11.10)

The dielectric constant κ_e determines the phase velocity of the wave. The value of the constant generally depends on the frequency of the radiation. With knowledge of this dependence, the frequency dependence of v_p follows from (10).

Index of refraction

The *index of refraction* η for a given material is defined such that

$$v_p = \frac{c}{\eta(\omega)}$$ (11.11)

is the *phase* velocity of an EM wave in that material. Comparison of (11) to (10) reveals that

$$\eta = \kappa_e^{1/2}.$$ (11.12)

Thus, the determination of $\kappa_e(\omega)$ for a particular material is tantamount to ascertaining the index of refraction. The former, and hence the latter, will generally be a function of frequency.

Dispersion relation

Because the speed (phase velocity) of any wave can also be given as ω/k (4), we have from (11) the *dispersion relation*

➡ $$\frac{\omega}{k} = \frac{c}{\eta(\omega)}.$$ (Dispersion relation)   (11.13)

This expression defines the relationship between the three variables, ω, k, and η. If a function $\eta(\omega)$ is inserted into this equation, one can solve for k in terms of ω, or vice versa. When η is a function of ω, the relation is no longer the simple linear relation for a vacuum, $\omega = kc$. In the cases of interest to us, η is close to unity and varies only slowly with ω; hence, ω/k is approximately, but not exactly, c.

Recall that light waves travel more slowly in glass or water than in a vacuum. For most types of glass, the index of refraction is $\eta \approx 1.5$, and for water it is $\eta = 1.3$. White light is dispersed into many beautiful colors in passing through glass with nonparallel sides (a prism) or through water droplets to create a rainbow. These effects arise because the phase velocity of light in the glass (or water) depends on the frequency (or equivalently wavelength or color) of the light waves.

The changed speed of light in matter occurs because the electric vector of the incoming EM wave forces the electrons in the matter to oscillate. In turn, because they are accelerating up and down, these electrons radiate their own EM radiation. This is reradiation of the energy that was absorbed from the incoming wave. This radiation will generally be displaced in phase from that of the incoming wave because of the inertia of the electrons. This phase shift is characteristic of forced harmonic motion.

The reradiated wave combines with the (unabsorbed part of the) incoming wave. The result of the interference between these two waves is a net shift in phase. This means that the

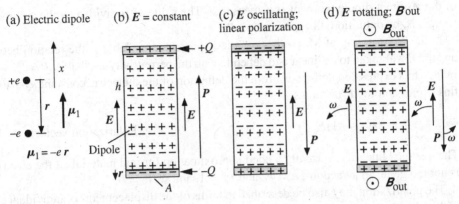

Figure 11.2. (a) Simple electric dipole and its dipole moment vector $\boldsymbol{\mu}_1$. (b) Charges in a rectangular volume displaced by a static electric field. This creates a net polarization of the medium resulting from a net excess of charges at the surfaces. The polarization vector $\boldsymbol{P}$, the dipole moment of unit volume, is parallel to $\boldsymbol{E}$. (c) Polarization $\boldsymbol{P}$ due to linearly polarized and vertically oscillating $\boldsymbol{E}$ field at an instant when $\boldsymbol{E}$ points upward. Their inertia and the high frequency (relative to resonance) cause the electrons to be displaced in the same direction as the $\boldsymbol{E}$ field at all times; hence, $\boldsymbol{P}$ is antiparallel to $\boldsymbol{E}$. (d) Polarization due to a rotating $\boldsymbol{E}$ vector (circular polarization) in the presence of a static magnetic field $\boldsymbol{B}$ directed out of the paper. The electrons are driven in a circular motion, and again $\boldsymbol{P}$ is antiparallel to $\boldsymbol{E}$.

peak of the wave would arrive at the next bit of matter slightly earlier or later than would be expected in the absence of matter. In this manner, the phase velocity in a medium can deviate from c.

The index of refraction η for a wave traversing a dilute plasma of electrons, protons, and magnetic fields surprisingly turns out (29) to be less than unity:

$$\eta(\omega) < 1.0. \qquad \text{(Index of refraction in plasma)} \qquad (11.14)$$

In this case, according to (11), the phase velocity is *greater* than the speed of light in a vacuum. This seems like a violation of a special relativity tenet. In fact, we will find below (46) that the *group* velocity is *less* than the speed of light; this resolves the dilemma.

Polarization of medium

The dielectric constant is a characteristic of a given medium. It depends on how susceptible the electrons in the medium are to being displaced by an external electric field. The quantity describing this displacement is the *polarization vector* $\boldsymbol{P}$, which is defined as the *dipole moment per unit volume*. If a single electron is displaced from a proton by distance r (Fig. 2a), the electric dipole moment, $\boldsymbol{\mu}_1 = -er$, points in the direction of positive charge displacement (upward in the figure). In a sample of many such dipoles (Fig. 2b), the sample as a whole exhibits its own dipole moment owing to the separation of the surface charges $+Q$ and $-Q$ (shaded areas in Fig. 2b); see also discussion leading to (17) below. Hence the medium has a polarization $\boldsymbol{P}$, which also points upward.

The dielectric constant for a given material is obtained, in principle, from a laboratory experiment in which a static field $\boldsymbol{E}$ is applied to a sample of the material and the strength of the resultant dipole moment $\boldsymbol{\mu}$ is measured. The value of $\boldsymbol{\mu}$ divided by the volume of the sample is, by definition, the polarization $\boldsymbol{P}$.

In the construction of Maxwell's equations presented in Table 1, the relation between $\boldsymbol{P}$ and the $\boldsymbol{E}$ was taken to be linear and dependent on the permittivity ε – that is, $\boldsymbol{P} = (\varepsilon - \varepsilon_0)\boldsymbol{E}$; in fact, this expression defines ε. From the definition of the dielectric constant $\kappa_e \equiv \varepsilon/\varepsilon_0$ (1), this expression for $\boldsymbol{P}$ becomes

➡️ $$\boldsymbol{P} = \varepsilon_0(\kappa_e - 1)\boldsymbol{E}. \qquad \text{(Polarization vector)} \qquad (11.15)$$

The assumed linearity is usually a good approximation for real materials if the electric field is not too great. In a vacuum, $\varepsilon = \varepsilon_0$ and $\kappa_e = 1$, and so $\boldsymbol{P} = 0$.

The polarization may also be described in terms of the displacements of individual charges within the medium. Apply a fixed electric field $\boldsymbol{E}$ to a sample with base area A and height h containing n_e electrons/m³ (Fig. 2b). The field will displace each electron a distance r from its companion positive ion. This separation of charges leads to a layer of positive charges at the "upper" surface of the volume (the surface in the direction of $\boldsymbol{E}$) and a layer of negative charges on the opposite surface.

Each charge inside the volume will find a charge of the opposite sign at its location, and so the charge density inside the volume is everywhere zero. The net result is that the volume is a macroscopic dipole with two layers of charge of area A separated by distance h. The total charge Q protruding from the surface is the charge density $n_e e$ times the separation distance r and the area A. The total dipole moment is thus

$$\boldsymbol{\mu} = -Qh\hat{\boldsymbol{r}} = -(n_e e A r)h\hat{\boldsymbol{r}}, \qquad \text{(Dipole moment of sample)} \qquad (11.16)$$

where $\hat{\boldsymbol{r}}$ is a unit vector in the displacement direction,

As stated, the polarization vector is the dipole moment per unit volume. Divide (16) by the volume Ah to obtain

➡️ $$\boldsymbol{P} = -n_e e \boldsymbol{r}, \qquad \text{(Polarization vector)} \qquad (11.17)$$

where we define the displacement vector $\boldsymbol{r} \equiv r\hat{\boldsymbol{r}}$. One could also obtain this directly by multiplying the dipole moment $\boldsymbol{\mu}_1 = -e\boldsymbol{r}$ by n_e. The vector $\boldsymbol{P}$ is directly proportional to the number of electrons per unit volume as well as to the displacement of each electron. If the electrons oscillate under the influence of an EM wave, the polarization $\boldsymbol{P}$ of the medium will also oscillate; it is not a static quantity.

Figure 2b represents the situation for a static electric field in which the electron displacement $\boldsymbol{r}$ is opposed to the direction of the applied field $\boldsymbol{E}$. Thus, from (17), $\boldsymbol{P}$ is parallel to $\boldsymbol{E}$. From (15), we find that, in this case, $\kappa_e > 1$ and hence $\eta > 1$ (12).

Our task in the next sections, in essence, is to determine the indices of refraction $\eta(\omega)$ for the oscillating electric field of an electromagnetic wave for different conditions (i.e., in a dilute plasma with and without an external static magnetic field and for circularly polarized radiation). Our procedure will be to evaluate, from first principles, the amplitude of the oscillating charge displacements in the presence of an EM wave. This provides the polarization $\boldsymbol{P}$, which in turn is related to $\kappa_e(\omega)$ through (15) and then to the index of refraction $\eta(\omega)$ through (12).

11.3 Dispersion

The phenomenon of dispersion refers to the variation with frequency of transmission speeds through a plasma. A pulse of radiation emitted by a radio pulsar will contain a wide range of radio frequencies. During the pulse's passage through the ISM, the higher frequencies propagate faster than the lower ones. A receiver on earth tuned to 100 MHz will detect the pulse substantially delayed relative to the arrival time at 400 MHz. Radio pulsars emit sequential pulses of radiation, which can be measured in this way. (One could not observe dispersion with a steady source; all frequencies would arrive at all times.) Here we relate the degree of dispersion to the electron density integrated along the line of sight to the pulsing source.

Polarization from equation of motion

Consider a linearly polarized plane EM wave (Fig. 1) passing through a plasma of protons and electrons. The instantaneous electric field E at some position displaces electrons, giving a displacement r and hence a polarization P (17). The field also acts on the protons in the opposite direction. However, the protons are so massive compared with the electrons that they move very little; thus, only the displacement of the electrons need be considered.

An electron in the plasma responds to the electric force of the EM wave,

$$F_E = qE = -eE_0 \cos(\omega t - kz)\hat{i}, \tag{11.18}$$

which we take to be sinusoidal (6). The electron charge e is taken, as usual, to be a positive definite number. The electron may be considered to be a *driven oscillator* under the influence of this force. We omit the negligible magnetic force $(qv \times B)$ due to B of the EM wave; see (3).

Write the equation of motion $(\Sigma F = ma)$ but with the electric force (18) only. The individual electrons in the plasma are assumed to be essentially free (i.e., the driving frequency is well above any resonant frequencies, and so the restoring and damping forces may be neglected). The electric field is in the transverse, x, direction, and so the electron motion is along the x-axis. Thus,

$$F_x = ma_x \tag{11.19}$$

$$qE_x(z, t) = m\ddot{x}. \tag{11.20}$$

The double dot notation represents the second time derivative. We will solve this for $x(t)$, the (transverse) electron motion.

Because the wave $E_x(z, t)$ varies sinusoidally, it is reasonable to adopt a sinusoidal trial solution for the motion, which we then differentiate twice:

$$x = x_0 \cos \omega t \tag{11.21}$$

$$\ddot{x} = -x_0\omega^2 \cos \omega t = -\omega^2 x. \tag{11.22}$$

Substitute into the equation of motion (20),

$$(-e)E_x = -m\omega^2 x, \tag{11.23}$$

and solve for the displacement x,

$$x = \frac{eE_x}{m\omega^2}.$$ (Displacement)   (11.24)

We find that x is in phase with E_x but 180° out of phase with the force $-eE_x$ on the electron. This is as expected for an oscillator driven at a frequency well above resonance.

The polarization vector follows from (17), where, in this case, $r = x\hat{i}$ and $E_x\hat{i} = E$:

$$\rightarrow \quad P = -\frac{n_e e^2}{m\omega^2}E.$$ (Polarization; negligible binding)   (11.25)

This is valid for negligible binding (i.e., for frequencies ω well in excess of the resonant frequency). The several vectors are illustrated in Fig. 2c; note that P is opposed to E in accord with (25). Compare this with the static field case, Fig. 2b, where they are parallel.

The polarization (25) is opposed to the electric field because the electron displacement is in phase with E (24), whereas, by definition, the vector P is directed opposite to the electron displacement; see (17). The amplitudes of both P and E oscillate sinusoidally, but they remain in fixed proportion to one another in this high-frequency limit.

Index of refraction and plasma frequency

Comparison of (25) with (15) immediately yields the dielectric constant

$$\kappa_e = 1 - \frac{n_e e^2}{\varepsilon_0 m \omega^2},$$ (11.26)

where m is the electron mass. According to (12), the square root of κ_e is the index of refraction,

$$\eta(\omega) = \left[1 - \frac{n_e e^2}{\varepsilon_0 m \omega^2}\right]^{1/2}.$$ (Index of refraction)   (11.27)

The variation of the index of refraction η with frequency ω results in a variation of phase velocity with frequency. This is the phenomenon of dispersion.

This expression can be simplified if we define the *plasma frequency* ω_p as

$$\rightarrow \quad \omega_p \equiv \left(\frac{n_e e^2}{\varepsilon_0 m}\right)^{1/2}$$ (Plasma frequency; m is election mass)   (11.28)

to obtain

$$\rightarrow \quad \eta(\omega) = \left[1 - \frac{\omega_p^2}{\omega^2}\right]^{1/2}.$$ (Index of refraction; low-density plasma; $\omega \gg \omega_{resonances}$)   (11.29)

Recall that η (29) was derived under the approximations that ω is much greater than any resonant frequency (e.g., ω_p), that (collisional) damping is negligible, and that the medium is isotropic and linear in its response to electric and magnetic fields. For our condition ($\omega \gg \omega_p$), the index of refraction is less than unity, as anticipated (14). The phase velocity (11) in this case is thus greater than c:

$$v_\mathrm{p} = \frac{c}{\eta(\omega)} > c. \qquad\qquad \text{(for } \eta < 1) \qquad (11.30)$$

The plasma frequency ω_p can be shown (Prob. 31) to be the resonant frequency arising from a small displacement of a volume element of electrons in the plasma from the ions in the same volume element. This frequency is a fundamental characteristic of a plasma.

Ionospheric cutoff

Rewrite the plasma frequency (28) with the numerical SI values of m, e, and ε_0. Because radio frequencies are generally given in hertz, use $\nu_\mathrm{p} = \omega_\mathrm{p}/2\pi$ to obtain

$$\nu_\mathrm{p} = 9.0(n_\mathrm{e})^{1/2}. \qquad\qquad \text{(Hz; } n_\mathrm{e} \text{ in m}^{-3}) \qquad (11.31)$$

In the ionosphere, the electron densities range up to $\sim 10^{12}$ m^{-3}. For this density,

$$\nu_\mathrm{p} = \omega_\mathrm{p}/2\pi = 9.0\,\mathrm{MHz}. \qquad (n_\mathrm{e} = 10^{12}\,\mathrm{m}^{-3}; \text{ ionosphere}) \qquad (11.32)$$

The ionospheric electron densities vary dramatically as a function of time of day as well as location.

A plasma will not transmit waves with frequencies below ν_p. The index of refraction is imaginary, and this leads to absorption of the waves; hence, radio astronomy from the earth's surface can not be carried out at frequencies $\lesssim 9$ MHz. Such radio waves (at $\lesssim 9$ MHz) from outer space are reflected back to space by the ionosphere.

You may know that AM radio can propagate long distances via such reflections from the bottom of the ionosphere. This is particularly effective at night when the relatively low "D" layer of the ionosphere becomes neutral; then reflection can take place at higher altitudes that have large values of n_e. In contrast, FM radio at ~ 100 MHz travels right through the ionosphere into outer space; it sounds better, but one has to be close enough to the station that the signal can be received more or less directly without ionospheric reflections.

Interstellar cutoff

In *interstellar space*, the number densities of ions are typically much less (i.e., 10^3 to 10^4 m^{-3}); thus,

$$\nu_\mathrm{p} \approx 300\text{--}1000\,\mathrm{Hz}. \qquad\qquad (11.33)$$

This is well below the frequencies that penetrate the ionosphere ($\nu_\mathrm{p} \gtrsim 9$ MHz). Consequently, radio waves that can penetrate the atmosphere can easily travel through much of interstellar space. This makes galactic and extragalactic radio astronomy possible.

Group velocity

The speed with which a pulse of radiation containing a band of frequencies will propagate through space is known as the *group velocity*. Here we derive its magnitude in terms of the index of refraction and find its dependence on frequency in a plasma.

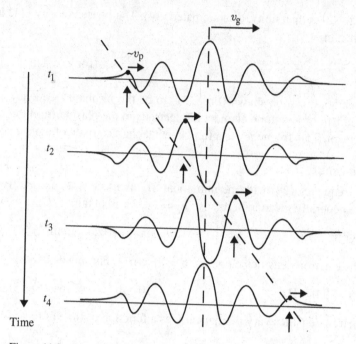

Figure 11.3. A wavelet or wave packet shown at four different times with each sketch centered on the packet centroid. The packet as a whole moves to the right at group velocity v_g. The interference between the component waves that make up the packet, with their different phase velocities, causes a single crest (arrows and dots) to overtake the centroid of the packet gradually. Relative to the medium, the overtaking crest moves at approximately the median phase velocity.

Phase and group velocities distinguished

The arrival time of a radio-wave pulse at a specified frequency ν depends on the *group velocity* of a band of frequencies $\Delta\nu$ at ν. The different frequencies travel together in a *wave packet*. Together they form a pulse. Recall that, in *Fourier analysis*, any waveform can be synthesized by summing sine (or cosine) waves with appropriate frequencies, amplitudes, and phases. Such a pulse can carry information such as "The redcoats are coming!"

In contrast, the phase velocity is the speed of a peak in a single-frequency wave. Such a wave is, strictly speaking, infinitely long and carries no information other than the fact that it is broadcasting. If the wave were finite in length it would contain other frequencies, and the group velocity would govern information transfer.

The group velocity phenomenon may be seen in the expanding circular ring of waves on a pond after a stone is thrown into it or in the group of waves sent out by a passing boat. In these cases, most of the frequencies in the packet are close to a common central frequency, and the envelope moves more slowly than does any given crest.

This is illustrated for the packet of Fig. 3. The arrow marks one peak as it moves toward the front of the moving packet. The individual peak moves at about the phase velocity of the central frequency in the group, and the centroid of the envelope moves at the group velocity. The information content moves at the latter speed.

The index of refraction in a plasma (29) was derived for $\omega \gg \omega_p$, and hence the phase velocity is greater than the velocity of light, as noted (30). This might be of some concern

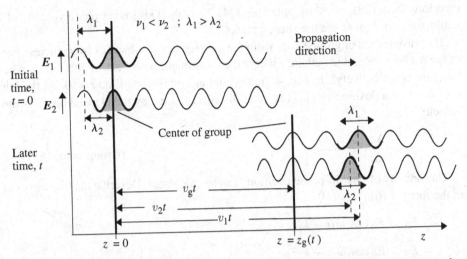

Figure 11.4. Phase and group velocities. Two frequency components ($\nu_1 < \nu_2$) of a wave packet are shown at time $t = 0$ (above) and at a later time t (below). The in-phase positions, marked with solid vertical lines, represent the packet centroids at the two times. Their speeds are thus the group velocities. The peaks, shown darkened at time $t = 0$, have moved to the positions shown for time t (again darkened). Their speeds represent the frequency-dependent phase velocities. The phase velocity of an EM wave in a dilute plasma is greater than c, but the group velocity is less than c. [Adapted from B. Rossi, *Optics*, Addison Wesley, 1965, p. 261]

because the speed of light in a vacuum is the maximum allowable speed in Einstein's special theory of relativity, but this limiting speed applies only to the speed of information transfer. Because no information is propagated at the phase velocity, there is no problem with its being greater than c. We will show that the group velocity in our plasma is less than c, as required by special relativity.

According to (29), when $\omega \gg \omega_p$, a large fractional change of frequency ω yields only a modest change of η (Prob. 32), which remains close to unity. Although v_p is greater than c, it is only slightly so, and it too changes only slowly with frequency.

General expression

The group velocity is a function of the frequency-dependent index of refraction $\eta(\omega)$. Consider two of the constituent waves in a group traveling in a plasma; let their wavelengths be quite close to one another with $\lambda_1 > \lambda_2$. The dispersion relation (13) and the index of refraction (29) tell us that, when $\lambda_1 > \lambda_2$, we have $\nu_1 < \nu_2$ (Prob. 32). In Fig. 4, these two waves are shown at two different times. Because the index of refraction is a function of frequency (or wavelength), the *phase* velocities of the two waves differ slightly; thus, from (30),

$$v_1 = c/\eta_1; \quad v_2 = c/\eta_2. \tag{11.34}$$

The crest of wave 1 moves a different distance in time t ($z_1 = v_1 t$) than does the crest of wave 2 ($z_2 = v_2 t$). For our example, $\nu_1 < \nu_2$ (or equivalently $\omega_1 < \omega_2$ because $\omega = 2\pi\nu$);

therefore, from (29), $\eta_1 < \eta_2$, and from (34), $v_1 > v_2$. This higher phase velocity of wave 1 is illustrated in Fig. 4; see the shaded peaks.

The wave packet or group arises from the interference of a band of frequencies. The center of the packet in Fig. 3 is the place where the constituent waves are in phase (i.e., where they interfere constructively). In Fig. 4, this occurs at $z = 0$ when time $t = 0$. At the later time t, it occurs at a position $z_g(t)$. This latter distance divided by the elapsed time t is the group velocity v_g,

$$v_g = \frac{\Delta z_g}{\Delta t} = \frac{z_g(t)}{t}. \qquad \text{(Group velocity)} \qquad (11.35)$$

The velocity of this "in-phase position" can be calculated. Describe the two waves of Fig. 4 in the form of (6):

$$E_1 = E_0 \cos(\omega_1 t - k_1 z) \qquad (11.36)$$

$$E_2 = E_0 \cos(\omega_2 t - k_2 z). \qquad (11.37)$$

These two waves are in phase at the places and times the arguments of the cosine functions are equal. This occurs at $z = t = 0$ and, in general, when

$$\omega_1 t - k_1 z_g = \omega_2 t - k_2 z_g, \qquad (11.38)$$

which yields

$$\frac{z_g}{t} = \frac{\omega_2 - \omega_1}{k_2 - k_1}. \qquad (11.39)$$

This is, therefore, the group velocity. Because the intervals are small, we can write

$$\Rightarrow \qquad v_g = \frac{d\omega}{dk}, \qquad \text{(Group velocity)} \qquad (11.40)$$

which is the desired analytical expression for the group velocity. This should not be confused with the similar appearing, but very different, expression for the *phase* velocity, $v_p = \omega/k$ (4).

Pulse speed in a plasma

The group velocity $d\omega/dk$ for a specific medium (a plasma, vacuum, etc.) is obtained from the dispersion relation $\omega(k)$ (13), which we can rewrite as

$$\omega = \frac{c}{\eta(\omega)} k. \qquad (11.41)$$

The derivative of (41) is the group velocity. For the specific case of a wave in a vacuum, where $\eta = 1$, this gives the expected value

$$v_g = d\omega/dk = c, \qquad \text{(Vacuum)} \qquad (11.42)$$

which is independent of frequency.

The group velocity for a plasma is simpler to calculate in terms of $1/v_g = dk/d\omega$. Rewrite (41) as

$$k = \omega \eta(\omega)/c \qquad (11.43)$$

and take the derivative of this product, keeping in mind that η is a function of ω:

$$\frac{1}{v_g} = \frac{dk}{d\omega} = \frac{d}{d\omega}\left(\frac{\omega\eta(\omega)}{c}\right) = \frac{\eta}{c} + \frac{\omega}{c}\frac{d\eta}{d\omega}. \tag{11.44}$$

One can now substitute any given index of refraction $\eta(\omega)$ into the right-hand term of (44) to obtain the inverse of the group velocity for that particular index of refraction. Substitute the index for our plasma (29) into (44) and carry out the differentiation and algebra to obtain (Prob. 33):

$$\frac{1}{v_g} = \frac{1}{c}\left(1 - \frac{\omega_p^2}{\omega^2}\right)^{-1/2} \tag{11.45}$$

or

$$v_g = c\left(1 - \frac{\omega_p^2}{\omega^2}\right)^{1/2}, \qquad \text{(Group velocity; low-density plasma;}$$
$$\omega \gg \omega_{\text{resonances}}) \tag{11.46}$$

which fortuitously equals $c\eta$ for this particular index of refraction. This expression (46) follows from the index of refraction (29), and so the same restrictions apply; see text after (29). For typical pulsar radio frequencies of $\gtrsim 100$ MHz in interstellar space where $\nu_p \lesssim 10^3$ Hz (33), the inequality $\omega \gg \omega_p$ is satisfied. The group velocity (46) is thus slightly less than c in accord with the requirements of special relativity.

According to (46), a group of low frequencies travels more slowly than does a group of high frequencies in our plasma. If the frequency is lowered toward ω_p, the group velocity decreases toward zero. For $\omega < \omega_p$, the index of refraction (29) becomes imaginary, leading to attenuation. (This is basisally correct even though (29) was derived for $\omega \gg \omega_p$.) In contrast, if the frequency ω is well above ω_p and increasing, the group velocity approaches c. This same limit is approached if n_e, and hence ω_p, decreases toward zero. As the medium becomes more closely a vacuum, the group velocity more closely approaches the vacuum velocity c.

Celestial source

Radio pulsars, make excellent dispersion probes of the medium along the line of sight. An individual pulse marks the beginning of the long transit to the observer, and this pulse will be delayed by the electrons along its path. How can one measure a dispersion delay if there is no observer at the source to record the time of emission? The answer lies in the frequency dependence of the index of refraction and hence of the delay.

Here again, the pulsar comes to the rescue because each pulse consists of a broad band of radio frequencies that are emitted simultaneously. Detection of different pulse arrival times at two different frequencies provides the evidence that there are free electrons along the path. It provides, as well, the column density of such electrons.

Time delay

The transit time for a pulse of radiation to propagate from a source at distance D to the earth is, from (46),

$$t = \frac{D}{v_g} \approx \frac{D}{c}\left(1 + \frac{1}{2}\frac{\omega_p^2}{\omega^2}\right), \tag{11.47}$$

where we invoked $\omega \gg \omega_p$ and used the expansion $(1 + x)^n \to 1 + nx$ for $x \ll 1$. For $\omega \to \infty$, the transit time is the same as in a vacuum, but it is greater for lower frequencies.

The difference in arrival times (47) of the pulse in two well-separated and relatively narrow frequency bands centered at $\nu_1 = \omega_1/2\pi$ and $\nu_2 = \omega_2/2\pi$ is

$$t_2 - t_1 = \frac{1}{8\pi^2} \frac{D}{c} \frac{n_e e^2}{\varepsilon_0 m} \left(\frac{1}{\nu_2^2} - \frac{1}{\nu_1^2} \right), \tag{11.48}$$

where we applied the definition of ω_p (28). If n_e is not constant along the path, one can write

$$
\begin{aligned}
t_2 - t_1 &= \frac{1}{8\pi^2 \varepsilon_0} \frac{e^2}{mc} \left(\frac{1}{\nu_2^2} - \frac{1}{\nu_1^2} \right) \int_0^D n_e \, ds \\
&= 1.345 \times 10^{-7} \left(\frac{1}{\nu_2^2} - \frac{1}{\nu_1^2} \right) \int_0^D n_e \, ds, \quad \text{(Time delay; s)} \tag{11.49}
\end{aligned}
$$

where n_e is the free electron density (m^{-3}), D is the distance to the source (m), and ν is the frequency of the radiation (Hz).

The measured time delay immediately gives a value of $\int n_e \, ds$, the integral of n_e over the distance D to the source. This integral is the number of electrons in a column of unit cross section (1 m^2) and length D (i.e., the *column density*).

Crab pulsar

Observations of the pulses from the Crab pulsar show the effect clearly; the delay between frequencies 100 and 400 MHz is 23.5 s, which immediately yields from (49) a value for the column density,

$$\int_0^D n_e \, ds = 1.86 \times 10^{24} \, m^{-2}. \quad \text{(Electron column density to Crab pulsar)} \tag{11.50}$$

Because the Crab pulsar is about 6000 LY distant (5.7×10^{19} m), the average electron density is

$$\langle n_e \rangle_{av} = \frac{1}{D} \int_0^D n_e \, ds = 3 \times 10^4 \, \text{electrons/m}^3. \quad \begin{array}{l} \text{(Average electron} \\ \text{density to Crab)} \end{array} \tag{11.51}$$

This is a relatively low density in the context of the ISM. It lies between the densities of the WIM and HIM (Table 10.2), which may not be surprising because the sun is embedded in an HIM region and the Crab is relatively local to the sun (distance 6000 LY at $\ell = 185°$).

The 23-s time lag from 400 to 100 MHz is very long compared with the 33-ms interval between pulses from the Crab pulsar. At the earth, each pulse is "heard" as a frequency that continuously decreases. This drags on and on for $\gtrsim 23$s, during which time another ~ 700 pulses are arriving at any fixed frequency. We show this schematically in Fig. 5. If the frequencies were in the audible range, this process might sound like a 700-person chorus singing the "dispersion glissando" in 700 parts.

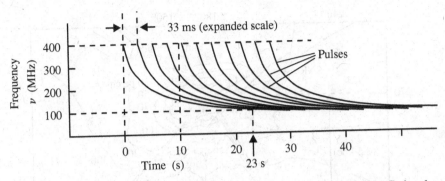

Figure 11.5. Sketch of frequency versus time at the earth of 11 pulses from the Crab pulsar. At any given time (e.g., at 10 s), multiple tones are detected. The 33-ms interval between pulses is greatly expanded. Actually, about 700 pulses would arrive at any specified frequency during the 23-s decay of a single pulse from 400 to 100 MHz.

Dispersion measure

The column density is often called the *dispersion measure* (DM) by radio astronomers:

$$\text{DM} \equiv \int_0^D n_e \, ds. \qquad \text{(Dispersion measure; m}^{-2}\text{, or LY/m}^3\text{)} \qquad (11.52)$$

The dimensions of this quantity are density (m^{-3}) times distance (m), or m^{-2}. Astronomers often use the mixed units pc cm^{-3} or, equivalently, LY m^{-3} (Prob. 34). The DM follows directly from a time-delay measurement (49) or equivalently from a measurement of $d\nu/dt$ (Prob. 35).

Galactic model of electron density

The number of known radio pulsars is now approaching 2000, and the DM has been measured for most of them. In some cases, the distance to the pulsar is known from independent evidence such as parallax measurements, H I absorption of radio waves (Chapter 10), or because the distance is associated with an object of known distance such as a supernova remnant. The DM in these cases provides an average value of n_e along the line of sight.

Radio waves can be deviated from straight-line paths by the electrons of the ISM. This results in temporal broadening of detected pulsar pulses because multiple paths contribute to the detected pulse. It also leads to angular broadening of galactic and extragalactic sources. These effects provide, in addition, information about the electron content of the ISM and thus complement the DM measurements.

These and other results have led to a model of the distribution of electron density n_e in the Galaxy. Ideally, the model would specify the electron density everywhere in the Galaxy. In actuality, of course, the model is only as good as the data used to construct it. A current model for the electron density n_e for the galactic plane is shown in Fig. 6. It is known as NE2001 because it is constructed from data available through the year 2001. The figure is a view from the north pole of the galaxy with contours that indicate the *integrated* densities, namely the DM, from the sun out to any given point in the galactic plane ($|b| \approx 0°$). A local

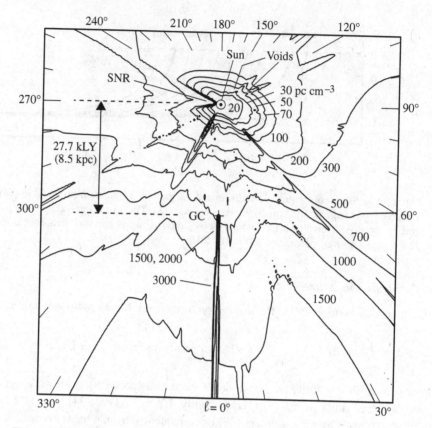

Figure 11.6. Contours of dispersion measure (DM) in units of pc cm^{-3} in the galactic plane viewed from above the plane (1 pc cm^{-3} = 3.262 × 10^6 LY m^{-3} = 3.086 × 10^{22} m^{-2}). The sun and galactic center (GC, cross) are marked; their separation is 27.7 LY (8.5 pc). View directions are labeled as galactic longitudes ℓ. The DM is the integral of n_e out to a chosen position and thus will generally increase with distance from the sun. Strong local enhancements of n_e, such as the GC at $\ell = 0$ or the Gum and Vela nebulae (SNR) at $\ell \approx 260°$, are reflected in the contours at distances beyond the enhancements. Similarly, voids such as those at $\ell \approx 120°$ produce "fingers" in the contours at greater distances. [J. Cordes and T. Lazio 2004, astro-ph 0207156]

density enhancement or deficiency will thus affect the contours farther from the sun at the same view direction. View directions from the sun are shown as galactic longitudes.

An interesting feature to note is that the sun is embedded in a hot, low-density plasma region, as indicated by the relatively large distance out to the first contour, 20 pc cm^{-3}, compared with the closer spacing of the next four contours in most directions. The contours bend dramatically inward toward the sun at longitudes near 260°. This is due to two nearby supernova remnants, Vela and the Gum nebula. Vela is ~800 LY distant and subtends about 10° on the sky. The even larger Gum nebula (~40°), an H II region about 1000 LY in diameter, lies in the same direction and is centered about 1600 LY from the sun. Both have enhanced electron densities that bring the first five contours dramatically inward toward the sun. Similarly, high densities in the galactic center (GC) region cause the contours outward of the GC to collapse toward it.

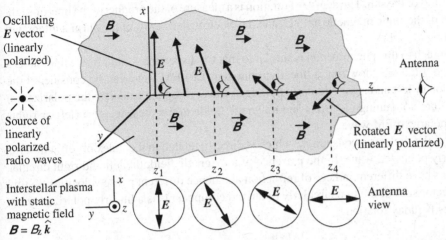

Figure 11.7. Faraday rotation. A linearly polarized radio wave passes from left to right through the electrons and magnetic field of interstellar space; the latter is parallel to the propagation direction. The direction of linear polarization (E vectors) rotates gradually as a function of position z; the directions of oscillation as viewed in the $-z$ direction are shown at four positions (inset antenna views). The gradual rotation assumes approximately uniform electron density and magnetic field. This pattern of oscillating vectors is stationary; it does *not* propagate to the right.

Elsewhere, the data indicate voids of electron density where the density is also relatively low, namely 1000–50 000 m^{-3}. For example, the model indicates voids at longitude ~114° at distance 8 kLY and 129° at 5 kLY. These cause the contours beyond the voids to project distinctly outward away from the sun at these longitudes.

Such a model can be used to find the distances to pulsars at unknown distances for which the DM has been measured. If the pulsar is in the galactic plane, one can enter Fig. 6 at the galactic longitude of the pulsar with the measured DM and read off the distance from the sun.

11.4 Faraday rotation

The continuum radio emission from discrete sources, such as a pulsar, a supernova remnant, or a distant quasar, is often found to be *linearly polarized*. The radiation from these sources often derives from synchrotron or curvature radiation (Chapter 8), which typically is partially polarized. Linear polarization of electromagnetic radiation is characterized by electric vectors lying in one direction orthogonal to the propagation direction, as in Fig. 1. (Do not confuse this polarization of the electric vector of an electromagnetic wave with the polarization P of the medium.)

Consider linearly polarized radiation traveling through an ISM containing a constant density of free electrons and a uniform static magnetic field B aligned parallel to the direction of travel. The linear polarization, as a function of position along the path, will gradually rotate about the propagation direction, as shown in Fig. 7. This phenomenon is called *Faraday rotation*.

For many astrophysical situations, the total rotation angle from source to observer is substantially less than a full rotation; however, it can be many rotations in some cases such as in observations of the sun. The degree of rotation is a measure of the product of the line-of-sight component of the static magnetic field B_z and the electron density n_e of the plasma integrated along the line of sight.

Our approach to this phenomenon is similar to that for dispersion – namely, we find indices of refraction. First we show that a linearly polarized (LP) EM wave can be considered the superposition of left and right circularly polarized waves. We then calculate the polarization P of a medium containing electrons and magnetic fields in the presence of a (left or right) circularly polarized EM wave.

This gives the dielectric constant κ_e, which is directly related to the index of refraction for the circularly polarized wave. In the presence of a magnetic field, the left and right circular polarizations have different indices of refraction and hence different phase velocities. In this case, the vector sum of the two circularly polarized components is a linearly polarized wave that exhibits Faraday rotation.

Rotation of linear polarization

Here, we examine the rotation phenomenon qualitatively.

Rotation with position

Refer again to Fig. 7. At the fixed position $z = z_1$ along the path, the E vector oscillates sinusoidally in amplitude along the x-axis; its direction remains fixed (see inset). This defines the radiation as being linearly polarized. Farther down the path, at z_3, the vector's direction is rotated $\sim45°$ about the line of sight, but here again it oscillates in amplitude along a fixed direction; here too, it is linearly polarized. At z_4, it is still linearly polarized but along the horizontal y direction. A stationary observer at this position detects linear polarization in a fixed direction $90°$ from that at z_1. This rotation as a function of position is not to be confused with circularly polarized light, described below in this section, wherein the electric vector at a fixed position rotates with the electromagnetic frequency ω.

Oscillating electrons

The origin of the rotation can be understood in a simplistic way by considering the motion of an electron in the plasma (Fig. 8). The sections of the figure illustrate three closely spaced positions along the path of the radiation. The plane of the paper is perpendicular to the propagation direction z with the wave approaching the reader. The wave is initially polarized with the transverse electric vector in the vertical direction (Fig. 8a). The electrons driven by this wave oscillate vertically.

In Fig. 8b, the wave encounters a magnetic field parallel to the propagation direction. The oscillating E field will accelerate an electron in the direction of the electric vector (up and down in the figure). As the electron moves, it senses a magnetic force, $(q v \times B)$ at right angles to its velocity, where $q = -e$. The electron path is bent to its left in both the up and down directions. This leads to a slight rotation of the electron's direction of oscillation.

These electrons reradiate EM radiation because they undergo acceleration as they oscillate back and forth along the slightly tilted path. The electric field vector radiated by an accelerated charge is parallel to the direction, projected perpendicular to the line of sight, of the

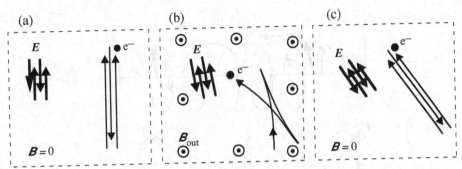

Figure 11.8. A simplistic view of the mechanism giving rise to Faraday rotation. The electric field and the electron motions are shown in sketches at three successive positions along the path of a radio wave, which emerges from the paper toward the reader. (a) Region of $B = 0$. (b) Region containing a magnetic field directed along the propagation direction. The $F = qv \times B$ force ($q = -e$) gives curvature to the electron trajectories, leading to rotated polarization of the reradiation. (c) Region of $B = 0$.

acceleration vector of the charge (Fig. 5.2a). For this reason, the reradiated electric vector will be parallel to the slightly rotated motion of the emitting electron.

This reradiated EM wave combines with the original wave to produce a resultant EM wave. The result is a slightly rotated E vector (Fig. 8c). This occurs in each segment of the path. Over a long path, the rotation of the electric vector can be substantial.

Circular polarization

Faraday rotation is properly understood in terms of the different phase velocities of "right" and "left" *circularly* polarized waves (RCP and LCP, respectively).

Rotating vector

A circularly polarized EM wave is one in which the tip of the E vector traces out a spiral in space if the wave is frozen at a given time (Fig. 9a). One cycle of the spiral occurs in the wavelength λ. Compare this with the linear polarization shown in Fig. 1, where the E vector lies solely in the x–z plane.

If time is now "turned on," the circularly polarized wave rushes to the right (z direction) at close to the speed of light; the corkscrew of Fig. 9a moves to the right *without rotating*. An observer at a fixed position z looking in the $-z$ direction detects E vector sample 1 first and then 2, 3, and 4 in sequence. The E vector thus rotates counterclockwise (Fig. 9a) one complete cycle for every wavelength that enters the eye. The vector rotates 100 million times a second for a 100-MHz radio wave.

Similarly, the linear polarization pattern of Fig. 1 rushes along to the right when time is turned on, and the eye sees 10^8 linear oscillations each second. Distinguish the patterns of Figs. 1 and 9 from the slow Faraday rotation of linear polarization shown in Fig. 7. The latter pattern extends over many wavelengths, and the polarization direction remains fixed at any given location as time progresses. We demonstrate below that this behavior may be understood as the sum of two circularly polarized waves (RCP and LCP) both of which are rushing to the right.

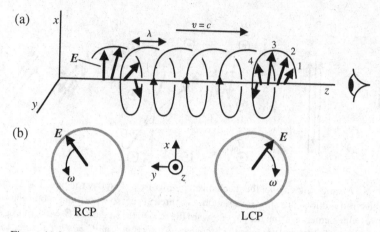

Figure 11.9. Circularly polarized EM waves. (a) Right circularly polarized (RCP) wave at a fixed time. The entire helix of E vectors moves as a rigid body to the right ($+z$ direction) as the wave propagates. The observer will detect E vector samples sequentially, as numbered (i.e., with counterclockwise rotation). (b) RCP and LCP waves at a fixed position z with waves emerging from paper. The IRE right–left naming convention is used.

Left–right naming convention

Consider the RCP wave at fixed position viewed in the $-z$ direction (Fig. 9b). If the thumb of a right hand points in the direction of propagation ($+z$), the fingers curl in the direction (counterclockwise) that the E vector rotates. An LCP wave rotates in the opposite direction. This definition of right–left is that of the Institute of Radio Engineers (IRE). The "classical" definition is the reverse of this. It is equally logical: If the wave is frozen in time, as in Fig. 9a, it forms a left-handed screw. We also use the IRE definition in Section 10.5 after (10.67) where we discuss the angular momentum carried by photons.

Components of E field

Mathematically, an RCP wave can be described as the sum of two LP waves at right angles and 90° out of phase with one another. For example, let one E wave oscillate and propagate in the x–z plane as in Fig. 1, and let the other lie in the y–z plane with a $\pi/2$ phase shift. We choose to write them as the x and y components of a resultant wave as follows:

$$E_x = E_0 \cos(\omega t - kz); \quad E_y = E_0 \cos(\omega t - kz - \pi/2). \tag{11.53}$$

Construction of the vector E from these two components as a function of z and t readily demonstrates that they do indeed form a propagating RCP wave. To see this, set $z = 0$ and note that E_x is at maximum, E_0, at $\omega t = 0$, whereas E_y at this time and position is zero. A quarter cycle later, at $\omega t = \pi/2$, and still at $z = 0$, E_y reaches maximum and $E_x = 0$. The vector E thus rotates 90° in the quarter cycle. Again, when viewed in the $-z$ direction, the rotation is counterclockwise; see the coordinate axes in Fig. 9b. If the phase shift in (53) were $+\pi/2$ instead of $-\pi/2$, the equations would describe LCP radiation. See Prob. 41 for an analytical approach.

Circularly polarized radiation is not just a mathematical construction; it is observed directly from celestial objects. Magnetic white dwarfs and "RS CVn type" stars emit circularly polarized radiation in the optical band.

How can such radiation come about? Let a single charge move in a circle in a magnetic field and view it from a position perpendicular to the plane of the motion. The observed radiation is circularly polarized because the instantaneous acceleration vector rotates with the particle and therefore so does the electric vector of the instantaneously emitted radiation. Observation of circularly or elliptically polarized radiation is a reliable indication of strong magnetic fields in the object giving rise to the radiation.

Superposition of RCP and LCP

It is now convenient to describe an LP wave as a sum of RCP and LCP waves, which is the converse of the exercise just described. An LP wave approaching the observer is shown in Fig. 10a with seven sequential snapshots of E at a fixed position ($z = 0$). The oscillation is vertical, as it is for the observer at z_1 of Fig. 7.

In Fig. 10b, the same oscillation is shown (at $z = 0$) as a vector sum E of the RCP and LCP vectors E_R and E_L at some time $t = t_0$:

$$E = E_R + E_L. \tag{11.54}$$

Equal amplitudes and frequencies and matched phases of E_R and E_L yield a vertically oscillating resultant E as the vectors rotate with time. The oscillating E vector is therefore an LP wave identical to that in the snapshots of Fig. 10a. At this position ($z = 0$), the horizontal (y) components perfectly cancel each other at all times.

The rotation angles of the vectors E_R and E_L at $z = 0$ and time t (Fig. 10b) are

$$\psi_R = +\omega t; \quad \psi_L = -\omega t. \qquad (z = 0) \tag{11.55}$$

This is the same ω that is the angular frequency of the EM wave ($\omega = 2\pi\nu$). The E_R and E_L vectors rotate a full circle once each full period of the wave. Because this is a traveling wave (right-moving helix of Fig. 9a), a given value of ψ_R moves down the $+z$-axis. Hence, the z dependence of the angle must be incorporated to describe a traveling wave as in (53),

$$\psi_R = +\left(\omega t - \frac{2\pi z}{\lambda_R}\right); \quad \psi_L = -\left(\omega t - \frac{2\pi z}{\lambda_L}\right), \qquad (z \text{ arbitrary}) \tag{11.56}$$

where $k \equiv 2\pi/\lambda$ and slightly different wavelengths for RCP and LCP are allowed. The different wavelengths do not affect our description of the wave at $z = 0$. Note, though, that the RCP and LCP must both have the same frequency ω to maintain the orientation of the summed vector at any fixed z (i.e., to yield LP radiation).

Rotated linear polarization

The effect of slightly different wavelengths for LCP and RCP (with equal frequencies) is evident from comparison of Figs. 10b and 10c, which are simultaneous views at two positions, $z = 0$ and $z = z_0$. The latter is somewhat farther along the propagation ($+z$) direction; it is closer to the observer by less than one wavelength. At $z = 0$ (Fig. 10b), the magnitudes of the LCP and RCP rotation angles are identical but opposite in sign, according to (56).

At $z = z_0$ (Fig. 10c), the vectors E_R and E_L are rotated backward toward the vertical compared with their positions at $z = 0$ in accord with (56). They are at an earlier phase

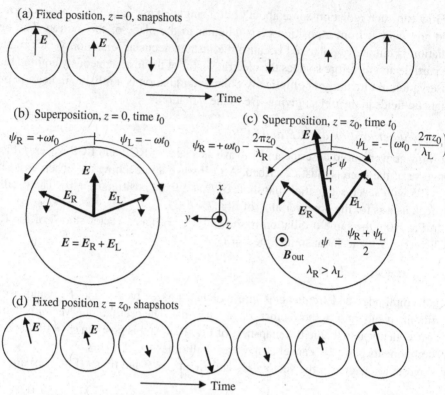

Figure 11.10. Sinusoidal linear polarized (LP) EM wave as the sum of left and right circularly polarized waves with identical frequencies for a wave propagating in the $+z$ direction as viewed in the $-z$ direction. (a) Electric vector E of an LP wave at fixed $z = 0$ at seven sequential times. (b) Vectors E_R and E_L emerging from page at $z = 0$ at time t_0 showing superposition that gives rise to the LP wave E of (a). (c) Vectors E_R and E_L and resultant E at a position slightly displaced in the $+z$ direction ($z = z_0$) at the same time t_0 as (b), where the LCP and RCP wavelengths differ slightly owing to the presence of a line-of-sight magnetic field. The plane of (linear) polarization is rotated counterclockwise by angle ψ. (d) LP wave of (c) at seven sequential times.

because the wave is traveling in the $+z$ direction and the vectors shown at $z = 0$ have not yet arrived at $z = +z_0$.

The different wavelengths ($\lambda_R > \lambda_L$ in this case) cause the vectors of Fig. 10c to be rotated backward by different amounts. For the longer λ_R, the given distance along the z-axis amounts to a smaller fraction of a wavelength; thus, the vector E_R rotates back less than E_L. In this case, the resultant vector, $E = E_R + E_L$, undergoes a net angular shift from the vertical ($\psi = 0$) to the new angular position,

$$\psi = \frac{(\psi_R + \psi_L)}{2}.\tag{11.57}$$

Substitute from (56) for ψ_R and ψ_L and set $z = D$ to obtain the net rotation angle over distance D in terms of the wavelengths λ_L and λ_R,

$$\Rightarrow\quad \psi = \pi D \left(\frac{1}{\lambda_L} - \frac{1}{\lambda_R} \right). \qquad (z = D;\ \text{rotation angle; rad})\tag{11.58}$$

Thus, if the source of the radiation is at $z = 0$ and if the position of the earth is at $z = D$, the plane of polarization is rotated by the amount given in (58). If $\lambda_R \neq \lambda_L$, the angle of rotation is nonzero.

The variation with time of the resultant E at $z = z_0$ is shown in Fig. 10d. During each cycle (period) of the wave, the resultant vector E oscillates back and forth once. The direction of the resultant will remain at the fixed angle ψ because there is no time dependence in (58). Thus, the result at z_0 is a linearly polarized wave with a rotated plane of polarization.

Similarly, at distance D (earth), one observes a rotated linear polarization. The amount of this rotation depends on the distance D and on the wavelengths for RCP and LCP in the presence of a magnetic field. The latter depend, in turn, on the indices of refraction for the two cases, which we now proceed to find.

Index of refraction

The calculation of the index of refraction $\eta(\omega)$ for a circularly polarized wave follows that for dispersion (Section 3) – that is, we solve the equation of motion for the electron displacement, which yields sequentially the polarization P, the dielectric constant κ_e, and η. Here the added effect of the interstellar magnetic field B is incorporated. As stated in the chapter footnote, we use the Helvetica font to distinguish the static magnetic field from that of the EM waves. We will find that it is the line-of-sight component of this magnetic field B_z that determines the Faraday rotation in interstellar space.

Circular motion postulated

The electrons in the plasma respond to the electric force of the EM wave and also to the $v \times B$ force due to the motion of the electron in the B field,

$$F = -e[E + (v \times B)]. \tag{11.59}$$

We will take this magnetic force to be significant but small compared with the electric force. We again omit the force due to the magnetic field B of the EM wave because it is negligible compared with the force of the E field if the electron is nonrelativistic; see (3).

Again, the electron may be considered to be a *driven oscillator* under the influence of this imposed force. In the usual manner, one writes the equation of motion ($\Sigma F = m\ddot{r}$) with the force (59) and tries to guess a steady-state solution. We consider, for now, the case in which the (static) interstellar magnetic field lies along the line of sight (z direction) so that $B = B_z\hat{k}$, where the hatted symbol is the unit vector in the z direction.

A reasonable, and correct, guess would be that an electron under the influence of RCP (or LCP) wave would follow a circular path with the same angular velocity as the rotating electric vector. With the aid of Fig. 11, we demonstrate that such a solution indeed satisfies the equation of motion.

Proceed by analyzing the forces on the electron. Figure 11 separately shows the forces for an electron driven by RCP or LCP radiation. In each case, at the instant shown, our (correct) trial solution has the electron at the top of a circular path when the rotating electric field, E_R or E_L, of the EM wave is directed upward (in the $+x$ direction). For an incident LP wave, the two component vectors E_R and E_L have the same frequency ω of rotation. The electron in the RCP case has velocity v to the left at the instant shown and thus follows the rotating electric field. The electric force $F_E = -eE_R$ is appropriately directed toward the center of rotation. The magnetic force $F_B = -ev \times B$, from (59), is also directed toward the center of rotation for $B_z > 0$; that is, $F_B = -e\omega B_z r_R$.

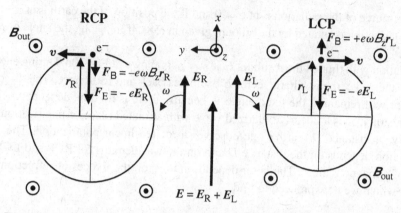

Figure 11.11. Circular motion of an electron driven by a circularly polarized EM wave in the presence of an interstellar line-of-sight magnetic field $\boldsymbol{B}$ for RCP and LCP. Both the $\boldsymbol{B}$ field and the wave propagation direction are directed out of the paper toward the reader. In the proposed solution, the electron rotates synchronously with the electric field $\boldsymbol{E}_R$ or $\boldsymbol{E}_L$. The force due to the interstellar magnetic field is oppositely directed in the two cases, leading to orbits of different radii.

These two forces continue to point radially inward throughout the postulated circular motion. The electric force $\boldsymbol{F}_E$ rotates synchronously with the electron, and the $\boldsymbol{v} \times \boldsymbol{B}$ force is always perpendicular to the $\boldsymbol{v}$ vector. A rotating net inward force of constant magnitude can give rise to the inward radial acceleration $\omega^2 r$ of circular motion. Thus, the radial component of Newton's second law ($\Sigma F_r = ma_r$) can be satisfied with this arrangement if the radius r has the correct value.

For the LCP case, the electron has to rotate in the opposite direction to stay in phase with $\boldsymbol{E}_L$, and therefore the magnetic force, $\boldsymbol{F}_B = -e\boldsymbol{v} \times \boldsymbol{B}$, is directed outward. Because the electric force dominates the magnetic force, the net force is again inward and constant in magnitude, and so Newton's second law can again be satisfied. However, the net inward force is reduced (relative to RCP), and this implies a smaller centripetal acceleration $\omega^2 r$. Because ω is fixed by the incident LP wave and is the same for both the RCP and LCP components, the LCP solution requires a smaller orbital radius, $r_L < r_R$.

Polarization vector

In our trial solutions (Fig. 11), the electrons are displaced a distance r relative to their equilibrium positions in the plasma, and so the plasma has a polarization $\boldsymbol{P} = -n_e e\boldsymbol{r}$ (17). The polarization vector $\boldsymbol{P}$ is smaller for the LCP case because of the smaller radius r of the electron orbit.

An expression for r, in terms of the physical parameters, is needed to obtain the polarization vector. Solve Newton's second law for both LCP and RCP:

$$\Sigma \boldsymbol{F} = m\boldsymbol{a}. \tag{11.60}$$

As usual, take the outward radial direction to be positive and continue to use $\boldsymbol{B} = B_z \hat{\boldsymbol{k}}$:

$$-e\boldsymbol{E}_{\mathrm{L,R}} \pm eB_z \omega r_{\mathrm{L,R}} = -m\omega^2 r_{\mathrm{L,R}} \qquad \text{(+ for LCP; – for RCP)} \tag{11.61}$$

$$r_{\mathrm{L,R}} = \frac{+(e/m)}{\omega^2 \pm (eB_z \omega/m)} \boldsymbol{E}_{\mathrm{L,R}}. \qquad \text{(Displacement)} \tag{11.62}$$

Here you can see that the LCP (+sign) has the smaller radius for $eB_z\omega/m \ll \omega^2$, which is the typical situation in interstellar space. This is in accord with our discussion of Fig. 11 just above. Note again that the electron at position r is displaced in the direction of the electric field or opposite to the force on the electron, as expected for forced oscillations at a frequency well above resonance.

Finally, substitute (62) into the expression for the polarization of the medium, $P = -n_e er$ (17):

$$\Rightarrow \quad P_{L,R} = \frac{-n_e e^2/m\omega^2}{1 \pm (eB_z/m\omega)} E_{L,R}. \quad \text{(Polarization vector; LCP (+), RCP (−))} \quad (11.63)$$

This provides the relation between the strength of the driving field and the induced polarization; they are antiparallel. Because E_L (or E_R) rotates at the EM frequency, so does P_L (or P_R). Figure 2d illustrates the situation at a fixed time. Compare the relative directions of E, P, and the electron displacement with the static and linearly oscillating cases (Figs. 2b,c).

Dielectric constant and the index

The polarization vector may now be used to obtain the desired index of refraction. Equate (63) and $P = \varepsilon_0(\kappa_e - 1)E$ (15) to obtain the dielectric constant κ_e,

$$\kappa_{e,L,R} = 1 - \frac{\omega_p^2/\omega^2}{1 \pm (\omega_{B_z}/\omega)}, \quad \text{(LCP (+), RCP (−))} \quad (11.64)$$

where the definition of the plasma frequency (28) has been invoked together with the non-relativistic *cyclotron* (or gyro) frequency (8.3) (Prob. 42) modified to be based on B_z,

$$\omega_{B_z} \equiv eB_z/m. \quad \text{(cyclotron frequency)} \quad (11.65)$$

The parameters ω_{B_z} and B_z are positive when the latter is directed toward the observer (out of the paper in our figures). The value ω_{B_z} will generally be less than, but of the same order of magnitude as, the true cyclotron frequency eB/m, where B is the magnitude of B.

The dielectric constant (64) is correct for *quasi-longitudinal propagation* in which the magnetic field is more or less in the propagation direction. *Quasi-transverse propagation* leads to a very different expression for the rotation, but in interstellar space (and at higher frequencies) the rotation is dominated by the quasi-longitudinal rotation. Thus, the latter adequately describes the rotation arising in interstellar space even when the magnetic field is not directed along the propagation direction.

The index of refraction η that we seek follows from $\eta = \kappa_e^{1/2}$ (12):

$$\Rightarrow \quad \eta_{L,R}^2 = \kappa_{e,L,R} = 1 - \frac{\omega_p^2/\omega^2}{1 \pm (\omega_{B_z}/\omega)}. \quad \text{(Index of refraction; LCP (+), RCP (−))} \quad (11.66)$$

For $B_z > 0$ (directed out of the paper toward observer), we have $\omega_{B_z} > 0$, and

$$\eta_R < \eta_L. \quad (11.67)$$

From the dispersion relation (13) and the definitions of $\omega = 2\pi\nu$ and $k = 2\pi/\lambda$,

$$\frac{c}{\eta_R} = \lambda_R \nu; \quad \frac{c}{\eta_L} = \lambda_L \nu, \quad (11.68)$$

which, with (67), yield

$$\lambda_R > \lambda_L.$$

(11.69)

This is the case illustrated in Fig. 10c.

Again we stress that the driving frequency ν is the same for both LCP and RCP. We will find that, for interstellar conditions, the magnetic-field term ω_{B_z}/ω in (66) is a small correction to the dispersion term in ω_p/ω, which in turn is a small correction to the vacuum value of unity for the index of refraction η. Finally, note that, if $B \to 0$ and hence $\omega_B \to 0$, the index of refraction (66) reduces to that for dispersion (29).

Cyclotron frequencies

The z-axis cyclotron frequency becomes, if one substitutes values for e and m into (65),

$$\nu_{B_z} = \omega_{B_z}/2\pi = 2.8 \times 10^{10} \, B_z.$$ (Cyclotron frequency, Hz;
ν_c in Hz; B_z in T)   (11.70)

In interstellar space with magnetic fields of order $B = 0.3$ nT and hence z components B_z of comparable magnitude, it is convenient to write ν_{B_z} for $B_z = 1$ nT as

$$\nu_{B_z} = \frac{eB_z}{2\pi m} = \left(\frac{B_z}{10^{-9}\,\text{T}}\right) 28.0 \, \text{Hz},$$

(11.71)

which is very small compared with frequencies in ground-based radio astronomy, which are $\gtrsim 10$ MHz (32). This leads to very small modifications to the indices of refraction (66), but, as will be seen, this will have a substantial effect on the rotation of the polarization vector because of the long distances traversed by astronomical radio waves.

In the relatively high magnetic field of the earth, $B_z \lesssim 10^{-4}$ T, the cyclotron frequency is

$$\nu_{B_z} \lesssim 2.8 \, \text{MHz}.$$ ($B_z \lesssim 10^{-4}$ T; earth's magnetic field)   (11.72)

This can be a substantial fraction of radio-astronomy frequencies ($\gtrsim 10$ MHz) and hence can yield significant Faraday rotation; see discussion under "Ionosphere" below.

Rotation angle

We are now in a position to calculate the rotation angle of the LP wave from the indices of refraction of its constituents, LCP and RCP. We have seen that the radio waves used in astronomy usually have frequencies ω significantly greater than either the plasma (33) or cyclotron (71) frequencies in most of the ISM. Thus,

$$\omega_p \ll \omega; \quad \omega_{B_z} \ll \omega.$$

(11.73)

Our expressions for the angle of Faraday rotation may be simplified with these two inequalities. We continue to assume uniform conditions along the line of sight.

Uniform conditions

Write the index of refraction, the square root of (66), taking into account the limiting expressions (73). Use twice the expansion $(1 + x)^n \to 1 + nx$ if $x \ll 1$:

$$\eta_{L,R} \approx 1 - \frac{1}{2}\frac{\omega_p^2}{\omega^2}\left(1 \mp \frac{\omega_{B_z}}{\omega}\right), \quad \text{(LCP } (-), \text{ RCP } (+); \, \omega_p \ll \omega; \, \omega_{B_z} \ll \omega)$$ (11.74)

where, again, the expressions for ω_p and ω_{B_z} are given in (28) and (65).

The rotation angle may now be obtained. First, rewrite the dispersion relations (68) as

$$\frac{1}{\lambda_R} = \frac{\nu \eta_R}{c}; \quad \frac{1}{\lambda_L} = \frac{\nu \eta_L}{c} \tag{11.75}$$

and substitute them into the rotation angle expression (58) to obtain

$$\psi = \pi D \frac{\nu}{c}(\eta_L - \eta_R). \qquad \text{(Rotation angle; rad)} \tag{11.76}$$

The difference of indices of refraction follows from (74), $\eta_L - \eta_R = \omega_p^2 \omega_{B_z}/\omega^3 = \omega_p^2 \omega_{B_z}/(2\pi\nu)^3$, and so we have

$$\psi = \frac{D}{8\pi^2 c} \frac{\omega_p^2 \omega_{B_z}}{\nu^2}. \tag{11.77}$$

Substitute into (77) the definitions of $\omega_p^2 = n_e e^2/\varepsilon_0 m$ (28) and $\omega_{B_z} = eB_z/m$ (65) to obtain the desired expression for Faraday rotation for uniform conditions along the line of sight:

$$\psi = \frac{1}{8\pi^2 \varepsilon_0} \frac{e^3}{m^2 c} n_e B_z D \frac{1}{\nu^2}. \qquad \begin{array}{l}\text{(Faraday rotation;}\\ \nu \gg \nu_p, \nu \gg \nu_{B_z}; \text{rad)}\end{array} \tag{11.78}$$

Note that the rotation angle is proportional to the product $n_e B_z$ and to D and that it varies with frequency as ν^{-2}. The rotation is most pronounced at low frequencies (long wavelengths).

Nonuniform conditions

If the **B** field or the electron density (or both) is not constant along the path, one must integrate along the path. Thus, from (78),

$$\psi = \frac{e^3}{8\pi^2 \varepsilon_0 m^2 c} \frac{1}{\nu^2} \int_0^D n_e B_z \, ds, \tag{11.79}$$

where ψ is the accumulated rotation angle from source to observer in radians, m is the electron mass, e is the electron charge, n_e is the electron number density, and B_z is the line-of-sight component of the interstellar magnetic field. The integration is over the distance D from the source to the antenna. Substitution of the values for the constants yields

$$\psi = 2.36 \times 10^4 \frac{1}{\nu^2} \int_0^D n_e B_z \, ds, \tag{11.80}$$

where the units are ψ (rad), n_e (m^{-3}), B_z (T), ν (Hz), D (m), and ds (m). In the local (to the sun) ISM, one might have a $D = 2 \times 10^3$ LY $\approx 2 \times 10^{19}$ m, $n_e \approx 10^4$ m^{-3}, and $B_z \approx 0.3$ nT. These values, if taken to be constant along the path, would lead to a rotation of 1.42 radians or $\sim 81°$ at $\nu = 1$ GHz.

In terms of the wavelength λ, the expression equivalent to (80) is, from $\nu = c/\lambda$,

$$\psi = 2.63 \times 10^{-13} \lambda^2 \int_0^D n_e B_z \, ds, \tag{11.81}$$

where the units of λ are meters. The use of $\nu = c/\lambda$ to obtain (81) is evidently an approximation because the correct expression is $\lambda = c/\nu\eta$. For our conditions ($\omega_p \ll \omega$ and $\omega_{B_z} \ll \omega$), one has $\eta \approx 1$ and hence $\lambda \approx c/\nu$.

The angle ψ is the total rotation of the LP wave along the entire path from source to observer. It is *not* the measured angle of polarization at the observer, which is conventionally measured on the celestial sphere counterclockwise from the direction to the north pole. One must also find the angle of polarization at the source and take the difference to obtain ψ.

The angle at the source is readily obtained if the polarization angle is measured at several frequencies ν and the measured angles are extrapolated to $\nu \to \infty$, where, according to (80), there is no rotation along the path. The angle at $\nu = \infty$ thus equals the polarization angle at the source. The rotation angle ψ at some ν is the difference between the angle measured at ν and that obtained by extrapolation to $\nu = \infty$.

Rotation measure

The commonly used quantity that incorporates the physical quantities n_e and B_z is the *rotation measure* (RM), which is defined by (81) to be

➡
$$\mathrm{RM} \equiv \frac{\psi}{\lambda^2} = 2.63 \times 10^{-13} \int_0^D n_e B_z \, ds, \quad \text{(Rotation measure; rad/m}^2\text{)} \qquad (11.82)$$

with the following units: RM (radians/m^2), n_e (m^{-3}), B_z (T), and D (m). For the Crab nebula, the rotation measure is RM ≈ -25 rad/m^2; see (85) below. The rotation measures for radio pulsars range from 0.1 to 1000 rad/m^2.

Keep in mind that Faraday rotation is a meaningful concept only if the source emits polarized radiation. Fortunately, many radio sources (galactic and extragalactic) are synchrotron sources that emit LP waves. As they travel to us, they probe the path they travel, and information about electron densities and magnetic fields is obtained.

The average of the product $n_e B_z$ is the integral in (82) divided by the distance D. Thus, (82) may be written as

$$\mathrm{RM} = 2.6 \times 10^{-13} \langle n_e \, B_z \rangle_{av} \, D. \qquad (n_e \text{ and } B_\parallel = \text{constant}) \qquad (11.83)$$

Knowledge of the RM for radiation from a source at a known distance D yields an average value of the product $n_e B_z$. The RM is obtained from a determination of the total rotation ψ of E at some λ (82). This requires measurements of the direction of linear polarization at several wavelengths, as pointed out just above.

Crab nebula

The Crab nebula is of special interest because it is such a bright radio source, its distance is known, and it houses a pulsar that provides a measure of the dispersion (Section 3). Early measurements of the Faraday rotation for the Crab nebula are illustrated in Fig. 12a. The polarization angle at zero wavelength is set to zero. The abscissa is linear in λ^2, and thus, according to (81), a straight line is expected.

The measured values (plotted points) of Fig. 12 agree well with a straight line (i.e., with the expected λ^2 dependence). The straight-line fit to the data shows that the polarization angle varies from 0° as $\lambda \to 0$ m and $-130°$ at $\lambda = 0.3$ m. The total rotation along the path

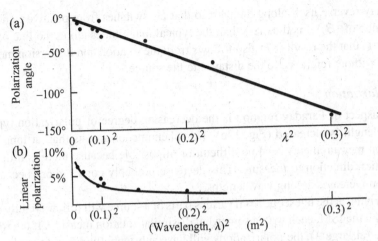

Figure 11.12. Early radio polarization measurements of the Crab nebula. (a) Polarization angle as a function of wavelength squared of the radio waves. The straight line is in accord with the λ^2 dependence of Faraday rotation (81). (b) Degree of linear polarization of the same radio waves. The degree of polarization P is derived from flux measurements at two orthogonal polarizations, $P(\%) \equiv 100 \times (S_2 - S_1)/(S_2 + S_1)$. [After J. D. Kraus, *Radio Astronomy*, 2nd Ed., Cygnus-Quasar Books, 1986, pp. 5–17]

at $\lambda = 0.30$ m is therefore $\psi = -130°$. This directly yields a value for the rotation measure from (82) of

$$\text{RM} = \frac{\psi}{\lambda^2} = \frac{-130}{(0.3)^2} = -1444 \text{ deg/m}^2, \tag{11.84}$$

or, if converted to radians,

$$\text{RM} = -25 \text{ rad/m}^2. \qquad \text{(Crab nebula)} \tag{11.85}$$

More recent measurements give RM ≈ -20 rad/m^2 for the ISM between the Crab and the sun after the variability of the RM over different parts of the nebula owing to rotation occurring within the nebula is taken into account.

The negative value of RM indicates that the rotation of the E vector polarization with z when viewing in the $-z$ direction is in the clockwise direction, which is the opposite of that shown in Figs. 7 and 10c. This tells us directly that the B field, on average, points away from, rather than toward (as in the figures), the observer. This value of RM (85) and the distance to the Crab, ~ 6000 LY, can be substituted into (83) to obtain a distance-weighted average value of the product $n_e B_z$ along the line of sight:

$$\langle n_e B_z \rangle_{\text{av}} \equiv \frac{\text{RM}}{2.6 \times 10^{-13} D} = -1.7 \times 10^{-6} \text{ T m}^{-3}. \qquad \text{(Crab)} \tag{11.86}$$

An average of B_z along the line of sight is obtained if a value of n_e is adopted and taken to be constant along the path. Adopt the average value of electron density obtained from dispersion to the Crab pulsar (51), $n_e = 3 \times 10^4$ m^{-3}, to find

$$\langle B_z \rangle_{\text{av}} \approx -0.06 \text{ nT}. \qquad \text{(Average } B_z \text{ on path to Crab nebula)} \tag{11.87}$$

The field **B** may reverse itself along the path so that B_z switches from positive to negative values. The value of $\langle B_z \rangle_{av}$ is thus less than the typical magnitudes along the line of sight. We will find (91) that the result (87) also follows from the rotation and dispersion measures taken together without reference to the distance to the source.

Depolarization

An important aspect of Faraday rotation is the decreasing degree of polarization typically found as wavelength is increased (Fig. 12b). This phenomenon means that, at long wavelengths, rotation measurements become difficult or impossible because there is little or no linear polarization direction to measure. This decrease probably comes about because the Faraday rotation increases at long wavelengths.

Radiation arriving at the telescope from various parts of a source will follow slightly different paths through the ISM, each with a somewhat dissimilar rotation measure. If the rotation angles are large (at large λ), the polarizations will have substantially different angles at the observer. Vector addition of the electric vectors results in a reduction of the net polarization. Depolarization can also occur if rotation takes place within the emission volume, as is the case for the Crab nebula.

Ionosphere

The rotation measure of the ionosphere turns out to be sufficiently large to merit careful subtraction from galactic measures of RM. Recall that the ionization in the ionosphere varies substantially with time – especially from day to night. For example, in an RM survey of the Galaxy from the Arecibo observatory (Fig. 13), the ionospheric rotation measure ranged from 0.5 to 4 rad/m^2. Signals from "RM-calibrated pulsars" were used to help remove this effect.

11.5   Galactic magnetic field

The average interstellar magnetic field (parallel component) between the earth and a pulsar can be obtained with no knowledge of the distance by invoking both the amount of Faraday rotation and the degree of dispersion.

Ratio of rotation and dispersion measures

The expressions (82) and (52) for the rotation and dispersion measures can be combined to yield

$$\frac{\int_0^D B_z n_e \, ds}{\int_0^D n_e \, ds} = 3.80 \times 10^{12} \frac{RM}{DM},$$

(Average magnetic field weighted by n_e; T)   (11.88)

where RM/DM is the ratio of the two measures. The ratio of integrals is recognized as the definition of the average of B_z weighted by the electron density, which surely is not uniform along the propagation path:

$$\langle B_z \rangle_{av} = 3.80 \times 10^{12} \frac{RM}{DM}.$$

(11.89)

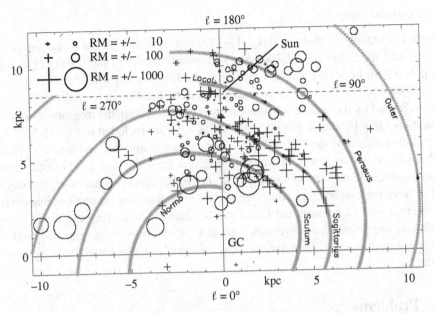

Figure 11.13. Rotation measures (RM) at 430 MHz in plane of Galaxy viewed from above the Galaxy and showing spiral arms as wide gray lines. Positions are given in rectilinear x, y coordinates with the galactic center at the origin $(0, 0)$ and the sun at the intersection $x = 0$, $y = 8.5$ kpc (27.7 kLY). Directions from the sun (galactic longitudes ℓ) are given. Each cross or circle indicates the measured RM to a specific pulsar, and the size denotes the magnitude in units of rad/m^2 (see inset code). The crosses (circles) indicate that the magnetic field points on average in the same (opposite) direction as the radio propagation direction, which is toward the sun. [J. Weisberg et al., ApJ Suppl. **150**, 317 (2004), Fig. 3]

Application of the DM from the Crab *pulsar* (50) and the RM from the Crab *nebula* (85),

$$DM = 1.86 \times 10^{24} \text{ m}^{-2} \qquad \text{(Crab pulsar)}$$
$$RM = -25 \text{ rad/m}^2, \qquad \text{(Crab nebula)} \qquad (11.90)$$

yields

$$\langle B_z \rangle_{av} = -0.051 \text{ nT.} \quad \text{(Average line-of sight magnetic field to Crab}$$
$$\text{nebula or pulsar; } \langle B_z \rangle_{av} \text{ directed toward Crab)} \qquad (11.91)$$

This agrees fairly well with the previous value (87) for this line of sight, but we note again that the typical magnitudes along the line of sight are likely to be substantially greater. We have further assumed, reasonably, that the dispersion or Faraday rotation along the portion of the path within the nebula is small compared with that in the ISM between the Crab nebula and the earth.

We chose to use the RM of the Crab nebula, not its associated pulsar, because it is very bright and yields a stronger signal. Most radio pulsars do not have such bright nebulae surrounding them, and the pulsed radiation is used for both the RM and the DM. The polarization direction typically varies through the duration of a given pulse, and so the RM must be determined independently for each phase interval of the pulse period.

Galactic map

Magnetic field mapping of the Galaxy (Fig. 13) has been carried out using pulsar rotation measures with the aid of the NE2001 electron density model of the Galaxy (Fig. 6). For a given pulsar with both DM and RM measures, the DM is used with the NE2001 model to obtain a distance. This then is used to mark the galactic location of the pulsar with a circle or cross, as dictated by the sign of the rotation, with size indicating the magnitude of the RM to that location. Only pulsars at galactic latitude $|b| < 9°$ are included in Fig. 13.

The plotted values of the RM in Fig. 13 are integral quantities, and thus a uniform field would appear as values that increase with distance from the sun. Because the integrated quantity can include regions with reversed sign of $n_e B_z$, caution is necessary in interpreting this plot. Nevertheless, one notes the large regions containing circles that lie adjacent to large areas of crosses (e.g., between the Local and Sagittarius spiral arms at $\ell \approx 60°$). Together they indicate large-scale field reversals. The small, dark arrows superimposed on the spiral arms (gray lines) indicate the magnetic field direction in those locations as deduced from these data.

Problems

11.2 Maxwell's equations and the index of refraction

Problem 11.21. (a) Demonstrate that the expression (3), $E_x = E_0 \cos(\omega t - kz)$, describes a wave traveling down the positive z-axis at speed ω/k. Evaluate this expression at times $t = 0$ and $t = P/8$, where P is the period. Note that $k = 2\pi/\lambda$. Sketch one cycle of $E_x(z, t)$ for each case and indicate the position of the first maximum in each. How far did the wave move and what was its speed? (b) Demonstrate analytically that the phase velocity of the wave is ω/k for the general case, $E_x = f(\omega t - kz)$, with a fuller explanation than in the text. Argue that setting the differential of $f(\omega t - kz)$ to zero is the proper procedure.

Problem 11.22. (a) Expand the divergence and curl operators in Table 1 and apply the simplifications given in the footnote to the table to obtain the component equations for a vacuum, as given in the table. What further restrictions are placed on the wave in doing so? (b) Use the component equations in Table 1 to obtain the vacuum wave equations (2); follow suggestions in the text. (c) Demonstrate that $E_x = f(\omega t - kz)$, where f is any arbitrary function, is a solution to the wave equation in E_x (2) with the constraint (5). (d) How do the divergence equations in Table 1 and the restrictions from part (a) contribute to the solution?

Problem 11.23. Carry out the following for a dilute plasma: (a) Verify that the wave equations (8) follow from the laws in Table 2 by operating on the component form of the laws with partial derivatives. (b) Verify that the function B_y (7) follows from E_x (6) and Faraday's law in Table 2. (c) Verify that E_x (6) and B_y (7) satisfy their respective wave equations (8) with the restriction (9).

11.3 Dispersion

Problem 11.31. A thin rectangular slab of thickness d and area D^2, where $d \ll D$, contains a plasma of protons and electrons with equal number densities, $n_e = n_p$, uniformly spread throughout the volume. Let the electrons be slightly displaced along the short axis a distance

$x \ll d$ such that a thin slab of negative charge lies along one face and a thin slab of positive charge along the opposite face. (a) Calculate the electric field within the overlapping volumes and the associated restoring force on an individual electron therein. (You will need Gauss's law.) (b) Find the frequency of oscillation if the electrons are released. You have just calculated the plasma frequency (28).

Problem 11.32. (a) Demonstrate that, for $\omega \gg \omega_p$, a fractional change in the frequency makes a much smaller fractional change in the index of refraction η (29). (b) Demonstrate that a crest of wave 1 moves farther than a crest of wave 2 (as shown in Fig. 4) if $\lambda_1 > \lambda_2$, $\omega > \omega_p$, and (29) describes the index of refraction. Use the dispersion relation (13) to justify the statement that $\nu_1 < \nu_2$.

Problem 11.33. (a) Evaluate the group velocity $v_g = d\omega/dk$ for our dilute plasma with index of refraction $\eta(\omega)$ given in (29) to obtain (46), without reference to the text. (b) Evaluate your expression for radio waves of frequency $\nu = 100$ MHz and electron density 3×10^4 m^{-3} and compare it with the speed of light in a vacuum. What is $c - v_g$? (c) For this electron density, what would be the time delay for a signal from the distance of the Crab nebula (6000 LY) relative to the transit time at speed c? [Ans. $-$; $\sim -10^{-10}\,c$; $\sim$20s]

Problem 11.34. (a) What is the dispersion measure (DM) for the Crab nebula in units of LY m^{-3} and in units of pc cm^{-3}? Use the measured value of DM given in (50). (b) Use your values of the DM to obtain $\langle n_e \rangle_{av}$ if you are given the distance to the Crab as 6000 LY or, alternatively, if you are given it as 1840 pc. [Ans. $\sim 10^8$ LY/m^3, $\sim$50 pc/cm^3; $\sim 10^4$ m^{-3}]

Problem 11.35. Find an expression for the rate of frequency change $d\nu/dt$ caused by dispersion as a function of ν, the dispersion measure DM and physical constants. Assume $\omega \gg \omega_p$ and start with (47). [Ans. $\propto -\nu^3/$DM]

11.4   Faraday rotation

Problem 11.41. Consider right circularly polarized radiation. (a) Demonstrate analytically that the components (53) yield a rotating E vector with magnitude that is independent of time t and propagation position z. What is the value of the magnitude? (b) Find the angle $\psi_R(z, t)$ of this vector in terms of z and t directly from the expressions (53). Hint: what is its tangent? Compare with the angle label for ψ_R in Fig. 10c. (c) Evaluate your answer to (b) at a fixed position z_0 while retaining the variation in t. Find, at this position, the angle ψ_R through which the vector rotates as t increases by one period $P = 2\pi/\omega$ of the wave. Specify the direction of rotation. (d) Evaluate your answer to (b) at a fixed time t_0 while retaining the variation in z. Find, at this time, the angle ψ_R through which the vector rotates as z increases by one wavelength $\lambda = 2\pi/k$. Again specify the direction of rotation. [Ans. $-$; $\omega t - kz$; $+2\pi$; -2π]

Problem 11.42 Apply Newton's second law to a free electron that is moving nonrelativistically normal to the field lines of a uniform magnetic field B in order to find the angular frequency ω_B of its circular motion; see (65). (There is *no* applied EM field in this case.) See Prob. 8.33 for the relativistic case.

Problem 11.43. (a) What is the expected angle of Faraday rotation for radiation at 3 GHz traversing the distance from the center of the Galaxy (25,000 LY away) if the free electron density is constant at 1×10^5 m^{-3} and the radial component of the magnetic field is constant at 0.1 nT? What is the rotation measure? (b) Repeat for the ionosphere. Adopt $n_e = 10^{12}$ m^{-3}, $B_z = 30$ μT, and $D = 300$ km. [Ans. ~360°, ~600 rad/m^2; ~1°, ~2 rad/m^2]

Problem 11.44. (a) Because we can not travel out to the source to measure the angle of the polarization at the beginning of the trip, how do we know how much the angle rotated en route to the earth? (b) Discuss the possibility, or lack thereof, of obtaining a measure of the column density of electrons along the line of sight with a dispersion measurement to a *steady* source (i.e., one that is not a pulsar).

11.5 Galactic magnetic field

Problem 11.51. Look up two articles in the *Annual Reviews of Astronomy and Astrophysics* pertaining to the magnetic field in the Galaxy and the interstellar medium and write a short paper on the roles the magnetic field plays in the interstellar medium. They are R. Beck *et al.*, "Galactic Magnetism: Recent developments and perspectives," *ARAA* **34**, 155 (1996) and D. P. Cox, "The three-phase interstellar medium revisited," *ARAA* **43**, 337 (2005). See especially Section 9 of the latter.

12

Gravitational lensing

What we learn in this chapter

Gravitational fields distort space, and light beams follow **null geodesics**, the "straight lines" of curved space. To a human observer, it appears that gravitational fields bend the paths of the rays. The bending angle of a light ray passing near a **point-mass gravitational "lens"** is deduced from **Newtonian** arguments, except that the proper result from **general relativity (GR)** is twice that so obtained, as was famously confirmed during the **solar eclipse of 1919**. Gravitational lensing phenomena now allow astronomers to detect otherwise invisible gravitational matter known as **dark matter**.

A **quasar** is a distant, luminous, and often variable **active galactic nucleus** (AGN). Electromagnetic rays from a quasar that pass through the gravitational field of an intervening galaxy can appear to an observer as multiple images of the quasar. The **twin quasar Q 0957+561** was the first discovered example in 1979. The **image positions** for **point-mass lensing** can be deduced from the **physical bending angle expression** along with the **geometric ray tracing equation**, which together yield a **lens equation**. The images may be either **magnified** or **demagnified**. Because the surface brightness of any image equals that of the source, an unresolved magnified (or demagnified) image will appear as a point source with increased (or decreased) flux density. The **magnification factor** of all images of a point source, taken together, **is always greater than unity**.

Two stars that temporarily become nearly aligned, one well behind the other, yield multiple images with milliarcsec separations. Although they may appear as a single unresolved image, their summed fluxes will be enhanced for the month or so the background star is in the vicinity of the foreground lensing star; the **background star** thus appears to **brighten temporarily**. Detections of such **microlensing** events have been used to place limits on the existence of **compact dark objects** in the halo of the Galaxy.

The image locations and magnifications for a quasar lensed by a **mass distribution** such as a galaxy can be determined and easily visualized through a **time-delay surface** and the application of **Fermat's principle**. A galaxy lens can produce more than two images, and four are not unusual. The different images have different photon transit times, both **gravitational and geometric**, from the source to the observer. If the source is variable, the differential times can be measured, and this leads to an independent measure of the **Hubble constant**.

Lensing also distorts background galaxies. One elliptical galaxy (MG1131+0456) images the lobe of a background radio-lobe galaxy into a tiny 3″ diameter ring of radiation, an **Einstein ring. Clusters of galaxies** image background galaxies into **large arcs** or slightly curved elongations called **arclets.** The arcs and arclets effectively **map the matter, both dark and luminous,** associated with the cluster (e.g., **Abell 2218**).

Gravitational structures on large scales in the absence of visible foreground clusters can be mapped with **weak gravitational lensing.** Barely perceptible systematic elongations of background galaxies over substantial portions of the sky can provide the **angular power spectrum of matter density fluctuations** out to $z \approx 0.2$.

12.1   Introduction

A powerful new tool for the study of astronomical objects became available to astronomers some 30 years ago with the realization that the gravitational bending of light rays from astronomical objects due to intervening matter (galaxies, clusters of galaxies, and stars) is a widespread, detectable phenomenon.

The observational result of gravitational lensing can be multiple images of a distant quasar or distortions of the shapes of distant galaxies. Given intensity variations of the lensed source, the Hubble constant can be determined without independent knowledge of the distance to the source. The lensing phenomenon also brightens faint objects that would not otherwise be detectable.

This chapter presents the lensing phenomenon in terms of both geometric ray tracing and Fermat's principle. Examples of the utility of this new branch of astronomy are presented.

12.2   Discovery

The bending of light by gravitational fields follows from Einstein's general theory of relativity, which is usually called "general relativity" (GR). The bending of starlight passing near the sun was found in 1919 to agree quantitatively with the 1.75″ deflection predicted by the theory; this made Einstein famous. Later, in 1936, Einstein examined the possibility that gravitational lensing might be detectable if one star intercepts the line of sight to a more distant star. He concluded that such an occurrence is highly unlikely and, even if it occurred, would have been impossible to detect with the then current instrumentation.

Astronomer Fritz Zwicky suggested in 1937 that galaxies themselves could serve as lenses to more distant objects. This was highly prescient, but discovery of the phenomenon awaited the detection of bright pointlike sources (quasars) and instrumentation with increased sensitivity and angular resolution. The phenomenon was brought forcefully to the attention of astronomers with the discovery in 1979 of the *twin quasar*, which turned out to be two images of a single quasar.

Quasars

A quasar is an unusually luminous nucleus of a galaxy viewed from a direction that provides a relatively unobscured view of the nucleus or of a jet emerging from it. Galactic nuclei that are less luminous or partially obscured by material orbiting the nucleus, for example, are known by other names (e.g., *Seyfert galaxies* of types 1 and 2). These nuclei of all types and luminosities are known generally as *active galactic nuclei* (AGN). The subclass quasars are now classified as optically violent variables (OVV), either radio loud or radio quiet. They are also known as *blazars* (Section 9.4). Radio images of quasars frequently reveal jets and lobes that indicate episodic ejections of material into intergalactic space (Section 7.6).

Only a tiny fraction of galaxies exhibit prominent AGN phenomena, but they are still quite numerous. Those brighter than magnitude $B = 22$ number about 100 deg^{-2}, yielding a total of $\sim 4 \times 10^6$ over the whole sky. Normal galaxies at this magnitude are ~ 100 times more numerous. The luminosity of an especially active nucleus can so outshine the stars in the rest of the galaxy that only the pointlike nucleus is visible on a photographic plate. Modern techniques can reveal the surrounding nebulosity of stars.

Quasars can be quite variable in intensity on time scales ranging from days to many years. This variability suggests source regions as small as a light day. The size of a flaring region is limited by the time it takes light to traverse it; otherwise, the several elements of the region could not coordinate their flaring. This argument for small sizes is relaxed by relativistic motions in the source; see "Superluminal motion" in Section 7.6.

The spectra of quasars show large redshifts of their spectral lines that are typically in the range $z = 0.2 - 2.5$, but a few reach beyond $z = 6$. The redshift parameter z is defined as

$$z \equiv \frac{\lambda - \lambda_0}{\lambda_0}, \qquad \text{(Definition of redshift)} \qquad (12.1)$$

where λ_0 is the rest wavelength of a spectral line and λ is its observed wavelength. A quasar at $z = 2$ would have its spectral lines shifted a factor of three; $\lambda/\lambda_0 = 1 + z = 3$. The large redshifts indicate relativistic recession velocities and hence distances that are large fractions of the Hubble distance, which is $c/H_0 = 15$ GLY for $H_0 = 20$ km s^{-1} MLY^{-1}. In turn the large distances imply huge luminosities of, for example, $\sim 10^{39}$ W ($\sim 10^{12}$ $L_\odot$). (The Hubble constant H_0 gives the recession speed of a galaxy in terms of its distance, $v = H_0 r$. At speed c, the distance is c/H_0. See AM, Chapter 9 and also Sections 6.3 and 9.5 of this text.)

The source of such huge luminosities from such small volumes has been called an "infernal machine." The difficulty in explaining the large luminosities led to much controversy as to whether the objects are really at the large distances indicated by their redshifts, but now we know that quasars are indeed very distant and very luminous.

The compactness and luminosity of these objects make it highly probable that they are *massive black holes* of order $\sim 10^8$ $M_\odot$ for the brighter examples. In this picture, the large luminosity arises from the capture of nearby stars or stellar material in the potential well of the black hole. The stellar gas loses potential energy and undergoes heating as it spirals around and toward the black hole. The hot gases radiate intensely as do jets of material ejected along the angular momentum axis (Fig. 7.7).

The radiation output of quasars spans all wavelength bands from the radio to the gamma ray. As bright compact objects, quasars are used as probes of the intergalactic medium between them and the earth. For example, cool, gaseous clouds in intergalactic space absorb the radiation at particular frequencies in quasar spectra – notably at the Lyman alpha line of hydrogen.

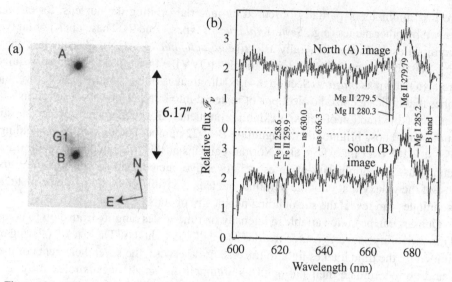

Figure 12.1. "Twin quasar" Q 0957+561. (a) Image taken with the WFPC2 camera on the Hubble Space Telescope. The lensing galaxy G1 is evident as a faint nebulosity. It is embedded in a cluster of galaxies that contribute to the lensing. The ray yielding image B passes deeper into the cluster potential than does the ray for image A. (b) Optical spectra of the two components of Q 0957+561. A broad emission line of Mg II is centered in each spectrum at $\lambda = 675.3$ nm. The rest wavelength of this line is $\lambda_0 = 279.79$ nm, which, from (1), gives a redshift of $z = 1.41$. Five narrow absorption lines of Fe II, Mg II, and Mg I appear at redshift $z = 1.39$ in both spectra. Atmospheric night-sky "ns" emission and B-band absorption are marked. [(a) G. Bernstein *et al.*, *ApJ* **483**, L79 (1997); (b) R. Weymann *et al.*, *ApJ* **233**, L43 (1979)]

One such cloud along the line of sight will have a distinct radial velocity relative to the earth observer owing to its recession velocity as part of the general expansion of the universe. Many such clouds at various distances, and hence at different recession velocities, lead to many absorption lines at different frequencies. The density of the absorption lines is such that this type of spectrum is known as the *Lyman alpha forest*. It is rich in information about the intergalactic medium.

The gravitational lensing of the radiation from quasars is another probe of the matter along the line of sight, whether it be dark or luminous.

Twin quasar Q 0957+561

Optical discovery

In 1979, two quasars (Fig. 1a) were discovered in optical images taken at the location of a radio source. They are of comparable B-band magnitudes ($m_B = 16.7$ and 17.0), the brighter designated "A" and the fainter "B," and are separated from each other by only $6''$. This is a highly improbable occurrence given the scarcity of quasars of this magnitude (one per a few square degrees). More than that, the emission-line spectra of the two quasars exhibit identical redshifts of $z_e = 1.4136$ (Fig. 1b).

The similarity of the two objects was further reinforced by the detection of a system of narrow absorption lines in each object with the same redshift, $z_a = 1.391$. This is normally indicative of an intervening cloud close to the quasar and very possibly ejected by it. Thus, one is forced to conclude that the two objects are, in some way, physically related to each other. This object is called Q 0957+561 after its celestial coordinates.

The case for gravitational lensing of Q 0957+561 was greatly strengthened by the discovery of an intervening galaxy at $z = 0.39$. This galaxy is a member of a cluster that also acts gravitationally on the passing light. The galaxy is the fuzzy extension (to the northeast) of the southern quasar image in Fig. 1a. The discovery of Q 0957+561 was thus a spectacular demonstration of the bending of light rays by gravitational fields and therefore of GR.

It is now well accepted that, for Q 0957+561 and many other quasar pairs, only one physical quasar exists and that the two images are formed by a *gravitational lens*. The lens in such cases is typically an intervening galaxy that is sometimes embedded in a cluster of galaxies.

True binary quasars are now known in which two quasars are in close proximity and gravitationally bound. As many as $\sim$20 of such binary quasars have become known to date. They most likely formed from the merging of two galaxies, each with its massive black-hole nucleus. In these cases, the spectra of the two distinct quasars can differ significantly, and they will most probably lack a visible foreground galaxy.

Radio imaging

Photons of all wavelengths follow the same (curved) path from source to observer as they traverse gravitational fields according to GR. They follow the same *geodesic*, the shortest line on a given surface between two points. Radio astronomers should therefore also see double images of Q 0957+561 because the quasar system is known to be a radio source. In fact, radio observations with the Very Large Array (VLA) of 27 large telescopes show emission from each of the twin quasars as intense pointlike sources labeled A and B (Fig. 2a).

The radio image also shows a jet extending from A to a lobe with hot spots C and D and another lobe E on the opposite side of A. Jets and lobes on opposing sides of a compact source are a common feature of extragalactic radio sources. The southern quasar image B does not show these jets and lobes. The imaging depends critically on the relative locations of the object and lens; the image of a large jet thus need not exactly mimic the imaging of the pointlike quasar. The radio image also shows emission from the position of the intervening galaxy seen in the optical band.

Radio studies of Q 0957+561 at very high resolution, 3.5 milliarcsec, using *very long baseline interferometry* (VLBI) show identical minijets in the two quasar images (Fig. 2b). Each extends only about 0.06″ from the pointlike core of the quasar and in the same direction. A jet this close to the quasar is much more likely to be imaged in the same way as the quasar. The two images also manifest similar polarization directions. This shows dramatically that, in the radio as well as the optical bands, the two quasar images are of a single quasar.

In the decade after 1979, several dozen examples of multiply imaged quasars were found. At this writing there are about 100 examples. The phenomenon of gravitational lensing is pervasive across the sky.

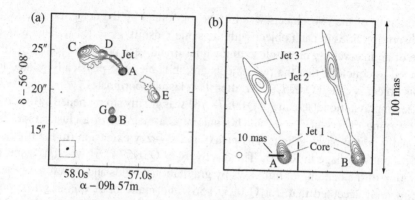

Figure 12.2. (a) Radio image at 5 GHz taken of the twin quasar Q 0957+561 with the VLA radio telescope array. It shows the two quasar images (A, B), a jet extending from the A quasar leading out to a lobe with hot spots C and D, lobe E on the opposite side of A, and the lensing galaxy G. Coordinates are epoch 1980 Dec. 16. (b) VLBI images of core regions set side-by-side for easy comparison. They show similar jets extending ~60 milliarcsec from the cores. [(a) D. Roberts *et al.*, *ApJ* **293**, 356 (1985); (b) M. Gorenstein *et al.*, *ApJ* **334**, 42 (1988)]

12.3   Point-mass lens

Gravitational fields are represented as curvatures of space in GR, and light beams follow "null geodesics" in space-time. For light just grazing the sun, the apparent deflection angle is, as stated, only 1.75″. This angle was a prediction of Einstein's general theory of relativity in 1915. A Newtonian treatment worked out below (see (6)) yields a deflection angle that is incorrect by a factor of two, or 0.87″.

During the solar eclipse of 1919, the bending of starlight was measured with photographic images and found to have a magnitude of 1.61″ ± 0.30″. This excluded the Newtonian value with marginally high confidence, 2.5σ, and Einstein became an instant celebrity. Decades later, in 1974–5, radio astronomers used interferometry to measure the bending of radio waves from quasars as they grazed the sun and found an angle of 1.761″ ± 0.016″, which is in excellent agreement with Einstein's value. Later work provides even better agreement.

In this section we find the effect of a pointlike gravitational lens on the light rays from a distant unresolved pointlike source. The bending region of a ray is taken to be small compared with the length of the entire light path. The bending angles are small, and so small-angle approximations may be used.

Bending angle

The bending of light from a source S (e.g., a quasar) by a gravitational body acting as a lens, L, is sketched in Fig. 3a for three observers, O_a, O_b, and O_c. The trajectory of photons en route to observer O_a remains sufficiently distant from L that the path is very close to a straight line. For observer O_b, the bending toward the gravitational mass is appreciable, and for observer O_c, it is very pronounced because the photons pass quite close to L. In this case, observer O_c intercepts a second beam that passes on the other side of L. In all three cases, the trajectories pass outside the galaxy, and so one can think (for now) of the lens as a point mass.

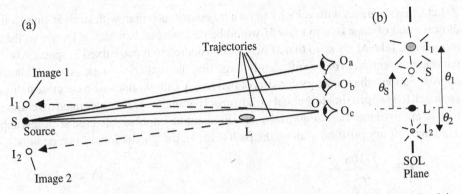

Figure 12.3. Gravitational lensing of a source such as a quasar. (a) Rays from source S arriving at observers O_a, O_b, and O_c who are at three different positions. Rays passing close to the lens L, a galaxy, are deflected so that the apparent source position in the sky is displaced. Observer O_c sees two rays and hence two displaced images I_1 and I_2. (b) Sky view as it appears to observer O_c. The images lie in the source–observer–lens (SOL) plane. Image I_1 is magnified, whereas I_2 is demagnified.

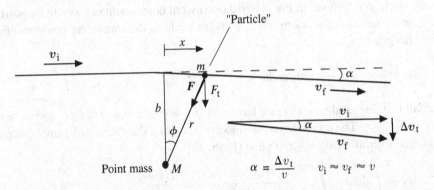

Figure 12.4. Force diagram for the Newtonian deflection of a "particle" of mass m passing a point mass M. The initial and final velocities, v_i and v_f, yield the transverse velocity Δv_t and hence the bending angle α. The deflection is assumed to be very small; the particle motion deviates only slightly from a straight line. The Newtonian scattering angle for a particle at speed c is a factor of two less than that derived from GR for the deflection of a photon.

For observers O_b and O_c, the photons arrive at the observer from a direction significantly displaced from that of the source. Each of the trajectories represents a bundle of nearly parallel rays that can be brought to a focus with an eye or a telescope. Thus, the observer detects the quasar at a displaced angular position. Observer O_c sees two such images, I_1 and I_2, one on either side of lens L. The view of O_c is shown in Fig. 3b.

Newtonian angle

The magnitude of the bending angle α for a beam passing near a point-mass lens may be derived in a Newtonian context in which we treat the photon as a particle of small mass m. The geometry is shown in Fig. 4. The result will be in error by a factor of two, but the functional dependence on the variables will be correct.

Let a mass m travel with velocity v_1 on a trajectory such that, with straight-line motion, the closest point of approach to mass M would be the distance b, which is known as the *impact parameter*. Take $M \gg m$ so that M may be considered to remain fixed in space. Also, let the particle momentum mv_1 be sufficiently large that the scattering angle is small. Under these approximations, the trajectory consists of two nearly straight lines with the closest distance of approach to M approximately equal to the impact parameter b. The integrated (over time) force may be approximated as a single transverse momentum impulse delivered at closest approach.

At an arbitrary position x along the path (Fig. 4), the gravitational force on m is

$$F = -\frac{GMm}{r^2}\,\hat{r}, \qquad \text{(Vector force)} \qquad (12.2)$$

where $\hat{r}$ is a unit vector in the radial direction (origin at M). The deflection will arise from the transverse (downward) component of the force F_t (i.e., the component perpendicular to the initial direction of motion of m),

$$F_t = \frac{GMm}{r^2}\cos\phi, \qquad (12.3)$$

where the angle ϕ is defined in Fig. 4 and a downward component is taken to be positive.

The related transverse acceleration $a_t = F_t/m$ leads to the transverse component of the velocity change:

$$dv_t = a_t\,dt = \frac{GM}{r^2}\cos\phi\,dt. \qquad (12.4)$$

For a small bending angle, $\alpha \ll 1$, we have $r \approx (b^2 + x^2)^{1/2}$, $\cos\phi \approx b/r$, and $dt \approx dx/v$, where $v \approx v_1 \approx v_2$. The total transverse velocity Δv_t along the entire path may be expressed as an integral over all positions x to yield (Prob. 33)

$$\Delta v_t = \int dv_t = \int_{-\infty}^{+\infty} \frac{GMb}{\left(b^2 + x^2\right)^{3/2}}\frac{dx}{v} = \frac{2GM}{bv}. \qquad (12.5)$$

This result may be inferred without integration if one assumes a force that is constant with its value at $x = 0$ over a distance of $\Delta x = 2b$.

The (Newtonian) angle of bending α_N for a point-lens attractor and small bending angle is thus (see Fig. 4 inset)

$$\alpha_N = \frac{\Delta v_t}{v} = \frac{2GM}{bv^2} \xrightarrow[v=c]{} \frac{2GM}{bc^2}, \qquad \begin{array}{l}\text{(Bending angle } \alpha \ll 1; \\ \text{Newtonian calculation)}\end{array} \qquad (12.6)$$

where the speed v is set to the speed of light c. This expression also applies to a spherical attractor (galaxy) if the Newtonian "particle" passes completely outside the galaxy.

General-relativity angle

It turns out that when GR is taken into account, the angle α for photon bending by a point-mass lens is a factor of two greater than the Newtonian result,

$$\alpha = \frac{4GM}{bc^2} \xrightarrow[b=R_\odot]{} 8.49 \times 10^{-6}\ \text{rad} \to 1.75'', \qquad \begin{array}{l}\text{(Bending angle for} \\ \text{photon } \alpha \ll 1; \text{GR result; rad)}\end{array} \qquad (12.7)$$

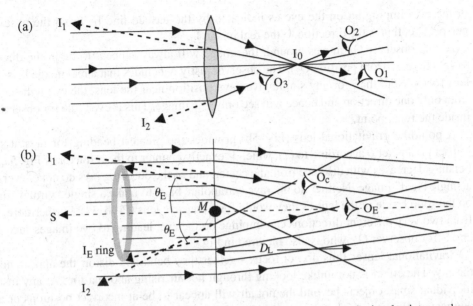

Figure 12.5. Comparison of (a) an ideal convex thin lens and (b) a gravitational point-mass lens, each for incident parallel rays from an infinitely distant source. The bending angle increases with distance from the lens center in the convex case and decreases inversely with distance in the point-mass case. Observers intercept only rays arriving at their local positions, and each such ray acts locally as a bundle of parallel rays focused to a single spot by the eye. Directions to images are indicated with heavy dashed lines. The Einstein ring (shaded) is at θ_E.

where G is the gravitational constant, M the lens mass, c the speed of light, and b the impact parameter as before. This expression is similarly valid for a spherical distribution of mass if the trajectory remains outside the spherical distribution of matter and the deflection angle is small. The equation is thus applicable to a beam of light just grazing the sun. The result is the aforementioned 1.75″.

The dimensionless ratio in (7) may be remembered as the ratio of two energies. If a photon had a mass m (which is not the case), the ratio of its gravitational potential energy GMm/b at radius b to its rest energy mc^2 would be functionally similar to (7).

Comparison with an ideal lens

It is instructive to compare gravitational lenses with the ideal thin lenses encountered in geometrical optics. Simply put, gravitational bodies make terrible lenses. They do not bring the rays of a pointlike source to a true focus at which a *real image* is created as does an ideal thin convex lens (Fig. 5a).

Let the pointlike source off the page to the left in Fig. 5a be at a great (infinite) distance so that the rays arriving at the lens form a parallel beam. Each eye in the figure intercepts a bundle of rays more or less parallel to the ray shown and will focus them to a point on the retina. The brain will thus interpret the flux as a point source in the direction of the ray entering the eye.

The ray in Fig. 5a passing through the center of the convex lens is not bent at all, but rays displaced from the center are bent through increasingly greater angles so that all rays are brought to a single focus at a "real" image I_0. Observer O_1 will see an image in the direction of the rays impinging on the eye as indicated by the dashed line I_1. From the observer's perspective, this is the direction to the real image I_0.

As the observer O_1 moves around, the image will always appear to be in the direction toward I_0 (e.g., from position O_2). The observer justly concludes that a real image I_0 is at the lens focus. Note that, for any single given eye position near the lens, the eye will see rays from only one direction and hence will see only one image. This is even true for observer O_3 inside the focal point.

A pointlike gravitational lens (Fig. 5b) provides the greatest bending for rays at small impact parameter b according to (7), which is exactly counter to the refractor of Fig. 5a. The bending decreases with distance from the center, and so the outgoing rays do not converge to a single "real" image. Neither do the rays extrapolate back to make a single "virtual" image on the source side of the lens as they would for a concave lens (Prob. 34). Furthermore, rays from two very different directions can impinge on the eye, leading to two images such as I_1 and I_2 for observer O_c, who is also pictured in Fig. 3a.

Gravitational lenses thus do not focus – rather they bend and distort the appearance of the sky. The effect is not unlike looking through hot air rising above a fire. At any instant, individual small objects beyond the hot air will appear to be at incorrect positions or even duplicated. Because the bending varies for different lines of sight, the appearance of large objects will be distorted. We will see below (under "Magnification and flux"), though, that the original surface brightness is preserved.

Einstein ring predicted

Observer O_E in Fig. 5b is directly in line with the point gravitational mass. Because the geometry is cylindrically symmetric about the lens direction, this observer would see a circular image. This was postulated by Einstein in 1936 and is known as the *Einstein ring*. A ring is observed from all positions along the central line to the right of the convergence point for rays just grazing mass M. A cone of rays converges on any such position. The half-angle of the cone represents the half-angle of the ring.

From the geometry of Fig. 5b, the angular radius of the ring will decrease as the observer moves to greater distances D_L from the lens. From the bending angle (7) and the geometry of Fig. 5b, one can find the angular radius of the ring θ_E. For small angle approximations and a parallel incoming beam, the result is (Prob. 31)

$$\theta_E = \left(\frac{4GM}{D_L c^2}\right)^{1/2}.$$

(Einstein ring angular radius; source at infinite distance; $\theta_E \ll 1$)   (12.8)

The more general expression for the ring radius for a source at a finite distance is derived below (15).

Image positions

Consider a quasar well behind, and at an angle θ_S from, a point-mass lens. The angular positions of its images are derived here. We will find that, in addition to the bending angle (7) discussed previously, a geometrical constraint equation is required. In turn, this leads to a "lens equation" that can be solved for image positions.

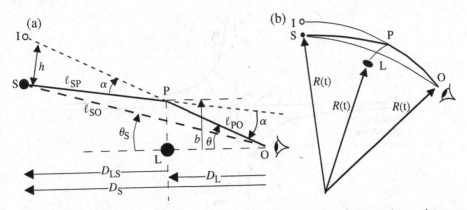

Figure 12.6. (a) Euclidean geometry for deflection of a ray by a galaxy taken to be a point-mass lens. The deflection is approximated as occurring at a single point P at a distance b from the lens. (b) The direct and bent paths of (a) on an expanding spherical surface, as in a closed universe with positive curvature.

Bending angle plot

The geometry for a single ray passing outside a lens is shown in Fig. 6a. The expression for the point-lens bending angle α (7) can be rewritten in terms of a hypothetical image position θ if the relation $b = D_L \theta$ is adopted from Fig. 6a, again for small angle bending,

➡
$$\alpha = \frac{4GM}{c^2 D_L \theta}. \qquad \text{(Bending angle as function of image angle; rad)} \qquad (12.9)$$

This is a useful form because the angle θ of an image is a measurable quantity if the lens direction is known. The lens distance D_L may be forthcoming from the redshift of the spectral lines of the lens galaxy spectrum and an assumed cosmology.

The bending function (9) is plotted in Fig. 7 for both positive and negative θ (solid curves). The bending angle α is the vertical axis, and the horizontal axis is the image position θ. Because θ is proportional to the impact parameter b, the horizontal axis is a measure of the impact parameter b. When the impact parameter is very large, there is very little bending ($\alpha \to 0$). If, on the other hand, the image is very close to the point-mass lens, the bending angle becomes large. Recall that our bending-angle formulas are valid only for small angles ($\ll 1$ rad). (The axes of Fig. 7 extend only to small angles.)

Ray-trace equation

The bending angle α given in (9) is a function of a hypothetical image angle θ. An actual image exists at θ only if the bent ray originates at a source of radiation. A geometrical *ray-trace equation* provides the additional information required to obtain the actual image positions for a source at a postulated distance D_S and (undeviated) angular position θ_S.

The angular displacement α (Fig. 6a) of the image viewed from the lens must subtend the same physical distance h at the source as does the angular displacement $\theta - \theta_S$ viewed from the observer. In the small-angle approximation, the angles α and $\theta - \theta_S$ are

$$\alpha = \frac{h}{D_{LS}} \qquad (12.10)$$

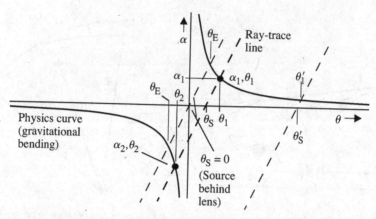

Figure 12.7. Graphical solution of lensing by a point-mass lens. The bending angle α is plotted versus image angular position θ, from (9), where the origin locates the lens. Dashed ray-trace lines (13) are drawn for three different angles θ_S of the source (measured from the lens), as indicated by the intercepts on the abscissa. Left line: source directly behind the lens at $\theta_S = 0$, which yields the Einstein ring. Central line: source displaced from lens by angle θ_S slightly less than θ_E, leading to images at θ_1 and θ_2, as in Fig. 3a,b for observer O_c. Right line: source far from lens and hence with very little bending (compare θ_S' and θ_1'), as in Fig. 3a for O_a.

and

$$\theta - \theta_S = \frac{h}{D_S}. \tag{12.11}$$

Their ratio yields the desired ray-trace equation,

$$\frac{\alpha}{\theta - \theta_S} = \frac{D_S}{D_{LS}} \tag{12.12}$$

➡ $$\alpha = \frac{D_S}{D_{LS}}(\theta - \theta_S), \qquad \text{(Ray-trace equation)} \tag{12.13}$$

where α is the bending angle, θ_S the source direction, θ the image direction, and D_{LS} and D_S the distances shown in Fig. 6a. The angles θ and θ_S are both measured relative to the lens.

This is a second, with (9), independent equation that defines α in terms of the geometric variables of the problem. It gives the bending angle required if the ray arriving at the selected impact distance $b = \theta D_L$ is to proceed on to the observer. Such a ray will exist only if the lens has the correct strength to bend it by the angle α; see (9).

Angular-diameter distance

The relation between the physical displacement of an object at cosmological distance and the associated angular displacement can vary as determined by the curvature of space-time and the chosen definition of distance. We have adopted, in (9), (10), and (11), Euclidian relations (e.g., $b = D_L \theta$) that are correct in relativistic cosmology if the distances are taken to be *angular-diameter distances*. Thus, the ray-trace equation (13) is generally valid if the distances D_S and D_{LS} are angular-diameter ones.

The distance D_S may be obtained from the measured source redshift z_S (1) by means of a relation that takes into account the geometry of the universe, We discuss this (Section 5) in the context of the Hubble constant. The lens distance D_L may similarly be obtained from z_L. The distance D_{LS}, however, does not so follow because the redshift z_{LS} of the source from an observer at the lens is unknown. Unfortunately, one can not use $D_{LS} = D_S - D_L$ because it is not valid in a geometry with curvature, but, with knowledge of the curvature, the distance D_{LS} can be calculated from z_L and z_S. We give an example in Section 5.

Graphical representation

On an α versus θ graph such as Fig. 7, the ray-trace expression (13) is a straight line of slope D_S/D_{LS} and abscissa intercept ($\alpha = 0$) at the source position $\theta = \theta_S$. This line is plotted in Fig. 7 (heavy dashed line). It intersects the solid curves at the angular coordinates α_1, θ_1 and α_2, θ_2. These are the positions that satisfy both the bending expression (9) and the ray-trace equation (13).

The plot thus tells us that an image should exist at each of the two positions θ_1 and θ_2. The former, at θ_1, is above the source in Fig. 3b with a moderate bending angle α_1. The latter, at θ_2, is on the opposite side of, and closer to, the lens with a larger (in magnitude) bending angle α_2 (θ_2 and α_2 are both negative).

Consider this graphical solution as the source position θ_S (the intercept) varies, and M, D_S, and D_{LS} are held fixed. In this case, the slope D_S/D_{LS} of the ray-trace line is held constant; see (13). If the source moves farther from the lens (the origin), the line must be moved to the right parallel to itself (e.g., to the light dashed line in Fig. 7). In this case, the new source direction θ_S' is close to the new image direction θ_1', and so the bending angle α is quite small.

This case (large θ_S') approaches the situation shown in Fig. 3a for observer O_a, where the lensing is quite weak. The image is nearly coincident with the source position. A second image exists in principle at very large negative α (off the plot) with a small negative image position θ (i.e., very close to the lens). Such images are expected to be very faint and hence are usually undetectable.

Now consider the source angular position to be closer to the lens than our original case. In this situation, the constraint line in Fig. 7 is moved to the left. As the source (intercept) moves to the left toward the lens (origin), the two images move to the left. The upper image moves more rapidly in θ than the lower one, which tends to make their positions more symmetric about the lens. When the source is coincident with the lens (left light dashed line), the images are perfectly symmetric about the lens, and the intersections with the bending-angle curve define the Einstein angle θ_E.

Lens equation

The positions of the two images are obtained analytically if the two expressions for the bending angle are equated, namely the physical (9) and the ray-trace (13):

$$\frac{4GM}{c^2 D_L \theta} = \frac{D_S}{D_{LS}}(\theta - \theta_S). \qquad \text{(Lens equation)} \qquad (12.14)$$

This is called the *lens equation*.

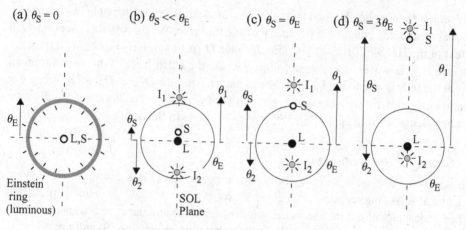

Figure 12.8. Angular positions of the images on the sky for four different positions θ_S of a point source relative to the point-mass lens position as viewed by the observer. The magnifications of the images are not shown. The "rays" indicate visible images; the (invisible) Einstein ring (thin line) is shown in (b), (c), and (d) for reference.

The lens equation readily yields the angular radius of the Einstein ring for a source at finite distance D_S. For the source directly behind the lens, $\theta_S = 0$, solve for $\theta(= \theta_E)$ to find

$$\theta_E = \left[\frac{4GM}{c^2}\frac{1}{D}\right]^{1/2} ; \quad D \equiv \frac{D_L D_S}{D_{LS}}, \qquad \text{(Einstein ring radius;} \qquad (12.15)$$
$$\text{point-mass lens; finite } D_S)$$

where M is the mass of the lens and D is a *distance parameter* based on D_L, D_S, and D_{LS}, which are the distances to the lens, to the source, and from lens to source, respectively. Note that, as D_S goes to infinity, $D_{LS} \approx D_S$, and the expression (15) reduces to our previous expression (8). The expression (15) applies to a point-mass lens and also to a lens consisting of a spherically symmetric mass distribution for rays passing completely outside the mass distribution.

The lens equation (14) can be simplified if we express it in terms of the Einstein angle (15) as

$$\theta^2 - \theta_S\theta - \theta_E^2 = 0. \qquad \text{(Lens equation simplified)} \qquad (12.16)$$

Analytic solution

Solve (16) with the quadratic formula to obtain the image positions θ:

$$\theta_{1,2} = \frac{\theta_S \pm \sqrt{\theta_S^2 + 4\theta_E^2}}{2}. \quad \text{(Image positions; point source; point-mass lens)} \quad (12.17)$$

Here again, we obtain the image positions if we know the source position and the quantities that make up θ_E. Hereafter we will associate the solution θ_1 with the "+" sign and θ_2 with the "−" sign.

The image positions for several values of θ_S obtained from (17) (Prob. 36) are given here. They are also shown in Fig. 8 as they would appear to the observer. The first is for $\theta_S = 0$

when the source is directly behind the point-mass lens, which is the location that would generate an Einstein ring. In this case, (17) yields

$$\theta_{1,2} = \pm\theta_E. \qquad \text{(Einstein ring; } \theta_S = 0) \qquad (12.18)$$

The result is the radius of the Einstein ring – namely the expression defined as θ_E in (15). As the source moves slightly off center, $0 < \theta_S \ll \theta_E$, the expression (17) yields

$$\theta_1 \approx \theta_E + \frac{\theta_S}{2}; \quad \theta_2 \approx -\left(\theta_E - \frac{\theta_S}{2}\right). \qquad (\theta_S \ll \theta_E) \qquad (12.19)$$

Two distinct images are formed (Fig. 8b). The Einstein ring is no longer visible; it is drawn for reference only. One image lies just outside the Einstein ring, and the other just inside in the negative direction. The figure shows only the image locations; magnifications and distortions are not shown.

When the source is exactly at the Einstein ring radius, $\theta_S = \theta_E$ (Fig. 8c), the images (17) are located at

$$\theta_1 = 1.62\,\theta_E; \qquad \theta_2 = -0.62\,\theta_E. \qquad (\theta_S = \theta_E) \qquad (12.20)$$

The upper image moved farther out to reach this position, but not as far as the source moved; the image and source are thus closer together than in Fig. 8b. The lower image moved closer to the lens. The solutions, θ_1 and θ_2, that appear in Figs. 3b and 7 approximate this condition.

The bending angles α_1, α_2 for this case can be calculated from (13) if one arbitrarily adopts the ratio $D_{LS}/D_S = 0.6$ and watches the signs:

$$\alpha_1 = 1.03\,\theta_E; \qquad \alpha_2 = -2.70\,\theta_E. \qquad \begin{array}{c}\text{(Bending angles for}\\ \theta_S = \theta_E, D_{LS}/D_S = 0.6)\end{array} \qquad (12.21)$$

The lower image undergoes a very large bending, as expected, because the ray passes close to the lens.

A final case is for $\theta_S \gg \theta_E$. The solutions are

$$\theta_1 \approx \theta_S\left(1 + \frac{\theta_E^2}{\theta_S^2}\right) \to \theta_S; \qquad \theta_2 \approx -\theta_S\frac{\theta_E^2}{\theta_S^2} \to 0. \qquad (\theta_S \gg \theta_E) \qquad (12.22)$$

Figure 8d is drawn for the case $\theta_S = 3\theta_E$, where one image lies very close to the source (i.e., very little bending) and the other (very faint) object lies close to the lens. The rays of the latter undergo large bending. In the limit of very large distances θ_S, the image θ_1 merges with the source position θ_S, and the image θ_2 merges with the lens. The latter also demagnifies to zero intensity, as will be seen below in the discussion of Fig. 10. The only image is thus coincident with the source; this is the elementary undeviated case with no intervening lens.

Determining system parameters

The image positions relative to the lens, θ_1 and θ_2, are measurable if the lens can be identified. With this information, one can find from the lens equation (16) (Prob. 35) or directly from the image positions (17) that

$$\begin{aligned}\theta_S &= \theta_1 + \theta_2\\ \theta_E &= \sqrt{-\theta_1\theta_2},\end{aligned} \qquad \text{(Source and Einstein ring angles)} \qquad (12.23)$$

where θ_2 is negative. The definition of θ_E (15) then yields the mass of the lens in terms of the Einstein angle and the distance parameter D,

$$M = \frac{c^2 \theta_E^2}{4G} D. \qquad \text{(Mass of the lens)} \qquad (12.24)$$

The value of θ_E is known from (23). The parameter $D \equiv D_L D_S / D_{LS}$ must be independently determined.

The parameter D, with units of distance (m), serves as an overall geometrical scale factor of the lensing geometry. It comes into play in our later discussion of the Hubble constant (Section 5). Note that for fixed geometry – that is, for fixed θ_1, θ_2 and hence fixed θ_E (23) – the lens mass M scales linearly with D.

Magnification and flux

The several images of a gravitationally lensed source do not in general appear equally bright; they will typically have different fluxes (W/m^2) or different optical magnitudes. These differences are best understood in terms of magnification and the conservation of specific intensity.

Conservation of specific intensity

We found (3.27) that the specific intensity of radiation I (W m^{-2} Hz^{-1} sr^{-1}) is conserved as it travels through space in the absence of absorption, scattering, or redshift and that this conservation is rooted in Liouville's theorem. The theorem states that the density of particles in phase space is conserved as one follows the particles' trajectory if the forces applied to the particles are "smooth" (e.g., no collisions) and independent of momentum or, if dependent on momentum, they do not change the momentum magnitudes of the particles (e.g., magnetic fields); see discussion preceding (3.27).

Gravity, viewed as a force field, is "smooth" because, for most of space, it is weak and is applied over long distances. The applied force is also independent of particle momentum, and much of space is transparent. It may therefore not be surprising, though it is hardly obvious, to find that, in GR, specific intensity is conserved in the gravitational deflection of light rays.

If one views an object of small but resolvable angular size such as a lit light bulb through the rising air above a fire, the bulb's location, shape and size will appear to vary (flicker) with time, but it will always have the same apparent surface brightness (i.e., flux per steradian or specific intensity). Similarly, the glints of sunlight seen from an irregularly shaped reflector (e.g., an automobile windshield) or through an irregularly shaped piece of glass will have the same apparent surface brightness as the sun if the reflection or transmission is perfectly efficient and independent of wavelength.

Magnification overview

The gravitationally lensed images of a source of finite angular size will be magnified or demagnified. If an image is magnified, it subtends a larger solid angle in the observer's view than would the unlensed source. If this larger image also has the same specific intensity (power per steradian) as the source, as we just argued, then the intensity integrated over the larger solid angle will be greater than that of the unlensed source. The observer will thus measure a greater spectral flux-density, $S = \int I \, d\Omega$ (W m^{-2} Hz^{-1}), or a greater flux density, $\mathscr{F} = \int I \, d\Omega \, d\nu$ (W/m^2). Similarly, if the image is demagnified, the product of the smaller solid angle and the invariant specific intensity yields a flux density $\mathscr{F}$ less than the unlensed flux.

Sources such as quasars, and their images, are too small to be resolved by current telescopes, and so the image magnifications can not be observed directly; however, their flux densities will still reflect their magnifications. Each image will appear as an unresolved (point) source, but its flux density will reveal its *magnification factor* μ.

Magnification M is usually a one-dimensional parameter. For example, 8X binoculars have $M = 8$ and produce an image of the full moon with angular diameter eight times that perceived by the naked eye. A magnification of unity means no change of angular size, whereas less than unity indicates demagnification. The magnification of binoculars is the same along the horizontal (x) and vertical (y) axes; hence, a circular full moon will yield a circular image. In this case, the solid angle and thus the flux density will increase by the factor $\mu = M^2$.

Gravitational lensing of a source gives a different magnification M_{rad} in the radial direction from the lens than in the azimuthal direction M_{az}. In this case, a circular source offset from the lens does not map to a circular image. If the source is small in radial and azimuthal extent relative to θ_S, the flux density increases approximately as the product $\mu \approx M_{\text{az}} M_{\text{rad}}$. If the source is large, the factor μ varies over the surface, and one must map many elements of the source to their image positions to find the overall magnification factor.

If there is more than one image after lensing and they are too close together to be resolved as separate images, the *total magnification factor* μ_{tot} comes into play. This is the sum of the solid angles of the several images divided by the solid angle of the source. In general, the total magnification factor for an imaged source is greater than unity. We now proceed to develop these concepts.

Extended source mapped

The mapping of an extended source into its images by a point-mass lens follows from the geometry we have already presented. Let the source be an angular segment of a circular band centered on the lens with constant surface brightness (Fig. 9). The Einstein ring (circle) is shown as a reference. Four corners of the source, a, b, c, d, are indicated. How do these four points image?

According to Fig. 3b, the two images of a source element must each fall on the line defined by the element itself and the lens L, and the distances from the lens must be in accord with (17), our analytical solution for point-source imaging. We are thus able to plot the two images of a, namely a′ and a″, as well as those for b, c, and d, as shown. If other points on the periphery of the source were similarly imaged, it is clear that the results would outline the circular band segments shown for the two images. Note that the θ_2 image is flipped horizontally but not vertically.

Magnification factor

Let the source size $\Delta\theta_S$ be much smaller than θ_S in both the radial and azimuthal directions. An analytic expression for the magnification factor $\mu \approx M_{\text{az}} M_{\text{rad}}$ for each of the two images then follows from the geometry of Fig. 9.

The magnification in the azimuthal direction is, by inspection of Fig. 9, proportional to the distance of the image from the lens, $M_{\text{az}} \propto \theta$. Because the azimuthal magnification should be unity for $\theta = \theta_S$, we have $M_{\text{az}} = \theta/\theta_S$. The magnification in the radial direction M_{rad} for the upper image in Fig. 9 is the ratio of the radial separation of image points b′ and a′

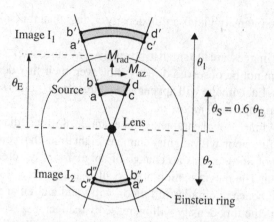

Figure 12.9. Magnification geometry for an arclike source at $\theta_S \approx 0.6\,\theta_E$ lensed by a point-mass lens as viewed on the sky by the observer. The points a–d of the source are imaged to the positions a′–d′ and a″–d″. The source and two images, I_1 and I_2, each have the same surface brightness. Both are magnified in the azimuthal direction and demagnified in the radial direction. The Einstein ring (circle) is shown as a reference.

to that for the source points b and a, which may be written as $M_{\text{rad}} = \Delta\theta/\Delta\theta_S \to d\theta/d\theta_S$ for infinitesimal elements. Visualize a small shift of a ray-trace line in Fig. 7 to estimate the relative radial motions of the source $\Delta\theta_S$ and an image $\Delta\theta$. For example, the image at θ_2, being on a steep portion of the bending curve, will move relatively little and hence $\Delta\theta/\Delta\theta_S < 1$. This image is thus demagnified in the radial direction. (It is also demagnified in the azimuthal direction because of its proximity to the lens.)

The overall magnification factor for a small source thus becomes

$$\mu \approx M_{\text{az}} M_{\text{rad}} = \frac{\theta}{\theta_S}\frac{d\theta}{d\theta_S}. \quad \text{(Magnification factor; small source; } \Delta\theta_S \ll \theta_S) \quad (12.25)$$

It is useful to write this expression in terms of θ, the observed image position, rather than θ_S. The azimuthal term θ/θ_S follows from the lens equation (16); solve it for θ_S:

$$\theta_S = \theta - \frac{\theta_E^2}{\theta} = \theta\left(1 - \frac{\theta_E^2}{\theta^2}\right). \quad (12.26)$$

Divide by θ and invert to obtain

$$M_{\text{az}} = \frac{\theta}{\theta_S} = \left(1 - \frac{\theta_E^2}{\theta^2}\right)^{-1}. \quad \text{(Azimuthal magnification)} \quad (12.27)$$

The radial term $d\theta/d\theta_S$ also follows from the lens equation in the form (26). Take the derivative $d\theta_S/d\theta$, which is the inverse of M_{rad},

$$M_{\text{rad}}^{-1} = \frac{d\theta_S}{d\theta} = \frac{d}{d\theta}\left(\theta - \frac{\theta_E^2}{\theta}\right) = 1 + \frac{\theta_E^2}{\theta^2}, \quad (12.28)$$

$$M_{\text{rad}} = \left(1 + \frac{\theta_E^2}{\theta^2}\right)^{-1}. \quad \text{(Radial magnification)} \quad (12.29)$$

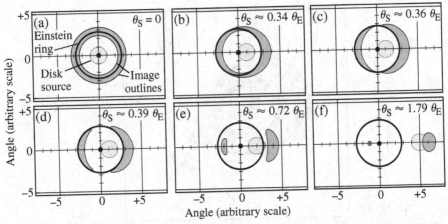

Figure 12.10. Image shapes and positions (dark shading) for circular disk source (light shading) imaged by a point-mass lens (black dot) for various positions θ_S of the source as seen by the observer. The heavy dark circle in each frame indicates the Einstein ring for a point source. The angular scale is arbitrary. The source radius is set to 1.0 unit, and the radius of the point-source Einstein ring to 2.75 units. As the source is offset from the lens, the broadened Einstein ring (a) separates into two arcs (c,d), and finally into arclets (e,f) of very different magnifications. [Adapted from E. Falco Acosta, Ph.D. thesis, MIT, 1985]

The magnification factor μ (25) is the product of (27) and (29),

$$\mu = M_{az} M_{rad} = \left(1 - \frac{\theta_E^4}{\theta^4}\right)^{-1}.$$

(Magnification; point-mass lens; source size $\Delta\theta_S \ll \theta_S$) (12.30)

The detected flux density of the image relative to that of the source is equal to the magnification,

$$\frac{\mathscr{F}}{\mathscr{F}_S} = \left(1 - \frac{\theta_E^4}{\theta^4}\right)^{-1}.$$

(Ratio of flux densities; $\Delta\theta_S \ll \theta_S$) (12.31)

Note that (31) gives the magnification in terms of the image position θ, not the source position θ_S. The quantitative image positions and magnifications in Fig. 9 (i.e., for $\theta_S = 0.6\theta_E$) are worked out in Prob. 37.

Let us look at the behavior of μ (30). First consider the image θ to be at the Einstein ring radius, $\theta = \theta_E$. This means that the source is directly behind the lens and an Einstein ring has formed. Here, the function (30) blows up, $\mu \to \infty$, because, in our small source approximation, $\Delta\theta_S \ll \theta_S$, a source element at the origin is infinitesimally small (Fig. 9).

A more realistic source model would be a small circular disk centered on the lens as shown lightly shaded in Fig. 10a. If the perimeter of the disk is mapped with (17), a broadened Einstein ring with magnification (dark shaded area) results.

As the disk source emerges from behind the lens (Fig. 10b,c), the image I_1 at θ_1 moves outward beyond θ_E, and the image I_2 at θ_2 moves inward. The magnification factors μ_1 and μ_2 from (30) are initially large, and they both decrease in magnitude as the source continues to move away from the lens (Fig. 10d,e). When $\theta_S \gg \theta_E$, the lens solution (17) gives $\theta_1 \gg \theta_E$, and the magnification μ_1 of image I_1 from (30) approaches unity (Fig. 10e,f). In

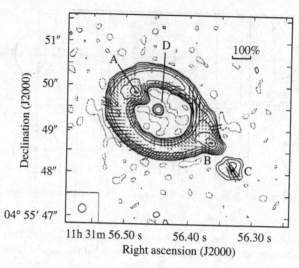

Figure 12.11. Einstein ring MG 1131+0456 obtained with the VLA at 8 GHz. The ~3″-diameter loop contains two hot spots, A and B, which are probably images of the active nucleus of a radio galaxy. The radio lobes of that galaxy likely give rise to the ring structure and the less pronounced hot spots C and D. The beam size is shown in the lower left. The line segments indicate polarization direction and degree (see scale). [G. Chen and J. Hewitt, *Astron J.* **106**, 1719 (1993)]

contrast, the magnification μ_2 of I_2 decreases toward zero and is negative, indicating the image is flipped. The image I_1 was always magnified, whereas I_2 initially was magnified and then became demagnified (Prob. 38) as the source moved outward from the lens. One can similarly check that the individual magnifications M_{az} and M_{rad} (27) and (29) are consistent with Figs. 9 and 10.

A candidate for the Einstein ring was first found in the lensing of the radio source MG1131+0456, which appears as a tiny loop of ~3″ diameter (Fig. 11). The ring is not due to the highly improbable circumstance postulated by Einstein – namely, the perfect alignment of two stars. Rather, it is most likely due to lensing by an elliptical galaxy that is more or less aligned with one of the lobes of a distant, typical radio galaxy containing a massive active nucleus and two lobes. The lobe behind the lens is distorted into the "ring," and the nucleus is doubly imaged into the two hot spots on the ring.

Total magnification factor

We now consider the case in which the two images are not resolved as separate images; the observer sees a single unresolved source made up of the two images I_1 and I_2, one magnified and the other either magnified or demagnified. In this case, it is necessary to calculate the summed or total magnification of the two images because the sum is proportional to the detected flux density of the unresolved source.

For this purpose, we first express the magnifications μ_1 and μ_2 (30) in terms of θ_S and θ_E rather than θ and θ_E. We accomplish this, in principle, by substituting the solutions for $\theta_{1,2}$ (17) into the magnification (30). The algebra is less onerous if one first rewrites (30) in terms of θ_S and θ with the aid of the lens equation (16) and invokes the definition $u \equiv \theta_S/\theta_E$ for the source position. The result after significant manipulation (Prob. 39) is

$$\mu_{1,2} = \frac{1}{2} \pm \frac{u^2 + 2}{2u\sqrt{u^2 + 4}}, \qquad \text{(Magnification factors as a function of } \theta_S; \quad u \equiv \theta_S/\theta_E) \qquad (12.32)$$

where, again, the θ_1 image is associated with the "+" sign and θ_2 with the "−" sign.

The right-hand fraction in (32) ranges from infinite at $u = 0$ (again owing to our small-source approximation) to $1/2$ for $u \gg 1$. The image I_1 (+ sign) thus always has magnification μ_1 that exceeds unity in accord with our discussion of Fig. 10 above.

In contrast, μ_2 ranges from a large negative number for small u to zero for large u. The image I_2 can thus be magnified or demagnified as determined by the distance of the source from the lens, which again is in accord with the discussion above. As before, the negative sign indicates the image is horizontally flipped.

The sum of the two *absolute values* of the magnifications (32) directly gives the total magnification. Because μ_2 is negative for all nonzero values of u, its sign must be changed in the addition. The result directly follows from (32):

$$\mu_{\text{tot}} = |\mu_1| + |\mu_2| = \frac{u^2 + 2}{u\sqrt{u^2 + 4}}. \qquad \text{(Total magnification factor;} \quad u \equiv \theta_S/\theta_E) \qquad (12.33)$$

The factor μ_{tot} has very large values when the source is near the lens ($u \ll 1$). For example, at $u = 0.1$, $|\mu_1| = +5.52$ and $|\mu_2| = +4.52$, giving $\mu_{\text{tot}} = 10.04$. When the source is at the Einstein-ring radius, $u = 1$, the two components are $|\mu_1| = 1.17$ and $|\mu_2| = 0.17$, giving $u_{\text{tot}} = 1.34$. A factor of 1.34 increase in flux density corresponds to an easily detectable magnitude change of $-2.5\log(1.34) = -0.32$ mag. Finally, when the source is far from the lens ($u \gg 1$), μ_{tot} approaches unity; this is the limit of no lensing.

The value of μ_{tot} (33) thus never falls below unity. The summed flux density of the two images of an unresolved source is always increased in the vicinity of a point-mass lens. The source brightens but never dims! Gravitational lensing means astronomers can see intrinsically fainter and more distant objects than they could without it.

Because the number of photons emanating from a source is conserved, unlensed lines of sight from these sources must experience a diminution of flux. The reduction, however, is spread over large solid angles and would therefore be minuscule in any given (unlensed) direction.

Microlensing

The magnification phenomenon together with modern technology has opened the field of gravitational *microlensing*, the lensing of one star by another star.

Projected stellar encounters

We have stated that stellar encounters (along a given line of sight) are unlikely occurrences because of the improbability of two stars' finding themselves almost exactly behind one other. Stellar proper motions would make such encounters transient events. Also, the image separations will generally be too small to detect. Lensing by galaxies ($\sim 10^{11}\,M_\odot$) over cosmological

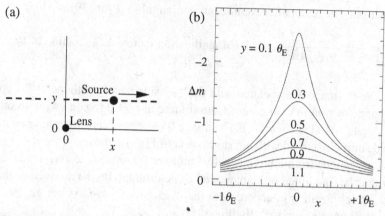

Figure 12.12. (a) Background star moving in the *x* direction at constant *y* with closest point of approach to a point-mass lens at $x = 0$. (b) Theoretical changes of magnitude Δm derived from the total magnification factor μ_{tot} (33) for trajectories at various values of *y*. If the relative velocity vector of the two stars remains constant, or nearly so, the curves represent the expected flux density changes as a function of time. [(b) R. Narayan, *Lectures on Gravitational Lensing*, astro-ph/9606001v2 (1997), Fig. 9]

distances (GLY) gives values of order $\theta_E \approx 1''$ (15), whereas lensing by stars ($\sim 1\ M_\odot$) over galactic distances (10^4 LY) gives values of only $\theta_E \approx 1$ milliarcsec.

These obstacles can be addressed (*i*) if one monitors the intensities of millions of stars over many years, hoping to find the relatively few line-of-sight encounters between a foreground and a more distant star, and (*ii*) one detects those encounters by virtue of the gravitational brightening of the background star rather than by the presence of multiple images. Stellar magnification events are expected to last typically a month or two, as a foreground star passes nearly in front of a distant background star.

We found from (33) that, for the encounter of two small objects such as two stars, the net flux change will always be positive. As the source angular position approaches that of the lens (Fig. 12a) to angles comparable to the Einstein angle, the flux density begins to increase (Fig. 12b). It reaches a maximum at the closest point of approach and thereafter diminishes as the stars separate in angular position. The uppermost curve tells us that, for a closest point of approach at $0.1\ \theta_E$, the star brightens by 2.5 magnitudes or a factor of 10.

The light curves (intensity versus time) of lensing events are expected in GR to be independent of wavelength. Also, they should be highly symmetric. In these two characteristics the light curves will differ from the many known types of variable stars in the sky.

Figure 13 shows an actual event in blue and red light with the difference flux shown in the lower panel. It truly is color independent. The event also exhibits left-right symmetry where the data are available; such events, of necessity, are discovered after they commence. The shape of the light curve also well matches the theoretically predicted response for such events.

MACHO project

Astronomers have carried out searches for microlensing events since the early 1990s. Optical telescopes with charge-coupled-device (CCD) detectors have automatically monitored, with frequent exposures (every day or so), regions of very high star density. This has been done

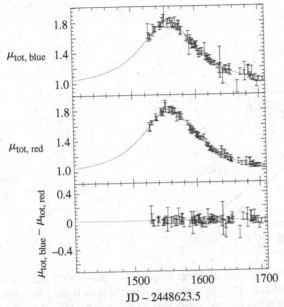

Figure 12.13. Microlensing event in the direction of the galactic bulge in two wavelength bands, blue and red. The measured fluxes relative to background are plotted against time in Julian days. The baseline ($\mu_{tot} = 1.0$) indicates background or no amplification. The difference in the two plots (bottom) is consistent with zero, indicating the color independence expected for gravitational lensing. The light curves represent a theoretical fit of the total magnification μ_{tot} (33) to the data. [MACHO project, P. Popowski *et al.*, *ApJ* **631**, 879 (2005)]

for the Large Magellanic Cloud (LMC), the satellite galaxy to the Galaxy. Objects in the halo of the Galaxy would serve as the intervening lenses.

Other studies have monitored stars in the bulge region of the Galaxy and in the nearby galaxy M31. In the former case, the lenses would be stars in the plane and bulge of the Galaxy (Fig. 14). In the latter, the lenses could be in either the Galaxy or M31. Fast computers make it possible to analyze millions of star images to look for intensity changes.

The primary objective of these studies has been to explore the matter density in the Galaxy – in particular to explore the matter density of nonluminous matter, or *dark matter*. Rotation studies of the Galaxy (and other galaxies) indicate the existence of large amounts of dark matter that could reside in a large, roughly spherical *halo* surrounding the Galaxy (Fig. 10.1 and Section 10.4). If the dark matter is in the form of compact and relatively massive objects, say 10^{-6} to $10^2 M_\odot$, they would serve as lenses for background stars. Such an object is called a massive astrophysical compact halo objects (MACHO).

A MACHO could be a black hole, neutron star, or brown dwarf, each of which is nonluminous or quite faint, but the expected numbers of these known types of objects do not begin to account for the gravitational masses suggested by the rotation studies. Thus, astrophysicists now generally favor the view that dark matter is an unknown type of matter. Microlensing can detect it without requiring it to be luminous if the lensing objects are relatively compact and massive.

Microlensing studies that have been, or are being, carried out include the EROS, MACHO, DUO, and OGLE projects. The MACHO project monitored 12 million stars in the LMC over 6 years and found about 15 microlensing events – significantly more than the 2 to 4 expected

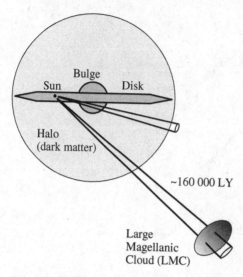

Figure 12.14. Schematic of lines of sight for Large Magellanic Cloud (LMC) and galactic bulge microlensing monitoring programs. The large numbers of bright stars in the LMC provide the potential for lensing by closer stars or compact dark matter objects in the halo of the Galaxy. For the bulge observations, the source stars are in the bulge and the lensing stars are in the disk or bulge.

from an estimate of known stellar populations. If the additional events are due to lensing by dark-matter halo objects of the Galaxy (and not the LMC), the group would conclude that between 8 and 50% of the halo matter could be in the form of MACHOs.

Follow-up studies of lensing events with, for example, the Hubble Space Telescope or radio VLBI have been finding that some of the excess events are due to previously known types of stars. This lowers the percentage just quoted. In addition, the EROS group has been quoting lower percentages. Thus, at present, there is no definitive evidence for MACHOs from these studies. Nevertheless, they dramatically demonstrate the microlensing phenomenon of GR.

The previously mentioned microlensing studies a few degrees from the galactic center (Fig. 14) probe the gravitational matter in the disk and bulge of the Galaxy. In this case, both the source and the lens reside in the Galaxy. More than 500 of these lensing events have been detected by the MACHO group. A small subset of about 40 selected bright microlensing events have been used to obtain a measure of column density of massive objects along the lines of sight. These studies constrain models of the Galaxy's structure. They appear to be in agreement with galactic models containing normal and compact stars without recourse to massive lensing objects of unknown types of dark matter.

12.4   Extended-mass lens

Until now we have discussed only point-mass lenses. In fact, the lensing of distant quasars is mostly done by galaxies or by clusters of galaxies, both of which have a measurable extent. Thus, in the lensing problem, one often encounters lenses that have finite extent and are not necessarily spherical. Rays that pass through the matter distribution as well as those that pass outside it must also be considered. It turns out, for example, that the imaging of

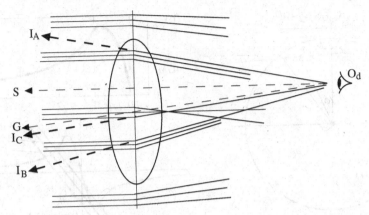

Figure 12.15. Extended-mass lens (e.g., a galaxy) refracting parallel rays arriving from the left. Near the galactic center, the rays converge, whereas, outside the mass distribution, they diverge as for a point-mass lens. In some intermediate region, the bending angle is relatively constant with position. The observer O_d potentially could see three images of the quasar, I_A, I_B, and I_C, though the latter is likely to be highly demagnified and hence not detectable. The direction of the center of the lens galaxy G and the direction of the true (unseen) source S are indicated.

Q 0957+561, the twin quasar, is materially affected by the cluster in which the lensing galaxy resides.

Galaxy as a lens

Within galaxies, absorption of optical light by interstellar grains (AM, Chapter 10) becomes an issue and must be a factor in the modeling of intensities. We ignore such absorption here. Elliptical galaxies are the main galactic contributors to lensing because they are more massive than spiral galaxies and are relatively free of dust – at least at our epoch. It is helpful, also, that the intergalactic space in a cluster of galaxies is comparatively free of dust.

Microlensing by individual stars in the lensing galaxy or galaxies leads to fluctuations in the magnifications of the individual images of a background quasar. Solutions for the underlying mass distribution thus rely more heavily on image positions than on the relative intensities of the images.

Here we present methods for the determination of image positions of point sources imaged by lenses consisting of mass distributions. We assume that all the lensing mass is localized along the line of sight over a distance small compared with the distance to the lens and that from the lens to the source.

Constant-density spheroidal lens

Ray patterns for a spheroidal extended-mass lens are shown, qualitatively, in Fig. 15. A parallel incoming beam (source at infinity) arrives aligned with the minor axis of the spheroid. A smooth mass distribution is assumed. Consider a uniform mass density, the antithesis of a point source.

The rays passing through the central part of the galaxy are focused much like a convex lens (Fig. 5a). A ray passing exactly through the center is not deflected ($\alpha = 0$) because the

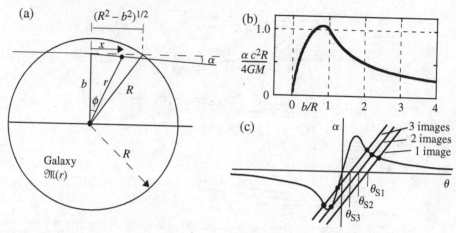

Figure 12.16. (a) Geometry for bending by a spherical mass distribution with enclosed mass $\mathfrak{M}(r)$ and radius R for $b < R$. A photon passes through or near the distribution at impact parameter b. (b) Plot of normalized bending angle α for a spherical mass distribution, $\rho(r) \propto r^{-1}$ for $r < R$ and $\rho(r) = 0$ for $r > R$, from (35) and (7). The bending angle is normalized to equal unity at $b/R = 1$. It reaches a maximum for photons passing through the outer part of the mass distribution. (c) Graphical solution for extended lens with no strong singularity at the center for three source positions; compare with Fig. 7. The lens is at the origin, $\theta = 0$. The black dots indicate image positions.

average transverse gravitational "force" on the photons passing through the center is zero. Just off axis, however, the bending increases with impact parameter (distance from the center) because of the increasing mass asymmetry relative to the trajectory; hence, rays passing through the center region will tend to converge as shown.

In contrast, rays passing completely outside the lens will experience deflection that decreases with impact parameter – as in the point-mass case – and hence will diverge. In some intermediate region, the rays must be deflected independently of impact parameter to reproduce parallel rays. These regimes are also illustrated in Fig. 15.

The observer O_d in Fig. 15 is immersed in rays emerging from the three local parts of the lens that happen to send rays in that direction. This leads to two images, I_A and I_B, from rays traversing the outer parts of the lens and a third, I_C, from close to the lens center. The latter would be highly demagnified and consequently might not be observable.

Bending angle

Consider an extended spherical galaxy of radius R with arbitrary, spherically symmetric mass distribution $\rho(r)$. The bending angle α for rays passing through the galaxy ($b < R$) can be calculated using an approach similar to the one for a gravitational point mass (Section 3). (Rays passing outside the galaxy ($b > R$) obey the point-mass bending function (7), namely $\alpha = 4GM/c^2 b$.) The geometry of interest for $b < R$ is shown in Fig. 16a. Again we take the deflection angle α of the photon to be very small.

In our Newtonian calculation, we place a mass m at radial position r within the mass distribution. For a spherical distribution, the gravitational force is the same as that due to a point mass at the origin with mass equal to the included mass $\mathfrak{M}(r)$. The function $\mathfrak{M}(r)$ is calculated from $\rho(r)$. It will replace the mass M in the integral (5), which yields the transverse velocity Δv_t imparted to the particle.

The required expression for the transverse velocity Δv_t consists of two integrals for $b \leq R$, one that sums from the midpoint to the edge of the mass distribution and the other from the edge to infinity,

$$\Delta v_t = 2 \int\limits_0^{(R^2-b^2)^{1/2}} \frac{G\mathfrak{M}(r)b}{\left(b^2 + x^2\right)^{3/2}} \frac{\mathrm{d}x}{v} + 2 \int\limits_{(R^2-b^2)^{1/2}}^{+\infty} \frac{GMb}{\left(b^2 + x^2\right)^{3/2}} \frac{\mathrm{d}x}{v}, \qquad (12.34)$$

(Newtonian transverse velocity in spherical galaxy
of radius R; small-angle deflection; $b \leq R$)

where $M \equiv \mathfrak{M}(R)$ is the mass of the entire galaxy. The integrands are the transverse component of the acceleration (F_t/m) times the time interval $\mathrm{d}x/v$. The radial distance to the particle is $r = (b^2 + x^2)^{1/2}$. The factors of 2 arise because the two integrals alone cover only one-half the path.

As an example, the expression (34) may be evaluated for a spherical mass distribution with mass density (kg m^{-3}) $\rho \propto r^{-1}$ for $r < R$ and $\rho = 0$ for $r > R$, where the coefficient of r^{-1} is set so that $\mathfrak{M}(R)$ will equal the total mass M. Invoke $\alpha = \Delta v_t/v$, the limit $v \to c$, and the factor-of-two correction for GR to obtain, in terms of $w \equiv b/R$ for $w \leq 1$ (Prob. 41a),

$$\Rightarrow \qquad \alpha = \frac{4GM}{c^2 R}\left[w \ln\frac{1 + (1 - w^2)^{1/2}}{w} + \frac{1 - (1 - w^2)^{1/2}}{w}\right], \qquad (0 \leq w \leq 1) \quad (12.35)$$

where $M \equiv \mathfrak{M}(R)$ is the galaxy mass. This function $\alpha(w)$ is plotted in Fig. 16b. It has features similar to those discussed qualitatively for a constant-density galaxy in Fig. 15. It is zero at the origin, increases as b increases, reaches a maximum, and finally, outside the distribution $(w > 1)$, decreases with the b^{-1} point-mass lens response of (7).

Contrast this to the point-mass lens of Fig. 7, where $b = D_L\theta$. The bending curve does not pass through zero; it gets larger and larger and then, at the origin, flips suddenly to the opposite sign. In Fig. 16c, we modify the curves of Fig. 7 to pass through zero bending angle at $\theta = 0$ to be in accord with the bending-angle curve for a distributed mass (Fig. 16b). The central rising portion of this revised curve represents rays passing through the central region of the extended galaxy, whereas the outer decreasing portions represent rays passing through the outer regions or completely outside the galaxy.

Singular isothermal sphere (SIS)

A stable, self-gravitating sphere of particles, all at one temperature, yields a bending angle that is independent of impact parameter if one assumes the sphere has no cutoff at large radii. This plausible, though highly idealized, model for a galaxy, which has a singularity (infinite density) at its center is known as a *singular isothermal sphere* (SIS).

One can show from the equation of hydrostatic equilibrium (2.12) and the nondegenerate equation of state (3.39) that the radial mass density distribution of a SIS is $\rho(r) \propto r^{-2}$ (Prob. 42a). This density distribution can be used in the manner leading to (35) to show (Prob. 42b) that the bending angle is indeed independent of b; that is, $\alpha(b) = $ constant.

It is interesting to note that the $\rho(r) \propto r^{-2}$ mass distribution of the SIS, with its flat bending curve, also leads to flat galactic rotation curves (10.46). Because rotation curves of real galaxies are generally flat, we might expect that observed bending curves would reflect these r^{-2} density distributions. Indeed they do, in general, albeit with deviations. Rotation curves and gravitational lensing each trace the total matter, both luminous and dark.

Image locations

The bending curve of Fig. 16c gives the characteristic bending for a galaxy as a function of impact parameter, $b = D_L\theta$, in a plane of directions that passes through the galaxy center. For such a bending curve, it is possible to obtain one, two, or three images as indicated by the intersections (black dots) with the three straight ray-trace lines representing three source locations; see (13). The distances D_S and D_{LS} are assumed to be the same for the three cases, and so the line slopes are identical.

The intersections give the values of α, θ for the several images. Finding two images would be improbable because the source must be located so the ray-trace line is just tangent to the deflection curve near its minimum or maximum. Thus, one usually would expect one or three intersections from a well-behaved extended mass distribution such as that represented by this plot.

If there is only one image (lower line), it will be displaced from the source position. If there are three intersections (upper line), the central image (black dot) will be demagnified because it is on a steep portion of the bending curve; hence, it may prove to be undetectable. The remaining two images in this case are the same two we have discussed for a point-mass lens; see Fig. 7. For a random alignment of source and lens, one typically finds either a single displaced image or an even number of images. See more on this in Section 5 under "Odd-number theorem."

Thin-screen approximation

The depth of the lensing mass along the line of sight is typically small compared with the observer–lens and lens–source distances. It is thus convenient to collapse (mathematically) the matter distribution into a single plane normal to the line of sight.

Lens plane

Project the three-dimensional (3-D) mass distribution onto a single *lens plane* normal to the line of sight (Fig. 17). If the z direction is along the undeviated line of sight, the 2-D density function $\Sigma \,(\text{kg/m}^2)$ may be defined as $\Sigma(\xi) = \int \rho(z, \xi)\,dz$, where ξ is the position vector in the plane. Position in the plane can also be expressed as a vector angular coordinate $\theta = \xi/D_L$.

For a relatively localized mass distribution, the bending of any ray will take place close to the lens plane and can be calculated as if all the mass were in the lens plane. For the large-scale geometry, one can also use the straight-line geometry of Fig. 6a. The total bending of a given ray is well approximated by an integral over the plane wherein each mass element contributes as a point source with the bending magnitude (7).

The lensing problem is thus broken into two parts: the projection onto a plane and the bending due to the resulting 2-D mass distribution. This is the *thin-screen approximation*. If the mass distribution is spherical, as in our examples, the projected density contours are circles and the vector ξ is replaced by the scalar radial coordinate ξ. For example, the SIS distribution, so projected, yields (Prob. 43a) a 2-D density distribution $\Sigma(\xi) \propto \xi^{-1}$.

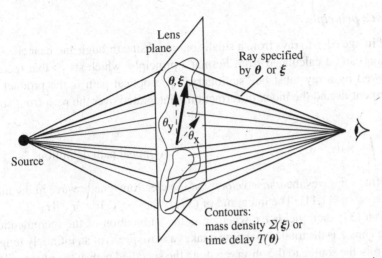

Figure 12.17. Lens plane and multiple optical paths for hypothetical rays passing through the lens plane. Actual paths must satisfy Fermat's principle. A given position in the plane or a particular ray is specified with the two-dimensional linear vector $\boldsymbol{\xi}$ (m) or the dimensionless angular vector $\boldsymbol{\theta} = \boldsymbol{\xi}/D_L$. The hypothetical contours could represent either constant mass density $\Sigma(\boldsymbol{\xi})$ (kg/m^2) or constant time delay $T(\boldsymbol{\theta})$, but not both.

Bending angle

One can integrate over the mass elements of such a 2-D circular-contour distribution, $\Sigma(\xi)$, to find $\mathcal{M}(\xi)$, the mass included within radius ξ, which is known as the *effective mass*. It turns out that the mass at radii $> \xi$ exerts no net force on a test particle at radius ξ in the lens plane, whereas the mass at lesser radii acts as if $\mathcal{M}(\xi)$ were a point source at the center of the distribution. This mimics the spherically symmetric 3-D case in which $\mathfrak{M}(r)$ acts on test particles as if it were a point mass at the center. One can therefore invoke the bending angle (7) to write

$$\alpha = \frac{4G}{c^2}\frac{\mathcal{M}(\xi)}{\xi} = \frac{4G}{c^2}\frac{\mathcal{M}(b)}{b} = \frac{4G}{c^2 D_L}\frac{\mathcal{M}(\theta)}{\theta}, \qquad \text{(Bending angle;}$$
$$\text{circular symmetry)} \qquad (12.36)$$

where we have substituted the impact parameter b for ξ, and then give b in terms of the observed angle θ, where $b = D_L \theta$. In each case, the center of the lens is the origin.

The effective mass $\mathcal{M}(\xi)$ for the $\rho \propto r^{-1}$ distribution can be found directly from a comparison of (35) and (36); Prob. 41b. The effective mass $\mathcal{M}(\xi)$ for the SIS distribution, $\rho \propto r^{-2}$, turns out (Prob. 43b) to be $\mathcal{M}(\xi) \propto \xi$. Hence, from (36), the bending angle is independent of the impact parameter b, as we found (Prob. 42) in a direct calculation. This applies if the ray passes through a perfect SIS distribution with no upper radial cutoff.

12.5 Fermat approach

The mass distributions that affect photon propagation can range from a compact (pointlike) star, to an ellipsoidal galaxy, to large clusters of galaxies. Visualization and calculation of image positions and magnifications due to complex mass distributions are greatly aided by the invocation of Fermat's principle together with the thin-screen approximation described in the preceding section.

Fermat's principle

The location of images due to rays from a small source passing through and near an arbitrary mass distribution can be calculated from Fermat's principle, which states that the *optical path* P_{opt} followed by a ray must be "stationary." The optical path is the product of the spatial path element ds and the index of refraction η integrated over the path from source S to observer O,

➡️
$$P_{opt} \equiv \int_S^O \eta \, ds. \qquad \text{(Optical path definition)} \qquad (12.37)$$

The index of refraction gives the *phase velocity* of the electromagnetic wave in the medium according to $v_p = c/\eta$ (11.11). The integrand of (37) is thus $(c/v_p)\,ds$ or $c\,d\tau$.

The integration (37) then yields $c\tau$ where τ is the summation of the incremental time elements ds/v_p. Thus τ is the total actual time it takes a wave peak (of an infinitely long wave train) to travel from the source to the observer along the specified path at the phase velocities of the medium. The optical path $c\tau$ is the hypothetical distance the wave peak would have covered in time τ had it traveled at speed c. For our purposes, though, the optical path $c\tau$ is simply a measure of the transit time τ.

In GR, the index of refraction η due to gravity acting on an electromagnetic wave is independent of frequency (see (38) below); in this case, the group velocity of an electromagnetic pulse equals the phase velocity c/η, $v_g = v_p$; see (11.44). Gravitational fields do not disperse electromagnetic signals.

There are many hypothetical paths between source and observer (Fig. 17). All the bending takes place in the near vicinity of the lens plane. Fermat's principle tells us that the only rays from a given source that can reach an observer are those with transit times that are stationary relative to the transit times of adjacent paths.

The transit time for such a path must therefore be at an extremum (maximum, minimum, or saddle point) of the transit time. Snell's law follows from this principle (Prob. 51). The problem of finding image positions thus begins with the determination of the transit times for all the possible ray paths that make a single bend at the lens plane.

Time delays

Electromagnetic signals can be delayed in two ways as they transit gravitational fields. First, there are geometric delays because the path deviates from a "straight-line" path from source to observer. Second, the gravitational fields themselves delay the signal, which is an effect of GR.

Effective index of refraction

The gravitational potential of a mass distribution slows a signal as if the potential had an effective index of refraction η. According to GR, this index is dependent on the depth of the gravitational potential Φ at each position along the track,

$$\eta = 1 - \frac{2}{c^2}\Phi, \qquad \text{(Effective index of refraction in potential; } \Phi \ll c^2\text{)} \qquad (12.38)$$

where the potential Φ is negative because it is set to zero at infinity. For a point-mass lens of mass M, the potential is $\Phi = -GM/r$. This expression (38) is valid only for small potentials,

$\Phi \ll c^2$, which is satisfied for most lensing. Note that both Φ and c^2 have the same units; multiplication of each by a mass gives an energy. This consequence of GR was confirmed in 1966–70 with radar reflections from Mercury and Venus.

The time a wavefront takes to pass from the source S to the observer O in the presence of such a gravitational potential (38) is, as just outlined, the optical path (37) divided by c:

$$\tau = \frac{1}{c} \int_S^O \left(1 - \frac{2}{c^2}\Phi\right) ds = \int_S^O \frac{ds}{c} + \int_S^O \frac{2}{c^3} |\Phi| \, ds \quad \text{(Transit time for} \quad (12.39)$$
$$\text{specified path)}$$
$$= (\tau_0 + \Delta\tau_{\text{geom}}) + \Delta\tau_{\text{grav}}.$$

We have expanded the integral into two terms. The first is the *geometric transit time*, the time required for the light wave to follow the bent path at speed c. It is expressed in the second line as the sum of the time τ_0 for an undeviated ray and the additional delay $\Delta\tau_{\text{geom}}$ due to the bending of the ray. The rightmost term is the delay $\Delta\tau_{\text{grav}}$ due to the gravitational potential. This is called the *gravitational delay* or the *Shapiro delay*.

Geometric delay

The geometric delay $\Delta\tau_{\text{geom}}$ for a hypothetical ray arises from the extra path length $\Delta\ell$ caused by the bend at the lens plane. This path length difference is (Fig. 6a)

$$\Delta\ell = (\ell_{\text{SP}} + \ell_{\text{PO}}) - \ell_{\text{SO}}. \quad \text{(Excess path length)} \quad (12.40)$$

This length difference in terms of the bending angle α for flat (Euclidean) space is (Prob. 52)

$$\Delta\ell = \frac{D_L D_{LS}}{D_S} \frac{\alpha^2}{2}. \quad (12.41)$$

This tells us that, among the paths possible for a given source and lens, larger bending angles yield greater excess-path lengths for fixed distances to the lens and source. The lower path to O_c in Fig. 3a thus has a greater excess path than the upper. At $\alpha = 0$ (no bending), the path is a straight line from the source to the observer, and there is no excess path. This expression can also be written in terms of the observed and source angles θ and θ_S by direct substitution from the ray-trace equation (13),

$$\Delta\ell = \frac{D}{2}(\theta - \theta_S)^2; \quad D \equiv \frac{D_L D_S}{D_{LS}}, \quad (12.42)$$

where again we adopt the distance parameter D (15).

The delay incurred by a signal traveling at speed c along this extra path length is

$$\Delta\tau_{\text{geom}} = (1 + z_L)\frac{\Delta\ell}{c}$$
$$= \frac{(1 + z_L)}{c} \frac{D}{2}(\theta - \theta_S)^2, \quad \text{(Geometric time delay; one ray)} \quad (12.43)$$

where we include a term $1 + z_L$. This can be thought of, somewhat imprecisely, as the redshift factor of the time interval as the signals travel from the lens to the observer. Recall from (1) that $1 + z$ is the ratio of wavelengths at emission and detection, and this, in turn, is equal to the ratio of times between wave peaks.

The result (43), in fact, follows from a proper treatment of the path difference (40) in an expanding universe with curvature, as illustrated in Fig. 6b. Our flat-space calculation (Prob. 52) gives the correct answer except for the missing $1 + z_L$ term.

The expression (43) is a very general and strictly geometric statement that gives the extra delay incurred by a hypothetical ray if it were to be observed at angle θ. This equation makes no reference to the physical effect (gravitational lensing) that could deviate a ray to this angle.

Gravitational delay

The gravitational term in (39) has its greatest value near the lens plane. The integrated value τ_{grav} is the transit time for the particular path under consideration, and the value will, in general, vary with the chosen path. In the thin-screen approximation, each path can be characterized by the position at which it crosses the lens plane (i.e., by the 2-D vector θ or its components θ_x, θ_y; see Fig. 17).

For the gravitational delay, it is convenient to define a dimensionless *gravitational delay function* $\psi(\theta)$ as

$$\psi(\theta) \equiv \frac{1}{D} \int_S^O \frac{2}{c^2} |\Phi| \, ds. \qquad \text{(Two-dimensional gravitational delay} \qquad (12.44)$$
$$\text{function; dimensionless)}$$

This is a two-dimensional function, where $D = D_L D_S / D_{LS}$ (15) and Φ is the gravitational potential. It may be viewed as a 3-D gravitational potential projected onto the two dimensions of the lens plane. We arrived at (44) by multiplying the last term of (39) by a factor of c/D to make ψ dimensionless and to simplify the final expression for the overall delay.

In essence, $\psi(\theta)$ is a dimensionless relative measure of the gravitational delay as a function of the path specified by θ or ξ. If Φ is zero everywhere on the path, $\psi(0) = 0$; there is no gravitational delay. If $|\Phi|/c^2 = 0.01$ over the entire path of length D_S, $\psi = 0.02 D_S / D = 0.02 D_{LS}/D_L$. The function ψ depends only on a ratio of distances; it thus contains no information about the overall scale size (e.g., D or D_L) of the geometry.

The gravitational delay term of (39) may thus be rewritten with (44) as

$$\Delta\tau_{grav} = D \frac{1 + z_L}{c} \psi(\theta), \qquad \text{(Gravitational time delay; one ray)} \qquad (12.45)$$

where again the expanding universe is taken into account with the $1 + z_L$ term. We find that the gravitational delay is proportional to D (45) as is the geometric delay (43).

Fermat potential

The overall time-delay function $T(\theta)$ is defined to be the sum of the gravitational and geometric delays. From (39) we have

$$T(\theta) = \tau - \tau_0 = \Delta\tau_{geom} + \Delta\tau_{grav}. \qquad (12.46)$$

Substitute from (43) and (45) to obtain

$$T(\theta) = \frac{(1 + z_L)}{c} D \left[\frac{1}{2}(\theta - \theta_s)^2 + \psi(\theta) \right]. \qquad \text{(Time-delay function;}$$
$$\text{one ray; s)} \qquad (12.47)$$

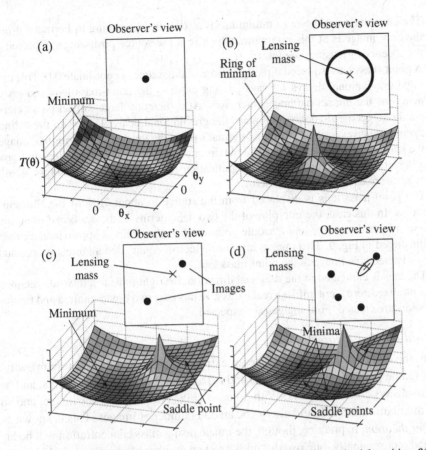

Figure 12.18. Examples of the time-delay function $T(\boldsymbol{\theta})$ versus celestial position $\boldsymbol{\theta}(x, y)$ for a point source fixed at position 0,0 as imaged by four mass distributions: (a) zero mass (no lens); (b) point-mass lens (cross) directly in front of source, yielding an Einstein ring; (c) point-mass lens offset from source; and (d) offset point-mass lens with extended component. The plotted function includes both geometric and gravitational delays. The observer's view (box) shows, in each case, image positions as large black dots; their magnifications are not indicated. Images appear at maxima, minima, and saddle points. [A. Cohen, Ph.D. thesis, MIT, 2000]

The delay function varies in the two dimensions of the lens plane. The angular positions of a hypothetical ray and the source are specified, respectively, with the 2-D vectors $\boldsymbol{\theta}$ and $\boldsymbol{\theta}_S$. This is the function to which Fermat's principle is applied to obtain the paths actually taken by rays. It is sometimes called the *Fermat potential*.

Four examples

Plots of $T(\boldsymbol{\theta})$ are shown in Fig. 18 for four different mass distributions. The delay is plotted vertically and the θ_x, θ_y coordinates of the lens plane are plotted horizontally. In each case, the distant unresolved source (e.g., a quasar) is at the origin at position 0,0. We now discuss the lensing by the four mass distributions in the figure as follows:

(a) There is no gravitational matter along the line of sight to the source. In this case the gravitational delay $\psi(\boldsymbol{\theta})$ is zero. The geometrical delay varies as the square of the offset from the source direction at the origin (47), and so the surface $T(\boldsymbol{\theta})$ is a paraboloid.

The only stationary point (a minimum) is at $\theta = \theta_S$. According to Fermat's principle, the only image is at this minimum, which is at the source position, as expected given the absence of lensing; see inset.

(b)   A point-mass lens is placed directly in front of the source at coordinate 0,0. This induces large gravitational delays for rays passing close to the point-mass lens and lessening delays as the impact parameter increases. At sufficiently large impact parameters, the parabolic geometric delay dominates the gravitational delay. This, with the cylindrical symmetry, gives a ring of minima. Fermat's principle thus dictates a ringlike image that we know as the Einstein ring. The maximum at the point-mass lens does not produce a detectable image because the magnification turns out to be vanishingly small; see "Curvature as magnification" below.

(c)   The point-mass lens is displaced from the source to about 70% of the Einstein ring radius. In this case, the interplay of the two delay terms leads to a broad minimum on one side of the source and a saddle point on the other. This is approximately the case illustrated in Fig. 9. Two images are thus detected. Again, demagnification precludes a third detectable image at the point-mass lens.

(d)   The lens is elongated in the diagonal direction, far right to near left, while retaining its centralized core – not unlike a real galaxy. In this case two broad minima and two saddle points are created. Four images are expected.

Odd-number theorem

In general, the addition of mass to create an extremum (maximum or minimum) will also produce a saddle point; compare cases (c) and (d) above. Thus, stationary points, and hence images, are always created in pairs. With no mass (Fig. 18a), there is one image, and so for any mass distribution there should always be an odd number of images. This is known as the *odd-number theorem*. In practice, though, the image near a mass concentration will be highly demagnified; consequently, one usually finds an even number of images, as in Fig. 18c,d.

The several images have delays that differ according to the heights of the Fermat surface at the stationary points. The highest points have the greatest delays. Thus, each image is a view of the source at a different time. The differential delays can range from weeks to months to more than a year for extragalactic lensing of quasars.

A real example of a quadruple lens very much like that of Fig. 18d is shown in Fig. 19. The two frames show different intensity cuts of the same image. The lensing galaxy is seen in the center of the cross in the deeper (darker) image.

Curvature as magnification

Consider Fig. 18c and let the peak be somewhat rounded to represent a slight diffuseness of the lens. In this case, an image would lie at its maximum nearly coincident with the lens. Now imagine that the source at the origin $(0, 0)$ is moved a small amount from the origin while the point-mass lens remains fixed at its off-axis position. The paraboloid contribution to the surface shape would be shifted with the source, but the peak position would move very little because it is quite well anchored to the stationary lens. The ratio of the two motions is the magnification, as we now explain.

Let the two source positions (shifted and unshifted) be elements at the outer edges of an extended source and the two image positions be the respective parts of the image. In our example, the image would be smaller than the source; the source is demagnified. See the similar argument preceding (25).

(a)   (b)

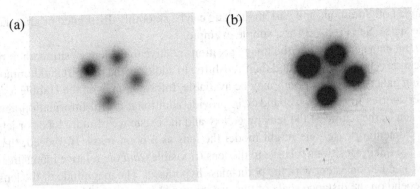

Figure 12.19. The quasar HE 0435–1223 at $z_S = 1.7$ imaged by a galaxy at $z_L \approx 0.4$ in the optical SDSS r' band with the Magellan Baade telescope. The two frames (8.8″ square) are the same image with different intensity cuts, with (b) showing the fainter features – notably the central lensing galaxy. This lensing pattern is similar to that of Fig. 18d. Another lensing example like this, Q 2237+0305, has been nicknamed the "Einstein cross." [L. Wisotzki, P. Schechter, H. Bradt (ahem!), J. Heinmüller, and D. Reimers, *A&A* **395**, 17 (2002)]

Now consider the image at the broad minimum in Fig. 18c. First consider source motion in the radial direction, away from the lens, to the lower left, and visualize the shift of the paraboloid with it. The Fermat surface is constructed of both the paraboloid and the lens peak. Thus, its minimum will move outward radially but not as far as the source; the lens peak restrains its motion. The image thus moves somewhat less than the source, giving a modest demagnification in the radial direction. This is in accord with the upper (I_1) image in Fig. 9.

Now consider the azimuthal motion of this source (Fig. 18c) about the lens. Again, it carries the no-lens paraboloid with it; the entire surface rotates about the lens peak. Because the image at the Fermat minimum is more distant from the axis of rotation than is the source (initially at 0, 0), the linear motion of the image is greater than that of the source. This gives magnification in the azimuthal direction, which again is in accord with the I_1 image of Fig. 9.

In these cases, the curvatures of the Fermat potential appear to indicate the degree of magnification or demagnification. The Fermat minimum in Fig. 18c is quite broad (low curvature) in the azimuthal direction in which we found magnification. In the radial direction, it has greater curvature but significantly less than that of the lens peak. This was associated with a modest demagnification.

It turns out, in general, that broad extrema (low curvature) in some direction yield large magnifications in that direction. In these cases, a source motion leads to a larger displacement of the image. Narrow extrema (high curvature), in contrast, yield lesser magnifications, which, if less than unity, is demagnification.

The Fermat potential is thus highly useful in that it not only contains time-delay information but also provides image positions and magnifications.

Modeling

The logic heretofore has been to determine image positions and magnifications from a specified lensing mass distribution. In practice the problem is usually inverted. One observes the lensed images and attempts to deduce the amount and distribution of the lensing mass. It can include both luminous and dark matter and can be smoothly distributed or granular (e.g., pointlike stars). See Prob. 53 for a simple example.

The input data could include the image positions relative to the lens position, their relative magnifications (flux densities), and their redshifts. In addition, the redshift and luminosity of the lensing galaxy, cluster, or both may be available. Information from the Hubble telescope and observatories in other wavebands can provide additional useful information about the system such as the nature of the lensing galaxy and the existence of radio lobes or jets.

In an elementary case, one might model the lens as a point mass. If, indeed, that were the case, the image positions relative to the lens (if visible) *and* the relative intensities of the images would be consistent with the point-mass hypothesis. The magnitude of the lens mass would depend on the distance scale of the system; see "Mass of lens" below.

In practice, there are no observational cases of multiple images of a quasar for a point-mass lens; all such lenses are galaxies often in association with a surrounding cluster. One thus adopts more complex mass distributions as trial models and tests them against the measured image positions with constraints provided by the image magnifications. The magnifications are subject to fluctuations due to microlensing by stars, as noted under "Galaxy as a lens" in Section 4. This and other limitations in the modeling can lead to less than definitive results for the mass distribution of the lensing matter.

Hubble constant

An important by-product of gravitational lensing is that a lens system consisting of a quasar and an intervening galaxy can yield a determination of the Hubble constant H_0, a fundamental cosmological parameter described above after (1). This is done by measuring the difference in occurrence times of intensity changes in the two images of the lensed quasar.

Quasars are known to vary in intensity in several wavelength bands. Intensity plots for the two images of a single quasar are expected to be similar but with a temporal offset because the signals from the two images travel along different paths to the observer and hence suffer different delays. We show here that the observed differential time delay $\Delta\tau$ between the arrival times of a particular variation (e.g., a sudden enhancement of intensity) yields the distance parameter $D \equiv D_S D_L / D_{LS}$ and hence the Hubble constant.

Distance–redshift relations

The three distances that make up the parameter D are angular-diameter distances, as noted previously after (13). They can be obtained from the redshifts, z_L and z_S of the lens and source, if the geometry of the universe is known. The redshift is defined as $z \equiv (\lambda - \lambda_0)/\lambda_0$ (1). In general, one can describe a distance-redshift relation (e.g., for the source) as

$$D_S = \frac{c}{H_0} f(z_S) \underset{z \ll 1}{\rightarrow} \frac{c}{H_0} z_S, \qquad \text{(Distance–redshift relation)} \qquad (12.48)$$

where D_S is the angular-diameter distance, H_0 (s^{-1}) is the Hubble constant, and c/H_0 is the *Hubble distance*, the distance a photon would travel at speed c in the Hubble time H_0^{-1}. The Hubble distance approximates the size of the observable universe. A similar expression relates D_L and z_L.

The function $f(z_S)$ depends on the chosen cosmological parameters – specifically the total matter density parameter Ω and the cosmological constant Λ in the usual description of the universe. For close objects, $z_S \ll 1$, the function $f(z_S)$ reduces to z_S. If the redshift is interpreted as a Doppler shift and the recession velocity v is nonrelativistic ($v/c \ll 1$), we then have $v = c\Delta\lambda/\lambda = cz$, and hence $z = v/c$. In this case, the $z \ll 1$ limit of (48) becomes $v_S = H_0 D_S$, the more familiar version of the Hubble relation. The units of H_0 (km s^{-1} MLY^{-1}, or more commonly, km s^{-1} Mpc^{-1}) reflect this view.

The relation for D_{LS} depends on both z_L and z_S:

$$D_{LS} = \frac{c}{H_0} g(z_L, z_S) \underset{z \ll 1}{\rightarrow} \frac{c}{H_0}(z_S - z_L), \tag{12.49}$$

where the function g depends, as for f, on the chosen cosmological parameters of the universe, which are hereafter referred to collectively as the chosen "cosmology." In the $z \ll 1$ limit, D_{LS} approaches the Euclidean value $D_S - D_L$; see (49) and (48). These several expressions yield the distance parameter D,

$$D \equiv \frac{D_L D_S}{D_{LS}} = \frac{c}{H_0} \frac{f(z_L)f(z_S)}{g(z_L, z_S)}, \qquad \text{(Scale factor–redshift relation)} \tag{12.50}$$

where all distances are angular-diameter distances. As an example, for a critical universe, with total matter-density parameter $\Omega = 1$ and cosmological parameter $\Lambda = 0$, the functions f and g are

$$f(z_S) = \frac{2}{1 + z_S}\left[1 - \frac{1}{(1 + z_S)^{1/2}}\right]; \qquad \text{(Redshift functions)} \tag{12.51}$$

$$g(z_L, z_S) = 2\left[\frac{(1 + z_S)^{-1}}{(1 + z_L)^{1/2}} - \frac{1}{(1 + z_S)^{3/2}}\right].$$

These expressions are the links between the Hubble distance, c/H_0, and the angular-diameter distances, through, for example, (49). The form of $f(z_S)$ may also be used for $f(z_L)$. The function $g(z_L, z_S)$ gives the angular-diameter distance of an object at z_S for an observer at z_L for the assumed cosmology. Thus, $f(z_S) = g(0, z_S)$.

System scale

The distances given in (48)–(50) for specified redshifts vary inversely with H_0. The Hubble constant thus sets the scale, or size, of the source–lens–observer system, whereas the functions f and g reflect geometry or curvature of the universe. The several values of redshift z locate objects within this universe.

The image positions, flux densities, and redshifts in a lensing system allow the observer to model the mass distribution to obtain a diagram such as that of Fig. 20a for an assumed H_0 and geometry. The observables, however, do not distinguish this case from that in Fig. 20b in which the entire system is larger but the recession velocities remain the same. In both cases the observed angles and redshifts are the same, but all distances are greater by a factor of 4/3 and H_0 is hence decreased by the same factor; see (48)–(50). In this comparison, we assume a fixed cosmology.

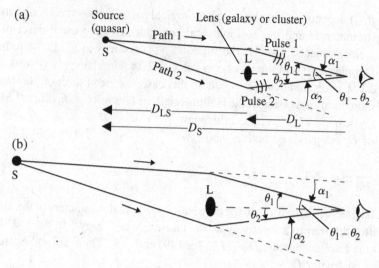

Figure 12.20. Geometry for differential time-delay measurement of the Hubble constant H_0. (a) Two geometrical paths for which signal transit times differ by a "differential delay" $\Delta\tau$ (53). The relative angles and distances represent approximately those given in (20) and (21) for the case $\theta_S = \theta_E$. Note that $\theta_2 < 0$. (b) Scaled-up (factor of 4/3) version of (a) with the same observable angles. The lens mass M and differential delay $\Delta\tau$ both scale up by the same factor. The measured value of $\Delta\tau$, together with the imaging geometry from modeling, yields the scale factor typified by $D \equiv D_L D_S / D_{LS}$. Adoption of cosmological parameters Ω and Λ then yields the Hubble constant through (50).

The scale factor, which we can describe with the parameter D (15), is thus not obtainable solely from the image positions and redshifts. It is, however, obtainable from the transit time differences for the various paths. Once obtained, D yields H_0; see (50).

Time difference (two paths)

Both the geometric and gravitational delays were found in (43) and (45) to scale linearly with D, and so their sum $T(\boldsymbol{\theta})$ also scales with D, as stated in (47). Unfortunately, the quantity $T(\boldsymbol{\theta})$ is not measurable; we do not know when the signal left the source. The difference in delays for two paths from source to the observer (Fig. 20), however, is measurable and it also scales with D. The difference can be written, from (47), as

$$\Delta\tau = T(\boldsymbol{\theta}_2) - T(\boldsymbol{\theta}_1) \qquad \text{(Differential delay for two images)} \qquad (12.52)$$

or

$$\Delta\tau = D\frac{(1+z_L)}{c}\left\{\left[\frac{1}{2}(\boldsymbol{\theta}_2 - \boldsymbol{\theta}_S)^2 + \psi(\boldsymbol{\theta}_2)\right] - \left[\frac{1}{2}(\boldsymbol{\theta}_1 - \boldsymbol{\theta}_S)^2 + \psi(\boldsymbol{\theta}_1)\right]\right\}, \qquad (12.53)$$

where $\boldsymbol{\theta}_1$ and $\boldsymbol{\theta}_2$ are the positions of the two rays in the lens plane, $\boldsymbol{\theta}_S$ is the source position, D is the angular-diameter distance parameter defined in (50), z_L is the lens redshift, and ψ is the gravitational delay function (44).

The factors in square brackets are, in principle, known from the modeling of the lens with the positional and intensity data. The dimensionless ψ terms require integration along the path from the source to the observer (44). The r^{-1} potential of a point-mass lens is sufficiently gradual that the photons are accumulating gravitational delay along the entire path. One can not approximate the delay as taking place only in the vicinity of the lens. (Placing the source or observer at infinity yields infinite delay!)

The distance parameter D is the unknown scale factor in the problem. A measure of $\Delta\tau$ provides, through (53), a numerical value for D, the scale factor of the system. The Hubble constant H_0 then follows from (50), where one uses the functions f and g appropriate to the assumed cosmology. The determination of D leads to values for D_L, D_S, and D_LS because the adopted model (e.g., Fig. 20) provides the distance ratios $D_\mathrm{L}/D_\mathrm{S}$ and $D_\mathrm{LS}/D_\mathrm{S}$.

Mass of lens

The mass M of the lens in Fig. 20a,b scales with D for a given set of observed image angles, as noted for a point-mass lens; see (24) and (23). This also follows from the bending angle's being proportional to M/b (7), where M is the lens mass and b the impact parameter. For the angles to remain the same in the scaling, this ratio must be conserved. Because b scales with D, so does M. The mass in Fig. 20b must therefore be $1/3$ larger than that of Fig. 20a. A measurement of $\Delta\tau$ thus determines M as well as D. By extension, this also applies to a system with a distributed-mass lens.

Example: point-mass lens

As an example, let us find the differential delay $\Delta\tau$ for a point-mass lens. Determine the gravitational delay function $\psi(\theta)$ by integration of the potential along the line of sight (44), where $\Phi = -GM/r$. The approximate result is (Prob. 54)

$$\psi(\theta) \approx \frac{2R_\mathrm{G}}{D} \ln\left(\frac{4D_\mathrm{LS}}{D_\mathrm{L}} \frac{1}{\theta^2}\right); \quad R_\mathrm{G} \equiv \frac{GM}{c^2}, \quad \text{(Gravitational delay function)} \quad (12.54)$$

where $\theta = b/D_\mathrm{L}$ is the image direction relative to the lens and R_G is the often-encountered factor GM/c^2 in GR. The latter has dimensions of distance, is one-half the Schwarzschild radius, and has magnitude 1.48 km for 1 $M_\odot$. The function (54) is an approximation for an impact parameter that is small compared with the path lengths D_LS and D_L and for gravity not excessively strong, $|\Phi(b)| \ll c^2$.

Because all detected rays are in the plane of the source–lens–observer (e.g., Fig. 20), we use signed angles, θ_1 and θ_2, rather than vectors. From our two expressions for θ_E, (15) and (23), we may write, recalling that θ_2 is negative,

$$\frac{2R_\mathrm{G}}{D} = \frac{|\theta_1\theta_2|}{2}. \quad (12.55)$$

Use this in (54), expand the logarithm, and use $\theta_\mathrm{S} = \theta_1 + \theta_2$ (23) to rewrite $\Delta\tau$ (53) in terms of the image angles (Prob. 55a) as

$$\Delta\tau = D\frac{1+z_\mathrm{L}}{c}\left[\frac{1}{2}\left(\theta_1^2 - \theta_2^2\right) + |\theta_1\theta_2| \ln\left|\frac{\theta_1}{\theta_2}\right|\right]. \quad (12.56)$$

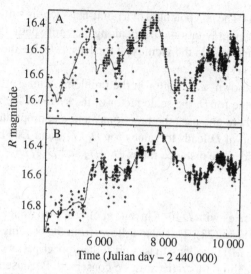

Figure 12.21. Radio flux curves for the two images, A and B, of the twin quasar Q 0957+561 (Fig. 1) over a period of 17 years. The two curves have similar broad features, but those of the B image are delayed by 417 days relative to those of the A image. The data points are precise to about 0.01 magnitudes (∼1%), but the scatter is about 10 times greater. This scatter and other differences between the two curves are probably due to microlensing by stars in the lensing galaxy or cluster that occurs on time scales of days to months. [J. Pelt *et al.*, *A&A* **336**, 829 (1998)]

If one has established that the lensing is truly by a point source, it is only necessary to enter the measured values of z_L, θ_1 and θ_2, and the time delay $\Delta\tau$ to obtain a value of the distance parameter D. Then, as before, with the functions f and g for the assumed cosmology, the Hubble constant H_0 is obtained from (50) (Prob. 55).

Q 0957+561

Of course, this can work only if the quasar cooperates by producing significant time variability of its measured flux density and if observations are made month after month to track the variability in both images. At present, about ten lens systems have yielded significantly similar variability in both images. The twin quasar Q 0957+561 has been monitored since 1979; it yields a delay of about 417 d (Fig. 21). This, together with the geometry, results in reasonable values of H_0. See Prob. 55b for an example with a 47-d delay.

What does this result tell us about the Hubble constant? Unfortunately the results do not yet have sufficient accuracy to improve on other methods because the mass distributions of the lens systems are not sufficiently determined through modeling of the imaging data. Even so, the measurements are a valuable check, for the method is completely independent of other methods and is based directly on GR in the well-tested weak-field limit. The large distance to a quasar is determined directly without recourse to a distance ladder that would require calibrations with its intrinsic uncertainties.

The small, rapid fluctuations in Fig. 21 are due, at least in part, to *microlensing* of the quasar images by individual stars in the intervening galaxy. The relative flux densities of the several images will thus not always have the values expected from a smooth galaxy model.

~90″

Figure 12.22. Image of galaxy cluster Abell 2218 taken with the WFPC2 instrument on the Hubble Space Telescope. The cluster is in the northern constellation Draco and is distant about 2 GLY ($z = 0.17$). Its two gravitational clusters distort background galaxies into giant arcs (strong lensing) or arclets (weak lensing). Magnification (flux-density enhancement) of background galaxies makes visible those that would otherwise be undetectable. North is ~20° clockwise from up and east is ~20° clockwise from left. [HST/NASA, A. Fruchter and the ERO Team (STScI); STScI-PRC00-08; Astronomy Picture of the Day (APOD), 2001 Oct. 7; also see APOD 2004 Jan 17]

If we presume dark matter to be smooth, this becomes a mechanism for separating out the dark matter and stellar contributions in the lensing galaxy.

12.6 Strong and weak lensing

Gravitational lensing is sensitive to gravitational matter whether it is luminous or dark. Lensing can thus be used to map out the matter content of the universe. In the material above, we have described lensing by individual stars (microlensing), galaxies, and clusters of galaxies. The light from a single background quasar can be modified by all three; a galaxy contains stars, and it may lie within a cluster. The lensing described heretofore produces multiple images, whether resolved or not, and is known as *strong lensing*.

A Hubble telescope image of a cluster of galaxies, Abell 2218, is shown in Fig. 22. It has two concentrations of mass indicated by clumps of visible light, one more prominent than the other. The much more distant background galaxies are dramatically distorted into ellipses and arcs. Close to the cluster, they are stretched out into *giant arcs* opening toward the cluster center, which are similar to the arcs in Fig. 10d. They are an indicator of matter, either luminous or dark. These arcs and pronounced ellipses are further examples of strong lensing.

Farther out from the cluster, the galaxies are only slightly elongated into *arclets*. They are ellipses with their long axis in the azimuthal direction, as in Fig. 10f. When the elongation is only a small fraction of the diameter, it is known as *weak lensing*. With appropriate statistical averaging of many such images in a given sky region, one can map out the gravitational matter associated with an intervening cluster. The cluster masses derived from such studies indicate large amounts of dark (nonluminous) matter more or less consistent with those from dynamical studies of galaxy velocities (Section 2.4) and x-ray studies of the intergalactic gas (Section 5.5).

Astronomers are beginning the study of galaxy elongations (weak lensing) in the absence of a visible cluster. The expected elongations are small (a few percent). To obtain a detectable signal, one must average the galaxy shapes over many galaxies in a given sky region.

Such studies can, in principle, map out large-scale gravitational matter distributions and thus determine the *angular power spectrum of matter fluctuations* in the nearby universe out to $z \approx 0.2$. This complements optical galaxy surveys that find the distribution of luminous matter only.

The temperature fluctuations of the cosmic microwave background (CMB) provide another snapshot of the fluctuations, but at a much earlier epoch ($z \approx 1000$). The two views can be compared with models of the evolution of the fluctuations. Weak lensing studies are potentially of great importance.

Problems

12.2 Discovery

Problem 12.21. (a) At what distance would a U. S. dime of diameter $d = 10$ mm be if it were to subtend an angle equal to that of the deflection of starlight just grazing the sun ($1.75''$)? (b) What is the angular size $\Delta\theta$ of an elliptical galaxy of diameter similar to the Galaxy, namely, $d = 1 \times 10^5$ LY, if its measured redshift is $z = 0.39$, the redshift of the lensing galaxy of Q 0957+561? Adopt a Hubble constant of $H_0 = 20$ km s^{-1} MLY^{-1} and the distance–redshift relation for a total matter-density parameter $\Omega = 1$ and cosmological constant $\Lambda = 0$; refer to (48) and (51). (c) Demonstrate from (48) and (51) that the distance-redshift relation reduces to $D = cz/H_0$ for close objects ($z \ll 1$) and, from this, obtain the Hubble relation $v = H_0 D$. State any assumptions you make. [Ans. $\sim$1 km; $\sim$5$''$;–]

12.3 Point-mass lens

Problem 12.31. (a) Formally derive the angular radius of Einstein's ring (8) for a source at infinite distance ($D_S \to \infty$) and a point-mass lens of mass M at distance D_L from the observer. Use small-angle approximations, the bending angle formula (7), and the geometry of Fig. 6a. (b) Use your expression to find a typical Einstein-ring radius for stellar lenses ($M \approx 1\ M_\odot$) at a distance comparable to the radius of the Galaxy, $\sim$25 000 LY. (c) Similarly find θ_E for extragalactic lenses of galactic mass ($M \approx 10^{11}\ M_\odot$) at a distance of $\sim$1 GLY. [Ans. –; $\sim$1 milliarcsec; $\sim$2$''$]

Problem 12.32. (a) Evaluate the bending angle expression (7) for a light wave just grazing the sun to obtain the $1.75''$ value given in the text. Maintain your calculation precise to about 0.1%. (b) Consider two parallel rays from a very distant star that just graze the opposite sides of the sun. At what distance D_g from the sun (in AU) do they converge? How does this compare with the planetary distances from the sun? (c) Make a qualitative sketch of the sun and rays from a distant star similar to Fig. 5b but with a larger sun and the grazing rays shown. Indicate the regions or locations from which an observer could see two images and from which an Einstein ring could be observed. Also show the earth's location during the 1919 test of GR. (d) An observer at the distance D_{AC} of Alpha Centauri (4.3 LY or 2.7×10^5 AU) is aligned (improbably) with the sun and a very distant star on the other side of the sun and hence detects an Einstein ring around the sun. Find the impact parameter b for the rays in the ring and also the half angle θ_E of the ring. Hint: define b in terms of the solar radius $b = a R_\odot$, and use geometry (with a sketch) and the bending angle (7) to solve for $D_{AC}(a)$ and $\theta(a)$. Check your answer with (8). Comment on the practicality of such a detection. [Ans. $-$; $\sim$500 AU; $-$; $\sim$20 $R_\odot$, $\sim$80 milliarcsec]

Problem 12.33. (a) Carry out the integration in (5) that leads to the scattering angle α in the Newtonian context (6). (b) Demonstrate how you might obtain this expression without evaluating an integral if you make certain approximations. What are the approximations?

Problem 12.34. Make an enlarged copy of Fig. 5b. (a) Measure the bending angles of the five rays on one side of the lens. How do these compare with the $1/b$ dependence expected from (7)? (b) Does the lens of Fig. 5b exhibit a real image or a virtual image? For the latter, extrapolate the bent rays backward and compare with an ideal thin concave lens (with sketch).

Problem 12.35. You measure the directions θ_1 and θ_2 of the two images of a quasar, where $\theta = 0$ is the direction of a visible lens galaxy, that may be taken to be a point mass. (a) Apply the lens equation (16) to each image to find expressions for the true direction of the quasar θ_S and the Einstein angle θ_E as a function of θ_1 and θ_2 only – that is, with no other constants or variables, as given in (23). (b) Let the two images of a quasar be at $\theta_1 = 5''$ and $\theta_2 = -1''$ relative to the lens (similar to the twin quasar; Fig. 1). What is the (true) direction θ_S of the quasar, and what is θ_E? (c) If our quasar is at $z_S = 1.40$ and the lens at $z_L = 0.4$, again like the twin quasar, what is the mass M of the galaxy? Use the Einstein angle expression (15), the distance–redshift relations given in (50) and (51), $H_0 = 20$ km s^{-1} MLY^{-1}, and a result from part (b). (d) For this system, what are the magnification factors μ_1 and μ_2 of the two images? How well does their ratio compare with the magnitude difference $\Delta m_B = 0.3$ of the two images of the twin quasar? What might explain the discrepancy? [Ans. $-$; $\sim+4''$, $\sim2''$; $\sim10^{12}\ M_\odot$; $\Delta m \approx 3.5$]

Problem 12.36. Find, from (17), the image positions for a point-mass lens as a function of θ_E (and θ_S if appropriate) for the following source positions: (i) $\theta_S = 0$, (ii) $0 < \theta_S \ll \theta_E$, (iii) $\theta_S = \theta_E$, and (iv) $\theta_S \gg \theta_E$, where θ_E is the radius of the Einstein ring. [Ans: See text, (18)–(22) and Fig. 8]

Problem 12.37. (a) Find the image locations $\theta_{1,2}$ in units of θ_E for a small background source located at $\theta_S = 0.6\theta_E$ from a point-mass lens. Compare your results with Figs. 9 and 10d,e. (b) For each image, what are the radial and azimuthal magnifications? Compare again with Figs. 9 and 10d,e. (c) For each image, use these results to obtain the approximate flux-density increase relative to the unlensed flux from the source. (d) Find the flux increase factor including both images. Compare with the result you would obtain from (33). [Ans. $\sim$1.3θ_E, $\sim -0.7\theta_E$; $-$; $\sim$1.4, ~-0.4; $\sim$1.9]

Problem 12.38. (a) For point-mass lensing, find the source direction θ_S in terms of θ_E, at which the azimuthal magnification is $M_{az} = -1$ for the image inside the Einstein ring; Fig. 9. (b) For this situation, find the angular positions of the two images and comment on how the source and image positions differ from Fig. 9. (c) Find the source and image positions that yield combined magnification $\mu = M_{az}M_{rad} = -1$ for the flipped image. [Ans. $\sim$0.7θ_E; $\sim$1.4θ_E, $\sim-0.7\theta_E$; $\theta_S \approx 0.35\theta_E$]

Problem 12.39. (a) Find from (30) the magnification factors $\mu_{1,2}$ (32) in terms of $u \equiv \theta_S/\theta_E$ for a point-mass lens and source size small compared with its distance from the lens, $\Delta\theta_S \ll \theta_S$. First, factor (30) and then use the lens equation (16) divided by θ^2 to rewrite it as

$$\mu = \left(\frac{\theta}{\theta_S}\right)^2 \left(2\frac{\theta}{\theta_S} - 1\right)^{-1}.$$

Then, rewrite the solution (17) in terms of u as

$$\frac{\theta_{1,2}}{\theta_S} = \frac{1}{2}\left(1 \pm \frac{\sqrt{u^2 + 4}}{u}\right),$$

which can be substituted into the previous expression to obtain, after some manipulation, $\mu_{1,2}(u)$ in the form (32). (b) Proceed on to obtain the total magnification μ_{tot} (33).

12.4 Extended-mass lens

Problem 12.41. (a) Derive for $b < R$ the bending angle expression (35) from (34) for a spherical lensing galaxy of mass M and mass density distribution $\rho(r) = M/(2\pi R^2 r)$ out to radius R and $\rho(r) = 0$ beyond that. Consider small-angle deflections as a function of $w \equiv b/R$, the ratio of the impact parameter to the galaxy radius. Proceed as follows: (*i*) Satisfy yourself that the integral expression (34) is correct for Newtonian deflections. (*ii*) Find the function $\mathfrak{M}(r)$ and demonstrate that $\mathfrak{M}(R) = M$. (*iii*) Substitute into (34) and convert the variable r to a function of x, the particle position (Fig. 16a). (*iv*) Carry out the integrations. (*v*) Calculate the desired function $\alpha(b/R)$. (*vi*) Make the ad hoc correction needed to take into account photons and GR and write your result in terms of the variable $w \equiv b/R$. (b) What is the effective (2-D) mass function $\mathcal{M}(w)$ for this postulated mass distribution? Use your result from (a). What is $\mathcal{M}(\theta)$? [Ans. $-$; $Mw f(w)$, where $f(w)$ is the bracketed expression in (35)]

Problem 12.42. (a) Show that a singular isothermal self-gravitating sphere (SIS) of identical particles of mass m satisfies both the equation of hydrostatic equilibrium (2.12) and the nondegenerate equation of state (3.39) if the mass-density distribution is $\rho = Cr^{-2}$ with no upper radial cutoff, where the constant C, a function of temperature T, is appropriately chosen. Find C in terms of T and m and also in terms of the measurable variance of the one-dimensional line-of-sight velocity, $\sigma_v^2 \equiv \langle v_r^2 \rangle_{av}$. Hints: use $\rho = Cr^{-2}$ as a trial solution, integrate (2.12) to find $P(r)$, and justify the equality $m\sigma_v^2 = kT$. (b) For the SIS mass distribution, find, from the integral (34), the Newtonian bending angle α for a particle of mass m and speed v as a function of σ_v^2 and impact parameter b. Make appropriate corrections to get the GR result for photons. Comment on the b dependence. [Ans. $C = \sigma_v^2/2\pi G$; $4\pi\sigma_v^2/c^2$]

Problem 12.43. (a) Find the two-dimensional (2-D) radial mass distribution $\Sigma(\xi)$ (kg/m²) for a spherical SIS 3-D mass distribution, $\rho = Cr^{-2}$, with no upper radial cutoff. Integrate along a column that runs parallel to the line of sight (z-axis), has cross section $A = dxdy$, and intercepts the y-axis at $x = 0$, $y = b$, where b is the impact parameter. Illustrate your geometry with appropriate sketches. (b) From your answer and the thin-screen bending formula (36), find the bending angle $\alpha(\xi)$ as a function of C and ξ. How does it depend on ξ? Compare with the answer of Prob. 42 if you did it. [Ans. $\pi C/\xi$; $8\pi^2 GC/c^2$]

12.5 Fermat approach

Problem 12.51. Use Fermat's principle to find the path of the ray that travels from point A to point B crossing a plane boundary separating two media. Specifically, minimize the optical path defined in (37),

$$P_{opt} = \int_A^B \eta(r)\, ds,$$

where $\eta(r)$ is the index of refraction at position r, by varying x, the position of the ray as it crosses the interface at $y = 0$. Use the geometry shown in the sketch and show that your calculation yields Snell's law, $\eta_1 \sin \theta_1 = \eta_2 \sin \theta_2$.

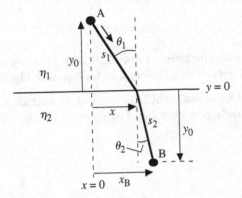

Problem 12.52. Use the Euclidean geometry of Fig. 6a to express the geometric path excess $\Delta\ell = \ell_{SP} + \ell_{PO} - \ell_{SO}$ (40) in terms of the bending angle α and distances for small-angle scattering and for a point-mass lens, as given in (41). Hints: use $\ell_{SP} = D_{LS}/\cos(\theta - \alpha)$ and similar expressions and eliminate θ_S with the ray-trace equation (13). Take all angles to be small relative to unity (e.g., $\theta \ll 1$), assume angular-diameter distances as before, and simplify (several times) with the Euclidean relation $D_{LS} = D_S - D_L$.

Problem 12.53. A pair of faint quasar images I_1 and I_2 are detected with no visible lens galaxy. The following two quantities are measured: the separation of the two images $\Delta\theta = \theta_1 - \theta_2$ and the ratio of flux densities $R \equiv \mathscr{F}_1/\mathscr{F}_2$. Spectra to determine redshifts are not obtained. (a) Model the system after Fig. 9 and assume the lens is a point-mass. Show that the quantities θ_1, θ_2, θ_S, and θ_E follow from the measured quantities as follows: write four independent equations in the unknowns, namely those of (23) and expressions for $\Delta\theta$ and R, find $R = |\mu_1/\mu_2|$ in terms of θ_1 and θ_2, and then find θ_2 as a function of $\Delta\theta$ and R. The other quantities follow directly. (b) Check your expressions with the example of (20) in which $\theta_S = \theta_E$, as follows: from the values (20), find $\Delta\theta$ and R and substitute them into your expressions from part (a). (c) What do the results tell you about the mass M of the lens and the distance parameter $D \equiv (D_L D_S/D_{LS})$? How does this relate to time delays and the Hubble constant? [Ans. $\theta_2 = -\Delta\theta(\sqrt{R} - 1)/(R - 1); -; -$]

Problem 12.54. Derive (54), the approximate gravitational delay function $\psi(\theta)$ for a point-source lens where the distances D_L and D_{LS} are much greater than the impact parameter $b = D_L\theta$. Start with the definition of ψ (44), adopt a straight-line trajectory from source to observer with impact parameter b (i.e., no bend at the lens), integrate from $x = -D_{LS}$ to $x = +D_L$, and simplify for $b \ll D_L$, and $b \ll D_{LS}$. Illustrate the geometry with a sketch.

Problem 12.55. (a) Show that the expression for the total differential delay (geometric plus gravitational) $\Delta\tau$ (56) for a point-mass lens follows from (53) and (54). (b) Use the 47-d delay between intensity variations in the two images of quasar B1600+434 to estimate the Hubble constant H_0 for the universe geometry represented by $\Omega = 1$, $\Lambda = 0$ (51) under the assumption of a point-mass lens. The lens has been identified, and so the angles θ_1 and θ_2 are known as are the redshifts of the source and lens: $\theta_1 = 1.14'' \pm 0.05''$, $\theta_2 = -0.25'' \pm 0.05''$, $z_S = 1.59$, $z_L = 0.42$, $\Delta\tau = 47 \pm 6\,\mathrm{d}$. The two images are on opposite sides of the lens, as expected for a point-mass lens. Express H_0 in units of s^{-1} and km s^{-1} MLY^{-1} and compare with the currently favored value of $\sim$22 km s^{-1} MLY^{-1}. [Ans. $\sim$30 km^{-1} s^{-1} MLY^{-1}]

12.6   *Strong and weak lensing*

Problem 12.61. Look up and closely examine the image of Abell 2218 (Fig. 22) on the Internet and scrutinize it for weak lensing features. Note the colors also. See caption to Fig. 22.

Problem 12.62. Look up the CASTLES Survey and MACHO Project on the Internet; both are gravitational lens projects. Write a brief report about one of them.

Credits, further reading, and references

Sources for the material in this volume include a diverse literature comprising individual published papers, general astronomical texts, conference proceedings, and specialized graduate and professional texts. Some works date to the 1970s, when I began teaching astrophysics courses. The derivations herein are rarely original with me; most have a long history of development and re-iteration. My contribution is the attempt to present them at a level appropriate to the upper-level undergraduate and beginning graduate student with appropriate amplifications, explanations, sketches, and exercises.

Some of the more useful texts and articles used over the years in preparing my notes are listed here with apologies to those whose works I used but do not now recall. The latest editions of which I am aware are listed. I do not include the general, mostly nonmathematical astronomy collegiate texts but do acknowledge the textbooks by Pasachoff and Kutner (*University Astromomy*) and Abell (*Exploration of the Universe*, 3rd Ed.) as important in acquainting me with the world of astronomy. The credits in the captions to figures are occasionally an indication of a source or sources used for the related portion of the text.

The references cited here are excellent starting points for further information on the covered topics. Following this, I call attention to other useful reference material. Figure credits are given in the figure captions. Personal credits are given in the acknowledgments.

Parenthetical numbers in the following lists indicate the chapters of this text that particularly benefited, at least in part, from the material in the reference.

Astrophysics – specialized topics

J. Binney and S. Tremaine, *Galactic Dynamics*, Princeton University Press, 1987 (10)

E. Böhm-Vitense, *Stellar Astrophysics*, two volumes, Cambridge University Press, 1989 (2,3,4)

B. Burke and F. Graham-Smith, *An Introduction to Radio Astronomy*, 2nd Ed., Cambridge University Press, 2002

D. Clayton, *Principles of Stellar Evolution and Nucleosynthesis*, McGraw-Hill, 1968 (2,3,4)

C. Hanson, S. Kawaler and V. Trimble, *Stellar Interiors*, 2nd Ed., Springer-Verlag, 2004 (2,3,4)

J. Kraus, *Radio Astronomy*, 2nd Ed., Cygnus-Quasar, 1986 (10)

J. Krolik, *Active Galactic Nuclei*, Princeton University Press, 1999 (7)

K. Lang, *Astrophysical Formulae*, 3rd Ed., Springer-Verlag, 1999

W. Rose, *Advanced Stellar Astrophysics*, Cambridge University Press, 1998

B. Rossi and S. Olbert, *Introduction to the Physics of Space*, McGraw-Hill, 1970 (5,8)

G. Rybicki and A. Lightman, *Radiative Processes in Astrophysics*, Wiley-Interscience, 1979 (5,6,7,8,9)

P. Schneider, J. Ehlers and E. Falco, *Gravitational Lenses*, Springer-Verlag, 1992 (12)

M. Schwarzschild, *Structure and Evolution of the Stars*, Princeton University Press, 1958 (2,3,4)

S. Shapiro and S. Teukolsky, *Black Holes, White Dwarfs, and Neutron Stars*, Wiley-Interscience, 1983 (4)

T. Swihart, *Astrophysics and Stellar Astronomy*, Wiley, 1968 (4)

W. Tucker, *Radiation Processes in Astrophysics*, MIT Press, 1975 (5,6,8,9)

G. Verschuur and K. Kellermann, *Radio Astronomy,* 2nd Ed., Springer-Verlag, 1988 (8, 10)

S. Weinberg, *Gravitation and Cosmology*, Wiley, 1972 (6, 12)

Physics textbooks

The following sources were useful to me. I make no attempt to list later, more contemporary texts. For the most part, the physics is well grounded and does not change.

G. Bekefi and A. Barrett, *Electromagnetic Vibrations, Waves, and Radiation*, MIT Press, 1977 (5, 11)

P. Bergmann, *Introduction to Special Relativity*, Prentice-Hall, 1942 (7)

C. Cohen-Tannoudji, B. Diu and F. Laloe, *Quantum Mechanics*, Wiley-Interscience, 1977 (10)

A. French, *Special Relativity*, Norton, 1968 (7)

S. Gasiorowicz, *Quantum Physics,* 3rd Ed., Wiley-Interscience, 2003 (10)

K. Symon, *Mechanics,* 3rd Ed., Addison Wesley, 1971 (1)

Review articles

The *Annual Review of Astronomy and Astrophysics* is an excellent resource through which to approach almost any astronomical topic. I list here a few of the recent review articles found there and elsewhere that were helpful to me. Some others that are recent and relevant to topics in this text are also included with, once again, chapter numbers of this book in parentheses.

J. Carlstrom, G. Holder, and E. Reese, "Cosmology with the S-Z effect," *ARAA* **40**, 643 (2002) (9)

L. Cowie and A. Songaila, "High-resolution optical and ultraviolet absorption-line studies of the interstellar gas," *ARAA* **24**, 499 (1986) (10)

D. Cox, "The three-phase interstellar medium revisited," *ARAA* **43**, 337 (2005) (10)

F. Melia and H. Falcke, "The supermassive black hole at the galactic center," *ARAA* **39**, 309 (2001) (1)

R. Narayan and M. Bartelmann, "Lectures on gravitational lensing," arXiv:astro-ph/9606001v2 (1997) (12)

A. Refregier, "Weak gravitational lensing by large-scale structure," *ARAA* **41**, 645 (2003) (12)

R. Remillard and J. McClintock, "X-ray properties of black-hole binaries," *ARAA* **44**, 49 (2006) (4)

P. Rosati, S. Borgani, and C. Norman, "The evolution of x-ray clusters of galaxies," *ARAA* **40**, 539 (2002) (9)

Y. Sofue and V. Rubin, "Rotation curves of spiral galaxies," *ARAA,* **39**, 137 (2001) (10)

T. Tauris and E. van den Heuvel, "Formation and evolution of compact stellar x-ray sources," in W. Lewin and M. van der Klis, eds., *Compact Stellar X-ray Sources*, Cambridge University Press, 2006 (4)

C. Urry and P. Padovani, "Unified schemes for radio-loud active galactic nuclei," *PASP*, **107**, 803 (1995) (7)

General astrophysics texts for science majors

These books are useful resources for students seeking another point of view. Each text in the following list has different emphases, and each finds its own solution to the dilemma that it is an overwhelming job to present the astronomical lore across the whole field while also giving careful derivations. Some books feature extensive textual discussion with a rather modest amount of mathematical derivations, whereas others stress the mathematical physics but frequently without the careful stepwise derivations appreciated by the student. One (Carroll and Ostlie) attempts both goals with some success but at the expense of a very large page count, though it is now available in a two-volume set.

B. Carroll and D. Ostlie, *An Introduction to Modern Astrophysics,* 2nd Ed., Addison Wesley, 2007
N. Duric, *Advanced Astrophysics*, Cambridge University Press, 2004
M. Harwit, *Astrophysical Concepts,* 4th Ed., Springer-Verlag, 2006
M. Longair, *The Cosmic Century*, Cambridge University Press, 2006
T. Padmanabhan, *An Invitation to Astrophysics*, World Scientific, 2006; see also his graduate texts (*Theoretical Astrophysics*)
W. Rose, *Astrophysics*, Holt, Rinehart, and Winston, 1973
F. Shu, *An Introduction to Astronomy*, University Science Books, 1982
A. Unsold and B. Baschek (W. Brewer, trans.), *The New Cosmos: An Introduction to Astronomy and Astrophysics*, 5th Ed., Springer-Verlag, 2002
M. Zeilik and S. Gregory, *Introductory Astronomy and Astrophysics*, 4th Ed., Saunders College Publishing, 1998
The Rose, Harwit, and Shu books were available in the early years of my astrophysics teaching career and were thus, in part, my introduction to the world of quantitative astrophysics.

Research results

Review articles in the *Annual Review of Astronomy and Astrophysics*; each volume contains a list of titles of the previous dozen or so volumes
Articles in amateur astronomy magazines such as *Sky and Telescope* (see "News Notes") and *Astronomy*
Occasional review articles in *American Scientist, Physics Today*, and *Scientific American*
NASA's *Astrophysical Data System* (ADS) on the Internet; search for review articles on a given topic or for a specific journal article.

General reference

A. Cox, ed., *Allen's Astrophysical Quantities*, 4th Ed., AIP, 2000
K. Lang, *Astrophysical Formulae* 3rd Ed., Astronomy and Astrophysics Library, 1999
S. Maran, *The Astronomy and Astrophysics Encyclopedia*, Van Nostrand/Reinhold, 1992
R. Meyers, ed., *Encyclopedia of Astronomy and Astrophysics*, AIP, 1989
J. Mitton, *Cambridge Dictionary of Astronomy*, Cambridge University Press, 2001
P. Murdin, *Encyclopedia of Astronomy and Astrophysics*, four volumes, Nature Publishing Group, 2001
I. Ridpath, *Norton's 2000.0 Star Atlas and Reference Handbook*, Longman, 1989
R. Zimmerman, *The Chronological Encyclopedia of Discoveries in Space*, Onyx, 2000

Glossary

1-D, 3-D	One-dimensional, three-dimensional
AGN	Active galactic nucleus or nuclei
AM	*Astronomy Methods* textbook
AU	Astronomical unit
BC	Barycenter or bolometric correction
B-E	Bose–Einstein (distribution)
Beppo SAX	Italian x-ray astronomy mission
CCD	Charge-coupled device
CGRO	Compton Gamma-Ray Observatory
Chandra	U.S. x-ray astronomy mission
CMB	Cosmic microwave background
CMD	Color-magnitude diagram
COBE	Cosmic Background Explorer (satellite)
CNM	Cold neutral medium (interstellar)
CNO	Carbon, nitrogen, oxygen chain
DM	Dispersion measure
DUO	Gravitational microlensing project
EAS	Extensive air shower
EGRET	Gamma-ray experiment on CGRO
EM	Electromagnetic (wave)
EOS	Equation of state
EROS	Gravitational microlensing project
F-D	Fermi–Dirac (distribution)
FO	Fundamental observer
GC	Galactic center
GLAST	Gamma-ray orbiting observatory
GR	General relativity
GRB	Gamma-ray burst
GZK	Greisen, Zatsepin, Kuzmin effect
HEGRA	TeV gamma-ray observatory
HESS	TeV gamma-ray observatory
HETE	High-Energy Transient Explorer (satellite)
H II	Ionized hydrogen (region)

HIM	Hot ionized medium (interstellar)
HMXB	High-mass x-ray binary
H-R	Hertzsprung–Russell (diagram)
HST	Hubble Space Telescope
H-T	Hulse–Taylor (pulsar)
IAU	International Astronomical Union
IC	Inverse Compton
IRE	Institute of Radio Engineers
ISM	Interstellar medium
LBV	Luminous Blue Variable (star)
LCP	Left circularly polarized
LMC	Large Magellanic Cloud
LMXB	Low-mass x-ray binary
LOS	Line of sight
LP	Linearly polarized
LSR	Local standard of rest
LTE	Local thermodynamic equilibrium
MACHO	Massive astrophysical compact halo object or gravitational microlensing project
M-B	Maxwell–Boltzmann (distribution)
LY	Light year
NGC	New General Catalog
OGLE	Gravitational microlensing project
OVV	Optically violent variable
pp	proton-proton nuclear chain
PSR	Pulsar
RCP	Right circularly polarized
RM	Rotation measure
RXTE	Rossi X-ray Timing Explorer
SED	Spectral energy distribution
SIS	Singular isothermal sphere of gravitational matter
SNR	Supernova remnant
SOHO	Solar and Heliospheric Observatory (satellite)
SSC	Synchrotron self-Compton
S-Z	Sunyaev–Zeldovich effect
UU, DD, UD	Up-up, down-down, etc.; spin states of hydrogen ground state
VERITAS	TeV gamma-ray observatory
VLA	Very Large Array (radio interferometer)
WIM	Warm ionized medium (interstellar)
WMAP	Wilkinson Microwave Anisotropy Probe (satellite)
WNM	Warm neutral medium (interstellar)
VLBI	Very long baseline interferometry
WFPC	Wide-field and planetary camera on HST
XMM-Newton	European x-ray astronomy observatory (satellite)
ZAMS	Zero-age main sequence

Appendix

Units, symbols, and values

Table A1. *Base units: Le Système International d'Unites (SI)[a]*

Quantity	Name	Symbol
Length	meter	m
Mass	kilogram	kg
Time	second	s
Electric current	ampere	A
Thermodynamic temperature	kelvin	K
Amount of substance	mole	mol
Luminous intensity	candela	cd

[a] National Institute of Standards and Technology (NIST), physics website. http://physics.nist.gov/cuu/Units/index.html

Table A2. *Some SI-derived units[a].*

Quantity	Name	Symbol	Equivalent units
Plane angle	radian	rad	$m/m = 1$
Solid angle	steradian	sr	$m^2/m^2 = 1$
Frequency	hertz	Hz	$cycles/s = s^{-1}$
Force	newton	N	$kg\, m\, s^{-2}$
Pressure	pascal	Pa	$N\, m^{-2}$
Energy	joule	J	$N\, m;\ kg\, m^2\, s^{-2}$
Power	watt	W	$J\, s^{-1}$
Electric charge	coulomb	C	$A\, s$
Electric potential	volt	V	$J\, C^{-1}$
Resistance	ohm	Ω	$V\, A^{-1}$
Magnetic flux density	tesla	T	$N\, A^{-1}\, m^{-1}$
Capacitance	farad	F	$C\, V^{-1}$

[a] NIST physics website, *ibid.*

Table A3. *SI prefixes*[a]

Log factor	Prefix	Symbol	Log factor	Prefix	Symbol
−3	milli	m	3	kilo	k
−6	micro	μ	6	mega	M
−9	nano	n	9	giga	G
−12	pico	p	12	tera	T
−15	femto	f	15	peta	P
−18	atto	a	18	exa	E
−21	zepto	z	21	zeta	Z
−24	yocto	y	24	yotta	Y

[a] NIST physics website, *ibid.*

Table A4. *Energy-related quantities*

Quantity	Symbol[a]	Unit	Definition[b]
Specific intensity	$I(\nu,\theta,\phi,t)$	$\mathrm{W\ m^{-2}\ Hz^{-1}\ sr^{-1}}$	
Spectral flux density	$S(\nu,t)$	$\mathrm{W\ m^{-2}\ Hz^{-1}}$	$S = \int I\,\mathrm{d}\Omega$
Flux density	$\mathcal{F}(t)$	$\mathrm{W\ m^{-2}}$	$\mathcal{F} = \int I\,\mathrm{d}\Omega\,\mathrm{d}\nu$
Power	$\mathcal{P}(t)$	W	$\mathcal{P} = \int I\,\mathrm{d}\Omega\,\mathrm{d}\nu\,\mathrm{d}A$
Energy	U	J	$U = \int I\,\mathrm{d}\Omega\,\mathrm{d}\nu\,\mathrm{d}A\,\mathrm{d}t$
Fluence	$\mathscr{E}$	$\mathrm{J\ m^{-2}}$	$\mathscr{E} = \int \mathcal{F}\,\mathrm{d}t$
Surface brightness	$B(\theta,\phi,\nu,t)$	$\mathrm{W\ m^{-2}\ Hz^{-1}\ sr^{-1}}$	$B = I$
Volume emissivity	$j_\nu(\mathbf{r},\nu,t)$	$\mathrm{W\ m^{-3}\ Hz^{-1}}$	$I = \int j_\nu\,\mathrm{d}r/4\pi$
Integrated volume emissivity	$j(\mathbf{r},t)$	$\mathrm{W\ m^{-3}}$	$j = \int j_\nu\mathrm{d}\nu$

[a] Directional quantities may be treated as vectors, but generally are used as scalers in this text.
[b] Definitions presume that area element traversed by flux is normal to the propagation direction.

Table A5. *Physical constants*[a,b]

Universal

Light speed in vacuum	$c = 2.997\,924\,58 \times 10^8$ m s^{-1}
Permeability of vacuum (magnetic constant)	$\mu_0 = 4\pi \times 10^{-7}$ kg m s^{-2} A^{-2}
Permittivity of vacuum (electric constant)	$\epsilon_0 = 1/\mu_0 c^2 = 8.854\,188 \times 10^{-12}$ m^{-3} kg^{-1} s^4 A^2
Gravitation constant	$G = 6.674 \times 10^{-11}$ m^3 kg^{-1} s^{-2}
Planck constant	$h = 6.626\,069 \times 10^{-34}$ J s $(= $ kg m^2 s$^{-1})$

Particle

Electron charge	$e = 1.602\,176\,5 \times 10^{-19}$ A s
Electron mass	$m_e = 9.109\,382 \times 10^{-31}$ kg $(\approx 10^{-30}$ kg$)$
Electron rest energy in eV $(m_e c^2/e)$	$= 0.510\,999$ MeV
Electron classical radius $(e^2/4\pi\,\varepsilon_0\,m_e\,c^2)$	$r_e = 2.817\,940 \times 10^{-15}$ m
Thomson cross section $(8\pi/3)r_e^2$	$\sigma_T = 6.652\,459 \times 10^{-29}$ m^2
Proton mass	$m_p = 1.672\,622 \times 10^{-27}$ kg
	$\rightarrow 938.272$ MeV
Neutron mass	$m_n = 1.674\,927 \times 10^{-27}$ kg
	$\rightarrow 939.565$ MeV

Atomic

Fine-structure constant	$\alpha = \mu_0 c e^2/2h = 7.297\,353 \times 10^{-3}$
	$= 1/137.036$
Bohr radius	$a_0 = 0.529\,177 \times 10^{-10}$ m

Physicochemical

Avogadro constant	$N_A = 6.022 \times 10^{23}$ mol^{-1}
Atomic mass constant $m(^{12}C)/12$	$m_u = 1.660\,539 \times 10^{-27}$ kg
	$\rightarrow 931.494$ MeV
Boltzmann constant	$k = 1.380\,650 \times 10^{-23}$ J K^{-1}
Molar gas constant	$R = 8.3145$ J mol^{-1} K^{-1}
Stefan–Boltzmann constant	$\sigma = 5.6704 \times 10^{-8}$ W m^{-2} K^{-4}
Radiation density constant $(4\sigma/c)$	$a = 7.5658 \times 10^{-16}$ J m^{-3} K^{-4}

Conversions

Electronvolt	1.0 eV $= 1.602 \times 10^{-19}$ J
Standard atmos. pressure	1.0 atm $= 101\,325$ Pa (N m^{-2})
Standard gravit. acceleration	$1.0\,g = 9.806\,65$ m s^{-2}
Temperature and energy kT(eV)	T(K) $= 11\,605 \times kT$(eV)
Photon frequency and energy E(eV)	ν(Hz) $= E$(eV) $\times 2.4180 \times 10^{14}$
Photon energy and wavelength	λ(nm) $\times E$(eV) $= 1\,239.842$

[a] Values are rounded to precision deemed useful for this text; most are more precisely known.
[b] P. Mohr, B. Taylor, and D. Newell, *The 2006 CODATA Recommended Values of Fundamental Physical Constants* (web version 5.0), NIST Web site, http://physics.nist.gov/constants

Table A6. *General astronomical constants*[a]

Quantity	Value
Astronomical unit (semimajor axis of earth orbit)	$1.0\ \text{AU} = 1.496 \times 10^{11}$ m
Light (Julian) year	$1.0\ \text{LY} = 9.461 \times 10^{15}$ m ($\approx 10^{16}$ m)
Parsec	$1.0\ \text{pc} = 3.086 \times 10^{16}$ m $= 3.262$ LY
Solar mass	$1.0\ M_\odot = 1.989 \times 10^{30}$ kg
Solar radius	$1.0\ R_\odot = 6.955 \times 10^{8}$ m
Solar luminosity	$1.0\ L_\odot = 3.845 \times 10^{26}$ W
Earth mass	$1.0\ M_\oplus = 5.974 \times 10^{24}$ kg
Earth radius, mean	$1.0\ R_\oplus = 6371.0$ km
Earth radius, equatorial	$1.0\ R_{\oplus,\text{eq.}} = 6378.1$ km
Moon mass	$1.0\ M_{\text{moon}} = 7.353 \times 10^{22}$ kg
Moon radius, mean	$1.0\ R_{\text{moon}} = 1738.2$ km
Moon orbit semimajor axis	$1.0\ a_{\text{moon}} = 384\,400$ km

[a] *Allen's Astrophysical Quantities*, 4th Ed., ed. A. N. Cox, AIP Press, 2000

Table A7. *Constants involving time*[a]

Day	
Mean solar day	86400.00 UT (mean solar) seconds
	$\approx 86\,400.002$ SI seconds
Sidereal day, relative to vernal equinox	86 164.09 UT seconds[b] $= 23$h 56m 04.09s UT
Year	
Julian year (exact)	365.25 d $= 31\,557\,600$ s (SI)
Julian century (exact)	36525 d
Tropical year (equinox to equinox)	365.2422 d
Sidereal year (fixed star to fixed star)	365.2564 d
Anomalistic year (perihelion to perihelion)	365.2596 d
Julian date	
1900 Jan. 0.5	JD 2 415 020.0
2000 Jan. 0.5	JD 2 451 544.0
2100 Jan. 0.5	JD 2 488 069.0
Standard Julian epochs	
J1900.0	JD 2 415 020.0 $= 1899$ Dec. 31, 12h TDB[c]
J2000.0	JD 2 451 545.0 $= 2000$ Jan. 1, 12h TDB
J2100.0	JD 2 488.070.0 $= 2100$ Jan. 1, 12h TDB

[a] *Allen's Astrophysical Quantities*, ibid.
[b] UT time is essentially mean solar time. The sidereal day has this same value in units of SI seconds within the given precision. See Eq. (4.13) in AM and the discussion that follows.
[c] TDB = Barycentric dynamical time

Index

Boldface page numbers indicate that there is a section or subsection on this topic beginning at the indicated page. A major section or chapter on this topic is denoted by a range of boldface numbers. An italic page number refers to a figure or a table.

Printed in the United States
By Bookmasters